Learning and Development for Innovation, Networking and Cohesion

Baltische Studien zur Erziehungs- und Sozialwissenschaft

Herausgegeben von Gerd-Bodo von Carlsburg, Irena Musteikienė und Airi Liimets

Band 12

PETER LANG
Frankfurt am Main · Berlin · Bern · Bruxelles · New York · Oxford · Wien

Gerd-Bodo von Carlsburg
Palmira Jucevičienė
Gediminas Merkys
(eds.)

Learning and Development for Innovation, Networking and Cohesion

PETER LANG
Europäischer Verlag der Wissenschaften

Bibliographic Information published by Die Deutsche Bibliothek
Die Deutsche Bibliothek lists this publication in the Deutsche Nationalbibliografie; detailed bibliographic data is available in the internet at <http://dnb.ddb.de>.

ISSN 1434-8748
ISBN 3-631-53426-4
US-ISBN 0-8204-7704-4

www.peterlang.de

Contents

Preface

The Institute of Educational Studies at Kaunas University of Technology in cooperation with Heidelberg University of Education presents the second book of scientific paper collection. The first book – „Towards the Learning Society“ – was published in 2002. As a continuation of the theme, we publish thirty articles discussing the problems of learning and development for innovation, networking and cohesion in the counties in-transition. The colleagues of the Institute of Educational Studies prepare most of the articles. One author represents Vilnius Pedagogical University. In this book, like in the previous one, the problem of innovation, networking and cohesion is looked upon through educational prism. Some papers though expand this attitude towards the development for innovation and networking by looking at the problem from the point of view of management and economics.

The book consists of three parts. The first part – „Innovation of and for learning and development“ – comprises twelve articles. The first paper "The enlarged European Union – towards an economic governance system for the collective learning and innovation?" (by Giedrius Jucevičius) is not only rather symbolic, reflecting the contemporary spirit of the European Union, but raising a question – what lies ahead? „The Importance of Educational Innovation Implementation in Achieving the Qualities of the Knowledge Society: What Barriers Can Emerge at the Initiation Phase and how to Identify Them?" (by Marija Jotautienė and Brigita Janiūnaitė) as if prolongs the previous article. Robertas Jucevičius' paper "Clusters as innovative form of regional development" brings the reader back to the field of managerial innovations. Eglė Katiliūtė joins both the managerial and educational attitudes and analyzes premises for development and implementation of the educational strategy in the knowledge society. Vilija Stanišauskienė keeps to a similar point of view when discussing the model of career competence and its expression in the contemporary labor world. Virginija Chreptavičienė continues with the considerations of competence problems paying great attention to the change of the conception of communicative competence in the context of the informative and multicultural society. Library as a preferred learning environment is analyzed with the help of the survey results of Gediminas Merkys, Palmira Jucevičienė and Gintarė Tautkevičienė's common research. Giedrė Adomavičienė and Liuda Šiaučiukėnienė, go deeper into the topic of learning environment by analyzing the creation of differentiated learning environment as a condition of liberal education aims. Nijolė Bankauskienė ir Ilona Skersienė build a bridge between the humanities and technical sciences by researching the cohesion of educating in philology and technological education.

Alicija Ramanauskaitė and Nida Stankevičienė deal with technological education as a means of reducing technological disjuncture. The article by Palmira Jucevičienė and Sandra Balevičienė is intended to analyze the necessity of natural and social sciences integration from the point of view of multidisciplinarity of the socially important theme of natural science in personal theory-in-use. These new ideas definitely require changes in the school activity, so Ona Visockienė and Vilija Dabrišienė analyze the tendencies in the secondary school of Lithuania.

The second part of the book is devoted to examining the issues of networking and cohesion.

The article by Viktorija Baršauskienė, Gediminas Merkys and Dalia Antinienė, discusses the problems of social network structure and presents a diagnostic study in the population of Lithuanian students. The contribution by Rūta Urbanskienė, Virginija Chreptavičienė and Rūta Janulevičienė focuses on creating closed partnership between business and science through social capital networks. Antanas Bagdonas approaches the problem of cohesion by analyzing the congruence of interests of different stakeholders of secondary school. Vilma Žydžiūnaitė deals with substantiation of a junction between teamwork competences and professional competence. Angelė Kaušylienė discusses characteristics of school's network in the article. Edita Štuopytė goes deep into peculiarities of socio-educational activities of the community centers.

The third part of the book is „University for development of new learning and mobility". Though dealing with the problems of the countries in-transition, these articles touch upon more general points of higher education. The twelve articles should catch the eye of those readers and researchers who are interested in the problems of higher education development associated with the conversion of educational paradigm.

Auksė Markevičienė analyses the contemporary university as a cultural phenomenon and identifies the paradoxes of being. The article by Palmira Jucevičienė and Violeta Karenauskaitė focuses on the role of university in liberalisation of society in the countries in transition. The approach taken to analyse this problem is quite unconventional: it addresses the potential of liberalisation on the content and methods of the natural sciences. Nijolė Zinkevičienė and Brigita Janiūnaitė analyse the university values from the point of view of the knowledge society and also identify two antithetic values: conservativism and innovativeness. The issue of the controversial nature of the university values lies at the background of the article.

Deep learning at university is the core of the article by Vida Stasiulionienė. The article aims at responding to the question whether deep learning is unambiguous quest for the truth. Rūta Klimašauskienė is preoccupied with the issue of knowledge mobility. Therefore, her article reflects the issues of knowledge dissemination from the university to the educational practice. Dalija Gudaitytė and Vida Vilkienė discuss the potential of co-operative learning encouraged by European Mobility Projects in teacher training college. Loreta Rudaitienė analyses the issue of self-directed development of emotional intelligence in professional performance; teacher professional performance is in focus. The article co-authored by Aldona Augustinienė and Nijolė Čiučiulkienė addresses the issue of developing students' decision-making competence in the context of problem-based learning. Vilma Žydžiūnaitė examines the potential of fostering the trans-cultural sensitivity of nurse students; she aims at identifying it in the field of curriculum development. Ernesta Račienė, a colleague from Vilnius Pedagogical University, focuses on the issue of interrelation between language and culture. Aistė Urbonienė and Irena Leliūgienė analyse the issues of professional motivation development in the study process. Regina Novikienė and Liuda Šiaučiukėnienė revisit the issue of deep learning: they address it in the context of learning mathematics at university. Ilona Adomavičiūtė refers to the chances of the Eastern enlargement in the EU-area. To reach the community in the area of the cultural work, getting to know other cultures offers many possibilities of an approach of developing new Europe together with the aim. The article of Inga Šukevičiūtė, European Parliament, deals with the special language in the economy communication himself and pursues the aim to show the unusual feature of the special language economy under the other special languages. Gerd-Bodo von Carlsburg, a colleague from Heidelberg University of Education, gives a view of the functions of school and society as well as a idea and vision of the 'good' teacher. At the beginning of his article he follows the question whether we study for the school or for the life.

We hope that the ideas presented in the articles of this book will stimulate new research. As the publications involve quite a lot of data on educational changes in Lithuania, this collection of articles may be useful to those researchers, teachers and students who are interested in comparative research. It may also be of interest to those who focus on the issues of lifelong learning, development of the knowledge and learning society, and stimulation of innovativeness its stakeholders (individual and organisations), especially those concerned with education.

Kaunas and Heidelberg,
November 28th, 2004

Gerd-Bodo von Carlsburg
Palmira Jucevičienė
Gediminas Merkys

Giedrius Jucevičius

The Enlarged European Union – Towards an Economic Governance System for the Collective Learning and Innovation?

This article seeks to analyze the emergence of the European economic governance system that can be defined through an increasing variety of national economic models that co-function and compete in the common European market space. We also seek to discuss the impact of the EU policies on the industrial competitiveness and innovation policies in the new member states as well as the scope of actual/potential *collective learning* inside the enlarged EU. Various authors have analyzed the aspects of the EU economic governance (Thompson, 2001; Rhodes and Apeldoorn, 1998) and collective learning (Storper et al., 1998), but very little research so far has been carried out regarding these two aspects combined, particularly in the light of the forthcoming enlargement.

The first question raised in the article is whether the EU can be considered as a specific economic governance system. We revisit the main integration and governance theories seeking to explain the nature of economic governance inside the EU. The second question raised by the analysis is whether the multiple policy conflicts should be considered as an obstacle or a characteristic of European economic governance. We then review the major national economic governance models inside the EU that determine the heterogeneity and many policy conflicts inside the Union. Finally, we provide insights into the Union's convergence vs. divergence dilemma in terms of the economic governance structures and try to evaluate it in the light of the *learning economy* environment as a precondition for Union's competitiveness and growth.

The European Union - market integration over market organization

The European economic system is primarily defined through the logics of the internal market. As Streeck and Crouch (1997) put it, the European integration has been and remains primarily about the market integration and the main loss of sovereignty of a European nation state so far has been not to the EU, but to the market (Tsoukalis and Rhodes, 1997). The creation of the European economic sphere is largely perceived as a regional response to the global economic challenges and neo-liberal nature of the international environment[1].

---

[1] The essence of neo-liberal ideology is the assumption that economic growth will be most rapid when the movement of goods, services and capital is unimpeded by government

Not surprisingly, overview of a bulk of academic literature on the European economy (Dyker, 1992, Pelkmans, 1997) reveals a pretty standard focus of analysis. It usually covers the theoretical considerations for market integration and Community regulation, integration dynamics, substance of the four freedoms and some implementing common policies (depending on the scope intended by the authors) as well as the Union's economic position in the global marketplace. Some authors also include the chapters on enlargement with rather superficial comparative overviews of candidate countries economies.

This approach dramatically differs from the usual structure of analysis used for explaining the *national* systems of market organization. Although both approaches share certain amount of historical discourse, the latter structures the analysis by the fundamental aspects of capitalism: nature of firm, economic role of the state, labor organization and industrial relations, corporate law and governance, educational system, established behavioral patterns and relationships across firms.

Such pattern of analysis is clearly absent on the European scale, which is more dominated by *market integration* than *market organization* perspective. Quite naturally, one should probably not expect from the European economy something that it has not been about – a clearly defined and naturally evolved socio-economic entity/model characterized by mutually complementary institutional elements and underlying uniform cultural values/behavioral patterns of social actors.

Nevertheless, fifty years of European integration suggest that there is something more to the European economy than a common marketplace. Some policy areas that only fifty years ago have been an indisputable part of national competence, today fall under a strong regulatory influence of the EU. According to Kapteyn (1996), 'states do not emerge without war and markets do not emerge without states' and refers to the EU as 'stateless market'. Indeed, there has always been much more ambiguity about the EU as a state than EU as a market. Whereas the national economic models are primarily described through state-market relationships, the status of European socio-economic entity was always subject to numerous questions.

Is there an emergence of European state in sight? Is it in principle possible that market integration leads towards the convergence of market organizations, or in other words, the emergence of a 'European' model of capitalism? What necessary elements should it have to obtain such status? What about the compatibility of the Union's economic policies? The search for answers to these questions calls for some more systemic and theoretically sound insights.

---

regulations (McEwan, A. (1999) 'Neo-liberalism or Democracy? Economic strategy, markets, and alternatives for the 21st century').

The entire process of European integration has been subject to various competing theories, especially in the field of political science and international relations. The *neo-functionalist* ideology, popular at the start of integration, assumed that a nation state gradually cedes its competences to the supranational structures through the policy spillovers. This approach has been backed up by the empirical evidence from the outset of integration, but later in the 70's, in the times of Euro-confusion, was confronted with counterviews (Wallace, 1996). The representatives of *liberal intergovernmental* approach argue that the Europeanization reflects the member states interests of economic modernization in the interdependent world and is shaped through intergovernmental negotiation (Moravcsik, 1999). Accordingly, the Europeanization is considered to be only the outcome of converging member states' preferences rather than the other way around.

Another approach, alternative to the traditional *intergovernmentalist-federalist* divide in the political science came from the domain of comparative politics and public administration (Hix, 1994, 1998). It was less concerned with the integration process *per se*, but with the nature of European political organization. The *new institutionalism* has reappraised the role of institutions in shaping the political actions. Its sub-streams such as *rational choice institutionalism* and *historical institutionalism* emphasized that the political choice was predetermined by the existing rules that, according to the latter, had to be considered in the historical perspective. Thus, at this stage, the construction of European polity was viewed as a process of the mutually reinforcing interplay between the national politics and pre-established Community institutions.

The last decade has seen a growing preoccupation with the issues of European governance, inspired by both of the above-mentioned approaches. The *multilevel governance* approach (Marks, 1993) explains the nature of European policy-making as a complex interaction of national, supranational and intra-national (regional, local) actors. It attacks the importance of national governments and independence of national political systems as advocated by the intergovernmentalists. However, unlike neo-functionalism, it fails to provide a more dynamic explanation of interactions among multiple actors in European policy-making (George, Bache, 2001). *Supranational governance* perspective regards the functioning of the EU as a series of regimes for different policy areas and holds that increased transactions over national borders favor the emergence of supranational regulation.

Thus, we can observe a real maze of the European integration and governance theories developed over the fifty years of integration. What is their relevance to our economic policy focus?

First of all, given a complex interplay of numerous interests and institutions at all levels of the policy-making plus continuous policy spillovers, it is really difficult to expect that the European integration could generate some kind of uniform economic model, explicable in conventional categories used for analysis of the nation-state economies.

Secondly, the aforementioned theories have emerged and been influential at different stages of the European integration. It is possible to observe that if the theories inspired by the political science (with emphasis on intergovernmentalist-federalist dilemma) were very dominating the early stages of integration, today we can notice a relative change of 'intellectual weight' towards the theories of governance and public administration. As the theories are closely intertwined with the European reality, it is possible to consider the emergence of European economic organization that does go beyond market integration.

National economic models vary primarily by the institutional solutions they offer to the state-market dilemma through the established systems of economic coordination and control (Whitley, 1992). Might Europe be not such a 'stateless market' after all?

Policy conflicts – characteristic or obstacle to the EU economic governance system?

One of the arguments against the EU as a consistent market organization is that its economic policies were '*ad hoc* created and not necessarily compatible' (Pitelis, 1994). It is really not surprising to find numerous policy conflicts in a multifaceted entity like EU, but what does it prove?

For example, when analyzing the EU industrial policy in a 'broad sense'[2] Darmer and Kuyper (2001), observe numerous EU policy conflicts. They mention the contradictory relationship between industrial policy (competitiveness perspective) and competition policy (fair competition perspective) or environmental policy (high environmental standards for firms), as well as between the technology policy (multi-annual nature of programs) and Community budgeting procedures. Faced with such policy dilemmas, Commission was sometimes producing ambiguous and contradictory positions (e.g. regarding the merger regulation).

However, a degree of policy conflict leading to certain inconsistencies in policy regimes can be equally observed on the national level. This gives no grounds whatsoever to doubt the existence of national varieties of capitalisms. For example, there can be as much functional/ideological disagreement between a na-

[2] Industrial policy in a 'broad sense' means any government policy with impact to business (Dar-mer and Kuyper, 2001)

tional Ministry of Economy and Competition Council as there is a potential clash between DG Enterprise and DG Competition. The difference is that member states have more flexibility in accommodating the national policy dilemma solutions into their existing institutional infrastructure, while Community solutions are more subject to rigid institutional framework (e.g. favoring the free market solutions over structural industrial intervention in the case of above mentioned policy clash). In some cases, the European institutional framework prevents from achieving systemic (and even desirable) solutions that would have been possible within national system. For example, in case of EMU, a potentially dangerous split that exists between the European monetary and fiscal competences was caused by Member states political interest to maintain fiscal independence that was supported by the unanimity requirement in the Council.

There is a general degree of controversy in evaluating the policy conflict and desirability of policy consistency. Some policy conflict is an inseparable part of life of any socio-political system, especially of such complex one as the EU. However, the consolidated national economic systems (German, Anglo-Saxon or French) have certain established institutional structures and underlying social conventions that accommodate the emerging policy conflicts and ensure a degree of policy consistency (sometimes, by the way, very resilient to the Europeanizing forces!). It is still a widely open question if, given the persisting diversity, the EU is on the way of developing some consistent socio-economic system of its own.

From the very start of economic integration, Community faced rather complex, if not contradictory, socio-economic objectives that included *harmonious economic development*, *continuous expansion*, *increased stability*, *rising and fair standard of living*. The common market was supported by the no less complex common economic policies, including the directly related external trade and competition policies, rhetoric common transport policy and socially-minded common agricultural policy. The further steps of integration saw both the deepening of existing policies and emergence of the new ones. The reasons for increasing scope of policies were rather diverse. The (re-)distributional structural policies appeared as a political precondition for Southern member states (and Ireland) to sign ambitious 1992-programme and as a reaction to formerly unfulfilled objective of 'harmonious development' (Pelkmans, 1997). The health and safety at work provisions emerged on the European level due to a commonly held consensus that was facilitated by a fixed QM voting requirement. The research and technological development policy was europeanized mainly by the push of the French lobby groups. The social policy, although with minimal direct EU regulatory (not re-distributional!) competence was a product of functional spillover of market integration. The environmental policy emerged as a result of preferences by numerous 'green' political environments in Europe of the 70's that have fixed the 'path dependence' for decades to come (Sbraggia, 2000). The EMU project emerged as a politically long-intended completion of the internal market with far-reaching policy spillovers on many of the aforementioned domains.

Apparently, these examples do not imply any uniform pattern in the Europeanization of different economic policy areas. To some extent, we can notice that the relevance of earlier discussed integration theories depended both on a specific policy area and the stage of integration.

Not only the emergence, but also the development of different policy areas was subject to pretty diverse institutional environment and member state preferences. Faster progress was achieved in the areas with the more converging member states interests and/or with the qualified majority voting (e.g. policy areas directly linked with the functioning of internal market). Whereas in the policy areas that had firm roots in national institutional infrastructure and/or with a unanimous vote requirement (e.g. social, environmental policy, corporate law), the progress in harmonization remained limited. The member states have cautiously sought to preserve the domestic policy areas that would otherwise misfit the European practices[3] (Cowles, Risse, 2001).

It is still unclear whether the policy solutions obtained on the European level are randomly negotiated compromises and 'lowest common denominator solutions' or clear-cut pieces fitting the 'puzzle' of emerging European economic system.

Majone (1996) analyzes the policy and regulation development in the EU through the traditional state functions: *redistribution*, *stabilization* and *regulatory*. He argues that the EU regulatory developments prove gradual transition towards the EU regulatory state. It is characterized by the dominating role of the regulatory function that is justified on the grounds of market efficiency (i.e. policy as an instrument for correcting the market failures) and implemented through non-elected expert bodies (or non-majoritarian institutions). Even the policy areas that traditionally belonged to the redistribution function of the state, became of an explicitly regulatory nature inside the EU (e.g. social policy). The rise in regulatory competences is mainly explained through the constraints facing Community re-distributional and stabilization functions.

If we define the Europeanization through such aspects as *emergence of formal EU legal structure* and *development of EU policies*, *transfer of decision making/policy competences to the European level*, *constructing systems of*

[3] The areas of misfit are referred to by the authors as the 'veto points'.

meaning and collective understandings (see e.g. Risse, Cowles, Caporaso, 2001), we will see that the progress in Europeanizing the fundamental elements that describe national capitalist systems (see Hall and Soskice, 2001, Whitley, 1992) – corporate law and governance, industrial relations, educational system – was very limited. There hardly exist one-size-fits-all regulations in any of these areas in Europe. Even the 30-year long providing a framework for national parties' voluntary agreement and a minimal common European denominator in case of disagreement solved Anglo-German dispute on European employee participation model in labour relations. The EC regulation in the fields of corporate law and educational system does not go much further than needed for implementing, respectively, freedom of establishment and free movement of labor. Thus, taking the usual framework for comparing capitalist systems and looking from European regulatory perspective, Europe truly consists of a great variety of capitalist systems and could hardly be called a distinct socio-economic entity.

Nevertheless such 'regulatory' approach has its limitations. Despite the minimalist European regulatory presence, the national sovereignty in the abovementioned areas has been dramatically eroded by the market integration processes as well as by global developments. Just like Leibfried and Pierson (2000) argue about the existence of huge restraints on a member state's autonomy in the field of social policy. Meanwhile, the emergence of EMU and the development of European stock markets pose serious spillover threats on continental corporate governance systems. Although the EU policy spillovers might not have produced an institutionally complementary model in its own right, their impact on *institutionally complementary* national systems (as proposed by Hall and Soskice, 2001) is becoming more than visible. It is still unclear, however, if the European impact on mutually complementary institutions leads to some kind of convergence across the systems, especially in the light of future enlargement that is likely to add to the socio-economic diversity in the Union.

National varieties of economic governance systems inside the EU

Classifications of economic systems

There are various possible classifications of business systems developed by different authors. The classification usually depends on the chosen variables.

Probably the most popular is *liberal-interventionist* divide (based on the *role of the state* variable) across national economic systems, going down to the classical Smith-Keynes dichotomy in the economic theory. The line is usually drawn between the Anglo-Saxon and continental European economic systems based on a degree of the government economic intervention. Interventionists hold views that national economic competitiveness lies on a well-established structural role of state carried out via industrial policy, whereas liberals rely primarily on the free interplay of market forces (via functioning competition policy), stable legal/macroeconomic environment, and justify state intervention only in the case of actual/potential market failure. According to Darmer and Kuyper (2001), the EU is relatively composed of five liberal and ten interventionist member states (market-oriented – Britain, the Netherlands, Denmark, Finland, Ireland; public sector intervention-oriented – France, Italy, Spain, Greece, Germany, Sweden, Austria, Belgium), although also admit that modern economies are too complex for bipolar generalizations (especially unclear are German or Swedish positions in such a divide).

The comparative micro-level research in business and management is dominated by the global *East-West* divide (e.g. Ouchi, 1981). The American researchers who were 'benchmarking' American management practices to the Japanese success story of the 80's produced most of the modern management theories. Such 'global' classification could still in a way accommodate the British and German management models, based on the ways they were respectively similar to the American and Japanese ones. However, it could not take into consideration some distinctive economic governance models, such as French, Swedish or Dutch. This divide is much less based on the role of the state (due to a different focus of the political economy and business studies), but on the organizational characteristics like individual vs. group relationship within/outside the organization, short vs. long-term orientation in economic activities, mechanistic vs. organic growth, analytical vs. integrative approach to the problem solving, emphasis on competition vs. emphasis on cooperation.

Hall and Soskice (2001) draw a line between *liberal* and *coordinated* market economies, based on five interlinked criteria: *financial system* or market for *corporate governance*, *internal structure of the firm*, *industrial relations, education and training systems,* and *inter-company relations*. Liberal market economies (Britain, USA) rely on the market forces as the main coordinator of the economic system (backed up by strong market regulations and deregulated labor market). Meanwhile, the coordinated market economies base their activities on the *strategic interaction* among different socio-economic actors (backed up by the recognized importance of network, long-term relations and reputation). According to the authors, the existing institutional environment is closely linked with the nature of national comparative economic advantage, it explains a lot why certain types of goods are produced in one country rather than another. Thus, not surprisingly, countries seek to preserve both their competitive edge and the corresponding institutional framework.

Whereas Whitley (1992) classifies business systems by *hierarchy-market* divide, Hall and Soskice (2001) rely on a more contemporary *network-market* relationship. The usefulness of dichotomies in classifying the economic systems may, however, be brought into question. Their use can be justified if it helps to locate divergent economic models on a continuum between the two extremes. However, the economic models are usually too complex for linear generalizations. In my opinion, from institutionalist point of view it still makes sense to distinguish at least three traditional kinds of economic systems: *market-based*, *network-based* and *hierarchy-based*. Although the emergence of the 'New Economy' is generally characterized by the shift from hierarchy to network relationships, it is difficult to expect that the latter is substituting the first, especially if viewed from the global scale (most of the third world economic system are still very hierarchy-based). Although culturalists (Hampden-Turner, Trompenaars, Hofstede, Laurent) measure international differences through the sets of numerous bipolar dimensions (e.g. individualism-collectivism scale), they too end up identifying at least three major value-based ways for solving organizational problems.

The trichotomy of economic coordination (*market, hierarchy, network*) was first proposed by G.B. Richardson (1972) and remains supported by numerous authors (Ozawa, 2002, Rhodes et al., 1997 etc.).

There is a degree of controversy in any static description of national socio-economic model, especially at times of intense economic internationalization. Thus, I will rely on the classical, somewhat simplified, characteristics of national models as described in literature on the comparative economic governance and organization studies (Hall and Soskice, 2002, Whitley, 1992, Rhodes and Apeldoorn, 1998, Hickson, 1993, Thompson, 2001, Hampden-Turner and Trompenaars, 1993, *et al.*). Only then will I discuss the actual/potential susceptibility of national economic models to the europeanizing and globalizing forces.

The discussion is based on the standard variables used in comparative studies of market economies: traditional role of the state and pattern of industrialization, nature of firm, financial system and corporate governance, education system, industrial relations and dominant management practices, corporate growth strategies and inter-company relations (see e.g. Rhodes and Apeldoorn, 1998).

British model – market-based economic governance system

Kapteyn as 'egalitarian centralism' describes the British economic model that very often falls into the same 'Anglo-Saxon basket' with its American counterpart. Although Britain has a long tradition of centralization, the *state* does not play an active role in economic policy making. It has always been more of a referee for the level playing field in trade and commerce than a coach of national economic players. This tradition is at least partly due to the country's *early industrialization* and establishment as the world's premium trading nation. The *firms* are regarded as independent economic units functioning in a clearly defined regulatory market environment ensuring adequate protection to the contracting parties. The firm's actions are directed towards increasing the shareholder value (i.e. that of the owner, investor) and accompanied by the corresponding legislative protection of shareholder rights. The *financial system* is based on the stock capital market rather than bank credits what means that listed companies are dependent on the short-term financing and have to deliver visible results (i.e. profits) in a short run than to invest in some long-term projects with unclear economic perspective. There is apparently little investment into education and training of employees/managers within the British companies, mainly due to the generally deregulated and contract-based labor market. It enables firms to attract the needed labor and intellectual resources 'from outside' for a required period of time and tasks. The manager's role primarily concerns optimal allocation of organizational resources (including the division of labor) rather than being a technical expert in some field. Manager's profile is supported by the generalist education system (MBA is a classical Anglo-Saxon invention). Thus, in British system, there is a clear division between the decision-makers and decision-implementers what also reflects in the pattern of *industrial relations*. Since Thatcherist policies, trade union movement remains fragmented and weak. The German attempt to europeanize the board-level employee participation (i.e. provide them with powers to influence the firm's policy) was a decisive 'veto point' in British policy for 30 years. Both inter-firm and intra-firm relationships remain highly contract-based, characterized by intense competition in the marketplace and encouraging individual short-term results (pay-for-performance motivation schemes). Such institutional setting favors more *primary* than *secondary innovations* (Hall and Soskice, 2001), and creates comparative economic advantages in new, massively produced, unsophisticated types of production (see Porter, 1990), however, it disfavors the areas where higher levels of technical training and organizational commitment are needed.

German model – network-based economic governance system

Germany's late emergence as a state as well as its federal structure has meant that there has traditionally been both limited state involvement in directing the activities of economic actors and political fragmentation of the market. Of all continental systems, the German business system is probably the most difficult one to categorize. In many aspects, it usually falls between the established dichotomies of the assessment criteria. However, it can be more easily contrasted to the Anglo-Saxon than other continental models of economic governance. Together with the Japanese model it comes closest to what is referred to as a *coordinated market economy* (Hall and Soskice, 2002). The German economic system is composed of numerous interactions among social and economic actors. The firm is perceived not as an individual competing profit-center (like in the

Anglo-Saxon world), but maintains a close dialogue with the stakeholders (labor, state and financial institutions). Consequently, the success of a German firm is measured by a more collective value it provides to the *stakeholders*. The relationship with the stakeholders is maintained not only through the established social conventions (importance of reputation!), inter-firm networks and associations, but also through the corresponding regulatory environment (i.e. high level of labor protection, prohibition of takeovers, protection of creditor rights etc. – Germany probably comes closest to be defined as 'regulatory state').

This close cooperative relationship with the stakeholders preconditions the *organic* growth of the company (in contrast to the mechanic growth in the Anglo-Saxon world). The German *financial system* remains bank-dominated, what means that companies can make use of relatively cheap capital for *long-term* investment into production, market or employee development without fears of being 'dismissed' by a stock-market (unlike in the Anglo-Saxon case). The German industrial organization is particularly known for the developed apprenticeship (on-job-training) schemes it offers to its employees. This is supported not only by the available financial instruments, but also by a highly specialized education system. The employee cross-company mobility (a fact of life in the Anglo-Saxon world) is not encouraged neither by firms' employment practices, nor by the regulatory environment. The German *industrial relations* are characterized by close and cooperative management-labor relationship (Germany has by far the lowest amount of strikes in the continental Europe) that is enabled through the supervisory function given to the stakeholder representatives in two-tier board system (in contrast to board-shop-floor divide in the Anglo-Saxon world). The German manager is neither a strategic resource allocator, nor a political figure, but rather an expert who is supposed to have answers to the technical questions arising from the shop floor level. Sometimes the German system is referred to as 'competence based' with no clear 'white collar' – 'blue collar' division. The cooperative nature of stakeholder relationship has traditionally marginalized the shareholder's impact on the company strategy. The company growth through mergers and especially hostile takeovers (i.e. discriminative tender for small shareholders shares in order to take over the control of the company) has not been tolerated by the economic and legal system.

The system, characterized by the abundance of long-term capital, high quality specialized and cooperative labor force naturally favors *secondary innovation* and the development of specialized high quality products, very often customized for the niche markets (Hall and Soskice, 2001). The disadvantages encompass lack of speed and flexibility in industrial restructuring, costly network support to the ailing industries, potential disregard of actual market needs due to the overemphasis on product development.

French (Mediterranean) model – hierarchy-based economic governance system

A long-standing tradition of the centralized state runs to the very core of the French socio-economic system. It has been characterized by an interventionist *(dirigiste)* role of the state in setting economic targets and plans. Over time, France has developed a degree of economic self-sufficiency, characterized by central allocation of resources, well functioning infrastructure and policy restraints on the free market solutions. There has been extensive public ownership of the large firms with former top civil servants occupying the managerial positions. As a result, the public companies have been first of all accountable to the politicians than to other stakeholders or the markets. However, the rationale of central policy measures has always been the creation of modern capitalist state and not some kind of centrally planned economy (Holmes, 1992). The public ownership of firms was in a way reinforced by underdeveloped financial markets, especially in the field of equity financing. Although the banking sector is generally more developed than the equity market, there is no established firm-bank relationship comparable to the German case. For example, the research and development projects were largely state-funded (in France) or close to absent (other Mediterranean countries). The *fragmented labour movements and high degree of labor-management conflict characterize the industrial relations system.* The worker representatives are only consulted on the issues directly related to their workplace, but have no possibility to influence the policy-making as in the German case. The French model has certain similarity to the Anglo-Saxon one in its clear separation of the decision-making (more precisely, plan-making!) and implementing functions. However, the underlying rationale is rather different and dictated by the hierarchical nature of societal relationships. The elitist character of French national education system and its Grandes ecoles, the graduates of which take high positions in civil service and enterprises without much prior working experience, reinforces these relationships. The system is much more based on formal academic excellence and social background than on professional competence and experience, thus managers in French organizations are more of the visionary political figures than field experts or task coordinators. The internal mobility inside/across the enterprises is very limited and 'class' boundaries relatively hard to overcome. According to Sorge (1993), the French economic system has always had more 'space on the top' than any other major systems, however, this space is reserved for the selected *ancients*. The *inter-company relationships* are usually maintained through the 'old boys' or family networks and not so much through the sectorial business associations as in Germany.

The main advantages of French economic governance model lie in the areas that need precision, careful planning and management of concentrated risk (aircrafts, TGV, nuclear power facilities etc.), clearly set responsibilities and their monitoring. However, the weaknesses are associated with the demotivated labour force, lack of managerial accountability, politicized and conflictual organizational environment, gap between the visionary thinking and implementing capabilities (Sorge, 1993). Moreover, Mediterranean economic systems traditionally suffered from the generally underdeveloped financial system and insufficient legal protection of investors/shareholders.

The new EU member states** and **economic governance system – mixture of influences

Due to the radical changes that recently occurred in their institutional environment, the new EU member states can hardly boast of any clearly defined model of economic governance or more or less fixed competitive advantages. Based on the case of Lithuania, Jucevičius (2003) presents the mixed nature of economic governance system in the transition environment. As shown in the Table 1, the aspects of each of the above mentioned governance systems can be found in the new member states.

Network-based	Market-based	Hierarchy-based
Diversity of stakeholders in corporate governance	Neo-liberal nature of transition and globalization	Hierarchical traditions and culture
Debt over equity	Neo-classical approach to privatization, competition-based restructuring	Renaissance of old structures
Continental investors	Short-term orientation and dynamic capabilities	Top-bottom approach in EU pre-accession process
Corporatist traditions	Power of executive	
Transformation of old networks		

Table 1: Factors favouring the emergence of certain socio-economic systems in the EU candidate countries

Žėruolis and Jucevičius (2002) claim that in the pre-accession process the European Commission has exercised a more top-bottom universalizing impact on the institutional structures of the new member states than it could otherwise have afforded in the case of the traditional EU member states under existing *acquis*. However, the sustainability of any uniform institutional solutions after enlargement remains under serious doubt both due to the limitations of the *acquis* and the increasing structural diversity inside the Union.

Convergence vs. divergence dilemma in economic governance

As already mentioned, there is an enduring controversy between the 'convergence vs. divergence' approaches to economic governance systems. The ailing convergence hypothesis, initially raised by Marxists and industrialists (e.g. Kerr, C., J.T. Dunlop, F.H. Harbison, C.A. Mayer), was brought back to life by the forces of globalization.

Global technological developments, trade and investment liberalization and growth of multinational enterprises with the annual turnovers exceeding that of annual GDP of some medium-sized nations have redefined some fundamental elements of national economic systems, such as the nature of firm, role of the state, approach to labor and its qualifications. Based on this, one can question the relevance of mentioned economic governance models and reconsider the viability of convergence hypothesis. In fact, the French state can no longer play its traditional interventionist role both due to the state aid restrictions within the EU and due to the increased scope for 'government failure' in more complex and volatile international market environment. The highly specialized German education system has recently accommodated a more generalist strand by in principle accepting the American-style MBA curricula what means that German managers will no longer be only specialized engineers. Besides, during the last few years Germany has experienced three hostile takeovers – something from the realm of fantasy only a decade ago! There is a generally spread argument about the Americanization of continental European systems, usually supported by the deregulatory neo-liberal nature of globalization that goes to the core of the market-based model. Nevertheless, one can also question the stability of the market-based Anglo-Saxon model that is not only exported to other countries, but also faces important challenges of survival (interestingly enough, many 'modern' American management theories borrow the ideas from radically divergent Japanese practices – what if not to compensate for the potential weaknesses of the domestic system?) Although these developments *per se* do not imply any sort of cross-national convergence, they do speak of the changes happening inside the domestic systems.

It is really problematic to draw a clear line between the impacts of globalization and European integration on national economic governance systems, especially as the latter is often presented as a response to the first. One line can possibly be drawn in the field of the regulatory impact. If globalization is more often perceived in terms of deregulation, so Europeanization is defined as a combination of deregulation *and* re-regulation (Majone, 1996). As already mentioned, the EU directs regulatory impact on traditional national institutional comparative advantages remains limited. Based on this assumption, it is possible to claim the existence of certain institutional/regulatory competition across the EU member states. The institutional/re-gulatory change that is happening within domestic systems is very much guided by their self-interest to adjust to the changing international environment. Dent (2001) claims that various national political and institutional impediments have so far prevented from achieving a singular European capitalist model, while responsibility for domestic market regulation remains largely in the hands of member states governments.

Majone (1995) claims that 'making the reconciliation of unity with diversity and competition with co-operation are the greatest challenges facing European-policy making'. Heritier (1996) analyses the process of accommodating the European diversity not only as a function of institutional setting, but also, and particularly, as a process driven by "deliberate strategies, such as the creative use of institutional channels, windows of opportunity, elements of surprise and 'a policy of stealth'. According to the author, if European policy-makers relied only on consensual institutional conditions, it would be practically impossible to accommodate the European diversity or the result would end up being the lowest common denominator. The question, however, remains if these theses could hold for the Europeanization of national economic governance systems (the policy areas covered by Heritier analysis (telecommunications, transport, environment, regional policy) fall somewhat outside the traditional economic governance focus).

I would like to argue that the potential Europeanization of economic governance systems is less likely to be driven by the deliberate strategies, even less so by the existing institutional framework. Most pressure on the national systems is likely to come in the form of policy spillover and structural adjustment needs of the member states in response to the competitive forces of globalization. As far as policy spillovers are concerned, probably the most marked pressure is influenced by the EMU project. The EMU convergence criteria have generally submitted the member states under strict neo-liberal macroeconomic agenda (Dent, 2001). The emergence and strengthening of the European stock market may lead to the reconsideration of firm performance appraisal criteria (i.e. more short-term orientation in network-based systems) with far-reaching governance implications. The role of the stock market should also be fostered by the growth of pension funds, dictated by the deteriorating demographics of the European welfare states (Morin, 2000 in Dent, 2001). However, it is still too early to predict the potential converging impacts of these developments. So far, it seems the greatest convergent pressures on national economic governance systems were driven by the forces of globalization, e.g. according to Rhodes and van Apeldoorn (1998) who define the nature of European corporate governance system as *embedded neo-liberalism* that is mainly shaped by the multinational companies and global financial capital flows.

The divergent' approach, on the other hand, is usually supported by several arguments

The first lies on the assumption that certain institutional framework renders certain comparative advantages that country seeks and should be able to preserve (see Hall and Soskice, 2001). According to Rhodes and Apeldoorn (1998), 'the efficiency and competitiveness of economies are linked to the microstructures of industries and firms whose incentives to use different organizational forms are shaped by particular national regulator policies: continued competitiveness will depend on *adjustment rather than abandonment* of those structures and policies'.

The second argument comes from the culturalist world. The culturalist perspective on the research of national economic systems has always been in a somewhat uncomfortable relationship with change in general and convergence in particular. According to culturalists, the socio-economic models are imbedded into the national value-systems, which remain particularly stable over time. Hofstede (1980) admits that cultures may change, but this change is very slow and can take hundreds of years. Moreover, even if cultures do change, they all change in similar direction, so as a result, differences across value-systems basically remain the same. (There is, however, limited empirical evidence to support these theses.)

However, the culturalist perspective may not be that fundamentally different from the institutionalist one. Culturalists in fact supplement the institutionally defined socio-economic models with the underlying values and informal social conventions (that are often ignored by the latter due to their problematic measurement). For example, very interesting findings were revealed during the survey of British, German and French students at the International Management School INSEAD (see Hofstede 1991). The students were presented with the problematic company situation and were asked to identify the major problem/solutions. Their answers were absolutely in accordance with the spirit of the above-discussed socio-economic models. The French students perceived organizational problem as a 'lack of hierarchy and unclear chain of command' (hierarchy-based system), the German students saw the problem of 'unclear responsibilities', i.e. functional specialization (network-based system), while the British identified the 'problem of communication and asymmetry of information' (market-based system). Hofstede comfortably locates all three models into a matrix by using two of his five 'cultural dimensions' (power distance and uncertainty avoidance).

Moreover, even Hofstede agrees that culture should not be used for explaining cross-national differences, if the institutional factors provide sufficient explanation. He also recognizes that culture is not inherited, but formed at the early age of person's life. A mutually reinforcing relationship between structure (institutions) and culture is also in principle recognized. The main point of disagreement concerns the perceived intensity of structural-cultural change. Culturalists like Hofstede particularly focus on the institutions that are less susceptible to change (e.g. family) than the more macro-level institutions that interest political economists.

A very interesting (though lacking empirical depth) perspective into comparative capitalists systems is proposed by Ch. Hampden-Turner and F. Trompen-aars (2000). It has gained popularity in the business world, however, remains virtually unknown among political economists. According to them, every capitalist model is resolving its own cultural dilemmas/paradoxes and, depending on the balance it finds between the two extremes, is either following 'virtuous' or trapped into 'vicious' circle. They analyze the traditional advantages and disadvantages of national models of organization (American, Japanese, French, German). The success of every system is very much determined by the way it perceives its own limitation and is capable of balancing it with another extreme. Every system that tends to rely on one extreme (e.g. individualist or collectivist) is doomed to enter the 'vicious circle'. Not surprisingly then, the American management methodologists are looking into the Japanese cases, in order to fight the 'inherent' weaknesses of the domestic system (e.g. to introduce more cooperative, long-term oriented work culture). Such perspective comes closest to the concept of the *collective learning* - whether on the social, or on the organizational level – that should be highly applicable in the enlarged European Union.

EU economic governance system for collective learning

The enlarged Europe faces at least two challenges – *accommodating diversity* and *accommodating change*. On the micro-level, these challenges are resolved through the process of *organizational learning* (e.g. see Senge, 1990). On the macro-level, one should look for the answers by creating the *learning economy* (OECD, 1996), a concept that goes even beyond *knowledge-based economy* (which EU seeks to become according to the Lisbon strategy).

The learning economy is defined as "the organizations of firms that facilitates the effective responses to technological change through the accumulation of know-how, continuous adaptation to new knowledge, and the pursuit of new, higher quality, and more effective production" (M. Storper *et al.*, 1998). The underlying rationale of contemporary industrial strategy is not to have a centrally formulated and formally adopted strategy document with a detailed course of action, but *an adequate coordination of micro and macro-level policies through the established social conventions and informal rules, in order to ensure the process of collective learning.*

Thus, the *learning economy* can hardly be built and sustained neither through regulatory measures, nor through mere market liberalization (both aspects especially characteristic of the europeanization process...). It primarily involves the development of new modes for policy co-ordination, distribution of learning capabilities, investment into both general and specific labor competences, and development of competence-based networks. It means that the national and supranational EU economic governance systems should be adjusted to the requirements of the learning economy and not under-estimate the *soft* aspects of the governance structures. In this context, it is also worth mentioning that some of the dominating Western (in fact, American) competitiveness methodologies, e.g. the influential M. Porter's 'Diamond model', show little consideration to the soft aspects of competitiveness, such as organizational culture and learning process (O'Shaughnessy, 1997). Their traditional focus of analysis is often subject to criticism for being limited on the developed world's leading economies, and, thus, for considering only the 'success factors' rather than a wider and more complex picture. Besides, there is some inherent cautiousness about industrial policy suggestions among the American methodologists, like M. Porter or P. Drucker, although this should be a logical evolution of many of their arguments (O'Shaughnessy, 1997). So, although often very useful tools in the policy making, the American theoretical models may not provide the universal explanations and guidelines, but should be adjusted to the local processes.

It has to be born in mind that the *learning economy* can hardly be built and sustained through the regulatory measures or mere market liberalization – on the contrary, such measures may simply erode the comparative institutional advantages of old and new member states. It is exactly in the light of the learning economy that one has to reconsider the role of the state institutions – both on national and European level. A 'strong' state is not the one that, in a *modernist* conception, has competence to impose centrally devised plans on economic actors. The neoliberal environment has already largely eroded this state power. Nor does a clearly defined, homogenous and static model of governance characterize a 'strong' state/organization. The boundaries between traditional models are becoming vague and overlapping, while the EU never had a conventional economic 'model' in its own right. Thus, a *"strong" state/organization is* a *dynamic system that suc-*

cessfully integrates diverging elements and ensures their synergic interaction (Jucevičius, 2003). The forthcoming EU enlargement creates excellent preconditions for such a paradigm shift.

Concluding remarks

The European Union is primarily characterized through the market integration processes. It can hardly claim to have a homogenous, institutionally complementary economic governance system that renders clear-cut institutional comparative advantages. It remains a set of diverse and mixed national economic governance systems/models that seek to preserve their domestic advantages and, if possible, to export them to the EU level. The Eastern enlargement supports the thesis of increasing institutional diversity inside the EU. The traditional EU regulatory approach in its economic policies will be increasingly less feasible in the changed and more heterogeneous institutional environment. It is neither by relying on the *hierarchical* regulatory solutions, nor merely on the *market* deregulation, that the EU can realize its ambitious Lisbon competitiveness objectives. Instead, the EU needs to adjust its structures to enable the *collective learning and innovations* among and inside the member states by creating and disseminating competence networks, benchmarking practices, promoting the innovation-friendly social conventions. The forthcoming enlargement will add to the external environment. The forthcoming enlargement will bring about an increased diversity and complexity inside the Union that can be turned either into a vicious or virtuous circle. To follow a later scenario, the Union needs to soften its traditional regulatory priorities towards softer instruments like learning partnerships, benchmarking, interchange and adaptation of the Community best practices etc. The policy climate in the EU should be such as to enable the member states to adequately develop their diverse national economic governance systems so that they are able meet the global competitiveness requirements and build/sustain their proper competitive advantages.

References

COWLES, M.G./ RISSE, TH.: 'Transforming Europe. Conclusions' in Cowles, M.G./Caporaso, J./Risse, Th. 'Europeanization and Domestic Change. Transforming Europe'. Ithaca and London 2001 (Cornelss University Press).

COWLES, M.G./CAPORASO, J./RISSE, TH.: Transforming Europe: Europeanization and Domestic Change. Ithaca, NY 2001 (Cornell University Press).

CROUCH, C./STREECK, W.: 'Institutional Capitalism: Diversity and Performance'. In: CROUCH, C./STREECK, W. (eds.): Modern Capitalism or Modern Capitalisms. London

1997 (Sage). (reference in M.Rhodes and B.van Apeldoorn, 'The Transformation of West European Capitalism?')

DARMER, M./ KUYPER, L. (eds.): Industry and the European Union. Cheltenham 2000 (Edward Elgar), p. 377.

DENT, CH.: 'Governing the EU economy as a whole'. In: THOMPSON, G. (ed.) Governing the European Economy. London 2001 (Sage Publishers).

DYKER, D. (ed.): 'The European economy'. Harlow, Essex, England, New York 1999 (Longman).

HALL, P.A./ SOSKICE, D.: 'Varieties of Capitalism. The Institutional Foundations of Comparative Advantage'. Oxford 2001 (Oxford University Press).

HAMPDEN-TURNER, CH./TROMPENAARS, F.: Building cross-cultural competence. How to create wealth from conflicting values. Chichester 2000 (John Wiley & Sons Ltd).

HICKSON, D. (ed.): Management in Western Europe. Society, culture and organizations in twelve nations. Berlin, New York 1993 (Gruyter).

HIX, S.: 'The Study of the European Union II: The "New Governance" Agenda and Its Rival'. Journal of European Public Policy 5(1994)1, pp. 38-65.

HIX, S.: "The Study of the European Community: The Challenge to Comparative Politics." West European Politics 17:1 (January), 1994, pp. 1-30.

HOFSTEDE, G.: Culture's Consequences: International Differences in work-related values. Beverly Hills, California 1980 (Sage Publications).

HOFSTEDE, G.: Cultures and Organizations: Software of the Mind. London 1991 (Mcgraw-Hill).

HOLMES, P.: The National Economies of Europe, France. DYKER, D.(ed.) London 1992 (Longman).

JUCEVIČIUS, G.: Hierarchy-Market-Network: the Patterns of Emerging Economic Governance System in an EU Pre-accession Country. Socialiniai mokslai 4(2003)41.

KAPTEYN, P.: 'The stateless market. The European Dilemma of Integration and Civilization', London, New York 1996 (Routledge).

PELKMANS, J.: European integration. Methods and Economic analysis. Netherlands Open University, Heerlen 1997 (Longman).

KERR, C./DUNLOP, J.T./HARBISON, F.H./MAYER, C.A.: Industrialism and Industrial Man. Cambridge 1960 (Harvard University Press).

LEIBFRIED, S./PIERSON, P.: 'Social policy' in Wallace. 'Policy-making in the EU'. Oxford 1996 (University Press).

MAJONE, G. (ed.): 'Regulating Europe'. London and New York 1996 (Routledge).

MARKS, G.: 'Structural Policy after Maastricht'. In: CAFRUNY A./ROSENTHAL G. (eds.): The State of the European Community. New York 1993 (Lynne Rienner), pp. 391-410.

MORAVCSIK, A.: The Choice for Europe. Social Purpose and State Power from Messina to Maastricht. Ithaca, New York 1999 (Cornell University Press).

OECD: Employment and Growth in the Knowledge-Based Economy. Paris 1996 (OECD).

O'SHAUGHNESSY, N.: The idea of competitive advantage and the ideas of Michael Porter. Strategic Change 6(1997), p. 73-83.

OUCHI, W.G.: Theory Z: How American business can meet the Japanese challenge, Reading, MA, 1981 (Addison-Wesley).

PITELIS, CH.: (1994) "Industrial Strategy: For Britain, in Europe and the World", Journal of Economic Studies 21 (5), pp. 3-92.

RHODES, M./VAN APELDOORN, B.: 1998 The Transformation of West European Capitalism? European University Institute, Working Paper, Robert Schuman Centre No. 60/1997.

RICHARDSON, G.B.: The Organization of Industry. The Economic Journal. Vol.82, n 327(1972), sept., pp.883-896.

SBRAGIA, A.: 'Environmental policy'. In: WALLACE, H. AND W. (eds.): 'Policy-making in the European Union'. Oxford 1996 (Oxford University Press).

SENGE, P.: The Fifth Discipline: The Art and Practice of the Learning Organization. New York 1990 (Currency Doubleday).

SORGE, A.: Management in France. In: HICKSON (ed.): Management in Western Europe. 1993 (Gruyter).

STEPHEN, G./BACHE, I.: Politics in the European Union. 1st ed. New York 2001 (Oxford University Press).

TSOUKALIS, L./RHODES, M.: Economic Integration and the Nation-State. In: RHODES, M./HEYWOOD, P/WRIGHT, V. (eds.): Developments in West European Politics. London 1997 (Macmillan), pp. 19-36.

WALLACE, H./WALLACE W. (eds.): Policy-making in the European Union. Brussels 1996.

WHITLEY, R.D. (ed.): European Business Systems. Firms and Markets in their National Contexts. London, Newbury Park, New Delhi 1992 (Sage Publications Ltd.).

WHITLEY, R.D. (ed.): 'European Business Systems. Firms and Markets in their National Context'. London, Newbury Park, New Delhi 1992 (Sage Publications Ltd.).

ŽĖRUOLIS, D./JUCEVIČIUS, G.: Strateginis matmuo Lietuvos ekonominėje politikoje – turinys ir lyginamoji analizė. Lietuvos metinė strateginė apžvalga, 2003 (Ge. J. Žemačio Lietuvos karo akademija ir Vilniaus universitetas). Vilnius 2003, pp. 237-258.

Marija Jotautienė, Brigita Janiūnaitė

The Importance of Educational Innovation Implementation in Achieving the Qualities of the Knowledge Society: What Barriers Can Emerge at the Initiation Phase and How to Identify Them?

Introduction

In the modern world the society lives the period that is distinguished by intensive striving for qualities of knowledge and information society as well as different technological, political, social and cultural changes. As Melnikas (2002) points out, in the common context of change, transformation processes accumulate and generate all fields of life that involve innovations, and topics of their implementation become very important and relevant in perceiving that in many cases effectiveness of innovations determine success of transformations and influence results of politics, social field and economical reforms.

Changes in society influence changes in education as well: in the system itself, institutional environment, level of an individual. The change in educational system gains particular importance as it is to be related not only to education, but also to social historic shift into the knowledge society, i.e. the educational system has to empower individuals to live and act in the knowledge society. One of the ways to induce a change in education is educational innovations. As a group of social innovations, they become especially important in the society development because, on the one hand, education influences society development, and on the other hand, social changes influence educational change.

Educational innovations as a research field are analyzed in the works of most researchers (Krainer, 2001; Dalin, Rolf, Kleekamp, 1999; Reilly, Tushman, 1997; Eadie, 1997, etc.), which reveal the concept of innovation, disclose processes of innovation implementation in various fields, their structures, analyze aspects of innovation classification as well as particularity of the management of innovation processes. Gilmanov (1997), Sovetova (2000), Zagvezinski (1998), Angelovski (1991), etc. analyze the nature of innovation implementation in the countries under transformations that are characteristic for the countries of Eastern and Middle Europe. Different aspects of educational innovations as a phenomenon and their implementation are disclosed in researchers' works (Stoll, Fink, 1998; Hopkins, Ainscow, West, 1998; Fullan, 2001; Stringfield, 1993, etc.).

Lithuanian researchers Janiūnaitė (in the period from 1999 to 2002), Tumėnienė (2002), Želvys (1999), Valiuškevičiūtė (2001) who analyze the importance of educational innovation implementation in the context of educational institutions modernization reveal peculiarities of teachers' innovative work and specific aspects of educational innovation implementation – classification, models of educational innovation implementation and teachers' innovation action research.

On the one hand, as the analysis of the above-mentioned researchers' works and practical experience shows that during the implementation of innovations as well as educational ones a lot of disturbances or, in other words, barriers appear. It is especially important to identify these barriers in the phase of initiation. Why? The success of innovation implementation depends on the success of this phase as during the phase of initiation one decides to implement an innovation, implementation environment, features of innovations, competence of implementers, etc. become important. It is also necessary to evaluate that every further phase is a concentric growth of the earlier phase, which can be treated as a continuation of the phase of initiation. The mentioned aspects become especially important in implementing external educational innovations, particularly during different international projects when the research expertise of these innovations does not take place all the time. So in the mentioned context the following questions of the research emerge: how to identify the barriers of educational innovation implementation during the phase of initiation as well as what barriers emerge in implementing systemic external educational innovation? The above mentioned issues make up the research problem of the article.

The ***research subject*** is the barriers of educational innovations in the phase of initiation.

The ***aim of the article*** is to reveal the barriers of educational innovation implementation at the phase of initiation in the context of striving for the knowledge society qualities.

The following ***research methods*** are applied in the article: analysis of research literature, case study, document analysis, and interview.

The ***article consists of four parts***. The *first part* substantiates the methodology of identification of the barriers of educational innovation implementation at the phase of initiation; the *second part* presents the methods and organization of the research; the *third part* defines the case of systemic external educational innovation; the *fourth part* presents the results of empirical research.

1. Methodology of identification of educational innovation implementation barriers at the phase of initiation

Multistructuralism, which is reflected by different structures and phases, is characteristic for the process of educational innovation implementation. Fullan

(1991) assumes that researchers refer to the first phase of innovation implementation in different ways: initiation or adoption; but its essence is the process during which one decides if the innovation will be implemented. At the phase of initiation the success depends on a lot of factors which researchers analyzing the phenomenon of phases present in certain order calling them determining factors of the success; however in the realization of the article's purpose – *to reveal barriers of educational innovations implementation* – it is possible to to think *that insufficient expression of these factors during the process of educational innovation implementation will become the barriers limiting the success of innovation implementation.* In order to substantiate the formulated statement, the analysis of research literature is performed following some logical sequence that the absence of factors determining the success is a theoretical basis to reveal presumptive barriers. Stoll, Fink (1998), referring to Fullan (1991), state that essential factors capable to influence the success of the phase of initiation are as follows: a) *necessity to renew improvement process;* b) *readiness of personnel to participate in innovative processes;* c) *accessible resources including both the time, and the support.* But these are not the only factors on which initiation depends; the success can be influenced by *the quality of the innovations already present* (Vonlanthen, 1995*), their accessibility, approval of teachers and internal as well as external environment, new policy and funds, direction of problem solution.* Fullan (1991) assumes that successful start/beginning of innovation implementation can be influenced by *innovation's relation to region's objectives and the needs; clear model of well-structured change; active supporters or defenders who understand the innovation and induce its implementation; active stimulation to begin the implementation of innovations; a good innovation.*

Fullan (1991) emphasizes that a lot of innovations influencing the change of value range can exist in educational institutions, but their dissemination in an institution depends on how pluralistic and heterogeneous the *culture of the organization* is.

The second factor – accessibility of the innovation – is to be related to the information about it. Fullan (1991) discloses that the dissemination can take place appropriately using personal communication and creating infrastructures in educational organizations and overreaching the educational institution. The author points out that the accessibility of the innovation depends on the infrastructure of overall relations. The phenomenon of the innovation at the phase of initiation is related to new knowledge. According to Stankevičiūtė (2002), the following aspects are important in the context of knowledge management infrastructure: a) effective application of different technological communication mechanisms; b) internal basis of information accessible to all organization members; c) necessary guarantee of material resources. Thus in the context of innovation accessibility *incomplete dissemination of information about the innovation, which emerges due to ineffective application of different communication technologies, closed internal basis of knowledge and insufficient resources at the phase of initiation* are treated as barriers.

Fullan (1991) supports the importance of the third factor by pointing out that in the environment of educational institutions the initiation of innovation is seldom successful without the support of the central administration (e.g. educational department, county administration), especially in evaluating that the educational institution has to acquire the right to act, i.e. in the phase of initiation the performance of institutions depends on taking particular decisions. Hence *central administration* is a very important source or *barrier* of support and defense of the innovation being initiated, *if in this level attitudes of the support are not expressed, or efforts of educational institutions* to initiate the implementation of innovations *are being ignored.* On the other hand, the role of the manager of the educational institution is also very important: he/she can become the source of overcharges in the organization if he/she tries to initiate a lot of innovations or block them using the support of the central administration. Consequently, *the manager can become a critical source of initiation* at the level of the institution.

According to Fullan (2001) teachers of educational institutions, even having less possibilities to find new ideas and not always enough time to realize them, still implement innovations. Teachers' innovatory work is explained by means of the search of self-realization ways, but as McLauglin and Talbert (2001) note the conditions of teachers' work largely are not favorable for teachers' innovatory work. Hargreaves, Fullan (1998) affirm that different associations of teachers can be important initiators of educational innovations implementation and this must be appropriately used as most teachers are in the mood to implement innovations into their work, but certain conditions have to be created for them: innovations have to be clear and of practical character, they have to be supported by the leaders of the organization, proper conditions for collaboration with other teachers have to be guaranteed, necessary resources for innovation implementation have to be granted and so on. Fullan (2001) points out that most teachers do not possess necessary information, time or just energy, so the innovations that they implement usually are not of large scope and it is hard to believe they will spread in the educational institution without the support of teachers' initiatives. Evaluating the remarks of the mentioned authors in the context of barrier formation it is possible to envisage that *having not guaranteed appropriate conditions for the expression of teachers' innovatory work, this will obtain* the expression of the *barrier* in the context of innovation implementation. The second barrier is *non-support of teachers' initiatives* in the background of this factor.

At the phase of initiation, community as a subject of education is important because the community in respect of an educational institution can use certain power in initiating the innovation implementation in it: to encourage and support them, reject them or simply to do nothing in the context of innovation initiation. This, according to the researcher, is influenced by particularity of the environment surrounding the educational institution in which the community has formed or even its certain exceptional features (e.g. demographic indicators).

A lot of new initiatives often emerge in the level of educational policy, which in a particular institution can be accepted or rejected. But one has to properly evaluate *bureaucratic orientation of problem solution*, which can influence the initiation and manifest itself as a barrier in the context of educational innovation implementation. This is explained so that local authorities can anticipate solving specific problems by means of additional funds. Then the initiation of innovations in respect of the institution can be formally delegated activity not necessarily expressed by characteristics of innovativeness.

Summarizing the factors influencing the success of the initiation phase, Fullan (2001) points out that the initiation depends on the variety of the application of the mentioned factors and their coordination: secondly, it is very important what we know about the initiation, i.e. what has to happen in mobilizing the teachers' community and how we are ready to plan changes. In this context the authors of the article formulate the insight that *insufficient capacity of educational institutions to use the factors influencing the initiation phase, emphasizing their variety and harmony, as well as insufficient knowledge about the initiation phase as the phenomenon of the process of the educational innovations implementation, evaluating the management aspects, are the essential barriers, negatively influencing the realization of the initiation phase.*

In the initiation phase of the educational innovation implementation the very *innovative idea* is an important aspect. The impulse of innovative ideas can occur from external or internal sources. In other words, the creation of ideas can depend on people and information flow between the institution and environment. Stoner (1999), stressing the importance of the sources, name them as follows: a) outside experts; in an educational organization they are researchers, consultants, experts, employees of the Ministry of Education, members of accreditation commissions, local and foreign partners, agents of educational innovation implementation and their competence; b) permanent members of the organization – school's administration, teachers (lecturers), learners, other employees.

Having not evaluated the importance of the named sources or the insufficient ability of the organization's members to use all possible sources for the search being present, the field of "idea search" becomes limited, and then it is treated as the barrier in the context of idea generation, as well as the initiation phase.

Thus the based barriers can be in general subdivided into external and internal as well as they can emerge in the levels of an individual, an organization and educational policy. This subdivision is the methodological basis for the identification of the barriers of the educational innovation implementation at the initiation phase (see Table 1, pp. 51-53).

2. Methods and organization of educational research on innovation barriers at the initiation phase

In order to carry out the research of the barriers in the implementation of educational innovations at the initiation phase, a ***case study*** was applied. Creswell (1998) points out the definition of the case and its ***context*** in the research structure of one case. For this purpose the method of the qualitative traditional analysis of the document content was applied. The main purpose of the realization of the chosen method is *to disclose the peculiarities of the systemic educational innovation implemented by the PHARE program, i.e. to disclose the research context.* For the realization of the purpose the official verbal documents were chosen, namely: information publications of the program of vocational training reform "Reform Steps/Phases" in 1998–2002, the conference "Towards Integration: Importance of Vocational Training Strategy for Economic and Social Development" (1998), "The White Paper" on vocational training. Another method is the interview. The *purposeful interview* type was applied for the respondent group – ***external experts***. It was chosen after the situation was evaluated during the pilot research, the protocols of institutions' external audit as well as having carried out the interview with the National expert; this enabled to cumulate certain quantity of the data before carrying out the interview with the respondents of this group, and during the latter interview the *subjective* evaluations of the *respondents* enabling to disclose the barriers at the initiation phase wider were very important.

The respondent group of ***external experts*** was formed considering the fact that on the basis of the PHARE program essential structural changes started in the system of Lithuanian vocational training – a new institutional link has formed: higher non-university institutions – colleges. The readiness to realize the programs of non-university studies is the beginning of the institutionalization of the innovations implemented by the PHARE program (of standards, systems of quality sponsorship, modular programs). On the other hand, with reference to the goal of activity delegated to external experts, to identify the preparation of the institutions to realize non-university study programs; this corresponds to the segments of the idea adoption of the initiation phase "making realization plans" and "decision to realize plans". In other words, the experience of these respondents helped to reveal the preparation of institutions to realize essential innovations – non-university study programs. The chosen respondents were of different levels and they had different flows of information at their disposition. One of them – the Chair of the External Council of Experts, the others represent the expert groups of different fields: technical-technological and social-humanities as, while identifying the preparation of the institutions to realize the non-university study programs, on the basis of such a principle the work groups of the external experts were formed.

The analysis of the data obtained during the interview is performed applying the method of *content analysis* and realizing the following steps:

1. While reading the stenographs, the manifestational essential aspects, which are reflected by phrases, statements, words, are distinguished;
2. While identifying the notional elements, the data is coded;
3. The structure of the data is built by grouping the notional elements into categories and subcategories;
4. The categories are integrated into the analyzed context. The integration of the obtained data takes place considering the statements of hermeneutics (Ricoeur, 2000), i.e. it is performed by the text-oriented interpretation.

3. The character of educational innovations implemented during the PHARE program

The aim of the PHARE program is the modernization of vocational training. In its broad sense, the aim reveals the possibilities of the implementation of European methodological attitudes – *to help the subsystem of vocational training to train professional, flexible and mobile workforce, which would be able to contribute to the development of the Lithuanian society and would be able to realize themselves in the intercultural environment.* It is possible to state that already at the initiation phase during the aim formulation the program's idea was directed to the attitudes warranting conditions, the essence of which is organization of the system; the attitudes directed to the personality stressing the improvement/perfection of a human being as the central figure as well as the premises enabling the implementation and realization of the previous attitudes. This is revealed by the objectives to be implemented and the choice of the objects that are implementing the program. In order to guarantee solidity/integrity of vocational training as well as the possibility of continuity, all links of vocational training were incorporated into the program – college, vocational and labor market training. Such choice disclosed the idea of bridges between the single links of the system. In the methodological sense, the PHARE program is important as it brought the modern methodology of the program implementation from the West; *its essence is that at the beginning the experiment takes place and only afterwards it is disseminated in all the whole the system,* as well as in *the systemic content*, which manifests in the realization of five internal projects during the program implementation, and this reflects the main idea of the program – the holistic modernization of the system, which stimulates the systemic reform of vocational training. These projects had specific aims, which caused considerable qualitative changes. Having decided to realize the PHARE program, the implementation of educational innovations was initiated in the following fields:

- ***Renewal of teaching/study programs.*** The goal of this project realization was to prepare and to accomplish *new programs of vocational training,* as well as the *model of standards* of national vocational training and employment. These were the essential innovations in the contemporary subsystem of vocational training. In preparing new or improving present programs *the attitudes of modality, flexibility, progressive specialization, relevance, monitoring and review were considered.*
- ***Training of teachers and leaders.*** The aim of this project initiation is to effectively train teachers, following modern teaching methodics, to train the leaders of educational institutions to manage education in a modern way. This activity was initiated referring to the analysis of the needs, thus the thematic scope was very diverse and wide. The training of teachers and leaders took place according to the designed programs of the qualification improvement.
- ***Resources.*** The project aim is to modernize the teaching material and relevant equipment of educational institutions in order to be able to implement new teaching programs. The EU experts, employed on a short turn basis, contributed to the accomplishment of the analysis of the project needs. The equipment provision was accomplished within the contest of equipment purchase in the countries of the EU PHARE program and Lithuania.
- ***Partnership with the educational institutions of the European Union.*** This project was initiated seeking to start the work contacts with the European educational institutions of the appropriate level, sharing the knowledge and experience that was accumulated in implementing the PHARE program: *the knowledge about the methodics of the study program design, the application of active teaching and learning methods as well as the preparation of new teaching material.*
- ***Formation of educational policy and dissemination of results.*** The project aims to create favorable conditions for continuous development of national and local educational policy and to circulate the results effectively. The dissemination of the experience is carried out using the scheme of the national partnership, implementing the *cascade model of* the *experience dissemination* in order to guarantee the program continuity in all possible ways.

The follow-up of the PHARE'94 program – was the PHARE'97. The aim of this program was to use the support of the EU while implementing the reform of the vocational training system in Lithuania.

At the initiation phase of the program the objectives disclosing the levels and objectives of the innovations to be implemented were formulated:

a) ***to extend the process of the institutional development*** (*the level of educational system,* the object – *the educational policy and management*);

b) ***to implement the national standards of vocational training and to design the system of quality security*** (the *level of an educational system*; the object – *a structural element of vocational training*; *the level of educational intuitions*, the object – *the subsystem of management*)

c) ***to disseminate and consolidate the experience of the PHARE'94 Vocational training reform*** (*the level of didactic/pedagogic system*, the object – *the teaching aims, curricula, methods, technologies; the level of an educational institution*, the object – *technical-technological and psychological subsystems*); d) ***to create the policy and strategies of vocational training considering the priorities of the integration into the European Union*** (*the level of an educational system,* the object – *the educational policy and management*). The strategy for the implementation of the program's aims and objectives was projected and it was decided to further follow the model of the active participation of the Program Coordinating Centre (PCC), which was created when implementing the PHARE'94.

In summary it is possible to assume that the PHARE program displays in the original structure of the content, which consists of different foreign educational innovations, which are implemented as single projects at the levels of the educational system (the formation of educational policy, the reformation of vocational training subsystem, the preparation of vocational training standards, regional development), educational institution (the improvement of management, the improvement of qualification, the development of material resources, the creation and implementation of quality sponsorship/voucher system) and didactic/pedagogical system (the modular programs of teaching/learning, the preparation of teaching/learning material, the implementation of active methods of teaching/learning). At the level of the educational system their implementation objects involve the educational policy, the structural elements of the educational system; at the level of the didactic/pedagogical system the objects of the educational innovation implementation are the educational aims, content, methods, at the organization's level – the management, structural and psychological subsystems. This allows naming the implemented innovations as a systemic educational innovation.

4. Analysis of the research results: the barriers of systemic educational innovation implementation at the initiation phase

While analyzing the external experts' survey, the main categories are distinguished referring to the methods of the external evaluation. The purpose of the external experts' activity (the expertise) was to identify the preparation to implement the non-university study programs, evaluating ten fields of the institutional activity. At the initial stage the activity fields are defined as categories; the subcategories revealed by the respondents are treated as particular barriers for the successful realization of every field. All obtained results are analyzed in the context of the initiation phase for the preparation to implement innovations is one of the activities of the initiation phase. Thus it is possible to treat the activity fields, which were evaluated by the external experts, in the context of the initiation phase of the innovation implementation process as one category – *the preparatory work*, and the subcategories are important so that by means of them the essential barriers are disclosed. The respondents of the interview distinguished 21 subcategories, which in the context of the implementation process of the educational innovations should be treated as the barriers, able to influence the successful preparation for the stage of the innovation implementation. From the plenty of the distinguished elements it is possible to make some insights that having summed all results of the respondents of this group at the initiation phase of the implementation process of educational innovations, the greatest barrier occurs in the field of management, which the respondents name as the subcategory of "management", "the organizational environment" would follow, and the subcategory "human resources" would become the third subcategory. Among the disparate subcategories the logical-notional links exist; consequently they are grouped into bigger logical-notional groups.

The subcategories distinguished by the respondents such as "perception", "self-government", "learning culture", "information accumulation and presentation", "disposition" are related to one logical-notional group of "organizational environment", the elements of which resolved according to the elements of the institutional environment distinguished by Vonlanthen (1995): employees' knowledge and skills, technical-material resources and the additional factor – the important attitudes and approaches of the participants. Then the distinguished category by the respondents can be generalized by the main category "organizational environment" and be treated as the main category.

Referring to the grouping of the obtained results during the interview, it is possible to assume that the organizational environment becomes the barrier due to *insufficient knowledge and skills of employees* at the initiation phase of educational innovation implementation. The employees' approaches and attitudes in the context of the organizational environment are also important, and the physical–material resources are one of the premises to discover the barrier, but that was not the essential one. Evaluating that the knowledge, skills and attitudes are the components of the competence structure, it is possible to draw a conclusion that in the organizational environment the barrier is the insufficient competence.

While analyzing the interview results, the management subcategory manifests as the essential barrier during the respondents' replies. Treating the management as one of the management elements (James et al., 1999), it is possible to relate the subcategories distinguished by the respondents integrating by the category of the "*management*" in the logical sense.

Then the category *"management"* would reflect the common combination of the subcategories presented by the respondents "***planning***", "***organizing***", "***leadership***", "***approaches and attitudes***", coded by the respondents in the concepts "***management***", "***leadership***", "***team management***", "***leaders' attitudes***". The respondents distinguish four subcategories, which are the management components: ***autonomy, team management, management, and leadership***. The management subcategory is named in two different fields of the institutions' activity: in the field of the management and quality sponsorship system implementation.

During the interview the respondents revealed the role of the institutions' leaders at the initiation phase. During the research the disclosed elements of the subcategories reveal certain essential but not insufficient competences at the initiation phase. The institutions aspire to acquire a new quality and to provide a new educational service of higher non-university education. With reference to Stoner et al. (1999), this kind of aspiration may be treated as a planned change: its wide scope involves both the system and teaching within institutions and structural changes in the subsystem of higher education. Thus the elements of the subcategories distinguished by the respondents "*incompetence to identify priorities, inability to understand the mission, incompetence to reason strategically*", etc. reveal that the essence of the planned change is not understood at all institutions, i.e. the managerial competence is insufficient in order to implement the planned changes. On the other hand, for the institutions where the planning is one of the reasons of the formation of the management barrier it is hard to hope for the competitive advantage in the educational market, as the reasons named by the respondents *"incompetence to reason strategically"*, *"incompetence to solve problems"* and so on can substantially influence the effectiveness of the institution activity. The respondents also disclose the insufficient understanding of the roles, the function of delegating, the information communication, etc. Referring to the respondents' statements and the groups of the barriers decoded from them, it is possible to assume that institutions are not oriented enough to the improvement of activity, using management in cooperation. In the context of the management the respondents distinguish an interesting notice in the aspect of the external influence – "the certain frame" formulated for organizations, i.e. the requirements, which oddly limit proactive ness characteristic for the initiation phase of the innovation implementation, formulated too strictly. Thus the approaches presented by the respondents inspire the insight that in the institutions getting ready to implement systemic innovations the basic barrier is the ***management*** at the initiation phase at the stage of the idea adoption (Thom, Zaugg, 2001), the elements highlighting it are ***insufficient competence of organizing, managing and planning.***

It is possible to split the resources as the barrier according to their structure into two groups: human and material. Non-university study programs and the applied research related to them are the essential innovations at the institutions visited by the experts. The elements, which were encoded in the respondents' replies during the interview, and in the context of the result interpretation reinforces the expression of the barriers, referring to Vonlanthen (1995), are grouped and two groups of the reasons become important, namely: c***ompetence,*** evaluating knowledge, abilities and attitudes, as well as ***organizational regulation***. Analyzing the organizational regulation in the sense of resources the managerial barrier – ***the competence of planning*** – is again unclosed. One of the respondents envisages also the ***insularity/closure of organizations*** as the element consolidating the expression of the barrier in the context of the resource barrier. He relates this to the search for the innovative ideas, argumenting that the opportunity of training at other institutions would be the source of innovative ideas, which also would bring the wider spectrum of innovations, while widening the aspect of competence, into the organization. In respect of the resources, referring to the respondents' approach, the scientific competence is very significant.

Insufficient research competence is to be related to the statements presented by the respondents already in the previous part of the interpretation: the insufficient evaluation of the mission. In terms of the managerial barrier, the ***insufficient competence of strategic forecasting,*** which was proved by one of the respondents by the statement "*one misses some clear strategy in the institutions. One limits oneself only by the present day"*, is envisaged. Evaluating the factors influencing the phase success defined by Fullan (1991), Stoll Fink (1998), as well as the fact that these researchers refer to the available resources and the preparation of human recourses to participate in innovation process as a factor, the barrier of insufficient resources is of particular relevance in the phase of initiation.

Summarizing the interview results, it was disclosed that quite a lot of attention is focused on relationship at the initiation phase. The respondents reveal this aspect by distinguishing the subcategories "international relations", "relationship with social partners", "relationship with educational policy makers", and "relationship with educational institutions". Referring to the earlier logical sequence of result generalization it is possible to assume the categories distinguished by the respondents in the context of the research. They can be treated as the barrier of the ***absence of relationship*** at the initiation phase of the educational innovation implementation. Treating the relationship as one of the barriers it is possible to state that the main factor – competence influences its formation equally.

With reference to the interview respondents' ideas and the factors of the initiation phase distinguished by Fullan (1991), one may assume that the barrier of relationship determines institutional isolation. This may constitute a threat as this may cause incomplete understanding of policy and problems of its implementation, ineffective use of funds and their improvement potential, and failure to use the support of central administration. In addition, according to Zagvezinski and Pilmanova (1998), the potential of external initiators of innovations (researchers, ministry, foreign partners) may be insufficiently employed. Thus as the relationship barrier exists in institutions, the external sources of ideas are partially used at the initiation phase because the creation of ideas depends on people and information flow between the institution and environment.

The respondents also distinguish certain specific subcategories, i.e. the ones that are related not only to the institutional environment: ***external pressure and political support of education***. According to Vonlanthen (1995), these two subcategories can be grouped into one: the factor of external environment as a barrier, and the elements of the subcategories involving the groups of external attitudes and decisions. Referring to the statements expressed by the respondents, the external attitudes, distinguishing unfavorable attitudes of educational politicians can be named as the premises for occurrence of the barrier of external environment at the initiation phase of the innovation implementation process. Reflecting the above-distinguished barrier "relationship", which, according to the respondents, develops at the institutional environment, the external barrier of the environment can be treated as the premise for the occurrence of the relationship barrier. This barrier, having evaluated the statement "*... the suppression of the innovative activity at institutions*", according to Fullan (1991), reflects *the bureaucratic orientation to problem solution*, and the group of barriers' reasons "educational policy" *("...they are not subject to take more flexible decision")* enables to state that the institutions wishing to implement innovations are not always supported by the central administration. Having revealed this barrier, at the same time the insufficient influence of the factor "community pressure, support and indifference" (Fullan, 1991), which becomes the barrier at the particular case for the implementation of educational innovations at institutions.

Summarizing the results obtained during the interview of external experts' group, it should be stated that this group of the respondents presented the most numerous amount of the information, which is particularly important in the context of the research.

Phase	THEORETICALLY BASED BARRIERS	EMPIRICAL RESEARCH
	Characteristics	***Barriers***
INITIATION	Lack of competence	Insufficient competence
	Lack of creativity	Non-innovative initiatives
	Non-innovative management	Non-innovative management Insufficient relations
	Unfavorable organizational environment	Insufficient generation and acceptance of ideas
	Existing low-quality innovations	Ill-prepared organizational environment
	Influence of educational politicians	Lack of educational politicians' support
	Influence of central administration	Unfavorable attitudes of central administration
	Influence of community	Unformed attitudes of society
	Unprepared	*Insufficient resources*
		Socio-cultural environment
		Inappropriate information presentation/communication

Table 2: The Congruence of the Research Methodology and Interview Results

The main barriers of the initiation phase, which were highlighted by the respondents, can be ranked as follows: management (elements of 27 subcategories), organizational environment (elements of 22 subcategories), resources (elements of 16 subcategories), external environment (elements of 6 subcategories), relations (elements of 5 subcategories). Thus in summarizing the data of the respondents group, it is possible to argue that the barriers at the initiation phase form in the internal organizational environment, i.e. the ***internal barriers prevail***.

As presented in Table 2, p. 48, the barriers distinguished by the respondents at the initiation phase are of wider range. In the analysis of research literature at the initiation phase the ***resources***, the lack of which could negatively influence the barriers of educational innovations' implementation were not emphasized. The theoretical model did not include the barrier ***of communication***, which was indicated by the respondents to the lack of information. The respondents disclosed the important aspect of sociocultural environment. These are the groups of internal barriers for which the theoretical model is extended.

At the initiation phase the identified barriers in the context of aspiration for the qualities of knowledge society signalize for the initiators of educational innovation implementation as one of the ways for the change. One has to be prepared for their implementation. At the institutional level it is necessary to design the system of educational innovation implementation, which would guarantee the functioning of innovative organizational environment as well as the effective implementation of educational innovations at the levels of the didactic/pedagogical system and an individual.

Conclusions

1. The PHARE program can be characterized by a specific structure of its content, which consists of different external educational innovations, implemented as separate projects at the levels of educational system, educational institution and teaching/pedagogical system.
2. In the level of educational system their implementation objects include educational policy, structural elements of educational system; at the level of didactic/pedagogical system the objects of the educational innovations are educational aims, curricula, methods, and at the level of the organization – managerial, structural and psychological subsystems.
3. At the initiation phase at the individual's level during the implementation of educational innovations the internal barriers of competence and lack of creativity lack, non-innovative management, unfavorable organizational environment and existing low-quality innovations as well as external barriers can be noticed. These barriers manifest at the level of educational policy (non-innovative educational policy, inappropriate strategies for the implementation of educational innovations, too strong use of political powers, etc.), at the level of central administration (bureaucratic orientation of problem solution, lack of support and maintenance, etc.) as well as at the level of community (the lack of the support and maintenance of initiatives, pressure of the community);
4. During the initiation phase of implementation of systemic educational innovations realized by the PHARE program, the following internal barriers were disclosed: at the level of individuals – the lack of the competence of educational innovations introducers as well as insufficient generation and acception of ideas, and at the level of organizations – non-innovative management, unprepared organization environment of innovations. The external barrier that manifested in this phase is non-innovative educational policy.

References

CRESWELL J.W.: Qualitative Inquiry and Research Design. Choosing Among Five Traditions. London 1998 (SAGE Publications Ltd.).

DALIN, P./ROLFF, H-G./KLEEKAMP, B.: Mokyklos kultūros kaita. Vilnius 1999 (Tyto alba).

EADIE, D.C.: Nurturing creative capacity in nonprofits. Eadie, D.C. Changing by design. San Francisco 1997 (Jossey-Bass Publishers).

FULLAN, M.: The New Meaning of Educational Change. London 1991 (Cassel).

FULLAN, M.: The New Meaning of Educational change. London 2001 (Routledge Falmer).

HARGREAVES, A.: Keičiasi mokytojai, keičiasi laikai. Vilnius 1999 (Tyto alba).

HOPKINS, D./AINSCOW, M./WEST, M.: Kaita ir mokyklos tobulinimas. Vilnius 1998.

JANIŪNAITĖ, B.: Edukacinių novacijų diegimo ypatumai Lietuvos švietimo sistemos modernizavimo kontekste: metodologiniai aspektai. Kaunas. Socialiniai mokslai 3(2000)24 (Technologija).

JANIŪNAITĖ, B.: Edukacinių novacijų tyrimo modelio pagrindimas. Kaunas. Socialiniai mokslai 4(1999)25 (Technologija).

KRAINER, K.: Ausgangspunkt und Grundidee von IMST. Endbericht zum Pilotprojekt IMST 2000/01. Klagenfurt 2001.

MCLAUGLIN, M.W./TALBERT, J.E.: Contexts that Matter for Teaching and Learning. Palo Alto 2001 (Center for research on the Context of Secondary School Teaching).

MELNIKAS, B.: Transformacijos. Vilnius 2002 (Vaga).

O'REILLY, A.A./TUSHMAN M.L.: Using culture for strategic advantage: promoting innovation through socal control. New York. Managing Strategic Innovation Change: A Collection of Readings 1997 (Oxfoed University Press).

"Reformos žingsniai". Informaciniai biuleteniai. 1997 – 2002.

RICOEUR, P.: Interpretacijos teorija. Diskursas ir reikšmės perteklius. Vilnius 2000 (Baltos lankos).

STANKEVIČIŪTĖ, J.: Organizacijos žinojimo didinimo metodologija. Daktaro disertacija. Kaunas, 2002 (Kauno technologijos universitetas).

STOLL, L./FINK, D.: Keičiame mokyklą. Vilnius 1998 (Margi raštai).

STONER, J.A.F./FREEMAN, R.E./GILBERT, D.G.: Vadyba. Kaunas 1999 (Poligrafija ir informatika).

STRINGFIELD, S.: Lessons being learned from research on educational improvement programs in the United States (mimeo). Paper presented at Tamtec Conference. Oslo 1993.

THOM, N./ZAUGG, R.J.: Excellence durch Personal-und Organisationskompetenz. Bern-Stuttgart-Wien 2001 (Verlag Paul Haupt).

TUMĖNIENĖ, V./JANIŪNAITĖ B.: Pedagogo veiklos pokyčiai pasaulinių ir europinių švietimo dimensijų kontekste. Kaunas. Socialiniai mokslai 2(2000)23.

TUMĖNIENĖ, V.: Pedagogo novacinės veiklos ypatumai Lietuvos bendrojo lavinimo mokyklose. Daktaro disertacija. Kaunas 2002. (Kauno technologijos universitetas).

VALIUŠKEVIČIŪTĖ, A.: Lateralinis mintijimas šiuolaikinėje kūrybiškumo ugdymo paradigmoje. Daktaro disertacija. Kaunas 2001 (Vytauto Didžiojo universitetas).

VONLANTHEN, J.M.: Innovations management in Schweizer Unternehmen., Bern 1995 (Peter Lang AG, Europäischer Verlag der Wissenschaften).

ŽELVYS, R.: Švietimo vadyba ir kaita. Vilnius 1999 (Garnelis).

АНГЕЛОВСКИ, К.: Учителя и инновции. Москва 1991 (Просвещение).

ЗАГВЯЗИНСКИЙ, В.И.: Разработка и реализация проектов развития образовательных учреждений. ГИЛМАНОВ, С. А. (РЕД) Москва 1998 (Издательство АСОПиР РФ).

СОВЕТОВА, О.С.: Основы социальной психологии инноваций. Санкт Петербург 2000 (Издательство С - Петербургского университета).

Internal barriers		External barriers	
Characteristics of barriers	***Criteria***	***Characteristics of barriers***	***Criteria***
Level of an individual			
Lack of competence	Non-innovative attitudes Traditional values versus innovative attitudes Low ability of idea generation Lack of professional knowledge Lack of knowledge from other fields Lack of reasoning at the level of organization policy	Barriers of the educational policy level	Non-innovative educational policy Inadequate strategies for the implementation of educational innovations Too intense use of political powers Too intense use of legal powers Too intense use of administrative powers Too intense use of economic powers Lack of consensus in implementing innovations „from above“ Lack of complementary funds
Lack of creativity	Stereotype thinking Incapability to identify problems Narrow viewpoint to problems Incapability to distinguish the most important information Incapability to see a problem from different perspectives Fear of risk Avoidance of inexplicit situations Priority to evaluate but not to generate ideas Low ability of idea gestation	Unfavorable attitudes of central administration (department of education, county, etc.) towards innovation implementation	Inadequate decisions Bureaucratic orientation to problem solution Lack of support and maintenance Disregard of the efforts of educational institutions to initiate innovations

Level of an Organization			
Non-innovative management	Inappropriate infrastructure of the dissemination of new knowledge Ineffective communication with the environment Lack of leadership Inappropriate coordination of innovation initiation Inflexible strategy Lack of correct information Inappropriate realization of the phases of a creative process Insufficient analysis of emerging difficulties Insufficient analysis of the state of educational process Insufficient analysis of controversial aspects Insufficient drawing-up of realization plans The team of professionals missing	Community influence	Lack of initiative support and maintenance Passive viewpoint in the aspect of innovation implementation Pressure of community

Unfavourable organisational environment	Inappropriate conditions for the expression of innovative activity Disregard of initiatives Insufficient ability to use the factors determining the success Insufficient knowledge about the initiation phase Lack of innovative culture Limitation of freedom in studying ideas and their use Unformed team of professionals		
Existing low-quality innovations	Inappropriate system of standards, aims, control and evaluation Inflexible curricula Inappropriate dissemination of the information about existing innovations Insufficient resources		

Table 1: Methodology of the Identification of the Barriers at the Initiation Phase of Educational Innovation Implementation

Robertas Jucevičius

Clusters as an Innovative Tool for Regional Development

During the 1990s the explosion of specialized and popular literature on industrial clusters gave them an unprecedented relevance across a range of areas, including business management and economic, political, public and social policy. There was also a degree of confusion over what the various authors mean – and do not mean – by industrial clusters. The answer still is not clear enough even if there are so many works done in that field of research. Clusters and clusterisation have become one of the priorities in contemporary studies on competitiveness, industrial policy and regional development. Nevertheless, a commonly accepted definition of cluster does not exist. There are different definitions of the word 'cluster' itself. French literature features the term 'local productive system', Italian literature – 'industrial district', German literature – 'networks of competence'. Researchers from different countries try to discover presence of clusters and their peculiarities in their regions. Various aspects of industrial clusters were analysed by Camagni (1991), Sternberg and Tamasy (1999), Etzkowitz et al. (2000), Lawton Smith and de Bernardy (2000), and others. But probably the most fundamental work in the field was of Porter (1990).

Innovation dynamics and knowledge creation and sharing in clusters are discussed by Morosini (2004), Scheel (2002), Hansen (2002), Furman (2002), Lorenz (1992, 1996), Lawson (2000), Lawson and Lorenz (1999), Keeble and Wilkinson (1999). Gordon and McCann (2000) emphasise the social nature of clusters, arguing the importance of social networks inside the clusters. Porter (1998), Morosini (2002) and Rabellotti (1999) are trying to understand the binding forces of clusters. An important feature of every cluster is geographic proximity of its actors. This aspect is discussed in the works of many authors (Porter, 1990, 1998), Scheel (2002), Morosini (2004), Feser and Bergman (2000), Simmie and Sennett (1999), and others. Despite a large number of studies published, many questions about the phenomenon of clusters and related forms of industrial cooperation remain unanswered.

The ***aim*** of the paper is to examine the importance of industrial clusters in the context of the development of small country or region in open economy.

The research ***methodology*** is based on applying Porter's diamond model on network-type co-operative structures and local business systems. Analysis of statistical data and calculation of economic indicators reflecting the notion of clusters was applied. The main instrument used to reveal the nature and logic of inter-firm relations was in-depth interview. Study results of two relatively successful Lithuanian industrial sectors were presented to demonstrate to what extent clusters of related or connected industries are important for the competitiveness of the whole industry and to show the trends towards cluster development in Lithuania. In this country it is difficult to find the evidence of well-developed clusters of the type described by Porter, so our study offers some insight about the potential of cluster development. It is concluded that Lithuanian industrial policy should be focused on developing the type of industry clusters described by Porter's model with the strong emphasis on international dimension.

In this paper, we also offer a knowledge-based framework to facilitate understanding of the key factors governing the way a cluster functions and its international success. Although there are many publications on the subject – or perhaps because of this – both the use of the term and the existing explanations of the underlying phenomena of clusters can be rather confusing. As a result, regional policy makers and businessmen often find it difficult to address the potential threats as well as the promising opportunities that these clusters provide.

Clusters: definition and policy pitfalls

There is a considerable variety of definitions of clusters in the literature. Porter (1990) says that a cluster consists of industries linked through vertical or horizontal relationships. Rabellotti (1995) describes the type of clusters found in industrial districts as being comprised of:

- Spatial and sectoral concentration of enterprises;
- Socio-cultural ties amongst local economic agents, creating a common code of behaviour;
- Intense vertical and horizontal linkages, based both on market and non-market exchanges of goods, services, information and people;
- A network of public and private local institutions supporting the enterprises in the district.

The working definition of clusters used in this article is as follows: *a socio-economic entity characterised by a social community of people and a population of economic agents localised in close proximity in a specific geographic region. Within an industrial cluster, a significant part of both the social community and the economic agents work together in economically linked activities, sharing and nurturing a common stock of product, technology and organisational knowledge in order to generate superior products and services in the market-place.*

Industries and firms within a cluster can have either direct interrelationships or indirect common resource needs that support their competitiveness on national and world markets.

This definition allowed for identification and analysis of different types of emerging clusters in Lithuania. First, it is possible to observe a classic, vertically integrated cluster, such as wood industry, with considerable evidence of subcontracting, facilitated by the existence of umbrella associations or personal networking. Second, one can study clusters, the component industries of which have a common resource base, but few logical buyer-supplier linkages, such as the cluster of tourism or the industry of mechanical engineering. In these clusters, component firms share a common transportation infrastructure and subset of the workforce. Finally, one can also examine "emerging" clusters, the linkages or common resource issues of which are plausible, but not currently well established.

In the rush by various local governments to employ clusters, some fundamental issues have been slighted, including appropriate research methods and even the definition of a cluster itself. Too often, this lack of clear direction manifests itself in one of several pitfalls that limit the usefulness of the subsequent cluster research for effective policy making. Some pitfalls can be identified.

Prescribe but don't identify. This common pitfall occurs when clusters are not formally identified or analyzed in any way. Rather, rushed policy makers, eager to show results, implement a set of cluster-type remedies - for example, network formation, targeted attraction efforts - with only a very limited understanding of the clusters' component industries or the commonalities and policy needs of these industries.

Identify but don't prescribe. This purely quantitative approach lists a set of clusters (or simply a set of seemingly unrelated industry groups) with little or no discussion of the component industries or the commonalities of these industries. In general, meaningful policy options do not flow from this type of work.

Say cluster but think industry. An impressive report with a list of clusters and related policy issues could be presented. Upon closer examination, however, one finds that no rationale for identification of the clusters is presented, and that any analysis presented is of the component industries treated individually, rather than of the cluster as a whole. This is probably the most common mistake.

Only the "classic" cluster will do. This pitfall, which can occur among academic researchers, is demonstrated by a lack of interest in any clusters not conforming to theoretical constructs about buyer-supplier activity or local demand conditions. These researchers focus on what might be called the "classic" vertically integrated cluster. Unfortunately, this research mindset, while of value, can inadvertently distract attention from those horizontal or emerging clusters most in need of immediate study and policy support. In many regions, the most interesting sectors may not be well-established ones, but rather clusters of firms whose incipient and emerging interrelationships can be strengthened through policy. Here we define horizontal clusters as those the component industries of which have a common resource base, but few logical buyer-supplier linkages. We define emerging clusters as those linkages or common resource issues are plausible, but not currently established.

What binds companies in a cluster together?

The definition of industrial clusters highlights that its members work together in related or linked business activities. The scale- and knowledge-based advantages generated within an industrial cluster stem from both the number and the nature of the particular linkages between its members. In a well-developed industrial cluster, these linkages can be numerous, unique and specialised to the industrial cluster, including (Morosini, 2004):

- common customers (both firms and individuals);
- common suppliers and service providers;
- common infrastructure, such as transportation, communications and utilities;
- common pool of human talent, such as skilled professionals or specialised labour;
- common educational, training and coaching facilities and approaches for workers;
- common university, research centre and technology institute specialisations;
- common risk capital markets.

The members of an industrial cluster must share and nurture a common stock of product, technology and organisational knowledge. Some authors described this critical characteristic of industrial clusters as constituting a 'social glue' (Porter, 1998) that binds a cluster together. Other authors refer to a 'common glue' or an 'organisational glue' that socially amalgamates diverse structural agents and integrates key knowledge across cultural, organisational and functional boundaries (Evans, 1993; Morosini, 2002).

Morosini (2002) identifies five key capabilities that need to be in place to build this 'common glue' which realises economic value within – and across – economic organisations. When applied to industrial clusters, these key capabilities can be characterised as follows:

(a) *Leadership.* Well-functioning industrial clusters are deliberately amalgamated by groups of key individuals with explicit roles fostering mutual cooperation, knowledge sharing, leadership coaching and arbitration of disputes that are seen as benefiting the common interests of the members.

(b) *Building blocks.* Well-functioning industrial clusters have typically developed a clear, common stock of organisational knowledge that is shared by all members, across functional, cultural and firm-specific boundaries.

(c) *Communication rituals.* Within well-developed industrial clusters, there are regular communication events, interactions and approaches that continuously foster a common sense of identity among all members of the cluster. Communication interactions that support the development of a common sense of identity in industrial clusters include proactive industry associations, common interest groups to lobby local or national governments and building a common image through public relations initiatives and the like.

(d) *Knowledge interactions.* Well-functioning industrial clusters have a series of regular, explicit and highly developed mechanisms for sharing key technological and business knowledge among all members. In well-developed industrial clusters, meso-level institutions, such as industry associations, usually play a key role as both initiators and managers of these coordination mechanisms. It is important to note, however, that this role is substantially different from the conventional collective bargaining, political lobbyism or contact and networking roles that these kinds of associations typically play within industries or inside less developed industrial clusters.

(e) *Professional rotations.* Within highly competitive industrial clusters, there is typically a significant pool of human talent specialised around business and technological knowledge that is specific to the cluster's main economic activities. The degree of cross-firm mobility of these professionals largely takes place within the geographic boundaries of the industrial cluster.

The ultimate goal of industrial clusters is to generate superior products and services that are valuable to customers in the marketplace. According to Marosini (2004), there are at least two crucial points to be made about this notion. Firstly, although an industrial cluster might benefit from some protective measures at the outset, in the long-term, selection mechanisms that reflect the dynamics of business competition must be in place. Otherwise, an industrial cluster will simply not survive, or it will do so as a result of economic transfers that are not necessarily market related, e.g. in the form of state subsidies or fiscal incentives.

Secondly, industrial clusters have not only proved successful at creating real economic value nearly everywhere but have also – during the second half of the 20th century – emerged as formidable global players in their own right across variety of industries. Therefore, although these industrial clusters are tightly confined geographically, their scope of competition is *increasingly global*, not local. The global scope achieved by industrial clusters during the 1990s went far beyond the export potential and international appeal of a specific product range that in the past typically made the fortunes of industrial clusters. At the dawn of the 21st century, industrial clusters were taking over entire areas of many global industries, such as manufacturing, R & D and product design. As a result, leading multinationals in different industries found themselves increasingly using industrial clusters to their benefit. These multinationals would typically involve industrial clusters either as leading suppliers or as key customers and innovators in the key areas of their value chain.

Emerging industrial clusters in Lithuania

Lithuania has been treated as one single region in our research. A big variety of industries was analysed: machinery and equipment, food, textile and clothing, wood and furniture, biotechnology and others. The main findings from the two – machinery and equipment and also wood and furniture industries will be discussed.

Machinery and Equipment Manufacturing Industry

There are over 700 companies in Lithuania which can be regarded as a part of the Machinery and Equipment Manufacturing (MEM) cluster. They employ over 41 thousand people and produce 17.9% of the domestic output of the processing industry. Their export constitutes 17% of the total production of the Lithuanian origin. Although this sector includes a rather large number of small firms, the larger ones (with over 250 employees) employ almost 53% of all sector's employees, while their sales constitute 60% of total sales of this sector.

The MEM industry in Lithuania can be divided into six key sub-sectors: electronics, electrical engineering, machinery, equipment and device manufacturing, transport vehicle manufacturing and metal items production. The most advanced in terms of clusterisation are Electronics, Machinery and Equipment Manufacturing, and Device subsectors.

The ***Electronics sub-sector*** is dominated by three large companies: *Ekranas* (Panevėžys), *Vilniaus Vingis* (Vilnius) and *Šiaulių Tauro Televizoriai* (Šiauliai). Together with several dozens of related firms they form a so-called 'shadow cluster'. It is not a mature cluster because the key companies are relating the major part of their activities with the firms functioning outside this implicit cluster, i.e. mostly those operating abroad. The mutual sales volumes of these local companies operating in the electronics sector are significant only in the category of the vertically technologically related products.

TV production cluster also includes a number of small firms which are involved in design and production of the TV technologies, as well as companies such as *Elinta* and *GTV* which are developing automation systems and equipment for such companies. However, identifying the boundaries and features of this cluster requires deeper research.

The sub-sector of the ***Machinery and Equipment Manufacturing*** is the largest in the industry. It involves about 130 companies, however only 9 of them are relatively large, employing over 500 people. Nevertheless these companies are producing about 70% of the entire sector's output. Companies of this sector are still looking for their place in the new economic conditions. There are difficulties in renewing the former relations and developing the new ones. The level of the managerial competence of these companies varies: along with those characterised by modern management and rather successful, there are quite a few with unclear future. Although the management of a number of companies begin to understand the importance of specialisation and core competence development, some of them still believe that their companies should be at least to some extent vertically integrated, have their own technological infrastructure and everything else needed for operation.

One of the problems is that many companies of this sub-sector are slowly renewing their technologies. So, their possibilities to produce and export high value added products are limited. As a result, many companies are focusing on the domestic market, producing equipment and other products for various economic sectors – light industry, wood and furniture, construction, food, chemistry, etc. From the perspective of clusterisation, this fact also has a positive aspect since these companies are integrating into clusters and similar agglomerations of other industries. However, the output of such companies is very varied, therefore it is difficult to identify some potential clusters based on common groups of products.

The study showed that the managers of the most successful companies in this sub-sector's are interested in expanding collaborative links. However, they are facing the problem of finding reliable partners with enough capacity and commitment to follow the contract on time. Faced with this difficulty, such companies are forced to integrate vertically, in this way slowing down their growth (due to the lack of investment funds) and in some cases increasing the operation costs.

The sub-sector of ***Device production*** is rather small, including only about 40 companies. From the perspective of clusterisation, this sub-sector is especially fragmented. Almost all companies are small and medium-sized and are involved in rather specialised activities. Except several firms, such as *Brown & Sharp – Precizika, Katra, Medelkom, Eksma, Ekspla, Accel Elektronika* and *Intersurgical*, most of the companies in this sub-sector are producing medium or low level devices and instrument. Due to high fragmentation of the activities, co-operation between the firms of this sub-sector as well as with other Lithuanian companies is minimal.

SWOT analysis of the Industry from the clusters perspective: (+) – strength and (-) - weakness

- ***Dominant strategies of firms:***
  (+) integration into international clusters;
  (-) most firms are little specialised;
  (-) obsolete technological equipment;
  (-) lack of trust between firms and institutions;
  (-) lack of entrepreneurship;
  (-) low managerial competence (in many firms).

- ***Local demand:***

  (-) small and fragmented local demand which is unfavourable for mass production;

  (+) high demand of local firms to modernize production;

  (+) increasing demand in the NIS market;

  (±) undemanding market;

  (-) no demand in the local market for significant number of products.

- ***Factors and conditions of operation:***

  (+) long experience of engineering

  (+) many specialists with high skills relevant for the needs of the economy

  (+) low labour cost

  (-) research laboratories and study programmes are inadequate to high requirements

  (-) system of specialists' training is does not meet the demand (at all levels)

  (-) inefficient innovation system

  (-) low investment into R & D on the part of firms

  (+) rapidly growing economy

  (+) favourable geographic location

- ***Clusterisation processes:***

  (+) mature demand for clusterisation

  (+) still existing social network of specialists

  (+) strengthening relations between industrial firms and research institutions

  (-) low cooperation between firms and with research & education institutions

  (+) industrial association unites firms of many sub-industries

  (+) strong lobbyism

  (-) lack (almost none) of network development and management specialists

  (-) low awareness of firms and research institutions about each others' possibilities and interests

Wood and Furniture Industry

The wood processing and furniture production industry can be divided into three sub-sectors: wood, furniture and paper production. In 2003, there were 873 companies operating in these areas in Lithuania. Their distribution was as follows: 647 companies were involved in wood processing, 190 in furniture production and 36 in paper production. Wood and furniture sub-sectors are dominated by small firms (up to 50 employees), while larger companies produce the major part of output. This is especially true about furniture and paper production enterprises. However, the actual picture is distorted by the fact that out of 3,559 registered companies only 873 submitted statistical data.

Wood and furniture sector is one of the 3-5 main sectors of the Lithuanian industry. The statistical analysis performed for this study confirms this fact. According to the number of people employed by this sector, wood and wood products industry is in the 3rd place among the processing industries and in 12th place among all industries involved in economic activities. Furniture production industry is in the 5th and 21st positions accordingly. According to the number of economic entities, these industries are in the 1st and 5th places among all processing industries.

In 1999-2002, furniture production doubled, while export grew by almost 3 times. The study of international competitiveness showed that Lithuanian wood and furniture companies are lagging behind the most successful EU enterprises by to the number or indicators. This is especially the case with value added and other parameters defining the operation efficiency, although some positive changes happened in the last few years. Deeper analysis of the situation in the specific companies showed that further specialisation and cooperation could play the key role in increasing efficiency.

From the perspective of values creation, the structure of wood and furniture industry is not very favourable. The dominating activities are rather basic, not requiring high skill and complicated technologies. This conclusion can be drawn based on the distribution of the surveyed firms. They included 343 sawmills, 163 firms producing woodwork for construction, 54 – wooden containers, 36 – paper, 27 - plywood and panels, and 190 firms producing furniture. The remaining 60 companies are involved in unclassified activities.

The range of products and the activities performed by enterprises as well as the structure of ownership and geographic distribution are favourable for clusterisation. Although most of the small firms and many large companies are autonomous, there are 4-5 groups of companies which are capable of implementing joint strategies. They involve *Libra Group, SBA Concern, Baltic Furniture Group, Western Wood Group* and *Baltic American Concord.* Among top 20 wood products companies, there are 10 companies belonging to these groups.

Analysis of the wood production value chain showed that in terms of production and technology Lithuanian firms are able to cover almost all value chain activities according to the demand. Much bigger problems are related to business infrastructure, which is still at the initial stage of development. The study showed that wood and furniture sector's links with companies of other sectors are minimal. They are buying most of the metal components abroad. The key reasons behind this fact are that there are no suppliers offering relevant quality and there is lack of trust between companies. On the other hand, discussion of this issue with machinery and equipment manufacturing companies showed certain scepticism on their part regarding the preparedness of wood sector firms to by high quality products. Speedier clusterisation processes could facilitate solutions to this problem because the collaboration potential is high.

A more complicated situation is in the area of technological equipment and chemical products. It is still unclear what realistic solutions can change this situation in the near future. To gain a better understanding of the problem, a detailed study should be conducted.

Furniture production companies integrated into IKEA cluster are using certain specialised services, which are organised by this corporation. First of all, these services include transport logistics, as specialised transport firms according to negotiated schedule supply most of the transport services. A very important element, which in a certain sense helps clusterisation of the wood and furniture companies, are business organisation services which involve expertise and recommendations of the Swedish specialists on the improvement of production organisation and cost reduction. This kind of help is unavailable to non-clusterised companies. In general it can be argued that those Lithuanian wood and furniture companies, which are IKEA partners, have already joined the international cluster, although company links in Lithuania do not suggest the existence of many elements of clusterisation. By expanding these links, companies would gain an extra stimulus for increasing operation efficiency.

Wood and furniture companies have much weaker links with equipment maintenance and repair firms. First of all, there is only small number of such firms and they can be hardly regarded as specialised. Besides, management has a deeply-rooted stereotype that these operations should remain within companies because they are rather complex. One solution to this situation could be looking for possibilities jointly with machinery and equipment sector companies.

SWOT analysis for the wood and furniture industry:

- ***Dominant strategies:***

  (+) focus on final product;

  (+) increasing specialisation;

  (-) low specialisation;

  (+) emphasis on export;

  (+) orientation towards international clusters;

  (-) focus on low-value added products;

  (+) increasing number of firms developing their brands.

- ***Local demand:***

  (+) growing domestic market;

  (-) stagnating EU market;

(+) growing Russian and Central European market;

(-) low purchasing power of local inhabitants;

(+) growing demand for furniture among offices and service firms (hotels, restaurants, etc.);

(-) growing demand for raw wood in the international markets;

(+) consumers turning back to natural products.

- ***Factors and conditions of operation:***

(+) traditional business;

(+) proximity to culturally similar markets;

(+) good logistics infrastructure;

(+) international experience;

(+) good stock of basic resources;

(-) technical/technological backwardness;

(+) inexpensive and skilled labour force;

(-) lack of well-educated specialists;

(-) increasing pressure of Russian producers in the IKEA segment;

(-) poor business infrastructure;

(+) great variety of firms in terms of size and activities;

(-) lack of modern training centres;

(+) active industrial association.

- ***Clusterisation processes:***

(+) natural relationship between consecutive members of the value chain;

(-) low, although increasing cooperation;

(+) high quality industrial publications;

(-) lack of professional information system;

(-) weak links with other industries.

Conclusions

1. Experience of developed countries proved that clusters are an important tool of enhancing the competitive advantage of regions and nations.
2. *Cluster* is a broader notion than just an agglomeration of economic agents: it is a socio-economic entity characterised by a social community of people and a population of economic agents localised in close proximity in a specific geographic region. Within an industrial cluster, a significant part of both the social community and the economic agents work together in economically linked activities, sharing and nurturing a common stock of product, technology and organisational knowledge in order to generate superior products and services in the marketplace.

3. At least five key capabilities need to be present to build a common ground for organisations in the cluster. These capabilities are:
   - leadership;
   - building blocks;
   - knowledge interactions;
   - professional rotations;
   - communication rituals.
4. When analysing clusters, statistical methods should be enriched by qualitative techniques. Lack of statistical data on input-output can be compensated by qualitative data that allow to reveal the nature of interaction of actors in a particular cluster.
5. The study of two Lithuanian sectors – Machinery and Equipment Manufacturing as well as Wood and Furniture Production – showed that some clusterisation processes are taking place in these industries.
   - In the case of the first industry, there are tendencies towards clusterisation in the Machinery and Equipment as well as Electronics sub-sector, although one of the problems indicated by management is difficulties to find local reliable partners.
   - In the case of Wood and Furniture sector, it is rather fragmented. The major signs of clusterisation can be found around IKEA operations in Lithuania, although in this case one should talk about the international cluster. The success of this agglomeration enables one to suggest that Lithuanian industrial policy should be focused on developing industrial clusters with the strong emphasis on international dimension.

References:

CAMAGNI, R.: Local 'Milieu', Uncertainty and Innovation Networks: Towards a New Dynamic Theory of Economic Space. In CAMAGNI R. (ed.): Innovation Networks: Spatial Perspectives. London 1991 (Belhaven Press), pp. 121-142.

ETZKOWITZ, H./WEBSTER, A./GEBHARDT, C./TERRA, B.: The Future of the University and the University of the Future: Evolution of Ivory Tower to Entrepreneurial Paradigm. Research Policy 29(2000), pp 313-330.

KEEBLE, D./WILKINSON, F.: Collective Learning and Knowledge Development in the Evolution of Regional Clusters of High Technology SMEs in Europe. Regional Studies 33(1999)4, pp. 295-303.

LAWSON, C.: Collective Learning, System Competences and Epistemically Significant Moments, In: KEEBLE, D./WILKINSON, F. (eds.): High-technology Clusters, Networking and Collective Learning in Europe. Aldershot 2000 (Ashgate), pp. 182-197.

LAWSON, C./LORENZ, E.: Collective Learning, Tacit Knowledge and Regional Innovative Capacity. Regional Studies 33(1999)4, pp. 305-317.

LAWTON SMITH, H./DE BERNARDY, M.: University and Public Research Institute Links with Regional High-Technology SMEs. In: KEEBLE, D./WILKINSON, F. (eds.): High-technology

Clusters, Networking and Collective Learning in Europe. Aldershot 2000 (Ashgate), pp. 91-117.

LORENZ, E.: Trust, Community and Co-operation: Toward a Theory of Industrial Districts. In: STORPER, M./SCOTT, A.J. (eds.): Pathways to Industrialisation and Regional Development., London 1992 (Routledge).

LORENZ, E.: Collective Learning Processes and the Regional Labour Market. Unpublished Research Note. European Network on Networks, Collective Learning and RTD in Regionally-Clustered High Technology SMEs. 1996.

MOROSINI, P.: Industrial clusters, Knowledge integration and performance. World Development, 32(2004)2, pp. 305–326.

MOROSINI, P.: Competing on social capabilities: a defining strategic challenge of the new millennium. IMD Working Paper 2002 – 1. Lousanne, 2002 (IMD).

PORTER, M.E.: The Competitive Advantage of Nations. New York 1990 (Free Press).

PORTER, M.E.: Clusters and the New Economics of Competition. Harward Business Review. November-December, 1998.

RABELLOTTI, R.: Recovery in Mexican cluster: devaluation bonanza or collective efficiency. World Development 27(1999)9, pp. 1571–1585.

STERNBERG, R./TAMASY, C.: Munich as Germany's No. 1 High Technology Region: Empirical Evidence, Theoretical Explanations and the Role of Small Firm/ Large Firm Relationships. Regional Studies 33(1999)4, pp. 367-377.

Eglė Katiliūtė

Premises for Development and Implementation of the Educational Strategy for the Knowledge Society

Introduction

Changing society and it's vision has been variously labeled by different researchers: postcapitalism (Drucker, 1993), the knowledge – value revolution (Sakaiya, 1991), the relationship age (Mollner, 1990), the learning society (Fullan, 1993; Fink, 1996), the intellectual society (Lyotard, 1993), postindustrialism (Abercrombie at all., 1992), "the third wave" society (Tofler, 1980), the postmodern society (Hargreave, 1999). Irrespective of naming today's and future society, it is clear that it distinguishes and will be distinguished by special relation to knowledge and science.

The key characteristic of the 'risk society' is uncertainly with respect to all those modernist institutions and belief systems rooted in the Enlightenment tenets of Truth, Progress, and Science. These 'old truths' have given way to doubts and anxieties about the nature and validity of natural scientific knowledge itself, the 'Big Science', which might be thought of as the core domain of the modern knowledge. Nowhere else this is more evident than in concerns about the questionable role of science and technology in relation to environmental degradation and ecological disaster. What was once considered a 'source of safety' has now become a 'source of risk' (Irwin, 1995, p. 61). It is not just that scientific knowledge can no longer be relied on as a secure base for decision-making, but these threats to the environment are actually produced by development in scientific and technological practices. We live in a period where uncertainty is 'manufactured' (Giddens, 1994, p. 4); that is to say, it originates not in 'natural' happenings but in the scientific and technological activity of society itself. The problem with the 'Big Science' is that it impedes the development of a more person-centred and community-oriented approach to the resolution of problems, which is an essential mode of operation in the risk society (Quicke, 2001, p. 113-114).

In the context of the changing society we could talk about new tendencies in education policy. As Young (1998) pointed out, 'more importance must be given to learning at work, learning outside school and in adult life generally. The crucial implication of this shift in emphasis for educational policy and theory is the new relationships between school and non-school learning which need to be developed. A further trend is the increasing emphasis in education policies on lifelong, work based and experiential learning – all forms of learning that take place outside institutions of formal education. The most successful private companies are seen to be becoming "learning organizations"; schools and colleges are encouraged to copy them, educational policy documents stress learner-centeredness and lifelong learning and research programmes are publicized as promoting a learning society'.

In this article I will make the assumption that ideally education policy includes coherence in explicit strategy of education, management of implementing of educational aims or objectives, and evaluation of achievements. Explicit strategy of education describes the process of aim or objective formation. I will touch upon only one level of education policy – strategy making. Speaking about education strategy level, I would like to mention problematic conditions of formulating a strategy or, in other words, problematic conditions of formulating educational aims.

The *aim* of this article is to analyze premises for development and implementation of the educational strategy in the knowledge society.

The argument in this article proceeds in three sections. The first one outlines the main concepts used in this article. In the second, I need to provide conditions of formulating a strategy: conditions that include what is 'good' (in formulating educational aims) and what is possible for strategy makers to formulate. In the final section, I present certain examples of educational aims.

I will make the assumption that curriculum is the core of educational policy; therefore I will talk about the changes of the curriculum. Creation and improvement of curriculum is related with all changes of education policy: development of teachers' qualification, need of lifelong learning, new directions of research, implementation of new programs of study and teaching, etc. For this purpose different work groups, changed structures of institutions are created. Usually the system of education quality is analyzed, and then the analysis of management of such system follows. Without curriculum, education policy is a bare document on the paper without its basis in education.

Descriptions of used concepts

Although in this concise article I will talk only about one level of education policy, I will give some descriptions concerning policy, education policy, strategy and curriculum.

One of the concise descriptions of education policy is as follows: *education policy* is the aims of education that are to be pursued. Having in mind lasting or long-range goals we could say that there is strategic planning. In strategic level policy makers should try to answer the following questions: what strategy should be implemented? What values and priorities should be observed in order to recruit and maintain the best employees and to create organizational culture, which is oriented towards continuous learning and professional development? Describing a strategy of education I'm talking about formation of education aims, values and priorities in the society level. Thus explicit *strategy of education* describes the process of formulating aims or objectives.

Ball and Bowe (1992) argue that the policy process is much more complex than this and that there is a dialectical process in which the 'moments' of legislation (the Act), documentation, and 'implementation' (the work of the teachers) may be more or less loosely coupled. Policy texts are not closed, their meanings are neither fixed nor clear, and the carry over of meanings from one policy arena and one educational site to another is subject to interpretational slippage and contestation. These texts are a part of a policy cycle consisting of significantly different arenas and sites within which a variety of interests are at stake. While it is the case that the Act as a whole aims to intervene in the workings of these arenas and sites in order to create a qualitatively different cycle with new conditions 'empowering' some at the expense of others, the different parts of the Act may well be taken up differently by particular schools and departments within schools, thus producing very different outcomes that may actually work against a National Curriculum (Ball, Bowe, 1992).

A new education policy often means a set of proposed initiatives, measures, or changes to the inputs and/or internal processes of the education system, with a hope of achieving some planned changes in education effects. According to Coombs (1994), education policy making is often characterized by extraordinary complexity (involving so many participants and stakeholders at different levels), visibility of the education system and most policy deliberations (i.e. almost all citizens have education experience and believe that they are "experts" in education), dispersion of authority in education policy making at different levels, ambiguity about goals in most educational settings, and labor-intensive processes involving the majority of the education budget on staffing (Cheng, Ng, Mok, 2002).

What about curriculum?

Before starting to talk about the conditions to formulate the education strategy, I would like to define the conception of curriculum and to show its link to education policy.

Curriculum is central in educating students to become certain kinds of people, individuals or citizens – whether such curriculum builds on students' own interests (Dewey, 1969) or is based on state-controlled outcomes (Wien and

Dudley-Marling, 1998). In both cases, there is an underlying assumption about what it means to learn and be 'educated'; indeed, who educators think students should become, frequently defines the aims and purpose of educational practices (Todd, 2001).

As Schelten (1991) points out, curriculum is a description of the whole of education processes i.e. its planning, performance and control.

Such characterization comprises full data about educational aims, its content, methods, means and control of aims achievement. It is a direct plan of teaching oriented to aims and its practical realization. Curriculum has been defined as wholeness of answers to the following questions: why is a certain process and means of teaching organized? Which aims and qualifications are pursued? Which material must be learned, imparted, experienced? Which learning experiences and teaching material are required? How are education aims pursued? Which forms of evaluating will give information about reaching the aims?

School subject content can be approached and conceptualized in very different ways. At one extreme, it is seen as unproblematized and given in accordance with educational policy aims. There are different essentialistic interpretations, and content can be seen as well in historically and socially changeable and interpretable terms. The phrase "educational content" is used to signify that the focus here is on the meaning and educational worth of what is being offered to students (Östman and Roberts, T. Englund, 1998).

The main point outlined in characterization of curriculum is that this wholeness of education aims, content, methods, teaching material, organization and control has been seen as an always changing and renewable system. Among all elements of curriculum aims play a special role. Changing of curriculum couldn't be meaningful without assessment. It defines how the change of one element will influence seeking attainments every time.

Young (1998) presents two points of view: 'curriculum as fact' and 'curriculum as practice'. He has suggested that 'curriculum as fact' expresses many of the prevailing assumptions of educational practitioners, teachers and policy makers. By accepting similar assumptions, curriculum theory does little more than re-describes a world that teachers already know, albeit sometimes in terms they find far from familiar. It also confirms for teachers both the irrelevance of theory for practical change in schools and their own passive role in such changes. Teachers have theories of knowledge, teaching and curriculum that are often remarkably like the ideas of curriculum 'theorists' and which, he shall argue later, play an important, albeit unconscious role in curriculum change or resistance to it. They continue to hold such views because they are congruent with views of the curriculum in the wider society and because, in part, they make some sense of the situation in which teachers find themselves. As a theory, however, 'curriculum as fact', though pervasive amongst academics, administrators and teachers, fails according to the criteria. It does not enhance the capacity of teachers to become aware of the possibilities of change and of gaining an understanding of the conditions of their own practice. 'Curriculum as fact' presents the curriculum as a thing, hiding the social relations between the teachers, students and curriculum policy makers who have historically and collectively produced it. What then does the 'curriculum as practice' have to offer, if it takes a quite opposite view?

The significance of 'curriculum as fact' is that it is not just a theory produced by academics but the basis on which education system is organized. In other words, it represents a part of the circumstances within which anyone concerned with changing educational practice has to work. 'Curriculum as practice' does not offer an adequate alternative to ideas about the curriculum defined in term of the structures of knowledge. Its weakness is in the limitations of its concept of practice. It replaces a notion of reality located in the structures of knowledge. Attempts by teachers to develop strategies for changing the curriculum derived from such a theory will confront with the limits of what they can do in their classroom practice. Teachers will also become aware of the limits of the theory that does not enable them to comprehend the origins of such limits or show them how their classroom practice does in part shape the external reality of the curriculum in meaningful ways for their students.

Speaking about the relation between curriculum and policy, I would like to give another quotation from Young (1998):

Although the burden on every school, college and university in the country, is to provide a vision for the next generation of learners, it cannot be done at the level of individual institutions alone. When we consider the curriculum as a whole, we need not only the visions of every school in the country but a definition of national educational purposes that goes beyond standards and is expressed in a democratically debated National Curriculum that applies to all students, at least until they are 18 and regardless of whether they are in school or college or of the status of the school (p.103-104).

Englund (1996) argues that we have to revitalize the citizenship concept as a criterion of education – a social citizenship right –, which means that we have to analyze education and curriculum as an ethical and political enterprise (Schilling, 1986). It is in the light of the way in which the schools relate to this problematic, and especially how it affects the content of education. Education is a fundamental (social) citizenship right in itself, but it is only when seen as an instrument for raising awareness and problematizing all the citizenship rights that the role of education can be analyzed (Englund, 1996).

Conditions of educational strategy formation

The first meaning of *condition* is 'a state or circumstances in which a person or thing is'. Thus if we are talking about conditions we could talk about environment or circumstances. In this article I will analyze conditions to formulate a strategy of education. Before starting to talk about environment we must take for granted that curriculum exists inescapable in certain circumstances.

Many of the key factors that should be taken into account when designing educational policies, as Sandi (1997) pointed out, are totally, or to a large degree, external to the system.

By my hypothetical assumptions, the main conditions of what is good for strategy makers to formulate are socio-cultural context. Education strategy makers should also evaluate cultural context, social objectives, aspects of ethic and values, psycho-anthropological and culture-anthropological issues and etc. I think that separately these fields are very extensive.

Before analyzing conditions of globalization I want to outline that the universalizing objectives in education strategy is a more difficult process than might seem. According to Taylor, Rizvi, Lingard and Henry (1997), 'educational reform proposals across nations bear remarkable similarity. Copying, which occurs across the boundaries of nation states leads to universalizing tendencies in educational reform'. It is important to stress that policy makers are usually copying or at least paying more attention to global tendencies of education based on UNESCO view on education or European dimensions of education. Is it good? To my mind it could be good if strategy makers evaluated socio-cultural context in a certain country and did not expect the same results after incorporating a new idea, which is copied from certain standards for education. According to Argyris (1985) 'policies are espoused theories, and true reduction of the mismatches will occur only if the officers and managers behave in accordance with the intend of these policies. Without a particular kind of re-education and without changes in the organizational culture, the intent will not be implemented'. Argyris (1985) talks about the organizational level, but we can transfer this theory to the society level. Then we will see a very similar situation. Strategy makers formulated a strategy in espoused theories and they acted in their theory-in-use. Thus formulated education objectives without changes in the social culture will not be implemented. According to the conception of curriculum strategy makers should evaluate their implementation.

Coming back to the socio-cultural context of the society and ethic conditions for formulating strategy. To my mind, it is important to outline Bernstein's (1996) point of view to the necessary and effective conditions for democracy. According to Bernstein (1996), 'parents and students must feel that they have a stake in the school and confidence that the arrangements in the school will realize or enhance this stake'. He suggests that 'if these conditions are to be realized in the schools then we will need to ensure that we have institutionalized three interrelated rights. The first right is the right to individual enhancement. Enhancement is not simply the right to be more personal, more intellectual, more social, more material, it is the right to the means of critical understanding and to new possibilities. This right is the condition for confidence, and operates at an individual level. The second right is the right to be included, socially, intellectually, culturally and personally. Inclusion is a condition for communities and this right operates at the social level. The third right is the right to participate in procedures whereby order is constructed, maintained and changed. Participation is the condition for civic practice, and operates at the level of politics'. I think that all the mentioned conditions - confidence, communitas and civic discourse - might be attributed to ethic conditions. Certain ethic aspects really existed but I do not want to believe that they are simply implemented. The problem is in evaluation and implementation of the mentioned rights.

Saljö (1992b) has developed what he calls a sociocultural perspective on the learning process. In contrast with research in which the learning problem is, as a rule, perceived in terms of the learners' common-sense views versus the adequate, scientific way of conceptualizing. Saljö sees the problem that scientific concepts – the epistemic content – are seldom confirmed in social environments other than those of the scientific communities (Saljö, 1995). Because of their specific scientific character, pupils often see these concepts as strange and difficult to understand. Simultaneously, due to our conceptualization of learning as learning scientific concepts – the metaphors of institutionalized schooling (Saljö, 1990) – we often see learning problems as psychological and not as communicative. In a sociocultural perspective there is an awareness that we are not living in an unambiguous "reality", but rather that there are always different perspectives ("meaning provinces") for looking at and understanding phenomena. There is a sociopolitical perspective that sees education and its content as the objects of struggling social forces. The ultimate power centre for this struggle is the state, but it is also a struggle at all levels over whose interpretation is to win out. The way in which the educational system manifests how reality is to be conceptualized and school knowledge constructed means that certain power relations are consolidated or transformed. This transmission of ideology is therefore subject to constant shifts, as power relations gradually change. At all levels there are permanent ideological conflicts, while at the same time a cohesive state power and state apparatus have the task of shaping political compromises out of the various political and ideological interests and making it look "rational" (e.g., through national evaluation programs). Educational policy documents such as national curricula and syllabi consequently are analyzed as interpretable political compromises (Östman and Roberts, T. Englund, 1998).

In a sociocultural perspective on human action (Saljö, 1992a, 1996b, Wertsch, 1985, 1991), the integration of artefacts – manufactured tools – into social practices is seen as basic starting point for the understanding of any human activity. To learn and to develop is to appropriate and master artefacts within meaningful social activities. The fundamental unit of analysis in such a perspective on human learning and development implies that individuals are conceived as operating in social situations with mental and physical artefacts as resources (Chaiklin & Lave, 1993; Goodwin, 1998; Hutchins, 1995; Hutchins & Palen, 1998). Such a formulation challenges the basic tenet of most modern dualist psychology in that it explicitly counter argues the feasibility of assuming that there is something inside the heads of individuals – cognition – that can be understood as separated from what is "outside" – culture and social life. On the contrary, a central assumption of a sociocultural perspective is that mind and culture co-constitute each other (Cole, 1991) and develop in close interrelationship. The link between cognition and culture is discourse.

At the cultural level the concrete differences in life-conditions between (and also within) societies result in radically different assumptions regarding how to think and act in social situations. What differ between societies in cultural terms are not just ideas and worldviews, but also the nature of artefacts that are used in daily life and the social organization of knowledge. In a sociocultural perspective, learning (and development) implies mastering the particular tools that are prevalent in society. This implies that in order to understand thinking and learning, "it is necessary to go beyond the limits of the organism and to search for the roots of these complex processes in the historically formed environment, in the communication of the child with adults" as well as in the "relations among objects, tools and language which have been laid down in the course of social history" (Luria, 1971). At a general level, tools can be described in two major categories: practical (or physical) tools and psychological (or intellectual) tools. The concept of tool is suggestive, and it deliberately puts the use of concepts and linguistically mediated knowledge into the same category as the use of practical tools and instruments; just as we do things with physical tools and instruments, we do things with words (Östman and Roberts, R. Säljö, 1998).

Speaking about socio – cultural conditions of formulating education aims, I would like to show links to values and social order. Social order is perceived as something very concrete, formed in precise terms. Although in pluralistic democratic society common agreement should prevail. The question is, whether the given concreteness (of social order) is the right for all? If the social order is not the right for all, it is not ethic. But the problem is, how to come to a common agreement on the same social order? Can the sole acceptance of the group be the basis for the moral norm? If we leave aside the objective norms with universal validity, wouldn't we fall in the use of force as the only solution to the conflicts between incompatible norms? Isn't ethical subjectivism simply the denial of the possibility of a harmonious coexistence in the sense that it does not necessarily recognize tolerance as a shared value?

The rational consensus seems to be the proposal in force in order to avoid the universalization of the norms, which respond to individual or group interests. The appeal to reason, to intelligent ethics, to "minimum agreed ethics" appears to be the tool used to avoid subjective irrationalism. It is a resource, which is not always effective as proved by our daily reality.

The socio-cultural changes, as Podest pointed out, assume political and economic transformations and affect the daily life of each one of us, and shake up in many cases the bases of what we have used to structure our existence, questioning the principles and values assumed in other times when other existential scenarios were faced.

According to Shapiro, we need an approach to moral education, which will help students develop values that will enrich their lives as individuals and as members of society. It should enable them to participate in a communal effort to find a balance between individual liberty, private property, market competition, and due process on the one hand and self-restraint and community obligation on the other.

Since we as a people have chosen pluralism and representative government over other forms of social organization (such as official moral orthodoxies or totalitarianism), it is incumbent upon our universities to prepare their students to live in a society where they will have to make their own moral choices, where they will have the capacity to help shape the moral contours of the society as a whole, and where their lives will be directly affected by the moral choices of others.

It is good when school creates various possibilities to develop personal qualities, yet it is very important to give the students estimation of values. The learning environment should be saturated with the values representing the social order and also other different values, coming from the students. Yet values brought by the students can exist in the learning environment only after having passed through the estimation of values. For example, if stealing is a virtue for a student, in the learning environment it should be estimated as a negative value.

Questions of a normative nature inevitably arise: who is it that we, as educators and citizens, desire people to become?

Philosophically speaking, exploring the place of ethics in education often begins from this normative point of departure: what values are to be invoked in educational encounters? How might students be educated to become better citizens, more responsible moral subjects, or people who can live and work better across social differences? Simon (1992) notes that pedagogy is not just about the teaching of morals, but is itself a 'moral vision'. Normative questions are often central in helping to define and reflect upon an educational project.

Shifting the focus from education as a scene where one ought to apply this or that principle, to a scene where the conditions or contingencies of ethicality may be found, means no longer simply thinking about education in relation to ethics; rather, it means thinking about ethics through education. This means exploring the day-to-day details of pedagogical encounters to see what they might offer in putting forth an understanding of education as a site of implied, rather than applied, ethics. To explore this idea of implied ethics more fully necessitates reading teaching-learning encounters for the way they promote conditions for ethicality as they promote conditions for being, both of which involve relationships between Self and Other (Todd, 2001).

Now I will turn back to the global conditions. As Young's (1998) points out, the first view of globalization sees it as an inexorable set of economic forces impinging on every country in the world. These forces, it is argued, diminish the potential role of national governments and privilege the interests of multinational corporations, which only have loyalties to them. Globalization, from this perspective, is about the power of the global market, which no individual country and no policy can avoid. In this view, the only policy option for a national government is to facilitate the entry of their country into global markets in as many sectors of the economy as possible. This, it is argued, will be achieved through the wholesale de-regulation of markets, the cutting of public expenditure, the extending of private enterprise into fields such as transport that in many countries were previously public and subjecting the remaining areas of public provision such as education to the rigours of markets This view has been powerful and pervasive and has been adhered to by the major international organizations such as the World Bank and the IMF and to varying degrees by the governments of the major industrial countries.

However, as Hirst and Thompson (1996) show clearly, this view of globalization, although plausible and much promoted, is at odds with much of the evidence; most key investment and other decisions are still national. Most countries share certain common circumstances and even a common interpretation of these circumstances; however, this does not make such circumstances into independent forces of history. Such a view of globalization is best seen as an ideology used by governments, usually of the Right, to justify taking certain decisions. Global reality is much more complex and is reflected in the increasingly wide differences, even among economists, about what changes can really be said to be global and what this means. These new circumstances, which have emerged since the 1970s, include the increasingly global markets for manufactured goods, the expansion and diversification of consumer demand, the emergence of the new communication technologies and the expansion of air transport (Castels, 1996), and the end of the Cold War ideological divide. All these changes are part of the new context within which national governments find themselves; how they are responded to and their implications for educational policy will depend on decisions made by governments and others which will reflect (a) the position of a country in the world economic order; (b) the national tradition of the country in question; and (c) the policy choices of specific governments.

The impact of globalisation on educational reform strategies

As Carnoy (1999) pointed out, changes in the world economy have provoked three kinds of responses:

Reforms that respond to shifting demand for skills in both the domestic and world labour markets can be called *competitiveness – driven reforms*. This kind of reforms is described in such characteristics as decentralization (an extension of such reforms is school choice and the privatization of educational delivery), standards, improved management of educational resources and improved teacher recruitment and training.

Reforms that respond to cuts in public-sector budgets and private company incomes, reducing public and private resources for education, can be called *finance – driven reforms*. The main characteristics or these reforms are: shifting public funding from higher to lower levels of education, the privatization of secondary and higher education and the reducing of cost per student at all schooling levels.

Reforms that attempt to improve education's role as a source of social mobility and social equalization can be called *equity – driven reforms*. In these reforms strategy makers are pursuing to reach the lowest-income groups with high-quality basic education, especially the large number of youth and adults who presently do not have access to basic skills, to reach certain groups, such as women and rural populations, that lag behind educationally.

In OECD countries, the equity-driven reforms are much more targeted towards particular "at risk" (low-income) and special needs children thought the education system, and focus on reforms that would increase their success rate in school.

Globalization tends to push governments away from equity-driven reforms. However, this does not mean that educational policy cannot pursue equity-driven reforms.

Analysis of the global conditions dissociated from economical forces is impossible. Before wider analysis of economy and education strategy formation I want to emphasize that socio-culture issues are the main conditions of what is good for strategy makers to formulate.

Trying to answer the second question: what are the conditions that include what is possible for strategy makers to formulate? I make a certain assumption, that there are all the same conditions that include what could be created by certain economic conditions. Therefore now I will talk more about economy. Strategy makers formulate educational policy primarily in response to national economic concerns. According to House (2000), 'there are at least four ways in which economic concerns influence educational policies:

- Economic conditions strongly influence educational policies.
- Educational policies are frequently formulated to reduce costs and increase the productivity of schools.
- Education and economic development presumed to be closely linked.
- Economic concepts and metaphors permeate educational thinking'.

It is inevitable that economic considerations must be entertained in education policy making. From an economic point of view, education policy needs to include economic considerations, such as the following:

- Meeting the short-term and/or long-term economic demands of society at different levels for education;
- Identifying, procuring, and allocating appropriate resources for inputs into the education system;
- Making appropriate education provisions (e.g. school places for students, number of trained teachers, schools, education facilities, levels of education, etc.);
- Changing the internal structures of the education system to meet different purposes in operation and education; and
- Enhancing the efficiency of internal processes of the system and its subsystems (such as schools, training institutions, etc.).

All these considerations and related efforts are aimed at improving educational services and practices, enhancing internal and external economic effectiveness and generating other social benefits to different levels of society (Behrman and Stacey, 1997; Solmon and Fagnano, 1994).

I agree that economic conditions have really more impact on possibilities but I disagree that it could be a panacea or answer. Bigger impact on possibilities is socio-cultural conditions. More issues could not bee soluted only by favourable economic conditions. More problems usually lie in social interaction, in other words, in relationship, cultural differences, values etc.

Objectives in education

By ending my article I will give some examples of formulated objectives in education. The critical analysis that has been made within the new sociology of education can be further developed in line with the Aristotelian starting point "that education is fundamentally not an epistemological but an *ethical and political enterprise*" (Schilling, 1986, p.12, cf. Aristotle, 1990). This perspective means that the content of education cannot be referenced to any definitive scientific-technical rationality. The content aims that are expressed and the tradition within which, science education is embedded (including its relationship to various scientific disciplines) have to be understood as historically and socially conditioned and always open to discussion.

With Aristotle and his modern interprets, Englund would put forward a communicative-practical rationality, stressing space and procedures for a rational discourse of value questions, as an alternative to the presumed adequacy of scientific-technical rationality. Once different conceptualizations of educational content have been acknowledged, a means is needed for discussing and debating their relative merits. This enterprise also can (given the aims set for the educational system today) be related to the question of values (companion meanings) to be communicated. Thus the educational philosophy that has focused on the relationship between democracy and education is placed centre stage (Östman and Roberts, Englund, 1998).

Standards for the science education system guide the policies that must be implemented and the alignment of the elements within education those policy makers and others must pursue to support science learning described in the standards. System standards also address the essential functions that serve to build and sustain the capacities demanded by the standards of teachers and school communities.

As Young (1998) points out, connectivity is a vision of a curriculum of the future but not only a vision. It can point to specific strategies for teachers, whether they are designing modules, recording the achievement of their students or assessing their work. It also recognizes the need to make explicit how such everyday practices are linked to the ways that schools and colleges are themselves part of a wider system and to debates about the kind of future society envisages for itself. The second premise is that if the importance of personal and social education is to be more than well intentioned rhetoric, a new approach to the curriculum is required that changes the relationship between personal development goals, the educational aims of school subjects and the whole curriculum goals of schools. Schools will need to develop their collective intelligence (Brown and Lauder, 1995) about those internal activities which involve the whole school – such as guidance and learner support as well as about activities such as marking and homework which in many schools are left to departments, and about their external links with primary schools, universities, employers and other secondary schools in their area.

As Young (1998) pointed out, the 1991 White Paper, Education and Training for the 21st Century clarified the two main elements of government strategy. The first was to establish a national tripartite qualifications framework based on an academic track, a full-time general vocational track, and work-based track. Within this framework, each track was defined in terms of an agreed set of levels; official policy was and remains that the three tracks should have parity of esteem at each level in relation to providing access to higher education. The second element of government strategy was to introduce a performance-related approach to funding that encouraged institutional competition as a way of getting value for money and raising standards'.

The appropriate objectives of educational policy, as McLaughlin (1996) pointed out, may be summarized as follows: first, to empower and liberate individuals so that they can exploit their own capacities, including their moral and judgmental capacities, to the fullest possible extent; second, to ensure that society as a whole will be a beneficiary of such general educational achievement and the socialization which accompanies it; third, to see to it, as well, that individual and social workplace needs will be a major consideration in designing specific curricula and educational structures.

Conclusions

Explicit *strategy* could be described as a process of formulating aims or objectives. The main point outlined in characterization of curriculum is that this wholeness of education aims, content, methods, teaching material, organization and control has been seen as an always changing and renewable system. Among all elements of curriculum the problem of aims is of special significance. By considering the curriculum as a whole a definition of national educational purposes should go beyond standards and be expressed in a democratically debated National Curriculum that applies to all students.

Summarizing the conditions which were analyzed I wanted to outline the main condition, that includes what is good and what is possible for strategy makers to formulate – it is socio-cultural context of society. Social order in a pluralistic democratic society would be ethical in the conditions of a common agreement. While social order is related to the values it comes out not easy to agree on what a social order has to be.

The curricular criteria must be tied to fundamental liberal notions of the autonomy and importance of the individual and of finding new and better ways to respect differences and reject domination. These are not commitments that are shared by everyone. However, these notions - together with the judicial and political systems and the many civic organizations designed to give them operational meaning - remain the greatest guarantee of our capacity to most fully realize, and give sustained meaning to, our human aspirations.

References

ARGYRIS, CH./PUTNAM, R./SMITH, D.M.: Action science. San Francisco, CA 1985 (Jossey-Bass).

AVIS, J.: Social difference and antagonism within the 16-19 core curriculum. British Journal of Sociology of Education 13(1992)3.

BALL, S.J./BOWE, R.: Subject departments and the 'implementation' of National Curriculum policy: an overview of the issues. Journal of Curriculum Studies 24(1992)2.

BAUMAN, Z.: Culture as praxis. London 1999 (Sage).

BERNSTEIN, B.: Pedagogy Symbolic Control and Identity. London 1996 (Taylor&Francis).

BEYER, L.E./ APPLE, M.W.: Values and Politics in the Curriculum. The Curriculum. Problems, Politics, and Possibilities. Albany 1988 (State University of New York Press).

BITINAS B. Ugdymo filosofija (Philosophy of Education). Vilnius 2000 (Enciklopedija).

CARLGREN, I.: National curriculum as social compromise or discursive politics? Some reflections on a curriculum – making process. Journal of Curriculum Studies 27(1995)4.

CHENG, Y.CH./ NG, K.H./ MOK, M.M.CH.: Economic considerations in education policy making: a simplified framework. The International Journal of Educational Management, 16(2002)1.

Defining the Curriculum. Histories and Ethnographies. GOODSON, I.F/ BALL, S.J. (ed). London and Philadelpia 1984 (The Falmer Press).

ENGLUND, T.: Education as a citizenship right – a concept in transition: Sweden related to other Western democracies and political philosophy. Journal of Curriculum Studies, 26 (1994)4.

ENGLUND, T.: The public and the text. Journal of Curriculum Studies 28(1996)1.

ENGLUND, T.: Towards a dynamic analysis of the content of schooling: narrow and broad didactics in Sweden. Journal of Curriculum Studies 29 (1997)3.

ENGLUND, T.: Rethinking democracy and education: towards an education of deliberative citizens. Journal of Curriculum Studies 32(2000)2.

GOODSON, I.F.: Studying curriculum: towards a social constructionist perspective. Journal of Curriculum Studies 22(1990)4.

McLaughlin, T.H.: Public Values, Private Values and Educational Responsibility. In: Values, Education and Responsibility. St. Andrews 1995, pp 19-32 (Centre for Philosophy and Public Affairs).

McLaughlin, T.H.: Education, Multiculturalism and the Demands of recognition. Amilburu, Maria G. (Ed) In Education, the State and the Multicultural Challenge. Pamplona Spain 1996, pp 135-158 (Ediciones Universidade de Navarra SA: EUNSA).

Portelli, J.P.: Exposing the hidden curriculum. Journal of Curriculum Studies, 25(1993)4, pp. 343-358.

Quicke, J.: The science curriculum and education for democracy in the risk society. Journal of Curriculum Studies, 33(2001)1, pp. 113-127.

Roberts, D.A./Östman, L.: Problems of Meaning in Science Curriculum. New York and London 1998 (Teachers College, Columbia University).

Rosenmund, M.: Approaches to international comparative research on curricula and curriculum – making processes. Journal of Curriculum Studies, 32(2000)5, pp. 599-606.

Sears, J.T./Marshall, J.D.: Generational influences on contemporary curriculum thought. Journal of Curriculum Studies 32(2000)2, pp. 119-214.

Shapiro, H.T.: Liberal Education, Moral Education. Can - and should - a university teach its students to be better citizens and better people? [cited March 12 2004]. Available from: http://www.princeton.edu/~paw/ archive_old/PAW98-99/08-0127/0127feat.html.

Teaching as a Reflective Practice. The German Didaktik Tradition. Mahwah, New Jersey, London 2000 (Lawrence erlbaum associates, Publishers).

The Curriculum. Problems, Politics, and Possibilities. Beyer, L.E./Apple, M.W. (Ed). Albany 1988 (State University of New York Press).

Todd, S.: 'Bringing more than I contain': ethics, curriculum and the pedagogical demand for altered egos. Journal of Curriculum Studies, 33(2001)4, pp. 431-450.

Witz, K.G.: Science with values and values for science education. Journal of Curriculum Studies 28(1996)5, pp. 597-612.

Young, M.F.D.: The Curriculum of the Future. London 1998 (Falmer Press).

Vilija Stanišauskienė

The Model of Carreer Competence and Its Expression in the Contemporary Labour World

Introduction

The contemporary world experiences deep political and social transformations. It faces an exceedingly wide range of interacting changes, starting from a different human thinking and self understanding as well as understanding of own activities, to global political, economic, demographic and other processes (Hargreaves, 1999; Fullan, 1998; Kavolis, 1991, 1996; Donskis, 1994).

These transformations significantly influence the change of labour world, which, according to Arnold (1997), has been noticed all the time, but became obvious in a current decade. The change of the work environment and work type stipulates new requirements for a person: he/she has to be able to act in a changing environment while assuming the responsibility for his/her own career. Thus, the question arises - what kind of competence should a person have while striving for a survival and successful performance in the contemporary labour world?

The ***aim*** of this article is *to base the model of carrier competence and reveal tendencies of its expression in the current labour world.*

The article consists of three parts. The first part presents the analysis of a change of characteristics of the contemporary labour world, which determines new requirements for a person in his/her career. The second part defines the concept of career competence and grounds the model of career competence. The third part introduces the analysis of tendencies of the expression of career competence model expression in the contemporary labour world.

1. Requirements to a person stipulated by changes of the labour world

It is important to emphasize essential characteristics of labour world while analyzing the change of the labour world. Having in mind that the labour world is a composite phenomenon, the multi-discipline attitude is necessary for its analysis and for the analysis of its changes. If to consider from the point of view of ***demography***, the problem of dynamics of the ***labour force*** or human resources becomes evident in the labour world. From the ***economic*** point of view, the dynamics of ***workload*** and ***salary*** as well as ***work places*** is very important. The ***work type***, being one of the characteristics of the labour world, changes fundamentally due to new ***technologies***. Thus, the aspect of technology changes has to be evaluated while analyzing the change of the labour world. From the point of view of management, changing ***organizations and/or organizational structures*** first of all reflects the change of the labour world.

Therefore, the change of four characteristics of the labour world, namely labour force, workload, work type and work organizations, will be analyzed hereinafter indepth with the emphasis on its influence on the change of the individual career conditions and requirements to personal competence arising due to this change. It should be noticed that the mentioned characteristics are particularly inter-related and the allocation of certain phenomena of change to one of these characteristics may be considered only as conditional.

Scientists (Loven, 2003, Arnold, 1997) notice essential changes of the composition of the **labour force**, the reason of which is cultural and social change (hereinafter the term "labour force" is used as a statistical index, describing a number of economically active population). Globalization, changed living conditions, achievements of medicine, which determine the general obsolescence of population, and women's emancipation stipulate cultural and ethnic diversity and dynamics of labour force, obsolescence of labour force and increased number of working women.

Population of working people gains an increasing cultural and ethnic diversity in many Western countries. Migration of employees is a natural phenomenon in the global labour market. From a cultural point of view, the homogeneity of labour force in the country mainly depends on economic indices and the performed employment policy in the specific country. Cultural ant ethnic diversity of labour force determines new requirements for employees. *Knowledge of foreign languages and wide cultural horizon* become a necessity. People not following these requirements face limited career opportunities.

Labour force is advancing in age in many Western countries. This factor is related to increasing life period of a productive person and lengthening period of young people's studying, which estrange the beginning of work career. The data of the analysis carried out Lithuanian organizations in 1998 (Kabaila, 1998) shows that employees advance in age, which becomes the most serious problem of personnel in the nearest future according to 39,9% of managers of investigated organizations. The general obsolescence of population influences economy of the country – on the one hand the proportion of people of employable age, i.e. tax payers, is decreasing; therefore the workload is changing. On the other hand, the competition for work positions is increasing, as people of retirement age may successfully compete with the youth because of the gained experience. Thus, the obsolescence of labour force, determining the increase of competition for work positions, impede the career development of an individual and emphasize the necessity of his/her correspondence with the contemporary requirements for career. *The experience in certain field of activities* gains great importance as a factor of competitive ability.

The change of the composition of labour force influences the change of other characteristics of labour world, namely workload. Flexible post-modern production technologies eliminate traditional work allotment: partially repetitive job descriptions and rotation appear. It becomes easier to change labour force while using temporary employment, part-time employment, and work in accordance with agreements etc. Possibilities of salaries become more flexible, as methods like payment for performed work, benefits at the discretion of the administration and payment for merits appear. Hargreaves (1999, p.79) states that companies may need twice less employees in the future, these employees will receive double salary and will have to work three times harder. What influence does it make on a person's career? First of all, *career safety and stability* become impossible. Secondly, such a workloade increase determines the decrease of job places and therefore unemployment, which impedes the career development of an individual.

The increase of unemployment is the essential feature of the contemporary career context, determining *the necessity of an individual to be ready to change the type of activities fundamentally or create his/her own work place*. It is undoubtedly related to the *preparation of an individual for life-long learning.*

Work type changes alongside with many changes happening in the society. The presentation "Re-designing Management Development in the New Europe" (1998, p.23) states that the movement of population from agricultural to industrial sector has been clearly observed in recent fifty years. Presently these people move again – to the sector of services. It shows that every generation collides with the movement of one third of that generation from one type of work to fundamentally different type of work. According to the authors, only a few Western European countries have been prepared for fast and successful migration of people from the sector of industry to the sector of services and great unemployment has been inevitable. Thus, the next generation will collide with analogous problem while moving from the sector of services to another sector, which still remains unknown.

The change of work type is related to changes of thinking and life style of individuals. The transition from the agricultural to industrial sector raised a demand to perceive and realize new concepts, namely the efficiency, time price, budget control etc. The movement from the sector of industry to the sector of services emphasizes new concepts, as the relationship between a client and a producer changes. It even influences the values of the society. On the other hand, new priorities appear in the field of career requirements, i.e. *social capacities become requisite for an employee of any field.*

A rapid development of technologies and particularly the development of information and telecommunication technologies influence essentially the change of a work type. Split-second means of communication and computerization allow more rapid and wide transfer of decisions. It becomes unnecessary to gather operations and employees in one place to make the work efficient. Subdivisions of enterprises become smaller and more flexible from the point of view of an entire structure of an enterprise. Places of factories are not traditionally established, they do not need numerous and loyal labour force. Factory towns become a relic of the past. Geographical space is economically flexible. Commitments to certain peculiarities of certain place are valid as long as they are guaranteed by local market, favourable land property and flexible, cheap labour force (Hargreaves, 1999, p.85).

Due to the progress of technologies, the number of people using the most modern communication technologies and working at home increase. This population might be described as teleworkers (Chapman et al., 1995): i.e., they use communication technologies as an alternative to moving to work place. It stipulates the necessity for almost all employees of all sectors *to be able to use modern information technologies.* With the sophistication of technologies, *requirements for the education of employees* undoubtedly rise.

All changes determined by the global change mentioned above promote the change of organizations in the labour world. One of the most common features of the change of organizations is the decrease of hierarchical layers of employees and organization (Wheatley, 1992, Sting, Vulf, 1994, Arnold, 1997, Jucevičius, 1998). Therefore, efforts for the promotion or greater responsibility become more complex. On the other hand, such changes in the structure of organizations stipulate lower control and greater individual responsibility for specific activities. The absence of clearly defined functions and hierarchical structure uplifts the necessity for capacities of *interpersonal* and *inter-group communication and cooperation* as well as emphasizes such features of employees as *initiative, independence and responsibility.*

The effectiveness of the organization activities and competence of an employee are determined by the contact with information, i.e. the most competitive employee is the one closest to the information source in the information society. Therefore, every member of an organization collides with the requirement *to be able to use information sources and apply the received information.*

The contemporary organization emphasizes the management in accordance with acute questions arising from the present situation (Jucevičius, 1998). That is why everyone has to *manage the analysis and the evaluation of a situation as well as naming goals. Critical thinking and innovation* are of high importance here.

Characteristics of labour world	Tendencies of the change of labour world characteristics	Arising requirements for a person
Labour force	Upgrowth of cultural and ethnic diversity;	Foreign languages and wide cultural horizon;
	Upgrowth of the age average of labour force;	Experience in a specific field of activities;
Workload	Upgrowth of workload;	Diligence;
	Popularization of temporary employment and part-time work; the emphasis on tendencies to have more than one work at a time;	Flexibility, ability to plan time and allocate attention;
	Upgrowth of unemployment degree	The preparation to change activities fundamentally, to find or create a work place; ability of constant learning;
Work type	Shift to service field;	Capacities of communication, cooperation and negotiations;
	Decay of enterprise towns, increase of persons working at home;	Ability to use contemporary information technologies;
	Modernization of technologies;	High education;
Organizations and/or organizational structures	Decrease of hierarchic layers	Interpersonal and inter-group communication and abilities of cooperation; initiative, independence, responsibility;
	Increased significance of information;	Ability to use sources of information and apply the received information;
	Management in accordance with acute questions arising from the present;	Ability to analyze and evaluate the situation, name aims; critical thinking and innovation;
	Individualized aims;	Set up and constantly verify goals, act in undefined changing situation
	Popularization of project teams;	Capacities of cooperation, activity planning and organization; creativity, responsibility and leader's features;
	Decay of organizational structures of career development	Individual responsibility for one's own career, capacities of self-education

Table 1: Tendencies of the labour world changes, determining the change of career conditions and new requirements for a person

Individualized aims and tasks characteristic for a contemporary organization promote person's individual creativity and flexibility. A person has to be able to *set up goals and constantly verify them, as well as act in an undefined and changing situation.*

One of the changes of organizations determining the more efficient work (as an answer to current requirements) is re-structurisation of organizations into project teams, which take the responsibility for certain tasks (Arnold, 1997, p.23). It is first of all the work with a number of new people for an individual, secondly, new tasks every time, demanding new skills, and thirdly, the possibility to find a new way (idea, relations, experience etc.) for the future. Therefore, *the necessity of cooperation, activity planning and organizational capacities arise.* Also personal *creativity, responsibility and leadership features* become important.

Another very important result of the change of organizations is the decay of organizational structures of career development. The career of an employee becomes his/her own business. The establishment of this attitude has been influenced by almost all above-mentioned changes, including the popularity of short-term agreements (it is unrewarding for an organization to spend money on the improvement of qualifications of temporary employees), division of wide organizations into independent subdivisions (it is ineffective for such subdivisions to have career development structure) etc. This tendency could be observed in Lithuania as well. The data of Kabaila analysis (1998) performed in 168 Lithuanian organizations shows that only 17,5 % of organizations have constant employees working with issues or personnel education. The majority of respondents mentioned that the qualification of an employee is considered during the employment procedures in their organizations, meaning that every employee *receives an individual responsibility for his/her own career. The significance of learning competence is emphasized too.*

Tendencies of the labour world changes and new requirements for a person determined by these changes are generalized in Table 1.

The Model of career competence

Requirements of the contemporary labour world raised for an individual stipulates the necessity to have certain capabilities and features that allow answering the mentioned requirements and overcoming obstacles arising during social transformation periods. In other words, ***a person needs to have career competence in the contemporary labour world.*** The concept of competence defined by Jucevičienė and Lepaitė (2000) are followed in this article, i.e. *competence is the expression of a person's qualification or the ability to act, determined by knowledge, know-how, skills, attitudes, personal features and values of an individual.* Capabilities and personal features enabling an individual to solve tasks arising in his/her career and follow requirements of the contemporary labour world may be named as the ones guaranteeing career competence.

The term *career competence* is used to define the preparation of a person for certain activities and the development of his/ her own career in many scientific works (Arnold, 1997; Jacobs, 1989 et al). Practical application of corresponding capacities is emphasized but not the knowing how to perform it, or the possession of formal qualification. For instance, Jacobs (1989, p.33), provides the following definition while talking about the management competence – "noticed capabilities to perform successfully managerial tasks". This definition could be applied easily to any field of activities, i.e. a person's competence could be judged by the fact whether he/she might apply possessed capacities for the effective performance of the work tasks.

Thus, *the career competence is a number of capacities and personal features necessary for a person in the contemporary labour work, which could be successfully applied in practical activities.*

The abundance and diversity of requirements for a person observed in the analysis of social and humanitarian scientific works (Loven, 2003; Hargreaves, 1999; Fullan, 1998; Brown, Scase, 1997, Kavolis, 1998; Shaw, Jovaiša, 1998; Jucevičius, 1998; Allison, 1998; Arnold, 1997 et al) stipulate the necessity of a certain theoretic construct that allows grouping and integrating these requirements. The scientific literature has been analyzed with the above mentioned purpose (Fullan, 1998; Franc, 1998; Brown and Scase,1997; Wats, 1994; Jovaiša, Shaw, 1998). It has been determined that the complexity of the current career competence and the integrity of composite parts are emphasized in the works of the researchers:

- Fullan (1998, p.31) names "four mediation capacities for changes" in the book "Change Forces" as requirements for a teacher in the context of change. It is ***"the creation of personal vision, interest, mastership and cooperation"***. These requirements are quite universal, therefore they may be applied to the definition of any activities in the context of the labour world.
- Franc (1998) states that the base of career is ***social, professional and continued self-education competence***. According to the author, social competence is necessary in any career course, as all activities of the contemporary labour world have social aspect: the communication with clients, work groups, relations with social partners. Professional competence, defined by Franc (1998) as specific professional knowledge and skills, is important in a specific chosen field of activities. This competence allows competition of specialists in the labour market. Basing on Franc (1998), educational competence is the capacity of continuous learning and updating of one's own professional and social competence. The system of these three competences provides a person with the ability to make a career in the contemporary labour world.
- Brown and Scase (1997), emphasize the following three groups of criteria while analyzing criteria of employee selection (comprehensive interview has been performed in sixteen organizations in Great Britain): ***suitability, capability and acceptability*** criteria. Suitability criteria are related to professional competence of an employee, i.e. specific professional capacities and general capacities, such as the capability to work in a team, manage projects etc. Capability criteria are defined as "the quality of intellect and thinking" (Brown, Scse, 1997, p.91). Acceptability criteria are related to a person's social competence – his/her capability to communicate and cooperate.

- Watts (1994) distinguishes the following three fields while analyzing issues of the preparation for career: ***personal / social*** (or learning how to live), ***educational*** (or learning how to learn), ***professional*** (or learning how to earn one's living). The author into the aims of general education, which essentially influence the curriculum, integrates aims for the preparation for career arising from these fields.
- After the analysis of the problems of the general education capacities, Lithuanian and foreign scientists (T. Jovaiša, Shaw, 1998; Laužackas, 1999), divide business competence into ***business, methodical and social competence*** in accordance with the object of activities of an employee: business competence corresponds the material object of activities, methodical – the object of the type of activities and social – the object of the relationship expression.
- Jarvis and Carlsen (1999), state the following while assuring that every person is responsible for own life and career: "in order to perform career tasks, schoolchildren acquire activity capacities uniting business, methodical and social competence, as the result of education. ***Business competence*** is capacities, skills and knowledge, necessary for the performance of professional tasks; ***methodic competence*** is first of all a capability to gain new knowledge and skills as well as to find ways independently for the solution of the work tasks; ***social competence*** is a capacity and preparation to work alongside and communicably with other people, participate in the creation of the work and environment.

Though scientists use different concepts, it is not difficult to envisage similarities between the groups of requirements emphasized for persons while comparing the meaning provided for these concepts. It is revealed in Table 2, p. 99.

After the analysis of the diversity of necessary capacities (requirements for a person arising from the labour world) and the generalization of attitudes introduced in works of the mentioned scientists, four fields are observed, to which the corresponding necessary capacities may be applied. The capacities applied to a certain field guarantee corresponding competence, which is an element of the integral career competence (see Figure 1, p. 99).

Thus, it might be stated that the *integral competence of a contemporary career is composed of the system of four competences, namely personal, social, learning and professional competences.*

Capacities ***of self-cognition and self-introduction*** are noted in the field of ***personal*** competence.

Capacities and features *harmonizing the relationship of a person with other people and social environment* are attributed to the field of ***social*** competence.

The field of ***learning*** competence includes capacities enabling a person for ***self-education***.

Professional competence consists of *special professional* capacities as well as of capacities and features related to the *working activities* of a person.

When analyzing every competence, it is worth mentioning that, firstly, capacities stipulating certain competence are related systematically, and, secondly, the set of the latter cannot be finite while evaluating the dynamics of the labour world.

Author	Requirements raised for a person			
Fullan (1998)	Ability to create personal vision	Abilities of cooperation	Abilities of interest	Abilities of mastership
Franc (1998)	Social competence		Competence of continued self-education	Professional competence
Brown, Scase (1997)	Criteria of acceptability		Criteria of capability	Criteria of suitability
Wats (1994)	Personal / social field		Educational field	Professional field
Jarvis, Carlsen (1999)	Social competence		Methodical competence	Business competence
Jovaiša, Shaw, 1998; Laužackas, 1999	Social competence		Methodical competence	Business competence

Table 2: Attitudes of scientists towards requirements arising to a person in his/her career

The attention has also to be paid to the term ***capability*** used in this article, which has been defined by Jovaiša (1993, p.62) as a physical or mental ability to perform certain action, activities. According to Jovaiša (1993), the pedagogic basis of capability is knowledge, know-how and skills. This term cannot be changed by a more common one, i.e. skills or know-how, as structural complexity of that capability is not revealed then.

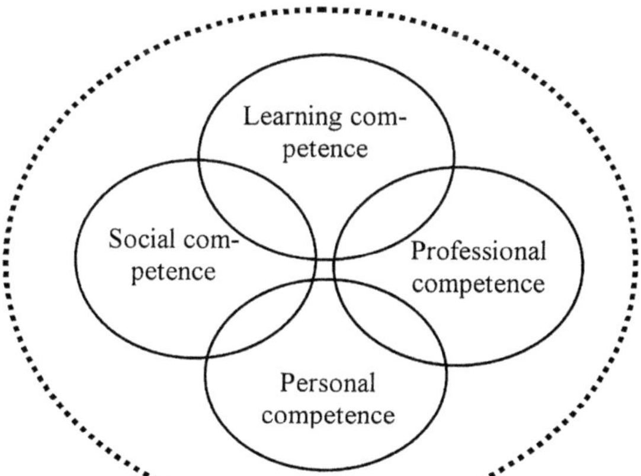

Figure 1: The model of career competence

Requirements for a person stipulated by changes of the labour world introduced in the first part of this article are divided into four fields of integral career competences (see Table 3). This division is conditional, as certain capabilities necessary for a person are related systematically and could be ascribed to one or more fields.

The analysis of the expression of tendencies of a career competence model in the contemporary labour world. The qualitative content analysis of work advertisements has been performed with the aim to reveal the reflection of the introduced career competence model (capability system stipulating career competence) in the contemporary labour world. The analysis of 220 advertisements with proposals of various work places selected at random in the national newspaper "Lietuvos rytas" in August has been performed.

The matrix of the analysis has been composed while basing on the above discussed division of capabilities and features stipulating the integral career competence in the fields of competences. Qualitative content analysis has been performed with the usage of SPSS software. To be specific, methods of descriptive statistics have been used.

Personal	Social	Learning	Professional
Self-cognition; skills of self-presentation; respect to yourself and others; independence; self-control; initiative; flexibility; personal style; abilities of self-control; responsibility; creativity; features of a leader	Trust in people; skills of interpersonal and inter-group communication and cooperation; capabilities of communication; capabilities of cooperation; skills of team and group work; social adaptation; wide social self-identification; acknowledgement of public processes and wide cultural horizon; social responsibility; knowledge of foreign languages	Preparation for self-education; critical thinking; innovation; abilities of the usage of information sources as well as transferring and aplication of information; computer literacy; high education	Preparation to act in an undefined, changing situation; project management; responsibility for own activity and its result; situation analysis and evaluation; set up of aims; decision; planning; professionalism in a specific field of activity; specific professional knowledge

Table 3: Allocation of capabilities and features certifying the integral career competence in the fields of competence

After grouping the requirements mentioned in work advertisements into four fields of career competence, the qualitative distribution of these requirements has been determined. (see Figure 2, p. 101). It is obvious, that these requirements are divided evenly with slight overweight of requirements for professional competence. The analysis of the division of the requirements for an individual in the order of rating in work advertisements has been introduced with the aim to reveal a more precise view of skills and features requested by employers (see Figure 3, p. 101).

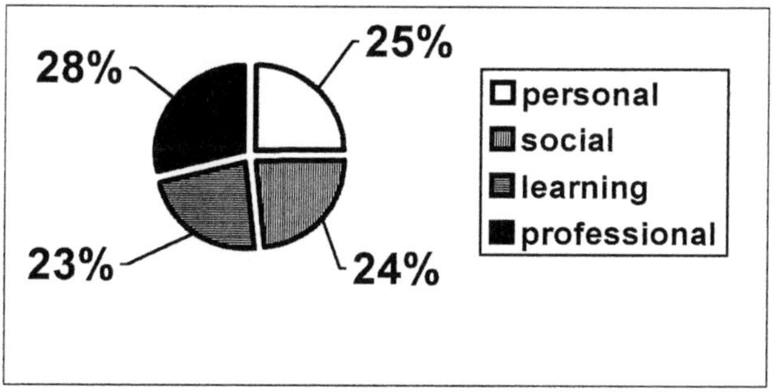

Figure 2: Allocation of requirements for candidates in the fields of carreer competence

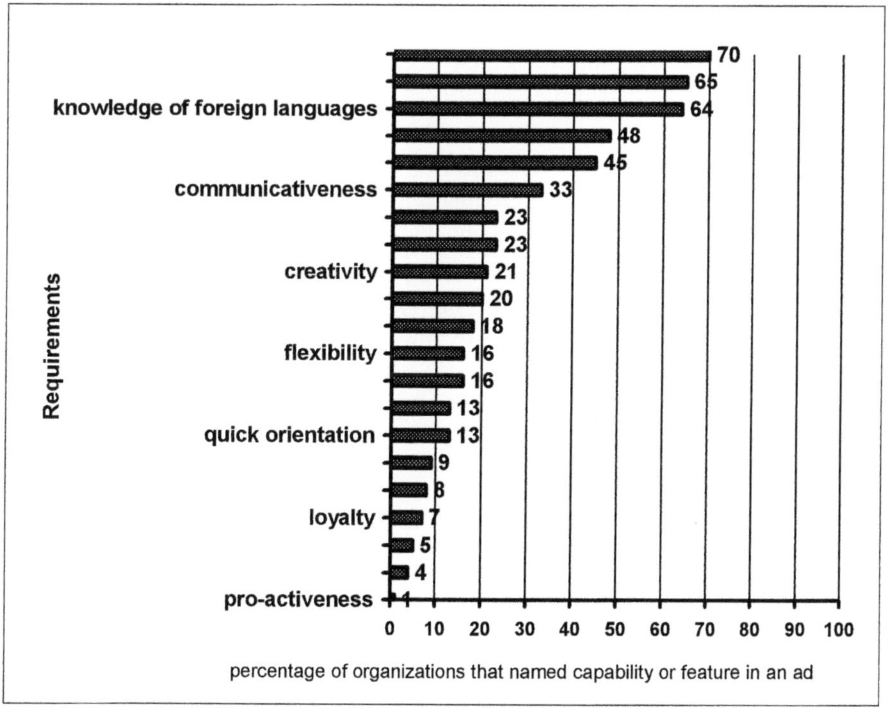

Figure 3: Rating of the requirements for candidates

It has to be acknowledged that skills and features guaranteeing career competence are divided unevenly in accordance with the frequency of references. The most frequently expressed requirements are the following: the experience in certain field of activities, higher education, knowledge of foreign languages and professional qualifications (with documents proving then). Meanwhile features as pro-activeness, organizational capacities, positivism, loyalty and honesty occupy the lowest positions in the rating. Their significance is not diminished, as the above-mentioned division is determined by the abundance and diversity of similar features. Certain requirements (as accuracy, orientation towards the result, enthusiasm, provenance, the ability of time planning and decision making, etc.) have been noticed once or twice in work advertisements and dropped out of the rating without making even 0,5%. Such structure of the rating allows concluding that work advertisements include quite standard requirements for candidates. Other requested features and skills only supplement this "framework" of the main requirements.

One third (36%) of the 220 analyzed ads for vacancies have been submitted by foreign companies working in Lithuania. The question has been paused: what are the differences among requirements named for future employees by Lithuanian companies and foreign companies working in Lithuania? The percentage of Lithuanian and foreign employers that raise at least one requirement for personal, social, learning and professional competences has been determined, while trying to answer the above paused question (see Figure 4). Obvious differences have been noticed, as advertisements of foreign companies working in Lithuania include and emphasize more requirements for personal and social competences of candidates. Such advertisements include the diversity of requested personal features, starting from enthusiasm, pro-activeness, flexibility, ability to perform a few kinds of work at a time to forthrightly expressed request for the candidate not to be "a gawk". Meanwhile ads of Lithuanian companies mainly emphasize requirements for professional competence. Requirements for personal and social competences are mainly formulated in a very stereotypic way: a communicative, creative and honest employee is requested.

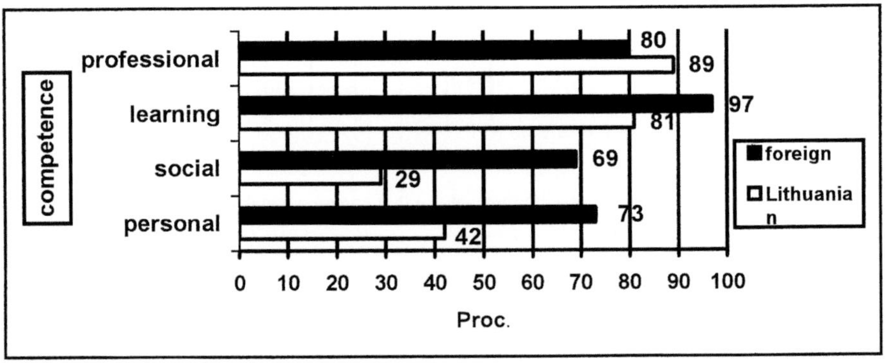

Figure 4: Requirements for personal, social, learning and professional competences of candidates in ads of Lithuanian companies and foreign companies working in Lithuania

Obvious differences have been noted while comparing specific requirements named for candidates in ads of Lithuanian companies and foreign companies working in Lithuania (see Figure 5, p. 103).

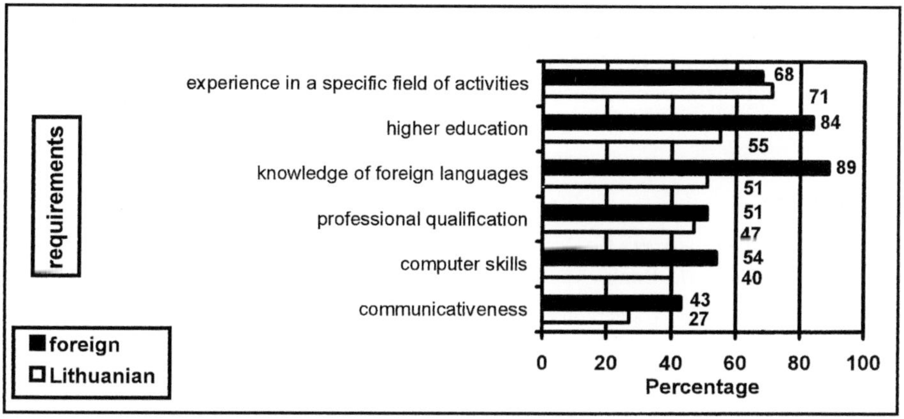

Figure 5: Requirements for candidates in ads of Lithuanian companies and foreign companies acting in Lithuania

Skills and features occupying the highest places in the scale of rating are evaluated differently by companies of these two categories: foreign companies emphasize good knowledge of foreign languages (it is undoubtedly nothing strange) and higher education; Lithuanian companies mainly emphasize the experience in a certain field of activities. It could be easily noted that foreign companies name more requirements for their future employees.

Conclusions

- Changes of the labour world, expressed by the changes of labour force, workload, work type and organizational structures, stipulate a number of new requirements to a person in his/her career life. Contemporary career competence is necessary to meet these changes.
- Career competence is defined as a system of capabilities and personal features necessary in the contemporary labour world that may be successfully applied by a person in practical activities. These capabilities and features are grouped into four fields of competences, namely personal, social, learning and professional competence. These are integral parts of career competence. Such model of career competence may be successfully applied while analyzing the specifics of the labour world.
- The analysis of the expression of the career competence model in the contemporary labour world revealed the following tendencies:
  - requirements named in ads for vacancies are gradually allocated in personal, social, educational and professional fields of competences;
  - ads for vacancies name quite standard requirements for a candidate, namely work experience, high education, knowledge of foreign languages, computer skills. Other required features and skills only supplement this "framework" of the main requirements;

- the comparison of the requirements for future employees named by Lithuanian companies and foreign companies working in Lithuania shows that foreign companies raise more requirements, which are more diverse; personal and social competence is more emphasized.

References

ALLISON, CH./RINGOLD, D.: Labor Markets in Transition in Central and Eastern Europe 1989-1995. 1996 (World Bank).

ARNOLD, J.: Managing Careers into the 21st Century. London 1997 (Paul Chapman Publishing Ltd).

BEACH, S.D.: Personnel: The Management of People at Work. New York 1985 (Macmillan Publishing Company).

BROWN, PH./SCASE, R.: Universities and Employers: Retoric and Reality. The Postmodern University? Contested Visions of Higher Education in Society. SMITH, A./WEBSTER, F. (Eds.). London 1997.

CHAPMAN, H./PARKMAN, G.: Steps Toward Implementing a Developmental Guidance Program. Journal of Connecticut Association for Counseling and Development 15(1988), pp. 17 –39.

DONSKIS, L.: Moderniosios sąmonės konfigūracijos. Kultūra tarp mito ir diskurso. Vilnius 1994 (Baltos lankos).

FULLAN, M.: Pokyčių jėgos. Vilnius 1998 (Tyto alba).

HARGREAVES, A.: Keičiasi mokytojai, keičiasi laikai. Vilnius 1999 (Tyto Alba).

HEATHCOTE, A.: Developing Personal Autonomy in continuing professional Development. Education, Autonomy and Democratic Citizenship Philosophy in an changing World. BRIDGES, D. (ed.). London and NEW York 1997 (Routledge), pp.165 – 176.

HODKINSON, PH.: Careership: The Individual, Choices and Markets in the Transition into Work. Konwledge and Nationhood. Education, Politics and Work London, New York 1996 (Cassell), pp. 121–139.

JOVAIŠA, L.: Pedagogikos terminai. Kaunas 1993 (Šviesa).

JOVAIŠA, T./SHAW, S.: Žvilgsnis į bendrųjų gebėjimų ugdymą Rytų ir Centrinėje Europoje. Profesinis rengimas: tyrimai ir realijos 1(1998), pp. 30–47.

JUCEVIČIENĖ, P./LEPAITĖ, D.: Kompetencijos sampratos erdvė. Socialiniai mokslai 1(2000), pp. 44-51.

JUCEVIČIUS, R.: Strateginis organizacijų vystymas. Kaunas 1998 (Pasaulio lietuvių kultūros, mokslo ir švietimo centras).

KABAILA, A.: Suaugusiųjų profesinio mokymo ypatumai Lietuvos regionuose ekonomikos reformos sąlygomis. Darbo rinkos profesinis mokymas – efektyvus darbo jėgos paklausos ir pasiūlos derinimo veiksnys. Vilnius 1998 (Agora).

KAVOLIS, V.: Civilizacijų analizė. Vilnius 1998.

LAUŽACKAS, R.: Sistemoteorinės profesinio rengimo kaitos tendencijos. Kaunas 1999 (Vytauto Didžiojo universitetas).

LOVEN, A.: The Paradigm Shift – Rhetoric or Reality? International Journal for Educational and Vocational Guidance vol. 3, No. 2(2003).

MATHIS, R./JACSON, J.: Human Resource Management. New York 1991 (West Publishing Company).

PARKMAN, G./NOEL, E.: Toward Developmental Guidance and Counseling in Connecticut. Dialog (13), Journal of the Connecticut Association for Counseling and Development 1989, pp. 37–45.

Re- dwsigning Management Development in the New Europe. Report of the Turino Group. 1998 (Europrean Training Foundation).

WATTS, A.G.: Lifelong Career Development. Cambridge 1994 (CRAC/ Hobson).

Virginija Chreptavičienė

The Change of the Conception of Communicative Competence in the Context of the Information and Multicultural Society

Introduction

The period of social transformations causes the changes of the interaction between environment and a person. If a person wants to be successful in the modern information or knowledge society and be able to solve its challenges, he must constantly renew his competence. In the information society, it is the especially urgent new qualitative characteristic, which determines the change of the conception of communicative competence – it is information communication technologies (ICT) because all human activity is based on information, knowledge and communication. In the information and knowledge society, as never earlier, there is a necessity for a person to obtain the competence of the use of the information and communication technologies and the competence of their creative application. The creation of information or knowledge, dissemination and usage become the main activity. Using ICT people can get information quicker, cheaper and become more educated. People's knowledge and education make a new knowing. However, in order to make ICT more effective and contributed constructively to the progress of a person, organization or society, the skills of using modern technologies are necessary. Thus one of the main factors influencing the change in the conception of communicative competence, which essentially opened new communication possibilities, is information communication technologies.

The other factor, which inspires the change of the conception of communicative competence and the space of the competence concept in the context of the information and multicultural society, is the changes of society structure. Contemporary communication does not occur culturally in the homogeneous but multicultural space. Multiculture is understood as "practice, which emphasizes equal groups of the society, particularly traditional minority groups, rights to satisfy the needs and to get help satisfying them" (Webster New World Dictionary, 1995). A democratic view to the variety in the society is revealed, the idea of equal possibilities, which encourages searching for the most proper ways to all group members of the society in order to make equal possibilities of the dialogue in the variety of cultural contexts. So in the multicultural society, the prominent idea of equal possibilities emphasizes the need to know other languages. It determines a new qualitative need for communicative competence as well as expands the space of the conception of communicative competence.

The problem of communicative competence in the modern world is equally important for business and educational organizations, so this has become the object of social research. The researchers of Western countries (Habermas, 1970, Hymes, 1972, Heberman, 1977, Harris and Cronen, 1979, Harris, 1979, Moirad, 1982, Allright, 1982, Littlejohn & Jabusch, 1982,

Monge, Bachman, Dillard, Eisenberg, 1982, Sypher & Sypher, 1983, Cooley & Roach, 1984, Sypher, 1984, Bostrom, 1984, Spitzberg & Cupach, 1984, Park, 1985, Jemeljanov, 1985, Gez, 1985, Henrici, 1986, Sypher & Zorn, 1986, Hammer, 1989, Spitzberg, 1989, van Ek, 1994, Fritzsche, 1994, Jablin et al, 1994 Sheils, 1995, Zorn Violanti, 1996, Hutmacher, 1996, Flusser 1998, etc.) pay a lot of attention to the analysis of this problem. In recent years the importance of this competence has been particularly emphasized in many EU documents (Barcelona European Council, March, 2002; Memorandum on Lifelong Learning, 2000; European Dimension In Secondary Education, 1997; Maastricht Agreement of European Union – article No 151, 1992; European Dimension in Education, 1993; White Paper on Education and training: Teaching and Learning Towards The Learning Society, 1995; European Council Resolution (December 16, 1997); Recommendation 1383, Linguistic Diversity, 1998, Passport to Mobility, 2001, etc.). The problem of communicative competence has been also analyzed in the space of Lithuanian social studies (Chreptavičienė (1997, 1999a, 1999b, 2000), Chreptavičienė, Urbanskienė (2000), some aspects of this problem were analyzed by Žibaitis (1996), the education of communicative competence in the native and other languages is particularly emphasized in the General Programs of Lithuania for School of General Education (1997), in General Programs for General Education and in Standards of Education development for XI-XII forms (2002), etc.

Although the problem of communicative competence, obviously, is urgent and this problem is investigated in the space of Western and Lithuanian social studies, in the context of information and multicultural society, the problem remains rather fresh and calls for sharp discussions. However, there is no clear and elaborate, acceptable by many people definition of the concept of communicative competence. The concept of communicative competence is poorly analyzed, it is not even distinguished as a separate concept in the space of common competence but named as the element of the other concept (social competence) (Stulpinas, 1996, Reetz, 1991, Franc, 1998, Jovaiša, Shaw, 1998, Jarvis and Carlsen, 1999, etc.) or it is attributed to general abilities (Leithwood, Chapman, Corson, 1996, Show, 1998, Kirby, Žydžiūnaitė, 1999, Toohey, 1999 etc.). A different interpretation of this concept often causes misunderstandings between the researchers and practitioners who investigate communicative competence on the theoretical level or form it in the process of education. On the other hand, as the structure of society changes, the multicultural society forms, information communication technologies are constantly being spread in all life spheres, the concept of communicative competence and the number of its structural elements also change. Global changes in the society make this concept more complicated, it becomes even more polysemous, contextual, and situational and gains new meanings and shares, thus it demands additional explanations and specific interpretations. Taking into consideration the demands of the time, a particular attention should be paid to this problem.

An urgent scientific problem is revealed in this context. It could be formulated into the following questions: What specific features does the concept of communicative competence distinguish in the development of the information and multicultural society? What sub-competencies does the concept of contemporary communicative competence incorporate?

The ***aim*** of this article is to reveal and ground the change in the conception of communicative competence in the development of the information and multicultural society.

The problem is discussed using the methods of scientific literature and documents analysis.

The article consists of two parts. The first part analyzes the factors which determine the change of the conception of communicative competence – the development of multicultural and information society. The second part reveals the space of the concept of communicative competence in the context of information and multicultural society.

Before analyzing the problem, it is necessary to explain the main definitions (concepts) of the article in order to avoid their ambiguous interpretation.

Information is the facts that have a meaning and an aim. Its purpose is to change the perception of an individual who receives this information, to influence his/her values and behaviour (Davenport and Prusak, 1998, p.2-30

Knowledge is the information which has a sense based on the previous knowledge, experience, values, beliefs, etc. (Stankevičiūtė, 2001).

Communication is a spiral process of information exchange between two or more people who exchange knowledge, information; it is not only a movement of communication but a accuracy and broadening of opinions and knowledge that people change (Chreptavičienė, 1997).

Competence is a person's expression or an ability to be active determined by his/her knowledge, skills, practice, attitudes, personal qualities and values (Jucevičienė, Lepaitė, 2001).

1. Factors determining the change of the concept of communicative competence

While the information and multicultural society is under development and ICT have been actively applied in all life spheres, a person's communicative competence is becoming important because the modern world is understood only as a process of information and communication, and life – as knowing. In the modern society, the main resources are not raw materials; these sources are more intangible, such as information and knowledge. Effective use of information becomes the means of increasing competition, the tool of control and even resources – assets and commodities. According to Gordon R. J., Gordon R.S. (1999 p. 8-9), "information as money, raw materials, equipment or time can serve as a resource, a kind of putting in, striving for some profit. Leaders can use information as capital to capital and work, this way mostly reducing the expenditure of an activity. Information can serve as wealth, which belongs to a person or an organization and gives profit. Using this attitude, leaders should value information as an investment, which later will allow acquiring the competitive advantage. An organization can sell information as a valuable commodity, as in the modern economics, where there is a sector of services, the number of companies is increasing, and those companies value any information as a commodity." New information, which creates new knowledge, is acquired only by communication. However, if there is a necessity to control the torrents of information effectively, one may need new qualitative strategic communicative competence. New qualitative communicative competence can be understood as a response to the changing environment as, according to Krogh (1996):

> "(I)n the information and knowledge society, the essential characteristics of which is the contact with information and knowledge, it would be very useful to make a distinction between communication related mostly to common daily work, based on the knowledge, which has been acquired earlier and now, i.e. the communication of work, and the other strategic communication, which creates new knowledge, during which a new space for finding something new is being tried to create (for finding decision, new knowledge, etc.)."

In the modern society, in which the main activity is the use of information or knowledge, creation and dissemination, a person's communicative competence becomes the most important part of holistic competence, which acquires a new quality in the information and knowledge society. Unlike ever before a necessity arises for a worker to acquire information and competence of using communication technologies as well as creative use, because lately ICT have become not only the most important means of activity, receiving information, cumulating and saving, but also a negotiator which is able to link separate elements into the system. Human activity is becoming less dependent on the distance between them, as ICT open wide, earlier unseen abilities to transfer, to accept and spread information, to distribute the torrents of knowledge and ideas despite the distance. ICT organizations can implement, what was impossible earlier. However, as ICT develop, organizations become more dependent on a person. According to Hope (1997, p. 65):

> "(T)he ability to collect with the help of computer, edit and distribute very structural information is not a miracle of the century, but organizations need people's intelligence and experience in order to transform that information into useful knowledge and good solutions."

In other words, it is very important to evaluate and select only necessary information from a big torrent of information, edit it and later create new qualitative information as an intellectual product. The value of the use of effective information urges for the usage of effective search information technologies, and mastering the latter demands special and so far only elite epistolary-cultural competence. Thus this ability of the members of society to navigate in the torrents of unlimited information and using modern ICT be able to select and use information as a valuable product can be called ICT competence, the meaning of which in the space of communicative competence is not discussed.

So, rapid development of ICT, their integration in all spheres of life and the improvement of communication ways has determined the changes of the space of communicative competence. If in the industrial society there have been important linguistic, sociolinguistic, discourse, strategic, socio-cultural, social competencies in the space of communicative competence, so in the information and knowledge society, ICT competence becomes very important in the space of communicative competence.

The other essential factor, which had influence on the space of communicative competence – global processes, which determined the structural changes of society, and at the same time inspired the birth of new multicultural society, distinguishing the variety of social, ethnic, political and other cultures. Realizing multiculture as "practice, which emphasizes equal rights of all society groups, especially minority groups, to satisfy the needs and to get help in satisfying them" (Webster New World Dictionary, 1995), a contemporary democratic view to the variety in the society is revealed. It consists not only of the individuals belonging to different nations, religions, cultures but also of the people belonging to different linguistic societies who have rights to satisfy their needs. Constant confrontation with "otherness" demands, tolerance and cultural knowledge from anyone who strives for intercultural dialogue, as in the democratic multi-cultural society different cultures should not live separately but together and with each other, they communicate in the area of a culture meeting (Saugėnienė, 2000, p. 28). A person's communicative competence becomes especially important for a successful inter-cultural dialogue, because through verbal and non-verbal communication the connection of vital importance between cultures starts. In the multi-cultural society the space of communicative competence changes essentially as the communicative space is not homogeneous. There is not only one linguistic community in such society, multicultural society is multilingual. So, the intercultural dialogue is possible only when an individual knows several languages and is able to construct the proper linguistic cultural context of the society, which corresponds to the discourse. On the other hand, every culture has its own values, the knowledge and acknowledgement of which also have some influence on a person's communicative competence. Obviously, the mastering of a foreign language is practically impossible without partial cultural adaptation. So, in the context of multicultural society, the space of linguistic, socio-linguistic, discourse, strategic, socio-cultural, social sub-competencies change essentially.

2. The concept of communicative competence and its space in the context of the information and multicultural society.

A lot of researchers have tried to define communicative competence; however, there is no steady universal definition till now. Striving to use this essential person's competence more effectively, it is important to make this conception more accurate: not only evaluate the issues of its changes, but also to highlight the sub-competencies of the concept of communicative competence.

Inquiring into the essence, meaning and characteristics of communicative competence, first of all it is important to find out the difference between the two often used synonymous definitions – communicative competence and linguistic competence, as their identification is one of the reasons of misunderstanding, especially on the theoretical level.

The main differences between the definitions of communicative and linguistic competencies have drawn attention of many researchers who analyzed linguistic and communicative competencies (Hymes, 1972, Heberman, 1977, Moirad, 1982, Allright, 1982, Gez, 1985, Henrici, 1986, van Ek, 1994, Fritzsche, 1994, Sheils, 1995, Flusser 1998, Chreptavičienė, 1997, Cesevičiūtė, Minkutė-Henrickson, 2001, etc.). However, although some of the researchers acknowledge that these terms mean different things, they tend to forget this difference themselves and continue to use these terms as synonyms again ("The General Programs of Lithuania for school of general education"(1997), "General Programs for General Education and standards of development for XI-XII forms (2002) or use the junction of linguistic and communicative competencies – linguistic communicative competence (Cesevičiūtė, Minkutė-Henrickson, 2001). Another phenomenon is being observed: the authors do not attach great importance to the differences of these conceptions, sometimes they are used as synonyms (General Programs and standards of development for XI–XII forms, 2002), and sometimes they are identified as communicative skills (Grebliauskienė, 1997).

Grebliauskienė (1997), presenting the educational system of phases non-written communicative skills, states that "a lot of authors, discussing communicative skills, use the term of communicative competence" and they do not go into details to the differences of these conceptions (communicative skills and communicative competence). Thus the author notices here only terminological differences that are not of great importance. Meanwhile to perceive and distinguish the differences of the concept of communicative competence and linguistic competence is not only important in the theoretical discussion.

So, how are linguistic and communicative competencies related? It was supposed that once linguistic competence has been brought up, the jump to communicative competence would occur automatically. Unfortunately, this is not so. The question is still open – how does the active use of a language occur? How does the language get its importance? Henrici (1986, p. 150) here distinguishes two components of communicative competence – an accumulated knowing and the use of this knowing. The mentioned author calls it "material" and "operating" things ("knowledge" and "ability to use this knowledge") and emphasizes the second component – ability to use knowledge according to the situation.

Fritzsche (1994), analyzing basic human competencies, refers to Habermas (1977). The latter author uses the term "competence" to express the structures, which improve the collision of the subject with the world, and argues that not every ability is a competence. Competence, according to the author, describes a person's main abilities, which can be named as three competencies: cognitive, interactive and linguistic. Each competence, according to the author, has the sphere of a reality, which is called "an environmental region".

Cognitive competence is related to the objective experience or the perception of things and situations. The highest form of cognitive competence is modern natural science. Cognitive competence helps to edit/evaluate information and to form a new idea, which, according to the author, can be called conclusion making.

Interactive competence includes an interhumane activity and a scale, according to which, we judge about a person's actions and activity. Another field occurs beside the cognitive one, i.e. society occurs beside nature as the other sphere of the reality. Not nature, but the laws of society, duties, rights, and responsibility, bad or good habits are valid there. While developing interactive competence (it is also called social), a child learns to distinguish between people and things. The difference between knowledge and interaction is problematic, as the interaction itself has cognitive components. We not only act, but estimate our actions. As Piaget (1932) states, when interacting, we consciously or unconsciously keep to some rules, and later – qualitatively different rules. The latter also match among themselves. Interactive competence occurs when the relations among people develop.

Linguistic competence. The basis of the use of a language is linguistic competence. Linguistic competence includes the rules, according to which we make words, sentences and texts, relate them with the reality. In the primary step, perceiving a language, we only learn words, and later imperceptibly begin to link them into sentences, memorize the rules, which are usually used unconsciously. It is said not only about grammar in a narrow sense (word-building and sentence-building), but about the use of a language (semantics and pragmatics) as well. When the rules are not used, the content of a language remains misunderstood. Misunderstandings occur not only because different meanings are attributed to the words, but because of the different understanding of a non-verbal language. Only purposeful language upbringing encourages growing and helps to form the basic competencies. Learning a language, morphology, syntax and semantic aspects play an important role in the development of linguistic competence, and the encouragement to think logically improves cognitive competence.

What should happen to get a social speaking/speech from grammatically correct sentence? The main task, according to Habermas (Gripp, 1986, p. 40) is to find out what is going on in a process of speaking to each other? How does the situation of possible understanding, making oneself understood originate?

According to Habermas (Gripp, 1986), every speaking situation means that the speaker and the listener get into two communicative spheres:

1. Inter-subject plane surface – the speaker and listener talk together, i.e. they exchange information (every time they exchange their roles during their conversation);
2. Things, phenomena, states, and individuals they are talking about. According to Flusser (1998), the space of linguistic and communicative competence could be described as a phenomenon when the change of signs is taking place, i.e. exchange of sounds, signs, gestures, symbols, colors. Some signs, from nodding of the head to literature, from Morse code to symbolic logics, from war trumpet to urbanization exist in that space. The competence of communicative theory expands into culture, thus into science, art, politics, economy, law, religion and other spheres of society.

According to American linguist Sapir's theory of language model/pattern, every person bears the main schemes of his own language organization, with the help of which he can express his thoughts (ideas), i.e. the models of his real possibilities the language gives him to ensure communication (Ivič, 1973). We construct our reality in the language; through our language we create the second world of meanings, which is common for us and other people. However, Scollon (1993) states, that a language is polysemous in the discourse levels of a word, a sentence. Humboldt's theory of the language flexibility helps us to understand important truth, that because of polysemy and versatility of the language it is sometimes difficult to understand each other, because everyone has different attitude towards the world; on the other hand, though, because of language flexibility and common regularity the language can be fitted to personal needs and be specific emancipation of spirit of some nation, surface manifestation revealing some attitude to the world (Ivič, 1973). So as for people to be able to talk to each other, to understand each other, they have to create communication community to fulfill their needs for communicative conditions, expression and content. The knowledge of language forms, rules and different possibilities of expression and the usage of these possibilities let people effectively exchange information. It is obvious that the skill to make language a helper gives possibility to be understood, which Habermas (1977) calls language competence. He states that a subject, being a member of specific linguistic community, together with his language, which he uses as a tool, is educated (acquired) by that language too. But the ability to make a language one's helper, according to Butkienė and Kepalaitė (1996, p. 97) gives not only a possibility to talk, but also helps to find the relations of causal phenomena, to relate the new with the well-known, to code the phenomena in such a way that we could use their inner reflections. So, linguistic competence is understood as keeping of production "inventory" of practical phonetics, grammar (morphology and syntax), lexis, word building, writing and fluent speech (monologue and dialogue) in mind (knowledge) and ability to use all this in the language (BPIS, pp. 109-110). It is obviously that the expression of thought (speech), i.e. the usage of language in speaking and writing is an individual linguistic act, the quality of which depends not only on the speaker's knowledge and abilities, but also on the personal features, attitudes and values. The development of linguistic sub-competence (linguistics in Latin *lingua* – language), as it is stated in the General Programs of Lithuanian schools and in the Standards of Education for XI-XII forms (2002, p. 123) does not end in itself: it is intended for the development of communicative competence and general cultural and linguistic education. The final aim and ambition of linguistic competence is communicative competence, which should be related with communication situations, sociolinguistic and linguistic competence. (The General Programs for General Education, 1997, p. 200). Three fields of competence – communicative, socio-cultural and linguistic – could not be named as phenomena of one level. Linguistic competence is the condition and assumption of communicative competence. van Ek (1994) sticks to the same opinion and states that linguistic sub-competence has to serve for the communicative competence and for general cultural and linguistic education.

In conclusion we can say that communicative competence and linguistic competence are not the same concepts. Linguistic competence is a component, a structural element of communicative competence. The identification of the concepts of communicative and linguistic competencies would narrow, simplify the concept of communicative competence and the other very important elements of communicative competence would be underestimated.

Switching to the analysis of the notion of communicative competence we need to say, that the complexity of communicative competence as a phenomenon explains the variety of attitudes towards the fact, that there is no exact and thorough definition so far.

Different definitions of communicative competence are found in social science literature; they reflect various aspects of these notions or just either duplicate or supplement each other. Some authors (Hamer, 1983, Hutmacher, 1996) give very general definitions of communicative competence; the others (Henrici,

1986, Kolomiceva, 1998) emphasize the linguistic aspect of communicative competence; the third group (Jemeljanov, 1985, Petrovskaja, 1989, 1996, Chreptavičienė, 1997) give a broader explanation putting stress not only on knowledge, proficiency, skills, experience, but also on attitudes of value in the space of communicative competence.

If we want to formulate and present the modern concept of communicative competence, to reveal its space in the context of information and multi-cultural society, it is meaningful to make the analysis of the definitions of communicative competence and models by various authors.

Hutmacher (1996) explains communicative competence as an ability to understand and speak more than one language, also an ability to hear out the speaker and take into account his opinion. Hamer (1983) and Henrici (1986) explain this concept of communicative competence in a similar way. Hamer (1983) says, "Communicative competence is a person's ability to exchange information on the basis of knowledge system", and Henrici (1986) admits that communicative competence is a ability to use language in a specific situation in the way as the partner hopes or to resist the partner with a reasonable idea. In this way, the author says, both speaker and listener understand each other and the linguistic activity is perceived according to motives, aims, choice of means (Henrici, 1986, p.148). It is obvious that with these definitions the authors emphasize linguistic but not communicative competence.

Kolomiceva (1996) also keeps to the same position, that communicative competence is an ability to communicate through the linguistic competence. The author identifies visibly the concepts of communicative and linguistic competencies, because, according to Stasiulevičiūtė, Narbutas, Januškevičius (1993, p.208), linguistic competence has to be understood as keeping in mind the linguistic "inventory"(words, their forms, phrases, graphic and phonic features), which is necessary for communicative activity, and the ability to operate with this "inventory"(to combine, inflect, link, paraphrase) using the consistencies of that language. The authors state that in the system of the elements of linguistic competence there is also a competence of grammar, phonetics, lexis and use of means. While explaining linguistic competence, van Ek and the authors of General Programs of Lithuania for School of General Education (1997) and General Programs for General Education and Standards of Education for XI – XII forms (2002) have the same opinion.

> van Ek says, that "when the linguistic sub-competence is being developed, it is necessary to know and to master the consistent of linguistic structure (grammar), to have a minimum vocabulary, to know alphabet, sounds and their relations (spelling and graphic), to pronounce and stress correctly, to be able to intone properly and put the words into meaningful sentences. The linguistic sub-competence has to serve communicative competence."

Other authors (Jamejanov, 1985, Petrovskaja, 1989, 1996, the Dictionary of Psychology, 1993) explain the concept of communicative competence broader, pointing out that its space consists not only of the knowledge and skills of an individual, but also includes the attitudes of value, personal features.

According to Petrovskaja (1989, 1996) "*communicative competence is use of knowledge, skills, experience and social regulation, i.e. creed and feelings, which help in some way to react to things, people and events*" (Myers, 2000, p. 633), in the interpersonal communication".

In the Dictionary of Psychology (1993) not only knowledge, skills, but also person's flexibility is stressed as a very important thing in communicative competence, i.e. ability to evaluate (estimate) different situations, orientate in them and use the knowledge, experience, skills, personal features, attitude talking to people in specific situations. Communicative competence is explained here as the whole of personal skills, knowledge and experience, giving a base for self-confidence, orientation in communication.

Jemeljanov (1991) gives a very clear definition of communicative competence and defines it as situational adaptation and development of social means (verbal and non-verbal). In his opinion the basis of communicative competence is not simple mastering of language and other codes of communication, but personal features – feelings, thinking and acting in a specific social context. Jemeljanov says a person's communicative competence is forming depending on his real social functions, which require to master specific part of the heritage of mankind culture. Situational adaptation should be understood as flexible practice of person's knowledge, skills, abilities, values, and personal features in specific situation. According to him, communicative competence is much more than linguistic competence.

Chreptavičienė, taking the idea of "information as a message, the code of which is artificially made with the help of linguistic signs, understandable for a person or a group of people" (1997) "as background, states that verbal communication is a process of exchanging information between two or more people, when that information has a code in specific language, as a system of linguistic signs" (1997, p. 41).

The author, after analyzing the differences between concepts – qualification and competence – states that "communicative competence is ability to perform communicative activity effectively, according to knowledge, skills, qualification, related to people's relationships, their functions, development (1997, p. 46). So the author emphasizes the knowledge and ability to understand other people's ideas (listening and reading) and to bring one's own ideas effectively to the others by speaking and writing not only in the native language, but in foreign language too, because the signs of that language can be written (spoken) in Lithuanian, Latin, Hebrew, Slavic, Mongolian and other graphic signs. It is important for the members of communicative process (the information sender and receiver) to know the same language and "to recognize and understand the message which is coded with artificially made signs." (Chreptavičienė, 1997, p. 14).

The effective performance of acting testifies the decision of a person to choose such methods of activity, which enable to reach the best results. The effective transmission of the idea is the ability to estimate the addressee, his expectations, situation and possible interference in the process of communication, and the ability to eliminate them by choosing the most effective means of information (thought) transmission. It is obvious that in the social communication in relation "man – man" next to the knowledge and ability, values allowing everybody to live together and support the social system (Mc. Laughlin, 1997) in relation "man – problem", personal qualities and standpoint causing efficiency of communication are also important.

Summing up the discussed standpoints, communicative competence in the context of the informative and multicultural society can be defined as an ability, connected with knowledge and skills, attitudes, personal qualities and values, to construct effectively, to pass, receive and interpret the discourse in various situations of intercultural interaction.

Still the above-mentioned concepts of communicative competence do not distinctly reveal the conception of communicative competence. In this point, the models of communicative competence, made by Moirand (1982) and van Ek (1994) in which some holistic structural elements of communicative competence (sub-competence) come to light, should be more obvious.

According to Hymes (1972), the communicative competence covers the person's who communicates ability of the right choice, when, why, where and with whom to talk (or write). The model of communicative competence, according to Moirand (1982), is made of the following components:

1. Linguistic component – the knowledge about the system of the language, its phonetic, lexical and grammatical models, the understanding and ability to use them in practice.
2. The component of discourse, inclosing the knowledge about the models and the ability to use them in various communicative situations.
3. Journal (referential) component, i.e. the experience and knowledge about the things and phenomena of the surrounding world and their connection.
4. Social cultural component or the knowledge of the interaction of social rules and standards, the knowledge of the user's culture and history. That is the knowledge of the socio-cultural context in which the language is used. So, in Moirand's (1982) opinion, the concept of communicative competence is a structural construct, incorporating linguistic, discourse, journal and social structural components.

van Ek (1994) points a little bit different elements of the model of the communicative competence. The communicative competence, according to the author, consists of 6 parts or subcomponents: linguistic, sociolinguistic, discourse, strategic, socio-cultural, and social.

So, in addition to the components of communicative competence mentioned above by Moirand (1982), van Ek (1994) introduces sociologic, strategic and social competence.

The author asserts that when developing the linguistic sub-competence one must perceive and gradually master the regularity of the language structure (grammar), accumulate minimal vocabulary, know the letters, sounds and their connection (spelling and graphic), be able to pronounce rather correctly, intonate and divide sentences into semantic groups. According to the author, the linguistic competence must serve the communicative competence and general cultural and linguistic education. When developing the sociolinguistic competence, according to van Ek (1994), one learns to use and interpret the language forms according the situation: the situation (who, with whom, about what, where and why is the communication taking place), settles the choice of linguistic forms; when developing the sub-competence of discourse, one learns to understand and form fluent and clear verbal speaking, and when developing the strategic competence one tries to become proficient to support a verbal language by non-verbal means and compensate insufficient (inadequate) knowing of the language".

The sociolinguistic competence mentioned by van Ek (1994) is not a separate component, and it shows the inner copula of linguistic and sociolinguistic competence, which is made of knowledge, knowing and skills connected with the usage of the language, verbal and non-verbal behavior, and the choice among its forms is determined by the situation.

The strategic competence is especially important for verbal spoken communication, at the same time it is not so important for written communication, even in the case when we speak about the professional written communication. The professional logbook of the document prepared according to arrangement of the documents and the execution rules requirements cannot be compensated by the lack of the knowledge of construction. The most important element in the professional documents is text. It must be accurate, concrete and clear. Otherwise, even if the exposition of the properties corresponds to the arrangement and execution rules, the addressee will not understand or will understand the information inadequately. Written, non-verbal communication only helps to understand written verbal communication, but it is not essential.

Developing the social competence, van Ek (1997) states, self-confidence and will to exchange information, to know how to behave in various social situations and ability to understand others are developed. Social (general) competence, according to Stulpinas (1996), means understanding and a high culture of association and communication of man-citizen, a member of the society. This competence is obviously very important in communication, because it determines the relationship of a society and a person. Social-cultural competence is the knowledge of that social-cultural context where the speech is used and the ability how to use that knowledge.

So, the given models of communicative competence evidentially affirm that communicative competence is an integral concept, which consists of six sub-competences such as linguistic, sociolinguistic, discourse, strategic, socio-cultural, and social. Certain knowledge, skills, abilities, personal qualities, point of view and values warrant every sub-competence as a sub-system.

Explaining the conception of today's communicative competence, it is particularly important to emphasize that communicative competence depends on a personality who is able to construct and interpret discourse not only in the mother tongue but also in some foreign languages.

Such regulations are revealed in many documents in EU. On the 31st of March the Resolution was carried by European Ministry of Education. There it is said that teaching and learning of foreign languages should become necessary and accessible for everyone. Learning and knowing of some foreign languages create conditions for a personality to disseminate and grow, help to communicate with the wide world, be acquainted with other cultures, and share the information and socio-cultural values with other people. One can broaden the outlook and develop the language culture (p. 213). Cultural and communicative skills are not inseparable from the development of communicative competence. The Resolution says that people who can speak several foreign languages are given plenty of opportunities: European citizens are given the right to live and work in any European country.

Knowing the languages gives an opportunity for a better job. Moreover, this Resolution sets two very important aims for European people: to understand each other if we want to collaborate and to retain our cultural identity. Thus, the first aim is to speak foreign languages, the second is to cherish our own language. The White book of the EU "White Paper on Education and Training: Teaching and Learning Towards The Learning Society" (1995) rises the aim to know three or more foreign languages. The EU documents mentioned above emphasize the necessity of this aim to necessity, priorities, means and possibilities. However, only the accent of the foreign language in the concept of linguistic sub-competence should testify that the communicative competence is corresponding to the foreign language. An individual should have a lot of various linguistic signs, which should ensure the freedom of the actions in all the space, not only in the international one. So, the knowing of one's native and some foreign languages only develops the space of linguistic sub-competence, though it doesn't make any separate sub-competence.

We think it is important to notice that in General Programs of Lithuania for school of General Education and in Standards for XI-XII forms (2002) the conception of the communicative competence is explained depending on the language (native, foreign), according to which the communicative competence is formed.

It is emphasized that "The most common purpose of Lithuanian is to develop the individual's communicative (linguistic, literary, cultural) competence..." (2002, p. 61). However, other languages, e.g. Russian, Polish, as a mother tongue have a very important aim of development. It is personal communicative competence, and it is already understood as an ability to perceive and create oral and written texts, taking into consideration the aim, addressee and situation. The communicative competence consists of several sub-competences: linguistic, discourse, socio-cultural, strategic (General Programs and Standards of development for XI-XII forms, 2002, pp. 81, 95). Speaking about the development of the communicative competence of Lithuanian as a state and foreign language, the communicative competence is understood in its broad sense. It is affirmed in the earlier mentioned programs and educational standards for XI-XII forms (2002, pp. 109, 123) that the purpose of learning of the state language is to develop sufficient communicative competence including the linguistic, socio-cultural, sociolinguistic, discourse, social and strategic competencies. However, speaking about the components of communicative competence, the concept such as a sub-competence is not used and that is why it does not reflect the inner unites of the components the communicative competence. The communicative competence of the foreign language is explained in the same way but here the concept of sub-competence is used, and not the concept of competence.

Thus, as it is affirmed in the earlier mentioned documents, both Lithuanian as a native language and other languages (Russian, Polish, Lithuanian state and foreign languages) have the basic aim. It is the communicative competence, though while developing the communicative competence of Lithuanian as a native language only linguistic and cultural sub-competence is important, but developing communicative competence of other native languages (Russian, Polish) linguistic, discourse social-cultural, strategic sub-competence is emphasized; and in Lithuanian state and foreign languages – linguistic, socio-cultural, sociolinguistic, discourse, social and strategic sub-competence is developed. We should think that while forming the communicative competence in Lithuanian or in some other languages in a Secondary School it is necessary to develop linguistic, socio-cultural, sociolinguistic, discourse, social and strategic sub-competences, though depending on the language that the communicative competence is developed, more attention should be paid to particular sub-competences, and less to the others. For example, while developing communicative sub-competence in Lithuanian as a mother tongue more attention should be paid to linguistic and cultural sub-competence, and to the other languages as native ones (Russian, Polish) -attention should be paid to the development of linguistic, discourse, and socio-cultural, strategic sub-competences.

Talking about the conception of contemporary communicative competence, obviously, it is necessary to add a structural element – ICT (sub-competence) to this model, as the realization of these technologies reveals new abilities of information obtaining, supplementing, storing and transferring. Only that person, who is able, according to the situation, to choose ICT properly and to use them as much as possible in the progress of communication, can ensure the efficiency of communication. As Anderson and Collis (1993) state, the relations between a computer and literacy involve the communicative skills that should be obtained by everyone whishing to act in a society.

Unfortunately, scientific literature and school educational programs have not evaluated the ICT sub-competence, as a structural element in the space of communicative competence. While analyzing the General Programs of Lithuania for General Education and in Standards of development its importance is emphasized only when it is spoken about the developing communicative competence in foreign language. When developing communicative competence in a foreign language it is indicated in the goals of the common (general) course and the broadened course "to learn to change information using modern information technologies" (pp. 129, 132). However, neither the programs of the native Lithuanian language nor the programs of the state Lithuanian language mention it. So, the question arises, whether the use of the ICT is important only in the international dialogue or only when the foreign-born "are talking"?

The answer to this question could be found in the following citation by Anderson and Collis (1993): "only the rational application of both local and global computer networks open new more effective possibilities for communicating".

Nevertheless, the perceived necessity of this rational use of computer networks should be attributed to the development of ICT competencies. Obviously, different professions demand a different level of ICT competence. The competence of one level is necessary for an administrator or an accountant, the competence of another level – for an administrator of a computer network. However, today all professions, according to Jucevičienė and Brazdeikis (2003, p. 79) need basic ICT competence, which can be defined by three characteristics:

- Ability to control and use ICT (technological literacy);
- Possession of informative abilities and their usage (informative literacy);
- Knowledge of social, ethnic and legal norms and ability to keep them using ITC in the activity (social literacy).

Today's communicative competence integrates seven sub-competencies: linguistic, socio-linguistic, discourse, strategic, socio-cultural, social and ITC. It is reflected picturesquely by an integral scheme (see Picture 1). The combination of these sub-competencies is especially important to a person's communicative competence.

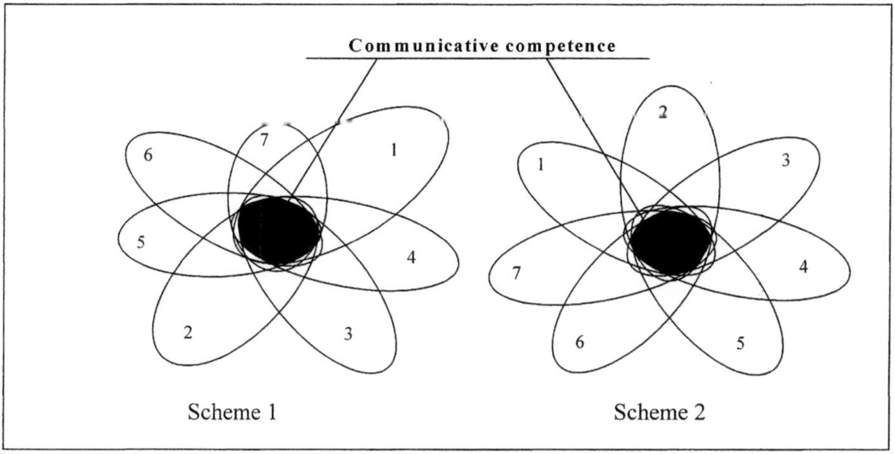

1. Linguistic component. **2.** Socio-linguistic component. **3.** Discourse component. **4.** Strategic component. **5.** Socio-cultural component. **6.** Social component. **7.** ITC component.

Picture 1: Model of Communication Competence

On the other hand, a person's ITC competence could be considered as a part of common competence (sub-competence), however, a person, striving for effective communication, must use ITC for various situations. It allows stating that it is an important part of communicative competence, an integral component (sub-competence).

In Picture 1 the model of communication competence (Scheme 2, p. 117) is symmetrical, which implies that all components are of equal importance. It is supposed that this is a rare case. Most often the components of personal communication competence are not equally well developed.

Every participant of the communication process does not possess ideal communication competence, since normally a person does not master all the subtleties (meanings, verbal expressions) of the native language, does not have sufficient skills in business language, skills in verbal and non-verbal business communication; our limited individual way of thinking cannot encompass a universal variety of communicative forms. In this case the model changes – it is strained (see Scheme 1, p. 117).

No doubt, personal communicative competence has a relative character. It also depends on the context of communication. This context is determined by the competence of partners and conditions, under which the exchange of information takes place.

Conclusions

1. The human competence in information and knowledge society can be relatively evaluated as the qualitative expression of accumulated knowledge, skills and abilities that is influenced by person's characteristics, attitudes and values. One of the main elements of contemporary human competence is communicative competence, since the modern world could be perceived as the processes of information and communication, and the life is understood as knowing.
2. Communicative competence in the context of the information and multicultural society is the ability related with knowledge, skills and abilities, regulations, personal characteristics and values which is able to construct, pass, accept and interpret the discourse effectively in the different cultural situations of the interaction.
3. Communicative competence – a structural combination incorporating seven sub-competencies: linguistic, socio-linguistic, discourse, strategic, socio-cultural, social and ICT. Every sub-competence, as subsystem, is guaranteed by some kind of knowledge, skills, abilities, personal characteristics, attitudes and values:

- *Linguistic sub-competence is a possession of the "inventory" (words, their forms, phrases, graphic and sound units) of the given language that is necessary for a communicative activity, and it is also an ability to operate with it (to combine, change, connect, and paraphrase) according to the regularities of that language. There should be a competence of the use of grammar, phonetics, and lexical means in the system of these factors of linguistic competence.*
- *Socio-linguistic sub-competence is an ability to use and interpret the forms of a language according to the situation: that situation (who, with who, about what, where and with what purpose communicate) decides the selection and use of the forms of a language.*
  - Discourse sub-competence is an ability to understand and form coherent and clear texts (not as an accidental combination of words and sentences, but as a coherent discourse).
  - Strategic sub-competence is an ability to use expediently the non-verbal strategies of a language passing and accepting information in order to achieve a communicative purpose.
  - Socio-cultural sub-competence is an ability to perceive a socio-cultural context and to construct the discourse which corresponds to that cultural context.
  - Social sub-competence is an ability to behave properly in different social situations and an ability to understand others.
  - ICT sub-competence is an ability to control and to use ICT (technological literacy); Possessing informative abilities and their use (informative literacy); The knowledge of social, ethnic, legitimate norms and the ability to keep them using ICT in the activity (social literacy).

References

ALLRIGHT, R.L.: Learning through communicative practice. In: The Communicative Approach to Language Teaching 1981 (OUP).

BOSTROM, R.N.: Competence in Communication. Beverly Hills, CA 1984 (Sage).

BUTKIENĖ, G./KEPALAITĖ, A.: Mokymasis ir asmenybės brendimas. Vilnius 1996 (Margi Raštai).

CHREPTAVIČIENĖ, V.: Dalykinės komunikacijos raštu mokėjimų ugdymas biznio administravimo universitetinėse studijose. Daktaro disertacija. Kaunas 1997 (Kauno technologijos universitetas).

COOLEY, R.C./ ROACH, D.A.: Theoretical approaches to communication competence: A conceptual framework. In: BOSTROM, R.N. (ed.): Competence in communication. Beverly Hill, CA 1984 (Sage).

DAVENPORT, T./JARVENPAA, S.L./BEERS, M.C.: Improving Knowledge Work Processes. Sloan Management Review, Summer, 1996, pp. 53-65.

Ek van ir kt.: Lietuvos bendrojo lavinimo mokyklos bendrosios programos. Vilnius 1994 (Leidybos centras).

Ek van ir kt.: Lietuvos bendrojo lavinimo mokyklos bendrosios programos. Vilnius 1997 (Leidybos centras).

FLUSSER, V.: Kommunikologie. Frankfurt/Main 1998 (Fischer Verlag).

FRITZSCHE, J.: Zur Didaktik und Methodik des Deutschunterrichts. Band 1. Grundlagen. Berlin 1994 (Klett).

GORDON, R.J./GORDON, R.S.: Information Systems – a Manager Approach. Second Edition. 1999 (The Dryden Perss Harcourt Brace Colledge Publischers).

GREBLIAUSKIENĖ, B.: ENKMUS: Etapinė nerašytinių komunikacinių mokėjimų ugdymo sistema. Vilnius 1999 (Žara).

HABERMAS, J.: Toward a theory of communicative competence. Recent Sociology, 2. 1970

HARRIS, L.: Communicative competence: An argument for a systemic: view. Paper presented at the annual conference of the International Communication Association. Philadelphia 1979.

HENRICI, G.: Studienbuch. Grundlagen fur den Unterricht im Fach Deutsch als Fremdsprache. Padeborn 1986 (Schöningh Verlag).

HYMES, D.: On communicative competence. In: PRIDE, B./HOLMES, R. (eds.): Sociolinguistics. London 1972.

HOPE, J./HOPE, T.: Competing in the Third Wave. Boston 1997 (Harvard Business School Press).

IVIČ, M.: Lingvistikos kryptys. Vilnius 1973 (Mintis).

JABLIN, F.M./CUDE, R.L./HOUSE/ A./LEE, J./ROTH, N.L.: Communication competence in organizations: Conceptualization and comparison across multiple levels of analysis. In: THAYER, L./BARNETT, G. (eds.): Organization Communication: Emerging perspectives Vol. 4. Norwood, NJ 1994 (Ablex).

ЕМЕЛЬЯНОВБ Ю. Н.: Обучение паритетному диалогу. Ленинград 1991 (ЛГУ).

JOHAN, R./KROGH, G.: Organizational Epistemology. 1995 (Macmillan).

JOVAIŠA, T/SHAW, S.: Žvilgsnis į bendrųjų gebėjimų ugdymą Rytų ir Centrinėje Europoje (partnerių šalyse). Profesinis rengimas: tyrimai ir realijos. Kaunas 1998 (VDU), pp. 30-47.

JUCEVIČIENĖ, P./LEPAITĖ, D.: Kompetencijos sampratos erdvė. Socialiniai mokslai 1(2000)22 pp. 44-50. Kaunas: Technologija.

KIRBY, J./ŽYDŽIŪNAITĖ, V. IR KT.: Bendrieji gebėjimai. Mokytojo knyga. Vilnius 1999.

LETTLEJOHN, W.W./JABUSCH, D.M.: Communication competence: Mode and application. Journal of Applied Communication Research, 10. 1982.

Lietuvos bendrojo lavinimo mokyklos bendrosios programos ir bendrojo išsilavinimo standartai XI – XII kl.). Vilnius 2002 (Švietimo plėtotės centras).

Lietuvos bendrojo lavinimo mokyklos bendrosios programos ir išsilavinimo standartai XI-XII klasėms. Vilnius 2002 (Leidybos centras).

MOIRAND, S.: Enseigner a communiquer en langue etrangere. Paris 1982 (Hachette).

Mokymas ir mokymasis – besimokančios visuomenės link. Europos Komisijos baltoji knyga. Brussels 1995.

MONGE, P.R./BACHMAN, S.G./DILLARD, J.P./EISENBERG, E.M.: Communicator competence in the workplace: Model testing and scale development. In: M. BURGOON (ed.): Communication yearbook 5. Newbury Park, CA 1982 (Sage).

PARK, M.R.: Interpersonal communication and the quest for personal competence. In: KNAPP, M.L./MILLER, G.R. (ęds.): Handbook of interpersonal communication. Beverly Hills, CA 1985 (Sage).

Psichologijos žodynas. Vilnius 1993 (Mokslo ir enciklopedijų leidykla), p.368.

SAUGĖNIENĖ, N.: Lietuvos tautinių mažumų švietimas multikultūriškumo vertybiniame kontekste. Daktaro disertacija. Kaunas 2000 (Kauno technologijos universitetas).

STANKEVIČIŪTĖ, J.: Organizacijos žinojmo didinimo metodologija. Daktaro disertacija. Kaunas 2002 (Kauno technologijos universitetas).

STASIULEVIČIŪTĖ, A./NARBUTAS, E./JANUŠKEVIČIUS, V.: Kokį kelią siūlome. Lietuvos švietimo reformos gairės. Vilnius 1993 (Valstybinis leidybos centras).

STULPINAS, T.: Autodidaktika. Kompetencijos ir labilumo principai. Kaunas 1996 (Vytauto Didžiojo universitetas).

Švietimo ministrų Tarybos 1995 m. kovo 31 d. Rezoliucija. [citet 1 May 2002]. Available from: <http://www.smm.lt/km/km_rezoliucija.htm>.

The European Dimension in Secondary Education. Council of Europe 1997.

Webster's New World Dictionary. Cleveland & New York 1988.

Webster's New World Dictionary. Packed Books paperback 1995.

Gediminas Merkys, Palmira. Jucevičienė, Gintarė. Tautkevičienė

Library as a Preferred Learning Environment

Introduction

The development of citizens' competence and permanent updating of knowledge are among the most important tasks in the knowledge society. Therefore, learning is to become an inseparable part of each member of society, and learning environments have to be accessible everywhere all the time. An old approach to teaching/learning as an activity, taking place only at the institutions of formal education (secondary schools, professional schools, institutions of higher education, professional development centers, etc.), is being supplanted by other views. According to these, learning is not just directly related only to education and teaching; learning activities underline supplying a learner with effective learning means and tools which match his/her learning style and demands (Marton, Hounsell & Entwistle, 1997). Adult learning (Roger, 2002, Field, Leicester, 2000), experiential learning (Kolb, 1984, Rogers, 2002), reflective learning (Schon, 2002; Boud, Walker, 2002) occupy an important part in the context of life-long learning. Even in institutionalized processes of education, the emphasis has been shifting from teaching to learning, from education based on the interaction between a student and a teacher to creating learning environments and making them work (Wilson, 1996; Barr & Tagg, 1995; Tuijnman, Kamp, 1992). As a matter of fact, Longworth (2000) refreshes the concept of life-long learning, considering it to be the development of human potential through the process of continuous support to an individual which stimulates him/her and enables the person to apply all the acquired knowledge, skills, values, and understanding in all the contexts of his/her life.

Aiming at securing the qualities of the knowledge society and at creating equal conditions for all citizens' learning (with no discrimination according to their age and social status), learning environments should be created not only in a formal educational system, but everywhere an individual could get acquire knowledge or could develop his/her skills. Learning can take place not only in a classroom but in the family (Moos, 1987), at a workplace (Boud & Garrick, 1999), in learning communities (Retallik, Cocklin & Coombe, 1999), at the library (McNicol, 2002; Simons, Young & Gibson, 2000; Bruce & Candy, 2000), in museums, on the streets, at the cinema, etc. Learning environments at both formal and informal institutions are interrelated and influence each other (Moss,

1987). On the one hand, knowledge, skills, values, and personal qualities, which have been formed in one environment, can be transferred into another; they can either facilitate learning or inhibit it. On the other hand, the same learning environment will probably have a different impact on an individual, stimulating his/her learning of subduing it. This depends on individual characteristics of each person: his/her experience, competence, values and attitudes, interests, motivation, etc.

Knowledge society puts special emphasis on new knowledge. Since the library secures accessibility to knowledge (Dowler, 1997), it is considered to be a learning environment and is given a special place in the knowledge society. Not a single educational institution is able to function properly, if it does not have a pool of information sources to satisfy teachers and learners' requirements. Therefore, it would be reasonable to treat the library as an intermediary between formal and non-formal learning environments. Nevertheless, libraries, for instance, public or municipal, are not primarily related to educational institutions and can be accessed by all the population to satisfy their learning needs. Thus, libraries provide learning environments to individuals belonging to different society strata, different age groups, and not necessarily involved in formal education (McNicol, 2002).

Literature analysis has shown that foreign scholars who investigate library activities (Kirby, Liddiard & Moore, 1998, Bruce, Candy, 2002) pay most attention to the development of information literacy and its integration into a study environment. Then, most research is carried out in libraries of educational institutions (universities, schools, colleges, etc.). Therefore, it could be concluded that there is a lack of research on learning environments in libraries of different types.

It was not possible to come across comprehensive empirical studies or projects focusing on the library, a learning environment, as an independent object of scientific research which would cover different types of libraries, various age groups, and diverse social strata. So far, the library has been investigated only as an academic unit (Juceviciene, Tautkeviciene, 2002), and this research has been directed towards the creation of an empowering learning environment. Therefore, it goes without saying that the investigation of learning environments, embracing various types of libraries, different age groups, and diverse social strata, would enable the evaluation of priorities in selecting a library for a learning environment, in comparison to other learning environments; it would assess the possibilities for its creation. This research is especially important in the context of a city, aiming at creating a learning society in it.

The *objective* of the present article is to carry out the analysis of priorities among the population in a country-in-transition, namely, Lithuania, when choosing the library as a learning environment, and to evaluate demographic factors influencing such preferences.

The research introduced in this article has been carried in the frame of a project "Kaunas as a Learning City". It should be noted that Kaunas specifically aims at moving its community into a learning city. The first part of the paper renders theoretical analysis of the problem, highlights the essence of the library as a learning environment, and introduces a social-political context of Lithuania. The second part presents research methodology, and the third part purveys the results of population survey as well as the analysis.

1. Theoretical analysis of the problem and a social-political context

In spite of the fact that each library has different potential and possibilities, they all can have more or less effective learning environments created inside. However, considering a learning setting, individual library visitors may recognize divergent learning environments. Many factors influence this discrepancy, and individuals' attitudes towards the library, as a potential learning environment, are one of them.

Throughout all the periods of society development, the library has been recognized as an important agent and an institution of culture, education, and information. The library has always tried to satisfy readers' demands for information, to serve their scientific, technical, artistic interests, and to organize educational work (Jovaisa, 1993). As a social institution, it has always been, and will remain, determined by the surrounding social environment (Shera, 1996). Therefore, changes in society influence the change in the activities of any library.

Libraries differ in many aspects: in their stock, composition of users, versatility of their purpose, and democracy. Throughout the history, different types of libraries emerged: scientific, academic, national, specialized, public, school, etc. However, the major functions of a traditional library have been similar, namely, mnemonic (storing human memory) and communicative (transmitting and spreading information). In addition, such functions as informational (providing answers to queries and informing), cultural, forming a public opinion, educating the user, etc., have become evident (Glosiene, 1998). These used to be quite stable functions, not subject to major changes. Only at present, they undergo smaller or bigger transformations. A mnemonic function is now carried out by different means, since the format of documentation has been changing. Newly created digital documents have to be stocked in magnetic or optical media, which requires additional equipment and technology to store and present them. A communicative function has experienced transformations due to new means in transmitting and publishing information, which requires new skills from library employees, and users in receiving, transmitting, and using information. Recently, major attention has been given not to providing answers to queries and informing the user, but to teaching users how to perform information search and to use search results. Thus, a new function of educating library users, or educational function, has been developing. In society, which fosters continuing education as a major principle, this function of a library becomes topical. In addition to educating, any library becomes a learning environment for the most part of the population.

A learning environment at the library is a space or location, physical or virtual, which influences each learner and is identified by him/her on the basis of one's own experience, competence, and learning goals. It encompasses the information, which matches person's learning goals, as well as means, subjects (peers, library assistants, etc.), and media of receiving and assimilating information (Juceviciene, Tautkeviciene, 2002.). It is individual for each learner and is formed on the grounds of an educational environment of the library. It is important to stress two aspects of a learning environment at the library, that is, contents, which reflects important learning information, and context, an educational environment of the library, perceived by each learner as an entity which determines person's learning process.

Features peculiar to both a traditional and a modern library distinguish a contemporary library. A traditional library is oriented towards stacks, hierarchy, and bureaucracy; it regards a document in a narrow sense. Such a library keeps to the equality principle for everyone; it recognizes such values as compliance, accurateness, knowledge of stacks and literature, and a neutral attitude towards work and users. A modern library is oriented towards the user, and it matches its management with the goals. Such a library keeps to a broad view towards the concepts of a document and service, and it is a functional and integral part of an organization or a community. Among the values recognized are initiative, openness to the novelty and change, skills of information management, responsibility and concern for work and for the user (Glosiene, 1998, Bakewell, 1997). Creation of empowering learning environments becomes one of the dominating aims of the library (Kirby, Liddiard, & Moore, 1998, Bruce, Candy, 2002). Learning at the library is related to such theoretical concepts as: resource-based, self-directed, discovery, or problem-based learning (Marcum, 2002).

The space of library activities has been expanded by such occurrences as the origin and development of news means to produce, store, and present information, and by constant perfection. Service has been provided not only in a real, or physical location, when visiting the library, but also in a virtual, or electronic space, that is, accessing a library catalogue via the Internet, directly connecting to databases, finding information on the Internet, or communicating with clients by e-mail. Virtual services and new channels of scientific information provide a possibility to better satisfy the need to get the necessary information. Possibilities of the library as a virtual learning environment allow to select time and place of learning, thus, increasing the flexibility of the learning environment. The merits a virtual library can provide make the library a more attractive learning environment than it used to be in the context of a traditional library.

A learning environment at a modern library provides a possibility for users to get print and electronic sources of information from local or worldwide information depositories (library catalogues on the Internet, databases, etc.), to use means of formatting a text (text editors, spreadsheets, text presentation programs, database creation and management, information preview and data transfer programs). The abundance of information sources and supply with technical and technological means are among the factors providing conditions for creating an educational environment at the library and, consequently, learning environments. Thus, rich and empowering educational environments can be created at the library and can be identified by an individual as comprehensible and useful learning environments. It is important to note that technical and technological means do not suffice for a library to be identified as a learning environment. An individual, who knows his/her learning goals, has to identify an environment at the library, which would enable implementation of these goals. In order to use this environment, the learner has to possess the necessary informational and meta-learning competences. As a matter of fact, the library has to have relevant data about its users.

Lithuania keeps to the standards of library statistics in collecting, processing, and publishing data about library activities. The standards reflect the following indicators: number of libraries in Lithuania, pool of documents, scope of loans, numbers of users and visitors, and number of library employees. This data is not compared and analyzed in statistical publications, since not all the libraries or their groups provide equally comprehensive data. Besides, the data in statistical publications emphasize the pool and the quota of users; it does not reflect the implementation of an educational function in libraries. There is no data available on trainings, seminars, presentations, or other kinds of user education activities carried out in libraries.

The research on library users, their needs and reading preferences in Lithuania is quite local or regional, and has been carried out episodically. There has been the investigation on the nature of reading among Lithuanian population (Rimsa, 1980), on the activities of public municipal libraries (2002), on readers' interest in periodicals and its satisfaction in public libraries of the Akmene region (Sinkeviciene, 2001), user survey at the Lithuanian National Martynas

Mazvydas Library (1999), etc. All the investigations mentioned above analyze library pool and user needs. Whereas the library as a learning environment has been only researched in a specific context of academic libraries (Juceviciene, Tautkeviciene, 2002).

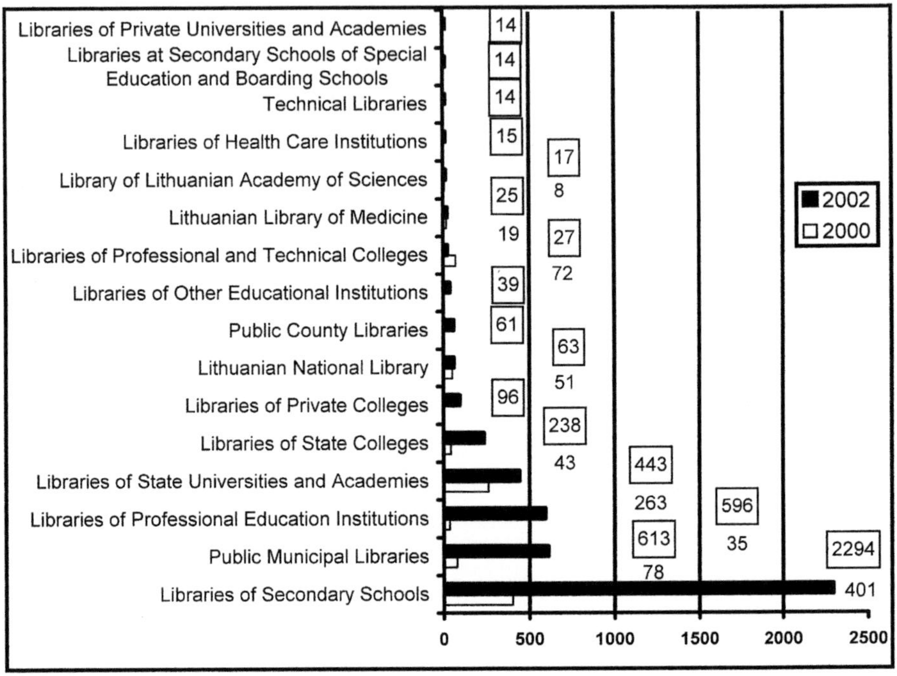

Figure 1: Comparison of Traditional and Computerized Workspaces in Libraries of Different Types (2000, 2002.).

According to the data provided by foreign researchers, (Hernon & Altman, 1996; Bruce & Candy, 2000), a computerized library is considered to be a more attractive learning environment. Therefore, it could be assumed that the creation and identification of learning environments among the population largely depends on the level of library computerization. The situation in Lithuania substantially differs subject to a library type. Table 1, p. 127, introduces comparative analysis of traditional and computerized workspaces in Lithuanian libraries of different types.

The data suggests that libraries of educational institutions (colleges, universities, professional education institutions, secondary schools) offer the largest number of computer workplaces. It could be concluded that educational institutions of all levels pay great attention to the creation of modern libraries and to securing the access to information worldwide.

Library Type	Number of Workspaces	Computerized Workspaces	Computerized Workspaces in Percent
Libraries of Private Colleges	187	96	51,3
Libraries of Professional Education Institutions	2111	596	28,2
Lithuanian Library of Medicine	126	25	19,8
Libraries of State Colleges	1498	238	15,9
Libraries of Secondary Schools	17183	2294	13,4
Libraries of Private Universities and Academies	112	14	12,5
Libraries of State Universities and Academies	4200	443	10,5
Lithuanian National Library	621	63	10,1
Lithuanian Libraries of Agriculture	10	1	10,0
Technical Libraries	156	14	9,0
Libraries of Other Educational Institutions (Methodical Rooms, etc.)	494	39	7,9
Libraries of Health Care Institutions	192	15	7,8
Libraries of Professional and Technical Colleges	395	27	6,8
Library of Lithuanian Academy of Sciences	279	17	6,1
Libraries at Children's Music, Art Schools	82	5	6,1
Public County Libraries	1009	61	6,0
Public Municipal Libraries	12205	613	5,0
Libraries of Other Institutions of Culture	20	1	5,0
Libraries of Agricultural Research Institutes	68	3	4,4
Libraries of Other Research Institutes	97	4	4,1
Libraries at Secondary Schools of Special Education and Boarding Schools	418	14	3,3
Lithuanian Technical Library	220	7	3,2
Libraries at Museums	42	1	2,4
Lithuanian Libraries of the Blind	122	2	1,6
Libraries of Children's Care Institutions	130	2	1,5
Libraries of Private Professional and Technical Colleges	48	0	0,0
Total:	**29820**	**4595**	**15,4**

Table 1: Traditional and Computerized Workspaces in Lithuanian Libraries (Data of 2002) (Statistical Data of Lithuanian Libraries 1991 through 2000. Vilnius, 2002).

In recent years, libraries of secondary schools have experienced the utmost change of computerized workspaces: in 2000 through 2002, the number of computers increased by 472 percent, rising from 401 to 2 294. In the institutions of professional education this number increased from 35 to 596 (213 percent), and in higher educational institutions – from 263 to 443 (68 percent). Professional and technical colleges, however, lost a substantial number of computerized workspaces, but this happened due to the educational reform, when some of these institutions became a part of tertiary education.

2. Research Methodology

After describing educational functions of the library in contemporary society, it is reasonable to investigate a practical implementation of these functions in social practices of the country. Naturally, this need can be satisfied only by the means of empirical studies, implemented in Lithuania. It is important to note that the data of similar studies carried out abroad would be useful and significant, but they should not be extrapolated without reservation due to cultural specifics of our country.

The lack of research on educational functions of the library suggests another way out: it is reasonable to study in more detail the data of other empirical investigations covering, even partially, the context of educational activities at the library. A project "Kaunas as a Learning City", carried out in Kaunas in 2001, could be considered as such. The investigation encompassed a written inquiry of the general population of the city, and it had the aim to research educational interests and educational preferences (especially those concerning the concepts of non-formal education and experiential learning) of the population. The sample amounted to 800; 729 questionnaires were processed. Demographic characteristics of the study participants varied from the age of 16 to 83; different social groups were covered. In order to compare the interests and preferences of Kaunas citizens, who come to represent a big city population, with the analogous data of the rural and periphery population, several "countrified" samples (about 30 percent of the whole sample) were represented.

The questionnaire embraced the following constructs:

1. Self-concept of social roles and competencies;
2. The interest to improve while performing social roles as well as implementing different competencies;
3. Reflectivity, openness, and rigidity (in the aspect of experiential learning);
4. Various educational preferences: (a) means and forms of learning, (b) places and environments for learning;
5. Educational interests;

6. Learning intentions (plans);
7. Attitudes towards paid educational services;
8. Unit of study participants' socio-demographic characteristics.

Table 2, p. 129, depicts a more detailed structure of the questionnaire.

It is important to note that all the scales (indexes) have been designed by the way of factorial validation, using Principal Components and VARIMAX Rotation; Cronbach-α coefficient was also calculated. Consequently, a psychometric quality of the indexes matched the estimation from "very good" to "satisfactory".

The construct of educational preferences (places and environments for learning) was reflected in 29 items (see Figure 2 and Table 2). Two of these items were directly related to the evaluation of the library as a learning environment. A general phrasing of the question for all 29 items was as follows: *"Which places and environment of learning are suitable for you? Where would you like to engage in studies in the areas important to you?"*. The items about the library had the following phrasing: 1) *"at a traditional library, without the access to the Internet"*; 2) *"at a modern library with the access to the Internet"*. The format of answers was defined in three stages (categories): "*no*", "*uncertain*", "*yes*".

Construct	**Scales (N)**	**Subscales (N)**	**Initial Items (N)**
Self-concept of Social Roles	1	4	16
Interest to Improve While Performing Social Roles	1	4	16
Openness/Reflectivity versus Rigidity	1	2	23
Learning/Cognitive Interests	1	11	106
Means of Learning as Preferences	1	4	22
Learning Environments and Learning Places as Preferences	1	5	29
Learning Intentions and Plans (open question)	-	-	-
Attitudes Towards Paid Educational Services	1	2	14
Socio-demographic Unit	-	-	10
Total	**7**	**32**	**236**

Table 2: Questionnaire Structure: Items, Sub-scales, and Scales

Having factorized the values of both answers of items, an index of the library as an educational preference was formed. Dispersion authorized by a factorial analysis model reached 60 percent; both items, mentioned above, reached a factorial lodging of 0,76. Thus, the index was considered to provisorily satisfy the requirement for the scale.

3. Results and Discussion

The preference has been clearly given to a library having access to the Internet, as seen in Figure 2, p. 130. A substantial sample (N = 729) has rendered reliability to the disparity between the following: Chi-Square = 368,42, df = 2, and p = 0.0000. Naturally, a sharp effect of inversion has manifested itself, as depicted in the below X-shape graph.

The so-called "yes-percentage" in estimating the value of a "traditional" library has reached 28,6 percent; in estimating a "non-traditional" library – 59,1 percent. The average of preferences of both items has equaled to 43,85 percent. Since there has been no "norm" of preference towards the library established, it is reasonable to compare this number with the results of other preferences (see Figure 2 and Table 2). Among 29 positions, a preference towards a library having access to the Internet occupies the 6 place and stands among the most favorable learning environments; a traditional library stands 16, and finds itself among mediocre learning preferences. In any case, it is delightful that the library, such a universally accessible learning space, escapes the description of an unattractive learning environment. It is symptomatic that even a "traditional" library, having less appeal than a modern, overtakes such a mass learning environment as a museum which has been favored only by 13,5 percent of the respondents (compared to almost 30 percent of traditional library preferences).

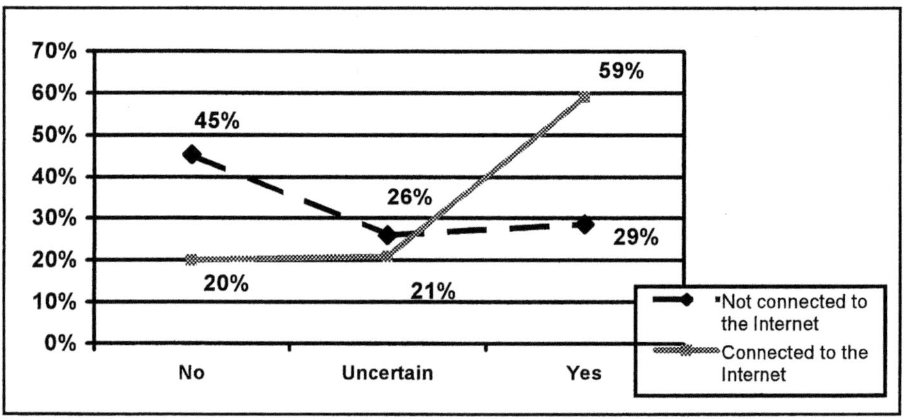

Figure 2: "Traditional" and "Modern" Library, as an Educational Preference. Would you like to study/learn at a library (a) connected to the Internet, (b) not connected to the Internet.

It could be hypothesized that the general population, judging from its attitudes, interests, and preferences (including educational preferences), is very composite. This thesis has been empirically grounded from the position of the two items specified above. Verification was carried out using a K-Mean Cluster Analysis method. After iterative testing of various classifications, theoretically, the most significant and suggestive appeared to be a model of four clusters.

Spaces and Environments Suitable for Learning	**Mean**	**SD**	**No %**	**Uncertain %**	**Yes %**
Workplace	1,55	0,76	16	12	72
Any place, where it is possible to exchange the experience with other individuals	1,49	0,70	12	26	61
Educational institutions and training courses established and run by foreign and international organizations	1,47	0,76	16	20	63
Institutions of higher education (university, academy)	1,43	0,82	22	14	64
Home	1,41	0,82	21	16	62
Modern library with the access to the Internet	1,40	0,79	19	21	59
Learning center which supports independent learning	1,38	0,78	18	25	57
Courses organized by municipalities or ministries	1,25	0,81	24	27	49
Non-formal environment (clubs, etc)	1,09	0,82	29	32	39
Institutions which organize trainings and courses (Kaunas Center of Culture, "Zinija" lyceum, etc.)	1,04	0,80	30	37	34
Courses organized by NGO'o or different associations	1,00	0,79	31	38	31
Internet cafes	0,94	0,82	36	33	31
Colleges	0,92	0,85	40	27	33
Fitness clubs, health centers	0,91	0,86	42	25	33
Traditional library without the access to the Internet	0,83	0,85	46	25	29
Prison (if found there)	0,79	0,83	47	27	26
Art or music school, or club	0,76	0,79	46	32	22
Voluntary service (The Red Cross, The Greens, Caritas)	0,71	0,75	47	35	18
Labor exchange	0,70	0,78	50	30	20
Professional or technical college	0,62	0,80	59	21	20
Local community centre	0,58	0,69	53	35	11
Night-school	0,52	0,76	64	20	16
Museum	0,51	0,72	63	24	14
Educational institutions established by the Lithuanian Catholic Church	0,50	0,71	62	26	12
Lithuanian Military service or National Defense	0,50	0,73	64	22	14
Educational institutions established and run by foreign Catholic Church	0,46	0,69	65	23	11
Educational institutions established and run other traditional religions in Lithuania (Lutheran, orthodox, etc.)	0,26	0,51	77	19	3
Educational institutions established and run by Christian non-traditional religions and denominations	0,22	0,51	82	13	4
Educational institutions established and run by non-Christian non-traditional religions and denominations	0,19	0,46	85	12	3

Table 3: Ratings of Learning Environment Alternatives

Qualitative interpretation and relative distribution of the obtained typological groups would be as follows:

- **Type 1:** "The liking" - population in favor of the library (42,2 percent) with the preferences towards the library (both traditional and modern) strongly expressed.
- **Type 2:** "The conservatives" – population with the lowest comparative lodging (6,6 percent). A very strong preference for a traditional library, and an extremely weak preference for a modern one. It could be concluded that, in general, this part of the population has negative attitudes towards the penetration of IT into daily life.
- **Type 3:** "The modern" – 33 percent of the population with the preference profile diametrically opposite to Type 2: a very positive attitude towards a modern library, and an extremely standoff view on a traditional library.
- **Type 4:** "The indifferent" – 13 percent of the total sample, with extremely low preferences towards both a traditional and a modern library.

Merits of the library, as an institution massively providing social educational services to the population, lie in the fact that most of the populations express an affirmative approach towards it; they tend to extremely positively treat the library as a learning environment. Theoretical interpretation suggests that Types 1 and 3 are characterized by functional attitudes towards the library. Consequently, about 80 percent of the population can be distinguished for positive attitudes towards the library.

It has emerged that the major socio-demographic variables, that is, gender, age, education, and place of residence (urban or rural), are statistically related to the preference investigated.

Unambiguously, educational preference towards the library is stronger among the female population than among the male (verified by *t*-test; $t = -2{,}74$; $df = 693$; $p = 0{,}006$). Furthermore, the impact of education on the preference was tested using Oneway ANOVA. The factor was divided into four categories, from the lowest stages of education to graduate studies. The leverage of education proved to be essential ($df = 3; 687; 690$; $F = 8{,}33$, $p \leq 0{,}000$). The comparison of the average of groups by Tukey test has proved that the part of the population, related to higher education[1], expressed more positive attitudes towards the library as a learning environment, if compared with categories of the population inferior in education[2]. It could be concluded, therefore, that the university is

[1] University graduates or undergraduate students (including dropouts) as well as current university graduate students, together with MA and PhD holders, fall into this category.

[2] High school dropouts, high school graduates, or technical school graduates apply to this category. No statistically significant differences have been determined among the above-mentioned groups. Consequently, two categories have been singled out: 1) population related to the university in their educational biography; 2) population not related to the university in their educational biography. The first group was relatively called "a group of high educational census", and the second – "a group of low educational census".

a cultural setting forming person's positive attitudes towards the library as a learning environment.

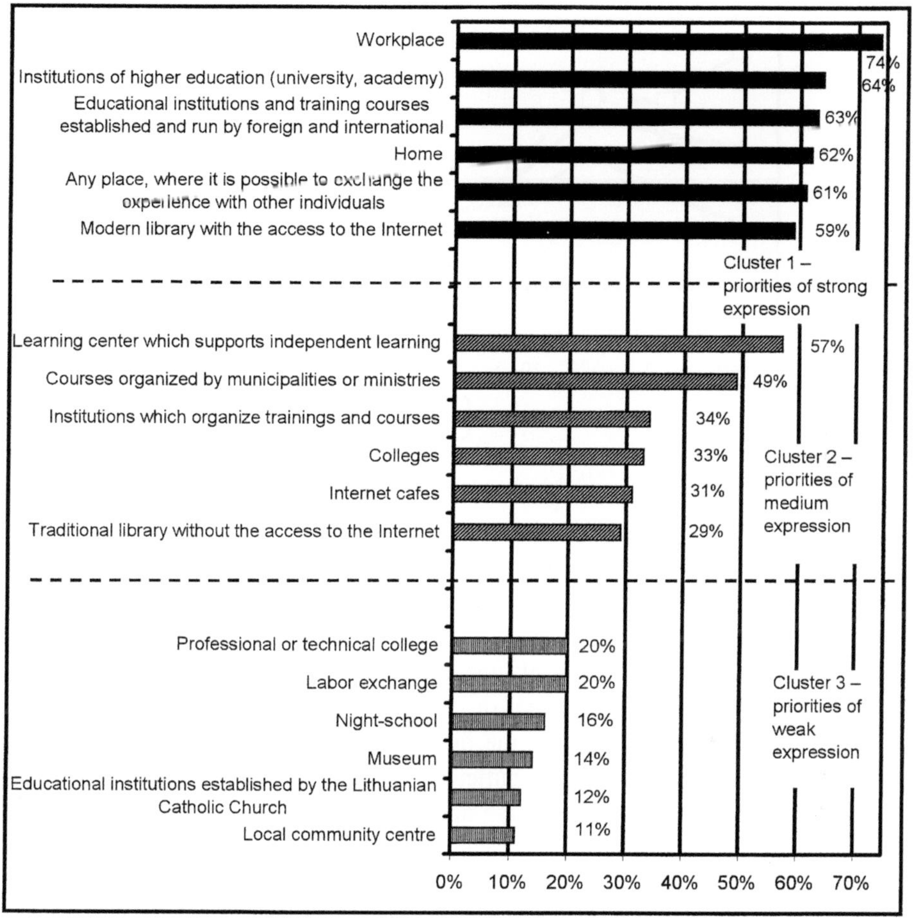

Figure 3. Ratings of Learning Environment Priorities, in percent.

ANOVA has also helped reveal the attitudes towards the library, influenced by the place of residence (df = 2; 696; 698; F = 3, 75, p = 0,024). The urban/metropolitan population than by the rural one has expressed more positive views towards the library as a learning environment. At the same time, however, no statistically significant differences have been noted between (a) metropolitan population and district center[3] population; (b) district center and rural population. In other words, statistically significant average differences have been noted only between the extreme categories, that is, between the metropolis and the country. Meaningful

[3] There are 20 administrative districts in Lithuania, each having a local government. A town with a municipality is a district center; it is usually a small town with the population of about 10 to 25 thousand people.

interpretation of this result could be rendered in the light of "disparity of the city and the country", a traditional sociological concept.

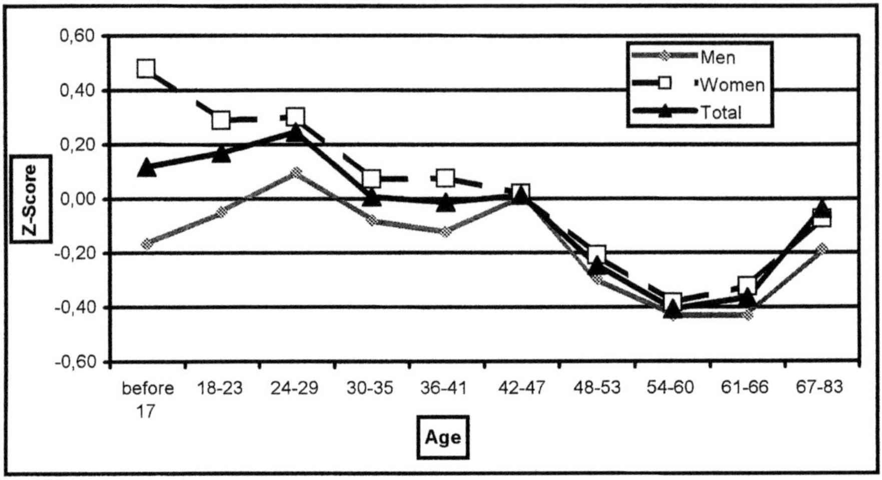

Figure 4: Selecting a Library as a Learning Environment with the Reliance on an Age Group and Gender.

Furthermore, it has been determined that the age makes a significant influence on the current educational preference. Therefore, the age between 16 and 83 has not been formally streamed in the matrix; rather, a human biography has been divided into 10 qualitatively meaningful categories. Having compared 10 age groups, statistically significant differences were obtained (df = 9; 683; 692; F = 2,08; p = 0,029). However, the division into six or five age categories has secured a considerably higher significance, 0,012 and 0,006, respectively. Still, the relation established between the age and attitudes towards the library does not possess features of the linear relation (see Figure 4). It is symptomatic that young people relatively have the most favorable attitude towards the library, which gradually declines according to the linear relation law. Having reached the lowermost point at the age of 55 to 65, it mounts up to the average, reaching it at the age of 66 to 83 which can be explained by a typical biography: in the young age, using a library is inseparable from studies at the university or at a professional college, or from a diploma and career-making. Afterwards, the period of a relative indifference towards the library follows, which also encompasses a critical pre-retirement age, easily explained by the psychosocial evolution of a personality. The same attitude renders clarification to the unexpected rise at the end of human life: most friends and contemporaries are dead, a social network shrinks, and a visit to the library becomes a daily cultural ritual for a senior citizen. Quite a large sample allows us to predict that the relation observed between the age and the attitude towards the library is not an artifact. Specifically, the initiation of a cross-cultural study would allow to determine if a non-linear nature of the relation is a really universal empirical law, resulting from the logics of human psychosocial development. The specifics of gender should be also mentioned. Namely, female population expresses a more positive attitude towards the library during all age phases, especially during the early stages, when studying.

Further consideration of the four types of the attitude towards the library, determined by the means of cluster analysis, reveals systematic influences of socio-demographic variables upon this attitude. The above enhance the assumption that the sample was mixed, non-homogeneous, and versatile in its attitudes. Therefore, investigation of the impact of several socio-democratic variables on the current preference stood to reason.

A special effect has emerged upon juxtaposition of three values: the preference investigated, age, and educational census, which are worth commenting (see Figure 5).

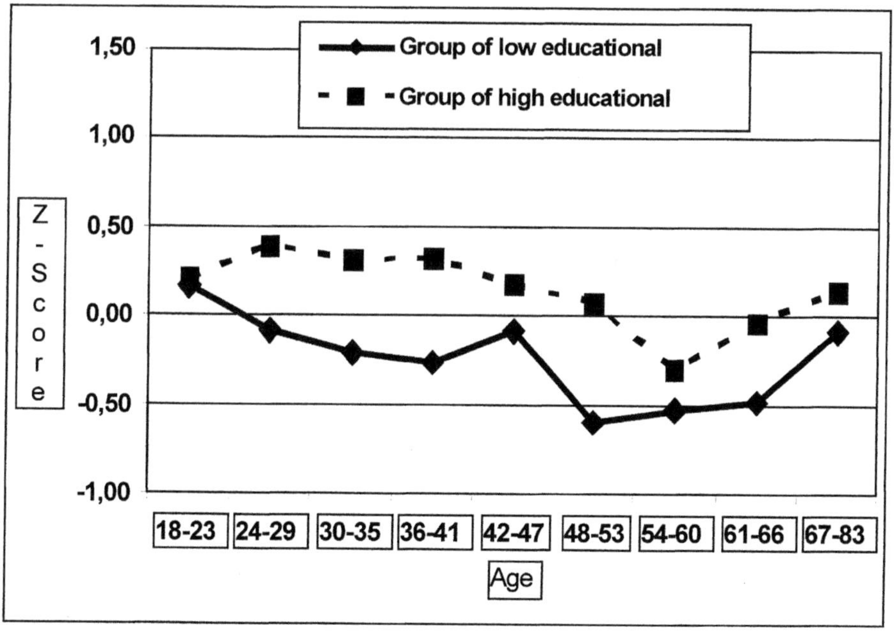

Figure 5. Reliance of Selecting the Library as a Learning Environment upon Educational Census in Different Age Groups

Both categories compared comply with the law, already established and discussed, namely: a positive attitude towards the library as an educational preference tends to decrease in time, but at the old age there appears a relative ascent.

of higher education express a more favorable preference; their positive consideration does not change even in the old age.

A similar effect of the dependence between the age and a socio-cultural status showed up when the comparison of social status, according to occupation, was considered (see Figure 5, p. 135). Having in mind that the sample represented a great professional and occupational variety, ranging from a housemaid to a lawyer, such an outcome could have been expected. By means of a qualitative analysis, the range of occupations was divided into three categories: "high status", "low status", and the "interference" category. Study participants whose professional status was difficult to ascribe to any marginal categories fell into this residual category. It is symptomatic that individuals belonging (or having belonged) to a "high status" category demonstrate a strongly expressed preference, even in their old age. In other words, their attitude towards the library as a learning environment is extremely positive.

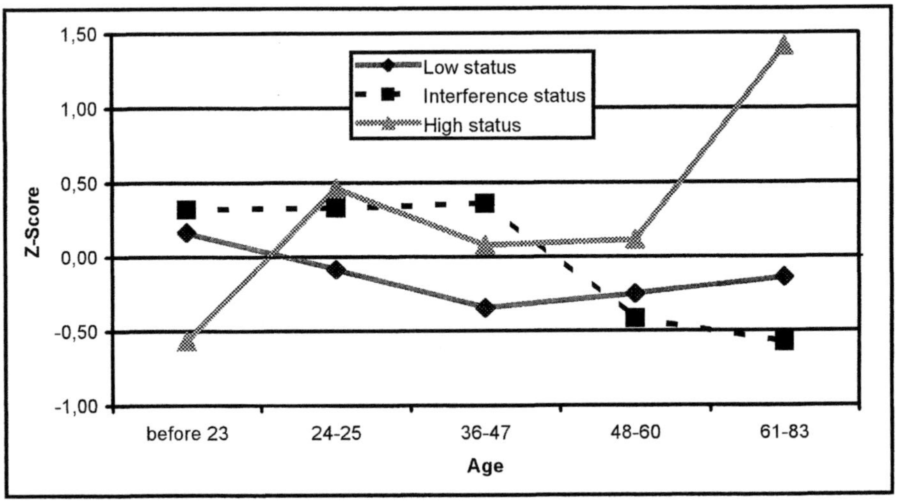

Figure 6: Reliance of Selecting the Library as a Learning Environment upon Occupational Status in Different Age Groups

Furthermore, a number of psychological personality constructs influencing the assessment of the library as a learning environment have become evident, namely: 1) interest; 2) reflectivity and rigidity; 3) self-conception of social roles and the interest to improve when performing social roles. Verification was carried out with the help of multidimensional regression. Having included all the above constructs into the model, the following results were obtained: Multiple R = 0,35; R Square = 0.12, $p \leq 0{,}000$. With a large number of independent variables in the regression model, due to missing values, the effect of matrix decrease usually appears. Therefore, the three groups (sub-scales) of independent variables were investigated separately in the model of multidimensional regression.

Eleven areas of the cognitive concern, included into the regression model, have produced such results: Multiple R=0,28; R Square=0.08, p≤0,000. The highest value of a standardized coefficient beta (β=0,23, p≤0,000.) has been obtained from a topical area of the cognitive interest, operationalized by the integrated index of computer literacy, foreign languages, and communication (10 items in total).

The subscales of openness – reflectivity and rigidity have acquired the meanings of beta coefficient 0,20 (p ≤ 0,000) and (minus) -0,13 (p ≤ 0,002) respectively. Negative correlation between rigidity and the library assessment is meaningful here. Other characteristics of the regression model obtained were R=0,27; R Square = 0.07, p ≤ 0,000.

Self-concept of social roles has lent the following results of the regression model: R = 0,26; R Square = 0.07, p ≤ 0,000. Both beta-coefficients are statistically reliable: 0,14 (p ≤ 0,000) for self-concept of social role implementation, and 0,19 (p ≤ 0,002) for the interest to improve when performing social roles.

It has become evident, however, that some of the constructs considered had no relation to the current preference. For instance, it has emerged that different sub-scales of respective attitudes, for instance, the attitude towards paid educational services, had no relation to statistical value of the library whatsoever. This example is especially interesting having in mind that 14 years ago paid educational services were non-existent in Lithuania, as well as in the Soviet Union and the Soviet block. At present, however, there are also no commercially operating libraries in the country.

Conclusions

It could be concluded that the results of this empirical investigation demonstrate the following:

1. The library has a potential to enable each person to individualize his/her learning environment, especially when a library, particularly a modern one, allows for a special attention towards creating an educational environment in it. Moreover, it is evident that each individual identifies his/her learning environment on the basis of his/her attitudes towards the library as an educational institution.
2. In the context of various learning environments, the library (especially a modern library with the access to the Internet) as an educational preference, ranks high.
3. Attitudes towards the library as a learning environment have been largely effected by the following: a) major socio-demographic characteristics: gender, age, education, and

place of residence; b) some psychological personal characteristics, such as interest, reflectivity and rigidity, self-concept of social roles, etc.

4. Results of the current survey have demonstrated the importance of the library as a learning environment and its potential in creating the knowledge and learning society. On the grounds of current research results, it could be suggested that the creation of learning environments in libraries is among the premises securing the creation of a learning city.

Due to the shortage of empirical research on the library in Lithuania, it stands to reason to carry out at least several empirical studies, which would define educational functions of the library as well as their manifestation as an autonomous object for research. Nevertheless, even the results of this modest investigation can serve the improvement of library work in many aspects, for instance, in individualizing the attitudes towards different categories of clients in the context of social service marketing.

References

BAKEWELL, K.G.: Managing User-centered Libraries and Information Services. 1997 (Mansell Publishing Ltd.).

BARR, R.B./TAGG, J.: From Teaching to Learning – a new Paradigm for Undergraduate Education. Change, November 01 1995.

BOUD, D./GARRICK, J.: Understanding Learning at Work. London, New York 1999 (Routledge).

BUŠIMIENĖ, S./CICĖNIENĖ, V.: Lietuvos nacionalinės Martyno Mažvydo bibliotekos skaitytojai. Vilnius 1999.

Constructivist Learning Environments: Case Studies in Instructional Design. WILSON, B.G.(ED.) Englewood Cliffs, N.J. 1996 (Educational Technology Publications).

DOWLER, L.: Gateways to Knowledge: the Role of Academic Libraries in Teaching, Learning, and Research. Cambridge, Mass 1997 (MIT Press).

HERNON, P./ALTMAN, E.: Service Quality in Academic Libraries. 1996 (Ablex Publishing Corporation).

Information Literacy Around the World: Advances in Programs and Research. BRUCE, C./CANDY, P. (eds.). Wagga Wagga 2000 (Centre for Information Studies, Charles Sturt University).

IVAŠKEVIČIENĖ, G.: Bibliotekos fondas: organizavimas, analizė, vertinimas. Vilnius 2001 (LNB).

JUCEVIČIENĖ, P./TAUTKEVIČIENĖ, G.: "Triangle of Competences" as one of the basic characteristics of the university library learning environment. European Conference on Educational Research: Lisbon, 11-14 September 2002. Leeds 2002, pp. [1-9]. [cited 14 February 2003]. Available from <http://www.leeds.ac.uk/educol/documents/00002343.htm>.

KIRBY, J./LIDDIARD, L./MOORE, K.: Empowering the Information User: New Ways into User Education. London 1998 (Library Association Publishing).

Learning Across the Lifespan: Theories, Research, Policies. TUIJNMAN, A.C./ KAMP, VAN DER M. (Eds.). 1992 (Pergamon Press).

Lietuvos bibliotekų 1991-2000 statistika. Vilnius 2002.

Life-long Learning: Education Across the Lifespan. FIELD, J./LEICESTER, M. (eds.). London and New York 2000 (Routledge Falmer).

LONGWORTH, N.: Making Lifelong Learning Work: Learning Cities for a Learning Century. London 1999 (Kogan Page).

MARCUM, J.W.: From Information Center to Discovery System: Next Step for Libraries? The Journal of Academic Librarianship. Vol. 27(2002)2, pp. 97-106.

MARTON, F./HOUNSELL, D./ENTWISTLE N.:. The Experience of Learning: Implications for Teaching and Studying in Higher Education. Edinburgh 1997.

MCNICOL, S.: Learning in Libraries: Lessons for Staff. New Library World vol. 103(2002)1178, pp. 251-258.

MOOS, R.H.: Learning environment in context: links between school, work, and family settings. In: FRASER, B. J. (Ed.): The Study of Learning Environments Vol.2. Perth, Western Australia 1987 (Curtin University of Technology), pp. 1-16.

MOOS, R.H.: Person-Environment Congruence in Work, Scholl, and Health Care Settings. Journal of Vocational Behavior. 31(1987), pp. 231-247.

Periodinių leidinių paplitimas ir skaitymas: tyrimo rezultatai. 1998 (LNB).

RETALLIK, J./COCKLIN, B./COOMBE, K.: Learning Communities in Education: Issues, Strategies and Contexts. 1999 (Routledge).

RIMŠA V.: Skaitymo tyrimai Lietuvoje: (analitinė 1918-1978 metų veiklos apžvalga ir būdai šiam darbui gerinti): metodika bibliotekoms, atliekančioms skaitymo tyrimus. Vilnius 1980.

Savivaldybių viešosios bibliotekos 1990-2000 metais: veiklos apžvalga. Vilnius 2002.

SCOTT, A.: Learning Centers. 1997 (Kogan Page Limited).

SIMONS, K./YOUNG, J./GIBSON, C.: The Learning Library in Context: Community, Integration, and Influence. Research Strategies 17 (2000), pp. 123-132.

SINKEVIČIENĖ Z. Skaitytojų interesai periodiniams leidiniams ir jų tenkinimas Akmenės rajono viešosiose bibliotekose: tyrimo rezultatai. Bibliotekininkystė 2000: bibliotekų fondai ir skaitytojai. 2001 (Lietuvos nacionalinė M. Mažvydo biblioteka).

Supporting Lifelong Learning: Perspectives on Learning. HARRISON, REEVE, F./HANSON, A./CLARKE, J. (eds.). London 2002 (The Open University).

WENGER, E.: Communities of Practice: Learning, Meaning and Identity. 1998 (Cambridge University Press).

Liuda Šiaučiukėnienė, Giedrė Adomavičienė

The Creation of Differentiated Learning Environment as the Expression of Liberal Education Features

Introduction

Since old times the educational science, its interaction with philosophy has not lacked attention of researchers. Most theories of education are based on philosophical viewpoints, i.e. the philosophy of education. Johan Dewey defined the relationship of the theory and philosophy of education most radically. According to him, the theory of education is the entirety of generalizations and abstractions related to education (Dewey, 1938).

Till the restoration of Independence in Lithuania, the educational process was of authoritarian nature under the influence of the Soviet pedagogy and state ideology, and it referred to the attitude that the learner must intercept the experience, which is given by the teacher; and the more significant the status of the teacher, the more intensive coercion is (Bitinas, 1996). Thus the construction and realization of educational process is greatly influenced by philosophical concept. It is known that the change of educational phenomenon occurs not only in the local but in the global context too. For Lithuania, where the variety of the development of educational science is distinct, it is important to which state of the world it gets now, what way of education it chooses while seeking to become a member of the European Union.

This issue is significant for all post-socialistic countries, as after having the authoritarian control/management system destructed, the main landmarks of education disappeared. It is necessary for the countries seeking for the qualities of knowledge and information (Drucker, 1993; Jucevičienė, 1997), to get oriented to methodological attitudes of the world and European dimensions.

This statement can be concretized by giving an opposite concept – the one of liberal education – against authoritarian pedagogy, the realization of which determines a free underlying position of a person as individuality's development in the society. In this sense differentiation and individualization as educational factors, which under the influence of macro system are important, and they are researched in the level both of an institution and micro system – a particular pedagogical process - as well as in the appropriate learning environment. These educational factors can be evaluated as original educational phenomenon, when at different levels the educational impact is organized so that human individuality – one's individual needs, interests, abilities, experience and motivation – take priority.

Thus the scientific problem *whether the creation of differentiated learning environment shows the expression of the features of liberal education in the contemporary teaching process arises.*

The research purpose is to reveal and found the environment of differentiated learning, disclosing its expression as the premise of liberal education.

The article consists of three parts. The *first* part reveals the features of liberal education and the possibilities of their realization. The *second* part analyzes the environment of differentiated learning in the context of liberal education; in the *third* part the expression of the premises of liberal education is based empirically in creating the environment of differentiated learning.

The Features of Liberal Learning and Their Expression in Educational Process

It is sought by means of liberal education to release the learner so that he/she would possess the intellectual and moral autonomy, a possibility to become free in choosing whom to believe and what to do (Bailey, 1984). Thus the aim of liberal education is the education and self-education of individuality, a free human being. Only being able to choose his/her goals, behavior freely and to decide independently, he/she discovers the most meaningful life values. Thus it is necessary to seek, to found and realize such education in Lithuania, at the didactic attitude hierarchy top of which the values of respect, tolerance, communication and help would occur, due to which not only the uniqueness and worth of every learner is recognized, but also rights, conditions, possibilities are created to influence the goals of teaching, content choice, its level and profile, teaching forms and methods, appropriate teaching institutions and teachers, to learn to react tolerantly to the choice of others and to take responsibility for one's own choice.

The system of differentiated education, referring to the cognition of a human being, united the statements and propositions of philosophers, psychologists, and pedagogues so that the curriculum and technology become concerted with the individuality of a human being. The goals of education ensure the right and freedom for the decision in the system of differentiated education for a learner. The pedagogue, working according to his/her individual viewpoint and referring to his/her professional competence as well as creative plans, forms particular didactic tasks/objectives, complements the precise curriculum, resounds the goals of liberal education and permits to form purposeful, oriented choice, intensive motivation of learning, when every learner orients in the activity and chooses according to his/her individual features and is able to behave according to subjective values.

While differentiating the curriculum, every learner chooses his/her profile, level or qualification degree corresponding the curriculum and learning according to individual programme. According to Ackermann, there is no such a constructive programme and its implementation, the methods of which can be applied to everyone who studies at the liberal school“ (1980, p.156); thus all programmes should predict the minimum of knowledge, and the gifted ones – the extended deepened volume of knowledge, and this ensures that more gifted children will not be suppressed, leveled, oriented towards the middle level. According to McLaughlin, the school based on the concept of general education does not have to avoid the forms of oneness (1997, p.146). Teaching process is organized through unlike level and volume tasks of differentiated independent work. Every learner can choose a task, its accomplishment proceeding according to one's individual possibilities. The time and forms of differentiated presentation and control as well as the right to choose the forms enable one to learn in his/her individual style of work.

These statements correspond to basic features and aims of liberal education, one of which is to strengthen the feeling of possibility to choose, to attain intellectual autonomy, to become free when choosing what to believe and what to do (Bailey, 1984).

The organization of differentiated teaching commences with the cognition of learners, their grouping according to corresponding individual attributes/features; it enables to realize the individual approach choosing for a small group different learning ways and forms.

Thus it is possible to state that theoretical statements of the differentiated teaching system create the significant offset for the former authoritarian educational system in Lithuania, their realization in school's practice is a democratic process as the premise of liberal education.

The Environment of Differentiated Learning and Its Characteristics

In the second half of the XX century external and internal contradictions in the teaching system challenged the transformation of educational paradigm. In the classical paradigm of learning a teacher is the main executive and conveyor of knowledge, and a pupil is only the object of teaching, a subordinate of the teacher, accumulating facts and information. In the modern paradigm of teaching a teacher is more a consultant, advisor or moderator for a pupil. He/she is a partner in educational process, who is constantly learning, gaining a lot of possibilities to grow through constant communication with a pupil. The role of a pupil also changes – the pupil himself/herself seeks for information resources, the teacher is his/her assistant, a senior colleague and the member of solving common problems.

Thus active interaction between a teacher and pupil occurs, and it shows that alongside the teaching process another important process – learning – emerges. This witnesses the transformation of educational paradigm, in which learning comes to the first place, and pupil's competence is defined as constructive activity (Ahlberg, 1997).

Various researchers are interested in the processes of teaching and learning and they analyze them (J.W.Brown, 1997; J.Martin, P.Mc.Keown, 1999; A. and J. Selden, 1998; MacFarlane, 1995; Arends, 1998; Flinck, 1996; Bond, 1993; M.Thrupp, 1999; P. Jucevičienė, 2000, B.Bitinas, 2000, etc.). The closest to the idea of differentiated learning are the statements of "self-directed learning" (Jarvis, 1999; Knowles, 1980) and the direction of constructive learning (M.Ahlberg, 1997, Lipman, Paul, 1998). Šiaučiukėnienė (2000) presents the comparison of differentiated teaching and learning (Table 1):

Differentiation of Teaching	**Differentiation of Learning**
Cognition of a pupil commences	It starts with the ability of a pupil to evaluate himself/herself correctly
A teacher makes an individual programme of teaching	A learner himself/herself makes his/her individual programmes of learning
It is most frequently offered to group learners according to their separate features	Learners themselves choose a group
A teacher chooses differentiated tasks according to the level of learners' readiness for future learning	A teacher offers learners tasks of different complexity, and they themselves choose their level, in this way learning to choose correctly

Table 1: The Comparison of Differentiated Teaching and Learning (L.Šiaučiukėnienė)

Referring to the statements of the above-mentioned researchers and the comparison of differentiated teaching and learning presented in Table 1, we would like to introduce the conception of differentiated learning.

Differentiated learning is individually chosen mental activity based on different intellect and experience. By interactive process between the search and interface with environment, it designs new understanding about oneself and the world. Through this kind of learning, the change and development of personal capacities are seen as enabling an individual:

- to assess himself/herself rightly;
- to learn and choose what is mostly needed for his/her harmonious development;

- to work by cooperating in team;
- to seek for higher quality of living by satisfying his/her needs.

This conception corresponds to the thoughts of Savin – Baden (2000), where the right of learners to choose, to correctly evaluate their abilities and possibilities is emphasized. The importance of differentiated learning is disclosed in the table of 10 qualification and competence levels, which is meant for national educational programme and certified by the European Council (Aktas, 1999).

No.	Levels	Environment (sphere) of influence	Result	Factor
1.	The level of educational system	Goals of differentiated teaching, conditioned by educational policy. Basic values, priorities, action strategy of differentiated teaching are predicted	Realization of differentiated educational system	Realization of differentiated educational system through profiled teaching.
2.	The level of institution	Specific goals of differentiated teaching, conditioned by administrative subdivisions that influence further programme of actions.	Realization of differentiated teaching system at an institution	Profiled teaching as the factor of the realization of external differentiation
3.	The level of a group or class	Operative goals of differentiated teaching, which directly related to teacher's work, activity, raise specific objectives for it	Differentiation of teaching within a class or group	Profiled teaching as the realization factor of internal differentiation
4.	The level of an individual	Specific, differentiated goals and objectives of learning of an individual.	The realization of learning differentiation	Profiled teaching as the environment of differentiated learning

Table 2: The Levels of Differentiated Teaching/Learning

Recently the content differentiation became important both in Western and Eastern countries (Jakimanskaja, 1995, McLaughlin, 1997). Thus in the system of differentiated teaching every learner has a possibility and right to choose a profile and level of content according to his/her interests and abilities, and the freedom of choice helps to form a positive viewpoint, stimulates corresponding emotions in the teaching process. At present the profiled teaching is introduced in Lithuania. Only having grasped and realized the process of change very well, one learns to manage innovations, as well as to implement particular curriculum and teaching methods. So the question arises what is the place of this teaching in the system of differentiated teaching. The statements in Table 2 give the answer to this question (Šaiučiukėnienė, Kaminskienė, 2001).

One of the innovations implemented in a modern school of Lithuania is the realization way of differentiated teaching – *the profiled teaching*, which is an important factor going towards the reformed school. *The profiled teaching at the individual's level should become the environment of differentiated learning* in upper classes.

The profiled teaching at the level of educational system is one of the factors of reform stage that realizes the system of differentiated teaching.

Darrison, 1993; Green, Collins, Resnick, 1996; Ahlberg, 19997; Knasel, Medd, Rossetti, 1999; Wilson, 1995; Arends, 1998 and other researchers analyzed learning environments.

The constructivists paid a lot of attention to the analysis of learning environments. Summarizing different constructivist definitions of learning environment (Garrison, 1993, Green, Collins, Resnick, 1996, Ahlberg, 1997, Bitinas, 2000), it is possible to state that *the learning environment is the place where learners can work together and keep up each other, applying different learning media and resources in order to achieve learning goals and to solve problems*. According to this statement, *a close learning environment is a class or group, in which a pupil learns.*

In this work close learning environments will be wider analyzed, i.e. among them a class or mobile group will be chosen. It is characterized by flexible and different structure. The concept of H. Kohl (1976) that offers creating flexible learning environment, to which not only openness, interconfidence but also undeviating continuity of learning is characteristic, proves this. This is the environment corresponding attitudes of liberal education, as the environment of differentiated learning has to ensure that conditions that a pupil, having chosen the learning content and activity that corresponds his/her needs and interests as well as skills, would experience certain satisfaction, would meet his/her needs and expectations, and would attain the set goals.

Referring to the statements of the above-mentioned authors about the environment of learning as well as to the presented concept of differentiated learning, it is possible to distinguish characteristics of differentiated learning environment, which are based on the following values of liberal education: autonomy of an individual, freedom and right to choose for every learner.

The following indicators reveal a free choice of the content: correct/fair self-evaluation, the know-how to choose, new experience of learning, creation of individual programme of learning, new information, the prospect of profile entirety. The following indicators will evaluate the strategy of differentiated learning environment: cooperation in learning in mobile groups, methods of active teaching and learning, evaluation of one's achievements, and reflection. The following indicators will assess the expected results: satisfaction of needs, attainment of the goal, and verification of expectations (Fig. 1, p. 147).

Having analyzed the concept of differentiated learning environment and its charakteristics, it is possible to point out that *the environment of differentiated learning is oriented towards a learner; it is liberal, informal, based on teacher's and learner's mutual understanding, colaboration and respect.* The programme of national education certified by the European Council predicts these requirements for learning environments (Aktas, 1999).

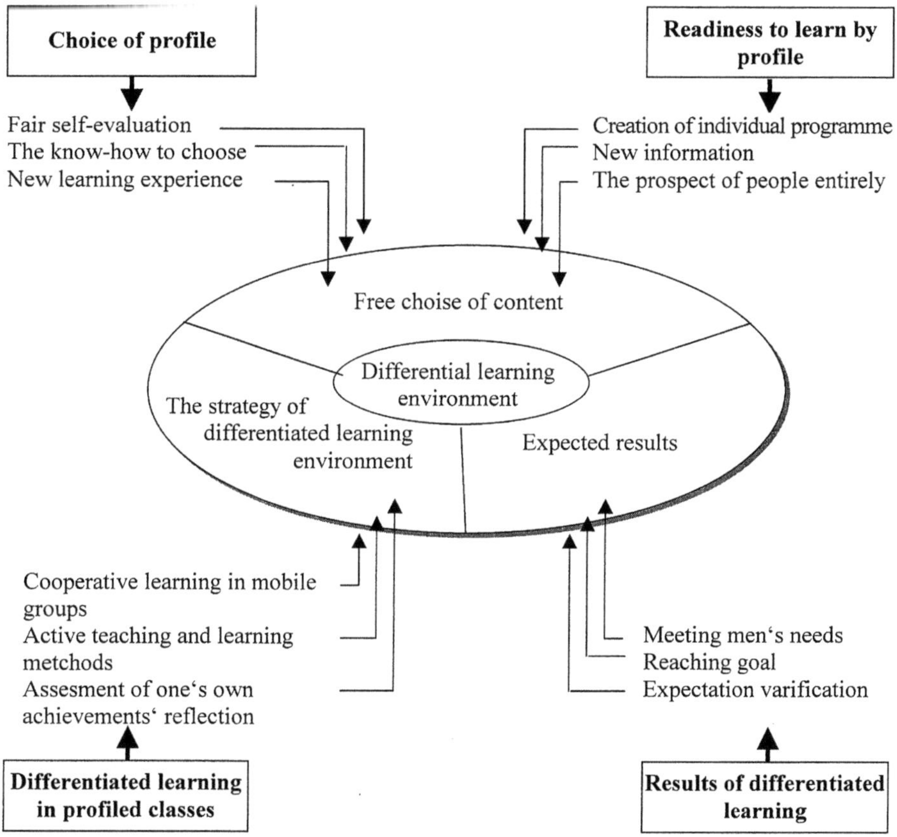

Figure 1: Characteristics of differentiated learning environments

The Research of the Expression of Differentiated Learning Environment

McLaughlin points out that only philosophical considerations alone are not able to solve the issues of educational policy and practice. It is also necessary to apply solutions of completely practical character (1997, p. 16). The values of education cannot be understood abstractly, they have to be interpreted and applied practically (Mclaughlin, 1997, p. 12). Fullan continues this thought by writing that they have to rise from one's activity but not to be forced by the activity (1993). In this article the theoretical statements based on the performed empirical research were presented (Šiaučiukėnienė, 1997).

In the presented research the premises for the realization of differentiated learning environment in upper classes will be analyzed referring to the indicators given in Figure 1.

The purpose of the research – is to empirically found the realization of differentiated learning environment in upper classes, as the premise of liberal education.

Referring to these indicators, in 2001 the quantitative research, the purpose of which was to envisage the premises of differentiated learning environment considering the answers of teachers and pupils. In order to attain this purpose the research instrument and questionnaire were constructed.

332 pupils of the 11th classes and 355 pupils of the 12th classes from different Kaunas secondary schools as well as 173 teachers teaching in upper classes, who were chosen by random selection, took part in the research.

The answers of the respondents are presented in three blocks, which are made according to the characteristics of differentiated learning: 1- free choice of content, 2 – strategy of differentiated learning environment, 3 – expected results.

The performed research on the expression of differentiated learning environment enables us to state that direct dependence between these criteria is observed:

- *correct self-evaluation and ability to choose (Kendall's tau-c = 0,286; StdE=0,036, p<0,05).*
- *the ability to choose and new experience of learning (Kendall's tau-c = 0,301; StdE=0,036, p<0,05)*
- *correct self-evaluation of the ability to choose and creation of individual teaching programme (Kendall's tau-c = 0,548; StdE=0,101;p<0,05);*
- *cooperation in learning in mobile groups and methods of active teaching and learning (Kendall's tau-c = 0,232; StdE=0,04, p<0,05);*
- *evaluation of one's new learning experience and achievements, reflection (Kendall's tau-c = 0,694; StdE=0,023, p<0,05);*
- *correct self-evaluation and expectation verification (Kendall's tau-c =0,320, Stde=0,032, p<0,05).*

Reverse dependence between learners' self-evaluation and responsibility development was observed, thus it is possible to draw the premise that pupils in upper classes did not possess developed responsibility for their learning, achieved results (Kendall's tau-c = -0,642; Stde = 0,022, p < 0,05).

Having evaluated the opinion of pupils in upper classes about the premises of differentiated learning environment, it is important to find out the opinion of teachers who work in upper classes about this process as in the theoretical model of differentiated learning environment there is a part related to teacher's activity, its peculiarities while working in mobile groups, thus it is impossible to limit only by pupils' answers, which can be not always competent.

Teachers gave the highest points for the statement (the medium of statement evaluation- 4,05), which proved the thought that profiled teaching stimulated teachers to orient their teaching methods to learning methods (SD -0,096), the lowest point was given to the statement (the medium of statement evaluation -2, 12) that during lessons teachers explained and pupils listened (SD -0,096).

By evaluating the statements, it was important to find out what teachers organizing lessons in upper classes pay their most attention to. The answers show that teachers understand that lesson's aims (R/No1) and pupils' psychological features (R/No2) are of the greatest importance while getting ready for a lesson and organizing it.

In differentiated learning environment teacher's activity, rights, possibilities, relations with other structural elements of teaching and learning process change as well, thus it was important to *find out how the teachers' activity changed while they worked in upper classes*. The presented data show that first teachers pointed out the creation of individual programmes for separate profiles and streams (for them it is the most important change while working with pupils of upper classes) (R/No1), as well as they emphasized that during lessons pupils' autonomy and responsibility is induced (R/No 2). It is possible to draw a premise that didactic approaches of teachers correspond to the concept of differentiated learning.

It was observed that teachers' answers are not related to their pedagogical work experience or possessed qualification category.

Summarizing the statements of teachers, it is possible to draw the following premises:

- The standpoint of most teachers is that the most important thing is to be oriented to teaching methods and this means that in upper classes the conditions to create differentiated learning environment are created.
- The medium of statement evaluation decreases much more when teachers have to evaluate their pupils, their responsibility, conscious and continuous evaluation of results and achievements in learning.
- While discussing the learning strategy, which teachers offer to their pupils, a positive phenomenon that only a small part of teachers still work using old classical methods (when pupils listen and teachers explain) emerged. But a part of teachers is not subject to stimulate differentiated learning because they give their pupils common, compulsory tasks, they do not arrange independent work, they do not choose tasks according to the learners' interests and skills.

While analyzing the answers of pupils and teachers, which are related to the characteristic of differentiated learning environment "Strategy of differentiated learning", it was noted that the answers of pupils and teachers differ.

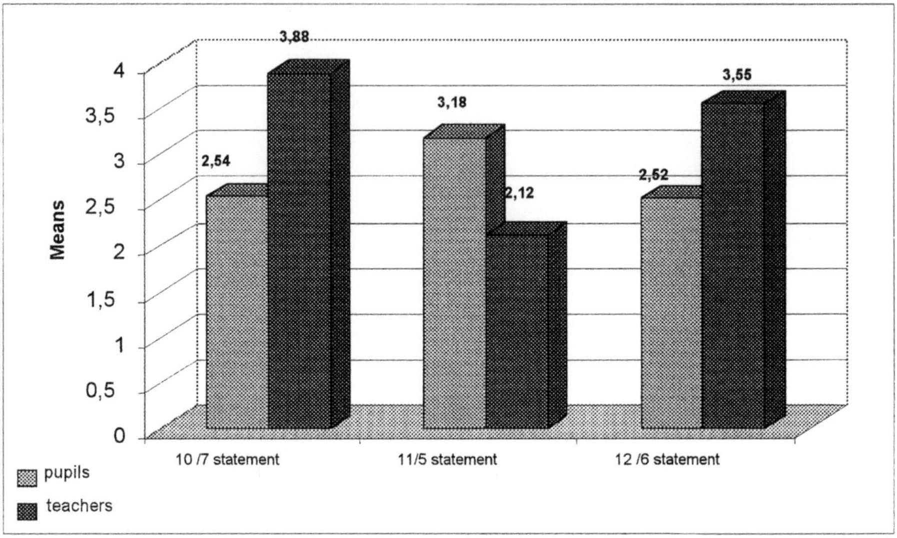

Figure 2: The comparison of pupils' and teachers' answers about teaching and learning methods

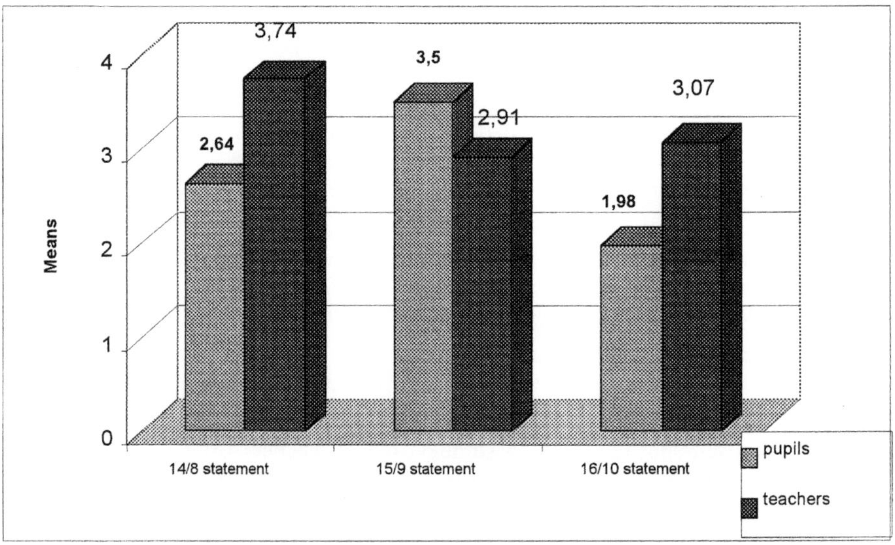

Figure 3: The comparison of pupils' and teachers' answers about differentiation of tasks

Referring to the data in Fig. 2, p. 150, it is possible to assume that the answers of pupils and teachers differ proportionally – pupils point out that traditional teaching methods are applied during their lessons, and teachers state that they most frequently apply active methods of teaching and learning during the lessons. Though the means of statement evaluation differ, applying the method of reliable intervals with $p < 0,05$ it is possible to state that statistically significant difference does not occur in the evaluation of pupils' and teachers' statements.

Similar ideas come to light while analyzing Figure 3, p. 150. Pupils state that most frequently during their lessons they do common and compulsory tasks; and teachers point out that the tasks they present are different, adjusted to pupils' learning skills and knowledge level. Though the means of statement evaluation differ, applying the method of reliable intervals with $p<0,05$, it is possible to state that statistically significant difference does not occur in the evaluation of the statements of pupils and teachers. However in the strategy of differentiated learning it is possible to envisage different tendency of pupils' and teachers' evaluation. This could be explained so that teachers, though knowing didactic approaches necessary for the realization of differentiated learning, not always apply them in their practice. Pupils' answers disclose this.

Conclusions

1. In the system of differentiated teaching (in the context of transformation of educational paradigm) the differentiation of learning uncloses at the level of an individual. The concept of differentiated learning corresponds the goals of liberal education in the educational process.
2. Profiled teaching is one of realization ways of differentiated teaching. Profiled teaching at the level of an individual should become the environment of differentiated learning in upper classes.
3. Profiled teaching in upper classes creates the environment of differentiated learning at the level of an individual. Profiled teaching at the level of educational system – is one of the factors of reform stage, realizing the system of differentiated teaching.
4. The environment of differentiated learning:
   - ensures conditions that a pupil, having chosen content and activity that correspond his/her needs and interests as well as capabilities, would experience certain satisfaction, meet their needs and expectations, as well as attain the set goals;
   - distinguishes in such characteristics: free choice of content, strategy of differentiated learning environment, and expected results.

References

AHLBERG, M.: Continual quality improvement as high quality learning in Finnissh.. Research Reports of the Faculty of Education. No.68. 1997 (University of Joensuu).

ARENDS, R.I.: Mokomės mokyti. Vilnius 1998 (Margi raštai).

BITINAS B.: Ugdymo filosofija. Vilnius 2000 (Enciklopedija).

BITINAS, B.: Ugdymo tyrimų metodologija. Vilnius 1998 (Alma Littera).

BROWN, PH./SCASE, R.: Universities and Employers: Retoric and Realety. The Postmodern University? In: SMITH, A./WEVSTER, F. (eds.): Contested Visions of Higher Education in Society. London 1997.

DEWEY, J.: Democracy and Education. New York 1996 Macmillan.

DRUCKER, P.: Post-capitalist society. 1993 (Butterworth-Heinemann).

GARRISON, D.: Self-directed learning: Toward a comprehensive model. Adult Education Quaterly. 1(1997)48.

GREEN, J./COLLINS, A./RESNICK, L.: Cognition and learning. In: BERLINER, D./CALFEE, R. (eds.): Handbook of Educational Psychology. London 1996 (Prentice Hall International).

JAKIMANSKAJA, K.: Требования к учебным программам, ориентированным на личное развитие школников// Вопросы психологии. Но.3(1995).

JARVIS, P.: Paradoxes of Learning. San Francisco 1999 (Jossey-Bass Publichers).

JUCEVIČIENĖ, P.: Ugdymo mokslo raida: nuo pedagogikos iki šiuolaikinės edukologijos. Kaunas 1997 (Technologija).

JUCEVIČIENĖ, P.: XXI amžiaus naujų mokymo teorijų paieškos: konstruktyvus aplinkos mokymas. Seminaro medžiaga LETA. 2000 02 03 (Edukacinės kompetencijos centras, Edukologijos institutas, KTU).

KNASEL, E/MEED, J./ROSSETTI, A.: Learn for your life. 1999 (Prentice Hall).

KNOWLES, M.S.: Self-directed learning. Cambridge 1980 (Adult Education).

KOHL, H.: On Teaching. New York 1976.

LIPPMAN, M./SHARP, A.M.: Markas: Mokytojo knyga. Vilnius 1999 (Kronta).

MACFARLANE, A.: Futur Patterns of Teaching and Learning. The Changing University? 1995 (SRHE).

MARTIN, J./SUGARMAN, J./MCNAMARA, J.: Classroom management. Calgary 2000 (Detselig Enterprises Ltd.).

MCLAUGHLIN, T.H.: Šiuolaikinė ugdymo filosofija. Kaunas 1997 (Technologija).

SELDEN, A.J.: Constructivism. MAA Online. F.Q. Gouvea (Ed.). 1998.

ŠIAUČIUKĖNIENĖ, L./KAMINSKIENĖ, G.: Diferencijuoto mokymosi pradų paieška profiliniame mokyme. LR Švietimo ir mokslo ministerija: Profilinis mokymas. Vilnius: 2001 (Presvika), pp. 5-12.

ŠIAUČIUKĖNIENNĖ, L.: Mokymo individualizavimas ir diferencijavimas. Kaunas 1997 (Technologija).

THRUPP, M.: Schools making a diference: Let's be realistic! School mix, shool effectiveness and the social limits of reform. 1999 (Open University Press).

WILSON, B.G.: Metaphors for instruction: why we talk about learning environments, Educational Technology, 35(1995)5, pp. 25-30.

Nijolė Bankauskienė, Ilona Skersienė

Cohesion of Educating in Philology and Technological Education

Introduction

The present Lithuania, its society, economics and policy move towards the knowledge society. New technologies are not its essence, but the people, who have a lot of knowledge, who are trying to renew them, who are creative and know how to use their knowledge and experience, are the background of such society. As V. Snitka (2001) states, economy of the state, which revive because of new knowledge, is competitive.

In this article we are trying to analyze a very relevant question of the cohesion of philology and technological education. Keeping in mind teaching / learning and preparation of youths for professional activity (vocational training), it is very important to answer the question, whether technological education of the society is the most important, in the start of the knowledge society. What is the role of humanitarian and philological education in this process?

F. Cajus (1995), M.J. de Vries (1999) discussed separate issues of technological education, P.A. Reed (2001), E.N. Israel (1994), L. Jewell (1995), D.L. Crory (1987), W.B. Woetjer (1995), T. Levis (1999) analyzed its curriculum. G. Gloechner, G. Adamson (1996), S. Petrina (1993), R. McCormick (1998) investigated the methods of technological education.

J.K. Galkauskas (2000), L. Šiaučiukėnienė (2002), A. Ramanauskaitė, N. Stankevičienė (2002), A. Ramanauskaitė (2001, 2002) analyzed the problems of technological education in Lithuania.

According to Sauter (1994), the rapid implementation of new technologies and huge amount of information make great demands of society members: they must be socially sensitive, they ought to have humanitarian, cultural and professional purposefulness, which is stipulated by wide general education.

It is natural that greater attention is paid to humanization and humanitarization of the system of education. The particular importance falls on philological educating of youth in this process.

L. Abraitytė (1999), Ž. Jackūnas (1993, 1994), D. Kuolys (1999), M. Lukšienė (1989, 1997), S. Matulaitienė (1991), N. Šervenikaitė (1999), V. Tapinienė (1999), N. Vėlius, D. Sauka (1990), G. Viliūnas (1998, 1999), V. Zaborskaitė (1993, 1999), S. Žukas (1995, 1999) analysed these processes. These questions also were discussed by N. Bankauskienė (1998, 1999, 2000, 2001) R. Brunevičiūtė (1998, 1999), R. Brunevičiūtė, N. Večkienė (1998), L.D. Barzdžiukienė (1999), D.

Drąsutavičienė (2002), Z. Nauckūnaitė (2000), L. Ruseckienė (1995), V. Salienė (2001), I. Skersienė, N. Bankauskienė (2001), K. Stoškus (1996), R. Tamošaitis (2000), Župerka (2001).

Lithuanian researchers also pay their attention to the new qualitative characteristic – to information and technological culture of the society (D.Urbonaitė, 2000). In certain cases they more emphasize technological means and the level of their development (A. Otas, 1997, Žintelis, 1999). Others (P. Jucevičienė, 1997; B. Janiūnaitė, 1998; P. Petrauskas, 1999) attach greater importance to human resources and to the characteristics of the personality. However we did not succeeded to find wider analysis of cohesion of technological and philological educating in the works of Lithuanian scientists.

We face ***a relevant problem***, whether it is possible to state that a cohesion of philological and technological education is being fixed in the process of modern education?

The ***object*** of our research is cohesion of educating in philology and technological education in gymnasium of technological type (such gymnasium was founded in the start of the knowledge society). The schools of vocational training turned into such gymnasiums naturally at the end of the XXth century. According to the White Paper (1998, p. 24), ***technological gymnasiums*** are learning institutions, where gifted pupils, oriented to practical activity are studying. The purpose of these institutions is to create preconditions to obtain professional qualification and to continue studies in a school of higher education. The pupils obtain secondary education with technological profile there.

As the "Memorandum of lifelong learning" (2001) states, the people are the main heroes in the knowledge society. A person's ability to use his/her knowledge effectively and reasonably in constantly changing environment is the most valuable. The essence is the high quality education in a comprehensive school for everyone. This stage of education and further vocational training must provide youngsters with general skills which they will need in the knowledge-based economy. It is necessary for them to know how to learn.

In the article we are trying to answer the following ***questions***:

- how are elementary philological education and its highest level – educating in philology – perceived?
- what essence of technological education come to light in the knowledge society?
- how do the experts value the cohesion of educating in philology and technological education in a gymnasium of technological type.

The main methodological principles of the article are based on the notion of educating in philology and philological education (Laužikas, 1943, 1993; Jovaiša, 1993, 1995; Alaunienė, 1994; Bankauskienė, 2002), on the notion of technological education (Ramanauskaitė, Stankevičienė, 2001; Ramanauskaitė, 2002) and on the conceptions of the knowledge society and life-long learning.

The article consists of three parts. In the first part the authors substantiate that educating in philology is the highest component of philological education. The authors emphasize the criteria of educating in philology in the knowledge society. The requirements, of the knowledge society for the technological profile gymnasiums, in which it is possible to find the display of educating in philology and technological education, are presented in the second part. The researches using the method of interview emphasize the opinion of the experts of the cohesion of educating in philology and technological education.

The ***methods*** used in this work are scientific literature analysis and interview.

1. Educating in philology in the knowledge society

When Lithuania is turning towards the knowledge society, it become very important and relevant in the whole Europe.

> One of them – "to guarantee universal and permanent access to learning, that the skills, which are necessary for taking part in the life of knowledge society, would be obtained and renewed" (Memorandum of life-long learning, 2001, p. 17).

First of all, a person obtains these issues through his/her native language. That is why, it is natural, that the particular attention should be paid to good knowledge of the native language, to the reading and writing skills. New and the most important skills, which are included into the conclusions of the European Board (Lisbon), point out the following priorities:

- ITT;
- Foreign languages;
- Technological culture;
- Business and social skills.

The importance of not only separate disciplines, but multidisciplinary knowledge and competences emphasized. So, the general skills often match with professional social skills in the sense of their content and function.

Philological educating is important not only in a comprehensive school, but in the person's future too, when he/she seeks for professional career. The notion of philological education is the most general conception, in which it is possible to see the essence of philological teaching, development and educating. Educating in philology is the highest level of philological education, in which integration of linguistic, literary and folk knowledge takes place. Such integration helps a person become more cultural, it creates possibility to develop his/her competence and intelligence, which are based on concrete and abstract thinking, to arise and solve problems. This integration enables to form wide orientation in the fields of cultural, artistic and social life; it allows a person to show his/her diverse and purposive communication (Bankauskienė, 1999a).

Educating in philology differs from the elementary philological teaching. Elementary philological teaching forms the bases of speaking, reading and writing. This is the necessary background for further and more complex philological education of a personality, which leads to educating in philology. According to Bankauskienė (1999b), elementary philological teaching develops the feeling of language, forms the intuitive and conscious experience of linguistic activity, skills of creative work, and forms the bases of written and spoken language. Meanwhile the whole view of the structure of a language is started to be formed during the process of educating in philology. A person is taught to perceive a linguistic system, its socio-cultural context. He/she learns to write correctly, to put the marks of punctuation in a text, develops skills of correct pronunciation and accentuation, skills of correct use of words, their forms and constructions. A person becomes acquainted with the most significant facts and phenomena of the language and literature. One of the criteria of educating in philology could be person's thoughtful ideas, which could be expressed in written or spoken form, using complex, stylish and sometimes artistic sentences. A person learns not only to retell and recall information, but also to analyze, interpret it, create a text, which is suitable for various situations, orient himself/herself well, use professional communication purposefully.

Elementary philological teaching takes place on the ground of one (mostly native) language. The criteria of educating in philology could be the person's ability to translate from/into the native language, perceive the similarities and differences of different linguistic systems. The person who is philologically educated, has knowledge and developed skills, is acquainted with treasures of his/her native language, literature and folklore. He/she can use these things in daily, practical activity and present them in theoretical level. Educating in philology develops thinking, emotions, imagination, intuition – the accumulation of personal characteristics, develops and enriches a personality. G. Viliūnas (1998) points out, that nowadays it is necessary to prepare pupils to live in a new culture, a new situation of values and communication. The Lithuanian language also has to carry out the role of continuation and preservation of traditions. It is not important to fight in the cultural market, but to prepare pupils for the life in the new society, to be able to combine verbal and visual language. Therefore the development of communication skills in a technological gymnasium is one of the most important goals in the process of personality development.

L. Ruseckienė (2001) states that the oratorical skills are also important, because a pupil of a technological gymnasium ought to manage to read, speak distinctly, prepare presentations, discuss. Therefore the teaching of elocution is a poly-functional and effective means of pupils' aesthetic education and development of emotions. It ennobles pupils' emotions, forms artistic taste, gives a possibility to communicate with work and audience creatively. Elocution is also a way to develop the sense of language: a pupil, who actively works with a text, remembers figurative words and expressions, and possesses reading skills, can make his/her speech more distinct.

Thus, educating in philology gives possibility to form a wide humanitarian view of general perception of the world to an individual, it helps to find and select appropriate information, which is necessary for verbal and written communication or searching for special knowledge of some field and putting it purposefully into the constantly changing world of knowledge and technologies. R.A. Slaughter's "New thinking for new millennium" (1999) pays great attention to humanitarian sciences, which teach a person to meet the future. J. Ogilvy (1996) points out the same ideas, emphasizing the importance of a language for culture, particularly its significance in the new millennium. Schwanitz (2001) emphasizes the significance of a language, as an instrument for better communication. Knowledge of linguistic forms, regulations and various possibilities of expression and use of these great possibilities are one of the main cultural practices, which enables a person to communicate with others and opens the way to treasures of the universal culture. We construct our own reality through language, the world of meaning is being created through language, this world helps people to communicate. *The mastering a language is a crucial condition, for a person to move freely in society*, because language style and vocabulary adjust to various environments and social fields. A person, who does not know "what he/she must say", feels himself/herself socially constrained. A person, who denies a language and advocates the priority of sincere purpose originality and deepness of unimpressionable things, becomes suspicious as a person, who slights what he/she does not know. A. Maceina, supporting L. Weisgerber, supplements this idea. He asserts that the purpose of teaching native language is a person, who is ready to participate in common affairs, which arise because of relationship with the community, using the same language. In other words, the purpose of native language teaching is not the knowledge of the language, but person's intention to take part in the life of his/her nation.

A. Maceina (1991, p. 179), talking about the significance of native language teaching, emphasizes three main goals:

- to define person's thinking in the aspect of nationality;
- to provide a person with a national world-view;
- to involve a person in the development of national culture.

The author concludes that native language teaching turns into creative intellectual development of personality.

Pupils' *educating in philology in a technological gymnasium is a particularly important part of general personality development, because a lot of personal abilities - philological possibilities of a person, ability to think, to use native language, make conclusions, evaluate a situation critically, analyze creatively and interpret – depend on educating in philology*. Educating in philology gets into other fields of human cultural activity and makes people stronger. A.J. Greimas (1991, p. 41-43) confirms this, by asserting that "a language and linguistics or science of language often serves as a model for other branches of science. (…) A language as a system of values, a system of meaning, is one of the most characteristic and the most significant systems in human life, because it is possible to translate all other signs' systems into natural languages. (…) Finally, a language is the most characteristic semiotic system, because it is an example that semiotic systems are functioning as socially compulsory systems. (…) Thus linguistics occupies a certain place of pilot research in the system of human sciences. Other human sciences have been borrowing methods from linguistics and applying them into special fields for more than twenty years".

After the analysis of scientific literature (Bankauskienė [1999a, 1999b], Daujotytė [2002], Greimas [1991], Maceina [1991], Ruseckienė [2001], Viliūnas [1998]), it is possible to mark out the following criteria of educating in philology, which are proper for pupils' philological education in the knowledge society:

1. A native language is the most valuable national wealth, one of the main conditions of nation existence.
2. The place of the Lithuanian language in the family of Indo-European languages and in the Baltic area.
3. Relations between mythology, folklore and literature.
4. Mastering of the rich vocabulary of the Lithuanian language, ability to compare native language with other languages, to point out the meanings of international words in a native language.
5. Ability to express one's ideas properly in a written or verbal form.
6. Orientation in the totality of world and Lithuanian literature, perceiving the role of writers' culture developers.
7. Ability to analyze and interpret literature works, perceive the main moments of their content and artistic expression and apply them in practice.
8. Ability to gather facts and prepare a presentation, read it distinctly, take part in discussions, debate, and prove one's opinion and present skills of rhetoric.
9. Knowledge acquired by extensive reading, which helps to evaluate belles letters and other texts and prove one's assessments rationally.
10. Ability to communicate, perception of language culture, ability to use and apply various terms.

11. Knowledge of at least one foreign language, ability to translate elementary texts of fiction/non fiction and edit them correctly, according to the main rules of language.
12. Ability to combine verbal and visual language, to adjust to living in a new cultural environment, in its new situation of values and communication.

2. Technological educating in the knowledge society

At the end of the XXth century we started to apply the term "technologies" not only to certain ways of production, but to other processes, too e.g., information technologies, humanitarian technologies, etc. The diversity of technological structures and relationships depend on the diversity of technologies. Ellulis, Schelsky (Huning, 1998) point out three fields, where technologies are used:

- production machinery, production of material goods;
- organization technology, social relationship regulation, methods;
- humanitarian technology, the purpose of which is to change, clean and form human internal world.

We can see that one of the aspects of ***"technologies"*** is oriented to wider humanitarian and technological interaction, which results in educating.

Pacey (1984) emphasizes the conceptions of "technology" and "technological practice". The author separates technical, organizational and cultural aspects of technology.

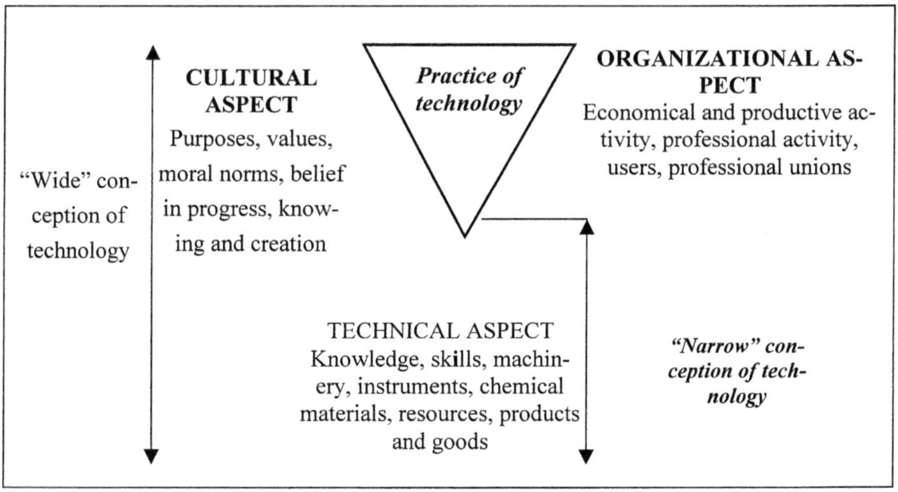

Figure 1: "Wide" and "narrow" perceptions of technology (Pacey, 1984)

In the Standards of Technological Literacy (2000), which were prepared by the International Association of Technological Education (ITEA), technology is defined in two aspects:

1. The innovative activity of humanity, which generates knowledge and processes, develops systems, solves problems and extends abilities of humanity.
2. The innovative change or modification of natural environment, to meet the needs and wishes of humanity.

Technology is modification of natural world. Such modification could be done according to the needs of humanity. The word "technology" in the ancient Greek language meant art, creativity or trade. The direct meaning of "technology" is the act of doing or creativity, but in the wider meaning it is changing different systems of processes or knowledge, which are used by people, with a purpose to meet their needs and wishes and to extend their abilities.

Humanity is turning into the information or knowledge society (Drucker, 1993), where knowledge becomes more important. According to the author, an educated person is the most significant in the knowledge society. Therefore nowadays the necessity of technological education is relevant in the institutions of education, because technique, as a means and way of material activity, is superior. The purpose of technological education, according to Huning (1998), is not only to develop skills of using technologies, but to develop ability to learn, to solve problems, communicate and cooperate, be ready for compromises, develop creativity and critical thinking.

The content of the term "technology" is changing in education: it becomes wider, covers the essential educational problems. M. Foucault (Lukšienė, 2000) pointed out four types of technologies, which are "matrices of practical mind":

- Technologies, related to production (manipulation with things);
- Technologies of signs' systems (manipulation with signs);
- Technologies of authority (manipulation with individuals);
- Technologies of oneself (I) or activity of individuals, directed to one's own thoughts, behavior, which enables them to change and reach the state of wisdom and perfection.

The invasion of technological innovations into life and activity arises specific purposes of technological education: it enables learning people to become technologically educated and literate, to be able to use new technologies consciously. But it is also necessary to evaluate responsibility and possible consequences.

"Computer literacy" (Eisenberg, Jonson, 1996), "technological literacy" (Fiddy, 1987), "information culture" (General Programmes, 1997) are some of the great number of terms, which define ability to orient in the chaos of information, to use successfully modern technologies in daily activity.

In educating pupils technologically, it is especially important to teach them to use technology of information and communication reasonably for competent decision-making and presentation of results. According to Jucevičienė (1997), the structure of usual knowledge is changing in the modern knowledge society, because knowledge and technologies are constantly developing, and work is changing. Baskas (2003) agrees with this idea. He states that modern technologies must make learning faster to enable a pupil learn at his/her personal speed and finish studies during the period of time, that is stipulated by his/her efforts and abilities. In that case every pupil would learn using his/her own way of learning. Technological educating creates possibility for pupils to become acquainted with the processes and knowledge related with technology, which is necessary for problems solving and development of human capacities. Technological educating is "about technology" and it is realised by the means of technology (Pannabecker, 1995).

Technological educating orients pupils to study processes, which help to create new technologies (critical thinking, ability to solve problems), fields of technology and the influence of technology on society and environment (Wright, 1992, quoted from Wicklein, 1997).

According to Ramanauskaitė (2002), technological educating enables learning people to become technologically literate, knowing how to obtain and renew knowledge, capacities, that help to understand, use and evaluate technologies. One must have a positive opinion on constant development of technologies towards "high" technologies, aiming for better life quality of society and its members' life. The author pointed out the following criteria of technological educating:

- to be able to understand and use technologies in one's own environment;
- to have a positive opinion on constant development of technologies;
- to be able to evaluate the significance of technologies ("high" technologies included);
- to be able to evaluate possible consequences of technologies for the quality of one's own and others' life;
- to know how to solve problems, think critically, creatively and cooperate.

According to Lukoševičius (2002), the constant search, requesting resources of human spirit and intellect, for volitional harmony that is called, as harmonious development could be a stabilizing factor. The notion of harmonious development rose from the determination of relationships between technologies and nature protection; this means that technologies could not develop at the expense of nature. The conception acquired the aspects and dimensions of time, space, and physical state: it is not possible to develop at the expense of our own future (time dimension), of our neighbors (space dimension), of our spirit (humanitarian dimension) or of our material welfare (physical state dimension).

Harmonious development of technologies and civilizations requires constant search for dynamic balance, changing the methods and ways of balance renewal in each stage of development. The society pays for the development of technologies in certain level of risk, and the harmonization of that risk requires control: prevision, identification, and evaluation of risk, information about it and means of its reduction. Educated society and harmonious policy of science and technologies are the conditions for the whole mastering of such risk (ibid). The guarantee, which can protect from negative consequences of technologies, is the competences of people and institutions, aiming at wide and deep knowledge, the complex of research and technological development.

According to Krištopaitis (1992, p. 7), "the intellect, which society will be able to develop, will affect events in socium". A knowing, thinking, educated society that has mastered technologies and innovations, can develop perspective, strong and stable economy, ensure social welfare, even if it does not have enough material resources (Kriščiūnas, Lukoševičius, 2002). The significance of educational system and relevant goals of the present period arise: to prepare young generation to live in the conditions of technological revolution, to form principles for universal life-long learning, as according to Mureika (1998), it would be a tremendous mistake to think that it is enough to be a good specialist or an honest person. These issues, taken separately, are not valuable preconditions for technological activity. Besides, all level Lithuanian educational programmes are not enough oriented towards humanitarian content, development of new branches of science and the place of technology in the modern culture. Aiming at coping with this misbalance, technological type gymnasiums are being founded. These gymnasiums are supposed to develop a good specialist and a cultural, spiritual, intellectual personality, who can work and adjust to constantly changing society.

In its sub-consciousness, society is still of the opinion that humanitarian and theoretical knowledge is more significant than activity oriented to practical success (Rapp, 1998), but in 1959 an English writer and physicist C.P. Snow noticed that the abyss between "literature-humanitarian" and "technical-natural sciences" intelligence of standpoints to culture and education still exists. Modern education declares the necessity and aspiration of diverse education. A person should manage to live in the culture of both humanitarian and natural sciences in the environment of information and knowledge society. Only the perception of the past and the present helps to develop self-perception, necessary in our days, understand the space of one's own freedom, responsibility and activity. Thus it is necessary to enrich technological educating with humanitarian disciplines, and include the analysis of technological culture, its structure, significance and consequences into humanitarian disciplines (Huning, 1998). As V. Adamkus (President of Lithuania) stressed in his presentation, made on 24 September 2003, it is very important to create conditions for harmony between humanitarian, social aspect and technologies in the process of globalization. Then technologies will not contradict human values.

In this context, it is necessary to emphasize the significance of pupils' computer literacy. As Otas and Telešius (2003) stated, there is a clear tendency for the age of information technology users to become younger. So it is necessary for young people to be clearly motivated for the use of information technologies. And the "General Standard of Pupils' Computer Literacy" (2001) points out the following principles:

- to understand the significance of learning for society life;
- to perceive the significance of information technologies for the realization of life-long learning principles;
- to perceive the importance of information technologies in professional activity and daily life;
- to understand not only the advantages of information technologies, but the dangers of universal computerization;
- to perceive that implementation of information technologies must be based on traditional values of a country and a person;
- to understand the significance of correct usage of the Lithuanian language in e-space and other technologies;
- to use technologies properly, to perceive the importance of ethics and communication norms in the field of technologies.

It is clear that valuable principles of technological education in technological gymnasiums complement ideas, which define an educated young person. The importance of influence of educating in philology in technological educating comes to light in technological gymnasiums. "The Strategy of Information and Communication Technology Implementation in Lithuanian education" (2000) points out that pupils must study in their native language. Thus, the importance of the Lithuanian language acquires a prior meaning in modern technological education, and educating in philology of pupils creates a possibility for them to achieve better results in technological education.

But it is necessary to stress that there is no state policy in Lithuania concerning the use of the Lithuanian language in computer programmes. It is not only a problem of educational system, but related with state institutions and private persons. Today we do not have any law or legal act, which would define the use of the Lithuanian language in e-space. The law of the state language confines itself only with general statements, and the inspection of the state language does not keep to the same requirements, which are applied for media and public notes for the Lithuanian language in e-space. The development of the Lithuanian language is lacking material of information technology – dictionaries, reference books. (Analysis of Implementation of technologies in the Education of Lithuania, 2002). ***So, the harmony of technology and philology is inevitable.*** Daujotytė (2002b) asserts that technological institutions of education should be the space of culture and humanitarian spirit. T.S. Eliot and other scholars foresaw that humanism would become stronger in XXIst century. But humanism must break out its way through heavy barriers. Technical, narrowly strong thinking applies its criteria in humanism. But a person lives more poetically than scientifically. If a person does not want to loose himself/herself, he/she must more often be in the environment of a poem, music, nature.

Philosopher Arvydas Šliogeris, who has technological university education, states (quoted Daujotytė, 2002b, p.7) that "the aim of humanism arises from the need, which is felt instinctively, to find an alternative for technological sterilization of a person and the world and finally the alternative for catastrophical destruction of reality". According to the opinion of the author, new technical and communication ideas will be raised by people, whose thoughts, emotions and imagination will be developed by creation, based on literary education, because things, which are not supposed and programmed, take place only in creation. Only creation encourages creativity. Experience of literature enables to perceive the language of technologies. The partner of a computer is allowed to be more active, too. A view also has its own language. If a person gets less experience and language from literature, it will be more difficult for him/her to understand a screen. As Daujotytė (2002b) asserts, humanistic culture bases itself on three main props: reading, writing, and speaking. The efforts and capacities of understanding and self-perception are fundamental. A person understands and is understood. Both things are not possible without language. To be understood means to develop the abilities to understand, not only to know narrowly. Therefore the dialogue between technologies and humanities is inevitable. Summarizing the expressed ideas, it is possible to identify the criteria of technological education:

1. The perception of influence of technology on culture and society, knowing how to use technological systems, information and computer programmes properly.
2. The ability to use technologies, especially applied computer programmes, for the search, selection and evaluation of information of various sources.
3. The knowledge, obtained in the field of technique and processes, which are related with it, the ability to apply knowledge into practice.
4. The knowledge of technological terms, clear expression of these words, using correct language.
5. The application of technology for creation, aiming to learn better and more productively, and development of positive opinion of life-long learning.

6. All-round cognition of the world, the use of technology for strategy preparation, task fulfillment and making competent decisions.
7. The ability of meeting the challenges of the modern technologies, emphasizing the importance of technical thinking and experiments.

3. The research of experts' opinion about the cohesion of educating in philology and technological educating

When criteria of educating in philology and technological educating were pointed out, and the necessity of influence of philology on technological educating was noticed, the opinion of pedagogues on how they evaluate cohesion of educating in philology and technological educating in new type institutions – in technological gymnasiums –, is of particular significance. The interview method was used and five questions were presented to the experts.

Five experts from philological and technological field were invited. They agreed to express their opinion, because they know the problems of philological and technological education. The experts were the professors and associated professors of Kaunas University of Technology, the pedagogues, who are working in Kaunas and have scientific doctor degrees, and the director of Kaunas Building Training Centre (the first technological type gymnasium classes were founded in this centre). The experience of pedagogical work of the selected pedagogues differs from 14 to 44 years.

The expert research helped to understand, how pedagogues with big experience of pedagogical work, perceive educating in philology and technological educating, where they can see the cohesion of these issues, what connections between educating in philology and technological educating they are able to find. It was interesting, what they think about technological gymnasiums' ability to develop a person, educated philologically and technologically, and capable to work and live in the knowledge society.

Answering the question, ***what educating in philology is***, all experts highly evaluate the significance of educating in philology in the knowledge society, but elementary things, such as scholarship, literacy, are enriched with new perception, which enables to communicate in concrete space and time (V expert), with knowledge of rhetoric (III expert), with ability to combine verbal and visual language. This complements Viliūnas' idea (quoted from Geda (1998)) that in educating pupils philologically, it is necessary to adjust to new reality and to prepare pupils to live in a new culture. The Lithuanian language must carry out the role of *continuation and preservation of traditions*.

As the majority of the experts are working in technological profile institutions of education, and notions of *educating in technology, technological educa-*

tion, technological literacy are often used in practical activity, they were asked to define these notions, pointing out the most significant ones.

During the discussions the specialists faced some misunderstanding concerning the above-mentioned concepts. Not all experts could define them and find essential differences between the concepts.

Expert N°1: "technological education, as a process, should begin in pre-school. Personality and technological culture, which we are especially lacking today, would be developed little by little. Talking about technological education, it is necessary to emphasize the necessity for everyone to work with a computer."

Expert N°2: "Educating in technology is striving for the top of professional activity (this suits for philologists too. This is the most important thing for a person)."

Expert N°3: "Educating in technology is ability to work with abundant technological equipment. The best humanitarian should be able to work with modern technologies. Computer literacy is various work with a computer. Educating in technology is the main and the most important concept."

Expert N°4: "Technological education is integration of systematic knowledge with technological knowledge and their practical use. Technological literacy – motivated use of one's own knowledge and practice, projecting, planning, organizing, producing and evaluating. Educating in technology is ability of thinking, creative and cultural personality to analyze environment, things and processes objectively, and ability to master simple processes of production and realization of situations. Educating in technology is the main and the most important notion."

Expert N°5: "Technological education is a process, when an ordinary personality turns into an educated personality. Educating in technology is the consequence of the process of education. Technological literacy is knowledge, abilities, skills, valuable ground, norms and regulations, which enable a personality to live and act in certain conditions. The widest and the most essential conception is educating in technology."

These ides conform to Croft's (1991) definition of a technologically educated person, who wants to have computer literacy: ability to evaluate technology, to make a certain decision; to develop the main skills, which are necessary for solving technological problems; ability to use technologies; ability to apply knowledge, means and skills for making society life better. That would help to achieve the aim of technological educating, which, according to Johnson, Gatz and Hicks (1997), is to help learning people to understand their technological future (quoted from Ramanauskaitė (2002)).

After the analysis of the notions of educating in technology and technological education, we asked the experts to answer the question, ***if educating in technology and technological education are compatible.***

Expert N°1: "Yes, these things are compatible, but one of them cannot exist at the expense of the other. It is necessary to be able to speak, write, perceive necessity of books reading, because only there one can see the beauty of the Lithuanian language. However educating in technology cannot be included in university programmes, because that education should be finished in the secondary school."

Expert N°2: "Yes. A nation is unhappy, if educating in technology has not any connections with educating in philology. It is possible to annihilate oneself without humanitarian culture."

Expert N°3: "Of course, they are compatible. The technologist cannot be "dry", he/she should have the bases of humanitarian education, because only such people are more inspired."

Expert N°4: "Yes, they are compatible, because intellect and practical activity are connected with things. The whole, educated personality develops, only interacting with all values of civilization."

Expert N°5: "Yes, they are compatible, because, in my opinion, general educating of personality depends on coordination between educating in philology and technological education."

All experts confirmed that educating in philology and technological education are compatible things, but they stress the scale of the cohesion differently. *Expert N°1* thinks that one of them cannot exist at the expense of the other. He is sure that educating in philology is not necessary in technological type higher institutions. That contradicts the results of Barzdžiukienė (1999) where she showed that the present system of humanitarian education in technological type higher schools is not able to realize its main aim – to develop a good specialist and a cultural, inspired personality, an intellectual. If educating in philology is left behind the doors of a higher school, further on only the training of specialists will be accentuated. Teachers of all specializations (the people, who prepare the programmes (modules) of teaching and educating and deans of faculties) should remember that aim. *Expert N°3* expert also contradicts this idea. He cannot agree that a technologist must be only a "dry" technological specialist, thus educating in philology helps to seek spirituality. Aspiration to responsibility and harmony, striving for cohesion between educating in philology and technological education are also important (*Expert N°2*). These ideas correspond to Lukoševičius' (2002) statements that it is possible to reduce the risk of technological development, striving for totality of competence, deep and wide knowledge and both individual and institutional research. Of course, humanitarian sciences are important here too. The remarks of *Experts N°4 and N°5* are also significant. They emphasize the harmony of educating in philology and technological education that develops general education of personality.

All experts agreed that educating in philology and technological education are compatible issues, so we tried to find out, ***what connections between educating in philology and technological education could be noticed.***

According to *Expert N°1*, one of such connections is self-perception in culture, but not at the expense of professional knowledge. *Expert N°1*emphasized a negative influence of mass culture and television, because they mangle the beauty of a language.

Expert N°2 stated that there is no balance between philology and technology, and technologies suppress philology more and more. But it is possible to notice the help of technologies to philosophy and vice versa.

Expert N°3 perceives the dangers of mass culture for harmony of educating in philology and technological education. The expert points out the abundance of information, computer technologies suppressing reading. In his opinion, a use of the English language is exaggerated, so in striving for cohesion of philology and technology it is necessary to create Lithuanian terms. Expert **N**°3 indicated the significance of rhetoric, because neither pupils nor students can speak fluently. One of the means, that helps to develop oratorical art, is the project method. According to *Expert N°4* in the change of the global approach towards technological education the connections between cultural intellectual self-expression are necessary, because a "technocrat" could not be called a personality, if his actions are automatic. The development of language culture, which is necessary for a philologically and technologically educated person, is an important connection.

Expert N°5 pointed out that the essential influence of educating in philology on technological education is an ability to express one's ideas reasonably and coherently, ability to communicate and "act" in conditions of technological "highways".

So, the experts' opinions showed that various connections emphasize positive and negative influence in striving for cohesion of educating in philology and technological educating.

The display of negative connections:

- Extension of mass culture and "audiovisual" (television) situation suppresses the beauty of a language, mangles imagination (Expert No1 and 2);
- Technology suppresses philology more and more (Expert No2);
- Abundance of information, computers keep from reading; it is difficult to orient in the flow of information, what to read and how to read; pupils' readiness to participate in the culture market, and to develop it (Expert No3);
- Exaggerated use of the English language (Expert No3).

The necessary preconditions of educating in philology, that positively influences technological educating:

- Creation of Lithuanian terms thesaurus (Expert N°3);
- Perception of basic rhetoric (Expert N°3);
- Development of cultural intellectual self-expression (Expert N°4);
- Development of language culture (Expert N°4);
- Communication and ability express one's ideas reasonably, using rich vocabulary, perceive concepts (Expert N°5).

Experts' opinions emphasized positive and negative preconditions of educating in philology on technological educating. This is natural, because the constantly changing and diverse world, development of technique, new technological possibilities (television, films, internet etc.) keep pupils from philological activity and encourage more interesting and requiring less intellectual efforts activity.

It is clear that new technological type gymnasiums are able to educate a person philologically and technologically, if:

- Pupils, interested in technologies, having high learning motivation, not identifying technological gymnasiums with schools of vocational training, will come to study at technological gymnasiums;
- Teachers of general education and technologies (professions), willing to improve their qualification, working productively, perceiving requirements of high level teaching, will work at these educational institutions;
- Teaching programmes, emphasizing academically level, will be prepared. They should not copy foreign educational experience, but adapt it to our economical, social and cultural environment;
- There is a strive for close integration of general and technological subjects;
- These gymnasiums will be able reach their goals.

Summarizing the opinions of experts, it is possible to affirm that cohesion of educating in philology and technological educating is noticed in new technological type gymnasiums and these educational institutions can be able to create proper conditions for pupils' cohesion of educating in philology and technological educating.

Conclusions

1. Elementary philological teaching provides a person with the basics of language, reading and writing. They are the necessary background for further more complex philological educating of personality, which leads to educating in philology. Educating in philology gives a possibility to compose a wide humanitarian view of general perception of the world, helps to find and select the necessary information, which is necessary in verbal and written communication or searching for special knowledge of some field and applying it reasonably in the constantly changing world of knowledge and technologies.
2. Technological educating gives possibility for learning people to become acquainted with the processes and knowledge related with technology. They are necessary for problem solving and enhancing human abilities. When educating pupils technologically, it is especially important to teach them to use information and communication technologies properly for competent decisions making and achieving results.
3. The native language, which is used in processes of education and development, becomes a priority, because pupils' educating in philology gives them possibility to reach better results in technological educating:
    - In the knowledge society, avoiding narrow specialization, it is necessary to create all-round conditions for cohesion of educating in philology and technological educating in technological type institutions;
    - All experts confirmed the possibility of cohesion of educating in philology and technological educating;
    - Positive and destructive connections of cohesion of educating in philology and technological educating were highlighted during the research;

- According to the opinion of experts, compatibility of educating in philology and technological educating is fixed in new technological gymnasiums, and these educational institutions are able to create good conditions for pupils' cohesion of educating in philology and technological educating.

References

ABRAITYTĖ, L.: Literatūrinis ugdymas šiandienos mokykloje – naujos galimybės ir paviršutiniškumo grėsmė. Mokykla 7(1999).

ADAMKUS, V.: Žodinis pranešimas (2003 09 24).

ALAUNIENĖ, Z.: Lietuvių klaba ir literatūra: tradicija ir naujovės mokykloje. Gimtasis žodis 1(1994).

Baltoji knyga. Profesinis rengimas. (1998). Projektas. Vilnius.

BANKAUSKIENĖ, N.: Aukštesniųjų klasių moksleivių filologinis išprusimas šiuolaikinio gimtosios kalbos mokymo požiūriu. Neskelbta daktaro disertacija. Kaunas 1999 (Vytauto Didžiojo universitetas)

BANKAUSKIENĖ, N.: Filologinio prusinimo ištakos Lietuvoje. Socialiniai mokslai 4(1999)21. Kaunas (Technologija).

BANKAUSKIENĖ, N.: Filologinio prusinimo ištakos. Socialiniai mokslai 2(1998)15. Kaunas (Technologija).

BANKAUSKIENĖ, N.: Filologinio ugdymo virsmas Lietuvoje integracinių kontroversijų fone: edukacinis žvilgsnis į XVI amžių. Socialiniai mokslai 3(1999)24. Kaunas (Technologija).

BANKAUSKIENĖ, N.: Filologinio ugdymo bei išprusimo tendencijos Lietuvos švietimo reformos kontekste: studentų požiūrių lyginamoji analaizė. Socialiniai mokslai 1(2001)27. Kaunas (Technologija).

BARZDŽIUKIENĖ, L.D.: Būsimasis inžinierius ir humanitarinė kultūra. Vilnius 1999 (VU).

BASKAS, A.: Dirbtinio proto įtaka mokymui. Kompiuterininkų dienos – 2003: LIKS suvažiavimo medžiaga. Vilnius 2003 (Žara), pp. 285-291.

BRUNEVIČIŪTĖ, R.: Klasikiniai humanitarinio ugdymo pagrindai. Lotynų klaba. Monografija. Kaunas 1999 (Technologija).

BRUNEVIČIŪTĖ, R./VEČKIENĖ, N.: Humanitarinių mokslų ir mokymo dalykų vieta šiuolaikiniame ugdyme. Edukologijos idėjos Lietuvos švietimo sistemos modernizavimui. Monografija. Kaunas 1998 (Technologija). pp. 47-59.

BRUNEVIČIŪTĖ, R./VEČKIENĖ, N.: Klasikiniai humanitarinio ugdymo pagrindai: šiuoolaikinės Europos tendencijos. Edukologijos idėjos Lietuvos švietimo sistemos modernizavimui. Monografija. Kaunas 1998 (Technologija). pp. 413-424.

DAUJOTYTĖ, V.: Kultūros šalys ir nuošalės. Kaunas 2002 (Technologija).

DRĄSUTAVIČIENĖ, D.: Literatūros kurso tęstinumo klausimai. Gimtasis žodis Nr.1(2002)157.

DRUCKER, P.: Post – capitalist Society. Oxford 1993 (Linacre House, Jordan Hill).

GREIMAS, J.: Bendrosios semiotikos problemos. Iš arti ir iš toli. Vilnius 1991 (Vaga).

HUNING, A.: Inžinieriaus kūryba/Sudarė J. Mureika. Technikos filosofijos įvadas. Vilnius 1998 (Alma littera), pp. 33-174.

Informacijos ir komunikacijos technologijos diegimo Lietuvos švietime strategija. Kompiuterininkų dienos – 2001: LIKS suvažiavimo medžiaga, praktiniai, metodiniai ir informaciniai pranešimai bei dokumentai. Lietuvos kompiuterininkų sąjunga. Vilnius 2001 (Žara), p. 184.

Informacinių technologijų diegimo Lietuvos švietime politikos analizė. Galutinė ataskaita. Vilnius 2002 (Demokratinės politikos institutes).

JACKŪNAS, Ž.: Ugdymo integracijos metmenys. Lietuvos švietimo gairės. Vilnius 1993 (Valstybinis leidimo centras).

JANIŪNAITĖ, B.: Šiuolaikinio lyginamosios edukologijos mokslo pasiekimų panaudojimo galimybės modernizuojant švietimo sistemą. Daktaro disertacija. Kaunas 1998 (Kauno technologijos universitetas).

JOHNSON, S.D./GATZ, E.F./HICKS D.: Expandig the Content Base of Technology Education: Technology Transfer as a Topick of Study. Journal of Technology Education. Vol. 8(1997)2.

JOVAIŠA, L: Edukologijos įvadas. Kaunas 1993 (Technologija).

JOVAIŠA, L: Hodegetika. Auklėjimo mokslas. Vilnius 1993 (Agora).

JUCEVIČIENĖ, P.: Ugdymo mokslo raida: nuo pedagogikos iki šiuolaikinės edukologijos. Kaunas 1997 (Technologija).

KRIKŠTOPAITIS, J.A.: Žmogus mokslo ir valstybės varžtuose. Filosofija, sociologija 1(1992)7, pp. 5-7.

KRIŠČIŪNAS, K./LUKOŠEVIČIUS, A.: Žinių ekonomikos Lietuvoje link: tyrimų ir technologijų plėtros strategija. KRIŠČIŪNAS, K. ir kt. (eds.). Mokslas, technologija ir visuomenė: harmoningos raidos paieškos. Kaunas 2002 (Technologija), pp. 241-270.

KUOLYS, D.: Kryptingos kaitos galimybės. Dialogas 37(1999)389.

LAUŽIKAS, J.: Pedagoginiai raštai. Kaunas 1993 (Šviesa).

LEVIS, T.: Content on Process as Approaches to Technology Curriculum: Does it Matter Come Monday Morning? Journal of Technology Education. Vol.11(1999)1

Lietuvos bendrojo lavinimo mokyklos bendrosios programos. Vilnius 1997 (Leidybos centras).

LUKSIENE, M.: Tautinė mokykla. Lietuvos vidurinės bendrojo lavinimo mokyklos koncepcija.

LUKSIENE, M.: Jungtys. Vilnius 2002 (Alma littera).

LUKOSEVICIUS, A.: Mokslo, technologijų ir inovacijų plėtra visuomenės gerovei. KRIŠČIŪNAS, K. ir kt. (eds.). Mokslas, technologija ir visuomenė: harmoningos raidos paieškos Kaunas 2002 (Technologija), pp. 205-241.

MACEINA, A.: Gimtosios kalbos mokymo prasmė. Tautinis auklėjimas. Kaunas 1991 (Šviesa).

MATULAITIENĖ, S.: Skubėkime apdairiai. Gimtasis žodis 2(1991), Vilnius.

Memorandum of life-long learning. Vilnius 2001 (LR Švietimo ir mokslo ministerija, Lietuvos suaugusiųjų švietimo asociacija).

MUREIKIA, J.: Techninė klutūra, žmogus ir technikas filosofija. Technikos filosofijos įvadas. Vilnius 1998 (Alma littera).

NAUCKŪNAITĖ, Z.: Iškalbos mokymas. Kaunas 1998 (Šviesa).

OGILVY, J.: Futures Studies and the Human Sciences: The Case for Normative Scenarios. In: Slaughter, R.A. (Ed.): New Thinking for a New Millenium. London and New York 1996 (Routledge), pp. 26-83.

OTAS, A./TELEŠIUS, E.: Technologinio ir profesinio kompiuterinio raštingumo ugdymo problemos. Informacijos mokslai t.26(2003), pp. 54-60.

PACEY, A.: The Culture of Technology. Cambridge 1984 (The MIT Press).

PANNABECKER, J.R.: For a History of Technology Education: Contexsts, Systems and Narratives. Journal of Technology Education Vol. 7(1995)1.

RAMANAUSKAITĖ, A.: Technologinio ugdymo prielaidos ir jų raiška Lietuvos vidurinėje mokykloje. Daktaro disertacija. Kaunas 2002 (Kauno technologijos universitetas).

RAPP, F.: Techninis ir humanistinis racionalizmas. Sudarė Mureika, J. Technikos filosofijos įvadas. Vilnius 1998 (Alma littera), pp. 335-344.

RUSECKIENĖ, L.: Literatūros pedagogikos praktikumas. Vilnius 1995 (Žodynas).

RUSECKIENĖ, L.: Literatūros pedagogikos studijos. Vilnius 2002 (Gimtasis žodis).

SALIENĖ, V.: Raštingumo samprata vidurinėje ir aukštojoje mokykloje. Aukštojo mokslo sistemos ir didaktika. Konferencijos pranešimų medžiaga. Kaunas 2001 (Technologija).

SAUTER, W.: Vom Vorgesetzten zum Coach der Mitarbeiter: Handlungsorientiere Entwicklung von Fuhrungskraften. Weinheim 1994 (Deutscher Studien Verlag).

SCHWANITZ, D.: Ką turi žinoti kiekvienas išsilavinęs žmogus. Vilnius 2001 (Tyto alba).

SKERSIENĖ, I./BANKAUSKIENĖ, N.: Technologinio profilio gimnazijos moksleivių leksinio išprusimo tyrimas. Aukštojo mokslo sistemos ir didaktika. Konferencijos pranešimų medžiaga. Kaunas 2001 (Technologija).

SLAUGHTER, R.A.: Critical Futures Study as an Educational Strategy. In: SLAUGHTER, R.A. (Ed.), New Thinking for a New Millenium. London and New York 1996 (Routledge), pp. 137-154.

SNITKA, V.: Informacinė ar žinių visuomenė? Studijų, mokslo ir švietimo informacinis leidinys Nr. 30(2001).

ŠERVENIKAITĖ, N.: Dalykiškesnio pokalbio elgesys. Mokykla 7(1999).

TAMOŠAITIS, R.: Literatūros egzaminas. Gimtasis žodis 9(2000)141.

TAPINIENĖ, V.: Literatūra, mokytojas, mokinys. Mokykla 7 (1999).

URBONAITĖ, D.: Mokinių informacinės kultūros ugdymo pedagoginės sistemos raiškos tipai (sąlygoti kompiuterių taikymu esant ribotiems ištekliams). Daktaro disertacija. Kaunas 2000 (Kauno technologijos universitetas).

VĖLIUS, N./SAUKA, D.: Etninės kultūros vieta tautinėje kultūroje. Lietuvos kultūros koncepcijos metmenys. Vilnius 1990.

VILIŪNAS, G.: Kultūrą integruojanti disciplina. Mokykla Nr.7. Vilnius 1999.

VILIŪNAS, G.: Teksto analizė mokykloje. Vilnius 1998 (Alma littera).

ZABORSKAITĖ, V.: Interviu. Gimtasis žodis 1(1993). Vilnius.

ZABORSKAITĖ, V.: Galimi literatūrinio ugdymo kontūrai. Mokykla 7(1999). Vilnius.

ŽUKAS, S.: Žmogaus vaizdavimas lietuvių literatūroje. Vilnius 1999 (Baltos lankos).

ŽUPERKA, K.: Teksto charakteristika. Šiauliai 2001 (ŠU).

Alicija Ramanauskaitė, Nida Stankevičienė

Technological Education as a Means of Reducing Technological Disjuncture

The ordinary structure of knowledge is changing in the modern knowledge society, together with the concept of an educated person, as knowledge and technologies are in constant development, and the work performed is changing too (Jucevičienė, 1997). According to J. Morkūnienė (2000), disability of a part of population to use the new technologies (because of lack of education or due to their social status) not only deprives them of the main source of development, but destroys human solidarity, tolerance, co-operation, i.e. increases inequality among people. Thus new forms of inadaptability and displacement from the society appear.

The approach to education as a general phenomenon, comprising the society members' development of general technological competencies concurrent with the needs of an individual is characteristic of the knowledge society. Thus for defining this sphere the term "technological education" is used. Like technology, education in its broad sense is associated with the civilization. Education is also affected by particular social structures, which are interested in education of their members, and by the quality of society living.

Technological education provides learners with a possibility to get acquainted with technology related processes and knowledge, necessary for problem solving and increasing the powers of the mankind. It is "about technology" and is realized with the help of technology (Pannabecker, 1995). Technological education orientates learners for studying processes (critical thinking, problem solving ability) with the help of which new technologies are created. (Wright, 1992, quoted from Wicklein, 1997). These assumptions lead to the problematic question:

Is technological education to be considered as a means of reducing technological disjuncture?

If, according to Johnson, Gatz and Hicks (1997), the far-off goal of technological education is to help the learners realize their technological future, the closest aim of technological education should be considered to be forming the skills of technological literacy, that make people active participants of technological reality. In order to determine the concept of technological education, technological literacy should be discussed. Hayden (1989, quoted from Waetjen, 1993) says that if technological literacy is knowledge and skills that allows selecting and applying technologies, suitable for a given situation. According to

Cajas (2000), the field of technological literacy comprises the relation between science, technologies and society, and that is expressed through designing, usage of control equipment, materials, production and resources. Hatch (1988) ascribes higher level problem solving skills – "technological method" – to technological literacy (Savage, Sterry, 1990), i.e., a systematic way of problem solving via thinking and acting.

The concept of technological literacy, comprising knowledge and technology, as well as understanding of its usage and skills, its means and assessment, is used by Steffens (ibid). When broadening this concept to technological literacy, Owen and Heywood (ibid) relate it to three components. It is ability to understand and use article production technology, organization technology, and information usage technology. Croft (1991) presents the characteristics describing a technologically literate person: he should be able to evaluate the technology when making a certain decision; possess the main skills, necessary for solving technological problem; be able to use technologies; to apply knowledge, means and skills for enriching society living ; to be able to describe the main technological systems of the society.

Dyrenfurth, Hatch, Jones, Kozak's (1991) definition of the concept of technological literacy might be considered to be the most general and disclosing the concept of the contents of technological literacy. The above-mentioned authors treat technological literacy a multi-dimensional conception that comprises:

- ability to use technologies (practical dimension);
- ability to understand technologies (civil- social dimension);
- ability to evaluate the importance of technologies (cultural dimension);

From the modern point of view, the definition is suggested to be supplemented in the following way: technological literacy is an ability to understand and use technologies in one's environment; positive attitude to constant development of technologies, ability to evaluate the importance of technologies (the "higher" ones included) and their possible consequences to the life quality of the present and future generations.

This concept of technological literacy as a result of technological education will open the way to defining technological education as such.

One of the social aims of technological education in the knowledge society is reducing social differentiation among the educated, what provide an individual with the possibilities of changes of self-realization and allows him/her to act in the environment of "higher" technologies (Krikštopaitis, 2002), and the non-educated who stay behind the knowledge society.

The aim of reducing social differentiation in technological education should be realized via understanding permanent development of technologies towards the 'higher" technologies and forming positive attitudes to technologies. All this is closely connected with the life-long learning paradigm.

By its contents, technological education is orientated towards the past, the present and the future. But the main direction of technological education is towards the perspective (Postman, 1992, Wicklein, 1997, Johnson, Gatz, Hicks, 1997), providing the learners with the harmony of the learning aim and meaning, as development of the modern society is concurrent with the development of technology that is progressing developing in exponential speed (Kriščiūnas, Lukoševičius, 1998). In such situation of "technological explosion" (ibid, p. 36) static contents of technological education looses its sense, as it does not fulfill its mission.

The principle of human responsibility is a very important part of technological education. Jonas (quoted from Lenk, 1998) considers it to be the ethics of future responsibility necessary for technological education. Technological education has to involve the principle of responsibility (as an axiological dimension of technological education); Jonas (quoted from Lenk, 1998) refers to it as to the ethics of responsibility for the future, necessary for the world of technology. He also points out the change in the effect of responsibility and its links with time, which takes place because of 'almost immeasurable technological power, extraordinary dynamic circumstances of life in industrial world and emerging danger for nature and living beings, included humans' (ibid, p.288). Responsibility for what has been done (for the outcomes) is transformed into preventive responsibility for the future; there is a transfer from 'responsibility for action' to 'responsibility for subsistence'. The author points out only the corporate possibility of technological activity and credits it not only to individual, but general responsibility too – responsibility of the team, the whole generation. The formation of this general preventive responsibility is one of the aims of technological education.

Ability to change and innovation are the relevant factors of technological education. The need of the ability to change is determined by the permanently and rapidly changing present situation. This ability is understood not as adaptation, readjustment to the situation, but a purposeful change for strategy development aims, the resources of which are considered to be intellect, knowledge, technologies, innovations and the potential of educated, motivated, creative youth. This need is inspired by the development of the "high" technologies that change not only the individual manner of living and thinking, but that of the society, too. Technological education should be directed towards the formation of the positive attitude towards such change.

In the contents of technological education, the concept of innovation as a significant factor of society development should be analyzed (Bansevičius, 1998,

Kriščiūnas, Lukoševičius, 1998), as well as the skills of problem solving, critical thinking, projecting, creative thinking.

Technological education by its characteristics correlates with the educational model of the rapidly changing society. (Targamadzė, Normantas et al, 1999) (Table 1).

Characteristics	Educational model of the rapidly changing society
Learning	Active learning, knowledge search is dominant. Learners form their knowledge by themselves. They create individual understanding.
Teaching	Learners are involved into the activity that requires critical thinking, active participation in discussions, ability to make fast decisions.
Curriculum	Fusion of teaching subjects
Environment	Integrated environment for self-study, where all is used – the learners, their interrelations, relations between teacher and learner, information technologies and special learning aids. Learning takes place through creation of common understanding of the problem discussed.
Assessment	Fulfillment of the task is assessed that requires not only facts, but problem solving and critical thinking. The whole teaching process – collecting information, assessment of alternative attitudes, searching for the best decision - is assessed.
Teaching aids and technologies	Special technologies for developing creativity and information search, structural presentation of information, subject games, situation analysis, modeling situations that are relative to the learners, discussions.

Table 1: Educational model of the rapidly changing society (Targamadzė, Normantas et al., 1999)

The table shows that its authors are keeping to the modern paradigm of transformation from teaching to learning, as the most important now is the learning process, newly understood teaching. Great attention is paid to environmental influence; the importance of information technologies and special learning aids is highlighted.

The literature analysis and the above mentioned thoughts lead to the conclusion that technological education should be defined as a part of holistic education enabling the learners to become technologically literate, able to acquire and constantly enrich knowledge, develop skills, that help them to understand, use and assess technologies and form a positive attitude to permanent development of technologies towards the "higher" technologies, towards better quality of the society and its members' life. Such individual attitude might be considered to be an assumption to one's high quality life in the modern world.

A research was carried out in order to find out Lithuanian secondary school teachers' attitude towards technological education. 309 teachers were questioned. There were two different cases of respondent selection applied in this research: the one of homogeneous sample (Bitinas, 1998) – teachers of all Lithuanian schools with technological gymnasium departments were questioned (97 teachers, 31% sample out of n = 309), and for comparative result analysis teachers of real, humanitarian, and technological profile schools were questioned according to sample of critical events. It describes minimal conditions for the phenomenon functioning that are the same for all possible objects of observation. (Bitinas, 1998). They were 212 teachers, 69% sample out of n = 309. The research data were statistically worked out by the package of SPSS software; hypotheses were checked by non-parametric statistical methods.

The respondents indicated that they had acquired pedagogical, technological and other kind of education (Figure 1)

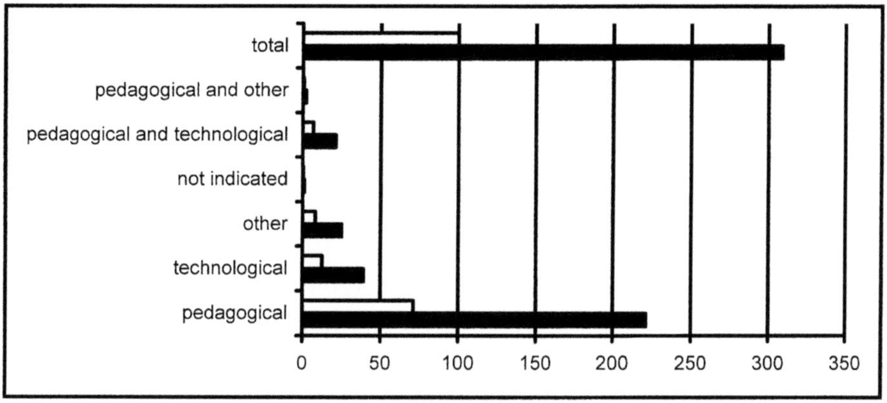

Figure 1: Teacher-respondents' distribution according to education

Most of the respondents 221 (71% out of n=309), have pedagogical education. 39 teachers have technological education (13% out of n=309). 21 teachers (7% out of n=309) indicated that they had both technological and pedagogical education. The reason why a big number of teachers have primary technological education is that they have graduated from vocational pedagogical schools or had acquired pedagogical education as their second one. This is characteristic of vocational school teachers (profession teachers).

Teachers were given some statements, which they had to assess within the range of 0-3 points (0 – I don't know, 1 – it's not important, 2 – it's not important, 3 – it's very important). In all statement evaluations scale indicators 0 ("I don't know) and 1 (It's not important) do not work, as probability of the indicated evaluation is lower 5 % (interval validity limits by applying mathematical statistics belong to interval [1,6%; 4,9%]). It shows that teachers have a positive attitude towards technological education. Evaluations "it's important" and "it's very important" are relevant to the research data analysis. While calculating media evaluations, the unanswered statements were not taken into consideration.

We were trying to find out whether teachers realize the impact of today's modern global tendencies on technological education in the educational practical activity of profiled schools. The mean calculations of each statement and standard deviations (Table 2) show that statements 1 and 4 got the highest points.

Statement	Mean	Standard deviation
1. Technology development has influenced the development of the modern world	2,58	0,039
2. The significance of high qualification specialists in the labor market is increasing	2,67	0,041
3. Technologies change the manner of production	2,53	0,051
4. Technological revolution influences peoples' manner of living, co-operation and work	2,46	0,038
5. The aim of the technological education should be directed towards creation of modern technology-intensive society	2,44	0,051

Table 2: Mean and standard deviations of answers

Teachers agree that the aim of the technological education should be directed towards creation of modern technology-intensive society (50% (154 out of n = 309) evaluated this statement as very important, 44% (136 out of n = 309) – as important). They understand that technological revolution influences peoples' manner of living, co-operation and work ("very important" – 49% (152 out of n=309), "important" – 48% (148 out of n = 309). But a number of the teachers doubt if the significance of high qualification specialists in the labor market is increasing.

We were also trying to clear up how teachers evaluate the possibilities of technological education that are determined by the tendencies of the Lithuanian economy development.

Statement	Mean	Standard deviation
1.Lithuania is striving to become one of the technologically preponderant states	1,88	0,063
2. Technology development in Lithuania is the major factor of the overall development of the country	1,64	0,054
3.The "high" technologies directly influence the living standard of the country	1,96	0,063
4. Lithuania is realizing the strategy of technological development	0,97	0,053
5.Lithuanian educational system fosters an individual's ability to participate and create a technologically developed society	1,62	0,056

Table 3: Mean and standard deviations of answers

Teachers understand (see figure 2) that the "high" technologies directly influence the living standard of the country (26% or 81 respondents out of n = 309 agree, 52%,161 out of n = 309 partially agree). The statement that Lithuania is striving to become one of the technologically preponderant states is positively evaluated (18% 56 out of n=309 agree, 56% ,174 out of n = 309 partially agree).

But a big number of answers "partially agree" show that teachers not fully agree with the statements. An assumption might be made that teachers are not enough familiar with this question.

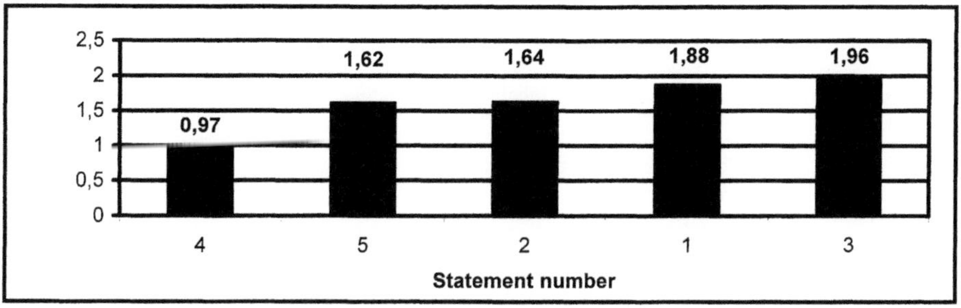

Figure 2: Comparative diagram of B1 statement evaluation

The lowest results can be seen concerning the statement that Lithuania is realizing the strategy of technological development (32%, 100 respondents out of n=309 do not agree with the statement, 41% , 127 out of n = 309, do not know). This result of the teachers' answers correlates with science, politics, business representatives' discussions on the questions of technological development strategy in Lithuania. (Lukoševičius, 1998, Bansevičius, 1998, Krikščiūnas, Lukoševičius, 1998 et al).

Teachers do not agree that they are not concerned with the globalization processes (64,5%, 200 respondents out of n = 309). This way they express their orientation towards the modern society (Table 4).

Statement	Mean	Standard deviation
1.Teachers are not concerned with the globalization processes	0,60	0,051
2.Technological knowledge should be imparted at universities, not secondary schools	0,68	0,053
3.Technological education is important only in teaching specialty	1,04	0,062
4.Technological education at secondary school is understood as integration of projection and production technologies	1,53	0,051

Table 4: Mean and standard deviations of answers

About one third of the respondents (37%) partially agree that technological education is important only in teaching specialty, though 60 % of the respondents point out that technological knowledge should be imparted at secondary schools. Thus they prove their "wide" point of view to technological education as a part of the holistic education.

Conclusions

1. From the modern point of view, technological education as a general phenomenon, should be orientated towards the development of technological literacy of the society members that is concurrent with the needs of an individual.

2. Ability to change and innovation are the factors relevant to technological education.

3. An assumption can be made that teachers consider technological education as a part of the holistic education, and they are positive to technological education.

4. The conclusions confirm that technological education can be a means of reducing technological disjuncture.

References

BANSEVIČIUS, R.: Technologijos mokslų problemos Lietuvoje. Technologijos mokslų plėtra. Lietuvos mokslų akademijos sesijos-konferencijos pranešimų medžiaga. Kaunas 1998 (Technologija), pp. 8-13.

BITINAS, B.: Ugdymo tyrimų metodologija. Vilnius 1998.

CAJAS, F.: Research in technology Education: What Are We Researching? A Response to Theodore Lewis. Journal of Technology Education. Vol. 11 (2000) N. 2.

JOHNSON, S.D./GATZ, E.F./ HICKS, D.: Expanding the Content Base of Technology Education: Technology Transfer as a Topick of Study. Journal of Technology Education. Vol. 8 (1997) N. 2.

JUCEVIČIENĖ, P.: Lyginamoji edukologija. Kaunas 1997 (Technologija), p. 28.

KRIKŠTOPAITIS, J.A.: Individualios raiškos problema informacinės globalizacijos tinkle. Filosofija,sociologija. Nr.1 (2002).

KRIŠČIŪNAS, K./LUKOŠEVIČIUS, A.: Technologijos specialistų rengimo Lietuvoje problemos Europos kontekste. Technologijos mokslų plėtra. Lietuvos mokslų akademijos sesijos-konferencijos pranešimų medžiaga. Kaunas 1998 (Technologija), pp. 27-50.

LUKOŠEVIČIUS, A.: Naujų technologijų reikšmė Lietuvai atsiveriant Europos erdvei. Lietuvos mokslas ir pramonė: Europos erdvės kontekstas. Konferencijų pranešimų medžiaga. Kaunas 1998 (Technologija), pp. 82-89.

MORKŪNIENĖ, J.: Informacinė visuomenė kaip technologinė trevoliucija ir kaip nauja visuomeninė sutartis. Filosofija,sociologija. Nr.4 (2000).

PANNABECKER, J.R.: For a History of Technology Education: Contexsts, Systems, and Narratives . Journal of Technology Education. Vol. 7 (1995) N.1.

TARGAMADZĖ, A./NORMANTAS, E./RUTKAUSKIENĖ, D. IR KT.: Naujos distancinio švietimo galimybės. Vilnius 1999 (Lietuvos neakivaizdinio švietimo centras).

WAETJEN, W.B.: Technological Literacy Reconsidered. Journal of Technology Education. Vol. 4 (1993) 2.

WICKLEIN, R.C.: Curriculum Focus for Technology Education. Journal of Technology Education. Vol. 8 (1997) 2.

WICKLEIN, R.C.: Curriculum Focus for Technology Education. Journal of Technology Education, Vol. 8 (1997)2.

Palmira Jucevičienė, Sandra Balevičienė

Multidisciplinarity of Socially Important Theme of Natural Science in Personal Theory-in-use

Introduction

In action science it is stressed that people have in themselves mental models referring to which they act in certain situations. These models include activity planning, realization and reflection methods. They exist in an individual's mentality as implicit cognitive structures. Agyris and Schon (1974) state that people usually follow these particular models and not those which are formed beforehand as "explicit knowledge". That is why there is often inadequacy between implied and actual activity carried by an individual (Agyris, 1980). Analyzing men's behavior, according to Agyris and Schon (1974), it is expedient to talk about two action theories. The first is espoused theory. Its essence is known. It stresses the model of perfect activity and shows a person's standpoint and values on which a man's imaginary activity is based. The second is theory-in-use that expresses standpoint and values about which are possible to judge from a person's behavior or fulfilled activity. This knowing it is so called implicit knowledge, which usually stays unrecognized, if an individual does not put addition effort. Knowing becomes explicit knowledge if it is expressed through a language in written form or orally. Usually it is carried out when a person reconsiders (reflects) his or her actions in the process of an activity or just after this activity has been finished. If some time passed after this activity had been done it is necessary to carry out the reconstruction of auto activity.

Used theories in men's activities of every day life can be reconstructed or developed acquiring new experience. Each new experience is interpreted in the context of previous ones. In men's general experience usually consolidate this knowledge or abilities that are constantly applied to in his or her practical life or professional activity (Jackunas 2003).

In a real world we come across with situations that are not divided into single disciplines. They are complex and require integration of knowledge and abilities from different areas. Therefore man's general experience and also action theories-in-use, which all human activities are based on, are more multidisciplinary than disciplinary in their nature.

Educationalists working out secondary school teaching programmes, traditionally, are guided by espoused theory model, according to which disciplinary proach to cognition of the world phenomena is acknowledged. In this way, life is artificially divided into "gratings" depending on disciplines. The knowledge and abilities that are acquired during such teaching sometimes virtually contradict the experience that a person already has. That is why such knowledge and abilities are quickly forgotten or exist in a person's mentality as isolated structures that do not play any important role in his or her practical activities or life. This problem is especially important in a secondary school presenting socially important topics of natural science subjects such as "Water", "Air", "Soil" and others. These topics are closely related with man's every day activities and implicate knowledge but also emotions, attitudes and values. Analysis of these topics at school requires knowledge not only in natural science but in social science and the humanities as well (De Paz, Pilo 1990).

Trying to avoid contradiction that arises due to the interaction of two theories – espoused theory and theory-in-use – in cognitive structure it is necessary to solve an urgent scientific problem. It deals with the question of what means and methods should be used taking into consideration the experience a person has in order to construct the contents structure of socially important topics of natural science.

Aim of the research is to ground the constructing mechanism of the contents structure of socially important topic of natural science. These objectives are sought using the following methodology:

- Theory of constructivism (Dewey 1972; Glaserfeld 1995; Novak Gowin, 1999), with the help of which it is stressed that a person interacting with social environment actively originates his or her own individual knowing on the basis of previous experience. Previous experience plays an important role in constructing new knowledge and understanding.
- Action theory (Argyris, Schon 1984) by which it is stressed that interacting with the environment a man originates certain mental models, which exist in the mentality of an individual as implicit cognitive structures. A man in real everyday activities follows these models, which are named as theories-in-use.
- Method of construction of concepts maps (Novak, Gowin 1999) that allows visually (by words, symbols) express earlier acquired understanding of a man about a certain object or phenomenon.

In the research methods of inquiry and content analysis are used.

Relation between the espoused and disciplinary approach on the socially important topics of natural science

Each person interacting with surrounding objects and taking part in everyday life activities forms his own perception of the world. A man constructs knowledge, attitudes and values that he follows in his further activities. The process of learning cannot be separated from everyday life and experience. A man coming to school already brings his own life experience, and it plays an important role in developing new knowledge and understanding (Balevičienė, Jucevičienė 2003).

Training specialists who acknowledge espoused disciplinary theory assert that those who study must acquire a "parcel" of knowledge and practical abilities systematically. According to them if one is not familiar with disciplines and their structure the threat to acquire fragmentary and chaotic knowledge arises. A person educated in such a way would hardly manage to deal with scientific, professional and even everyday life problems efficiently and with competence.

The supporters of espoused multidisciplinary theory think that single isolated discipline can highlight its relationship with reality in an insufficient way. It is not possible to systematize knowledge about the world phenomena choosing only one field because the aim of teaching is to show existing relationships in the world to those who study. Every taught subject is significant in its own way but, in order it could be possible to use knowledge in a real life, it is necessary to combine disciplines with each other ant not to limit oneself only to "pure" subject. As an example the authors give global ecological problems that can hardly be solved by biologist, physicist or chemist alone. The solution of the problem can be found with the help of joining knowledge, abilities and methods from different spheres (Johansen, Rathe 1999).

In spite of recognizing the importance of multidisciplinary teaching one must take into consideration already existing thesauruses that remain in disciplinary teaching. Therefore a young person should know which teaching area thesaurus to advice if he or she wants to solve a particular problem.

We suggest applying binary approach to contents formation of natural science teaching that would be based on two theories – disciplinary and multidisciplinary espoused theories. Moreover, we raise another significant question that is what relationship between these approaches should be constructing the contents structure of separate topics.

It is likely that analyzing individual topics this relation is not going to have a single meaning. Analyzing some of the topics disciplinary attitude will be stressed more while in others multidisciplinary one is going to be emphasized. There could also be such topics where both approaches will be equally expressed. Thus natural question arises which approach should be referred to more and which less. And also which of them should be followed making contents structure for one or another topic.

Earlier gained experience in some topic or question has a great influence for the development for new knowledge. Learning is understood as cherishing and developing of this experience in a proper direction. Then, establishing which approach is going to be followed more in making the structure of the topic contents, it is purposeful to pay attention to theories used by people.

When a person's experience in relation to the chosen topic is expressed as structures of knowledge of various disciplines then compiling the contents of such a topic it is expedient to refer to multidisciplinary espoused theory. On the contrary, when experience is expressed through the knowledge of single discipline then the disciplinary approach has to be applied.

Constructing of the contents structure of socially important topics of natural science based on the "Water" topic example

For construction of the contents structure of socially important topics of natural science the topic "Water" has been chosen, which in everyday life is perceived not only as natural but also as social problem. The analysis of the topic at school should help pupils not only to become acquainted with water characteristics but also to understand that clean water problem takes a global scale. Furthermore, rational water consumption and understanding of management principles as well as their application in everyday life become our duty.

Establishing social standpoint to the topic "Water" it was necessary to find out which spheres of life related with this topic are the most important to people. Do people perceive water as social problem? How much it is urgent to them? In what logical succession the sub themes of the topic "Water" should be given that in the process of learning a person would be motivated most?

In order to achieve this goal it was necessary to choose such method of questioning that could help obtain personally the most important knowledge about water to people. Also compiling the contents of the topic "Water" not only knowledge is significant but its structure and relationships as well, which have special meaning forming logical contents of the topic.

The methodics of the research

As it was mentioned earlier that action theories-in-use or experience exist in person's mentality as implicit cognitive structures. Therefore paying attention to these structures it would be good to express them as certain fixed forms. According to Argyris and Schon (1974) it is possible to express such theories by words, sentences and symbols applying reconsideration of the activity.

Using common research methods such as open and closed questionnaires it is possible to know only this knowledge of an individual to which specific questions are directed. On the other hand according to the received answers it is difficult to get a general view of how the person understands the whole of phenomenon and in what way this knowledge is organized (Tautkevičienė, Jucevičienė 2003). Thus to solve the research problem the method of construction of concept maps has been chosen (Novak, Gowin 1984). This method allows visually express not only person's understanding of phenomena but also the structure of man's knowledge. According to Edwards, Kraser (1983), Wandersee

(1990), Kinchin (1999; 2000), West, Park, Pomeroy, Sandoval and others using this method it is possible to get much more information about person's knowledge and about interrelationships among different pieces of that knowledge than using traditional research methods (for example interview, questionnaire, etc.). Therefore in our work the concept map was used as the main research method.

Trying to get more knowledge related with the experts' values, feelings and one's philosophy on the topic "Water" and also taking into account the specific character of the method (drawing of a concept map takes much time) the expert inquiry method was used (Merkys 1995). We considered that each interrogated person who represented typical profession or population group and who was able to explain what role water plays in his or her life could be recognized as an expert. In this research we present Lithuanian context (Lithuanian experts were inquired).

Sampling. Constructing the contents of natural science topic "Water", which would be urgent for those who study and also related with their everyday life, it is best to refer to that structure of knowledge that is conditioned by the general life experience of people. Thus in the research those citizens of the society that are in the closest social environment were interrogated.

The sampling method of this research is non-functional (probability) and objective. The sampling (N = 30) of experts was made taking into consideration the following criteria: age, education, field of studies and social status. The distribution of the experts according to these criteria is shown in the Table 1.

Organization and course of the research

Before the research the experts were acquainted with the method of construction of concept maps. Later they were asked to draw a concept map on the topic "Water in my life". The duration of the interrogation was not limited.

Criteria	N	%	Criteria	N	%
Age:			***Field of studies:***		
From 17 to 35	10	30.0	Not established	5	16.7
From 36 to 45	12	43.3	Biomedical	10	33.3
From 46 to 64	8	26.7	Technical	3	10.0
			Social	8	26.7
			Humanities	4	13.3
Education:			***Social status:***		
Incomplete secondary	1	3.3	White collar worker	11	36.7
Secondary	9	30.0	Blue collar worker	4	13.3
Vocational	4	13.3	Unemployed	2	6.7
College	3	10.0	Student	8	26.7
Higher	13	43.3	Pupil	2	6.710.0
			Pensioner	3	

Table 1: Sample criteria and their distribution

Analysis of concept maps

The analysis of the contents of concept maps of the experts was carries out combining quantitative and qualitative research (Kardelis 2000).

By the *quantitative* analysis of the contents of concept maps it was sought to establish the contents structural elements and their significance to the topic "Water". Thus obtained information was classified into categories. The number (n) and percentage of respondents having chosen a distinguished category were established. Also repetition frequency and its standard deviation were established.

The evaluation *of* contents structural elements of concept maps from the quantitative point of view shows only the importance of these elements, but does not reveal their contents. In order to reveal this it is necessary to use qualitative research methods. In this case contents analysis has been chosen.

Qualitative *contents* analysis of concept maps was carried out and obtained information was classified into categories and subcategories to which keywords were attributed (Table 2).

Carrying out qualitative analysis of the contents existence or non-existence of different conceptions in the concept maps has been chosen for identification. Before the analysis started is was decided that the conceptions having the same meaning are to be attributed to a specific category or subcategory in spite of being expresses in different words.

Respondent's statement	**Keywords**		**Researcher's remarks**
	Category	***Subcategory***	
"Water is used for preparation of food and beverages"	Household	Food	e.g. Water is used for preparation of soups, baked ware, tea, coffee, etc.

Table 2: The example of distinguishing categories and subcategories given in the concept map

Later it was sought to establish how the respondents construct contents structure of the topic "Water". Which categories are characteristic to all the groups and which to the groups of experts of different age, education and field of studies? Also attempts were made to evaluate if water problems are important to the experts, and in what way they are related to other conceptions characterizing social and natural environment.

After carrying out quantitative and qualitative analysis of the contents of concept maps it was sought to make a logical contents structure of socially important topic of natural science "Water".

The research results and their interpretation

Quantitatively analyzing the contents of the experts' concept maps 17 categories were singled out (Table 3). These categories were conditionally divided into two areas: 1) water expression in social environment; 2) water expression in natural environment.

The categories corresponding to the social sphere of water were identified in all (N = 30) experts' concept maps while the categories corresponding to the natural sphere were found in 26 ones (N = 26).

Although life experience and competence of all the experts were very much different it is possible to distinguish certain categories that are common to the most of the experts. In the sphere of social environment irrespectively of age, speciality, education and profession most of the experts marked "household", "industry", "agriculture" and "recreation" categories. And the smallest number of respondents marked "science", "means" and "traditions" ones. In the natural sphere similar number of respondents marked "physical" and "biotic environment" categories.

Categories	N	%	Mean	SD
Social environment				
Household	30	100	4.60	3.18
Industry	26	86.6	2.67	1.73
Agriculture	24	50.0	1067	1.03
Recreation	24	80.0	1.20	0.89
Sport	14	46.7	1.07	1.60
Values	12	40.0	0.60	1.04
Emotions	12	40.0	0.83	1.53
Transport	11	36.7	0.37	0.49
Medicine	10	33.3	0.50	0.82
Problems	7	23.3	0.53	0.97
Art	7	23.3	0.53	0.97
Religion	3	10.0	0.13	0.43
Science	2	6.7	0.07	0.25
Means	2	6.7	0.10	0.40
Traditions	2	6.7	0.07	0.25
Natural environment				
Physical environment	23	76.7	3.03	2.76
Biotic environment	21	70.0	1.60	1.57

Table 3: Quantitative analysis of contents categories expressing water concept

From 30 concept maps in which the category of "household" was marked in each of them there are approximately 4.6 (Table 3, p. 187) statements of this type. Comparing categories of "agriculture" and "recreation", which were marked by 24 experts, the former is more significant as its approximate repetition is greater. Comparing "problem" and "art" categories, which were marked by 7 experts, the latter was more important to the experts.

By the qualitative research it was established that in the group of the experts whose age is under 35 such categories as "transport", "religion", "traditions" "means" and "science" have not been identified (Table 3). The experts whose age is from 36 to 45 have marked all the distinguished categories. In the group of experts whose age is over 45 only the category "traditions" has not been marked.

Comparing existence and non-existence of the categories in the experts' concept maps according to the discipline it was established that representatives of biomedical and technical science have not marked "religion" and "traditions" categories.

All the respondents that marked "household" category mainly relate it with the subcategories of "food" and "hygiene" (Table 4, see pp. 191-192).

The experts relate "food" subcategory with such human activities as slaking one's thirst, preparation of food and beverages. "Hygiene" subcategory in concept maps is illustrated with the examples of cleaning of living quarters and person's hygiene activities. Table 4 shows that analyzing the experts' concept maps, in which the category "household" is marked, it is evident that the subcategory of "care of pets and animals" is marked only by one third and "heating" only by one fourth of all the experts. It seems that these subcategories were not as important as "food" and "hygiene" to the experts.

Analyzing experts' concept maps, in which "industry" category was indicated, from three subcategories related to it the subcategory of "light industry" was pointed out by the largest number of experts. The experts relate this one with water usage in food, beverages, and building materials production and in other industrial production of products that are important to men's life.

In Table 4 it is seen that from the most significant "household", "industry", "agriculture" and "recreation" categories, which characterize water expression in a social environment, the most important subcategories are: "food", "hygiene", "light industry", "plant-growing", "cattle-breeding" and "recreation".

Qualitative contents analysis has shown that, for example, water needs in medicine were marked more often and more comprehensively (pointing out the relationships between different structure segments) by the representatives of biomedical and technical sciences. On the other hand, the representatives of social science and humanities more comprehensively marked water as the factor that can influence man's emotional state, its significance to creative work and artistic activities. The representatives of biomedical science analyzed the problems of water environment more widely than the representatives of other science fields. What is more, relating these notions with other conceptions of social and natural fields they pointed out not only the consequences of these problems but also the ways of solving them. Meanwhile, representatives of humanities, social and technical sciences pointed out only consequences cause by water environment problems to a man and nature. For comparison the concept maps of experts of different age, education, science field and profession on the topic "Water in my life" (see Figure 1-4, pp. 193-196).

On the basis of quantitative and qualitative analysis of the contents of concept maps of researched experts the contents of the socially important topic "Water" was compiled (see Figure 1, p. 193). The data show that the contents of the topic "Water" mainly implies those structural elements that are closely related with social environment. On a dark gray background the categories of "household", "industry", "agriculture", "recreation", "physical environment" and "biotic environment" are pointed out. These categories were identified as the most significant ones because more than 20 experts indicated them. Less important categories such as "sport", "values", "emotions", "transport" and "medicine" are shown on a light gray background. They were identified in more than one third of concept maps of the experts. The rest, the least significant conceptions, which were indicated by less than 10 experts, are shown on a white background. The concept map (see Figure 1, p. 193) shows the relationships that go into the conception of "problems" and reveal, according to the experts' point of view, the most important man's activity spheres that have influence on the appearance of water environment problems.

Summarizing the research results it is possible to assert that people relate water conception not only with natural but also with social environment. In experts' concept maps the categories that are closely related with men basic needs and his or her everyday life are identified and repeat most often. The experts define water conception by categories corresponding to knowledge not only of single but also of different branches of science. To one fourth of interrogated persons global problems of water environment are not significant.

Conclusions

- Analysis of concept maps once more showed that it is efficient means that allows identifying people's knowledge on one or another question. This method also allows establishing the contents of knowledge and its relationships. The experts of different age and science fields pointed out different relationships among various notions.

- The experts' experience related with the topic "Water" was not defined by the knowledge of a single discipline. Due to this, constructing the contents of socially important topics of natural science, it is advisable to refer to constructivism theory presenting these topics at school. Also constructing their contents structure it should be referred to multidisciplinary espoused theory.
- Seeking that those who study would become acquainted with structures of certain subjects and presenting the socially important topics of natural science it is necessary to distinguish disciplinary knowledge together with its structures and later to extend it.
- On the basis of quantitative and qualitative analysis of the contents of concept maps, which were obtained from the researched experts, the contents of socially important topic of natural science "Water" has been compiled. Taking into account significance of the categories and relationships between notions a logical structure of this topic has been formed.
- The research results show that water environment problems are not significant to the inquired individuals. Thus, in order to excite the curiosity of those who study water environment problems and their solutions it is expedient to analyze the topic "Water" at school relating it with such most urgent spheres of human life as "household", "industry", "recreation" and others.

References

AGYRIS, CH.: Inner contradictions of rigorous research. New York 1980 (Academic Press).

ARGYRIS, CH./SCHON, D.A.: Theory in practise: increasing professional effectiveness. 1974 (Jossey – Bass Publishers).

BALEVIČIENĖ, S./JUCEVIČIENĖ, P.: Vanduo aplinkoje ir žmogaus gyvenime. Eksperimentinė medžiaga. Kaunas 2003.

DE PAZ, P./PILO, M.: European Project for Environmental Education. A Curriculum for European Schools. Genova 1999.

DEWEY J.: Experience and Education. New York 1938 (Macmillan).

EDWARDS, J./FRASER, K.: Concept maps as reflectors of conceptual understanding. Research in Science Education 13(1983), pp. 19-26.

GLASERSFELD, E.: Radical Conctructivism: A Way of Knowing and learning. London 1995 (Falmer Press).

JACKŪNAS, Ž.: Šiuolaikinė ugdymo turinio kaita prof. J. Laužiko integracijos idėjų atžvilgiu. Pedagogika 68(2003), pp. 19–23.

JOHANSEN, B.A./RATHE, A.L./RATHE, J.: Vaiko galimybės ir mokykla. Pedagoginė apybraiža – nuo idėjos iki tikrovės. Vilnius 1999 (Margi raštai).

JUCEVIČIENĖ, P./TAUTKEVIČIENĖ, G.: Academic library as a learning environment: How do students perceive it? The European Conference on Educational Research, Hamburg, 17 – 20 September, 2003/European Educational Research Association. Leeds: University of

Leeds. p. [1-13]. [cited:] Available from: <http://www.leeds.ac.uk/educol/documents/00003274.htm>.

KARDELIS, K.: Mokslinių tyrimų metodologija ir metodai. Kaunas 2002 (Judex leidykla).

KINCHIN, I.M.: Concept cones. School Science Review, 80(1999)293, pp. 107-108.

KINCHIN, I.M.: Using concept maps to reveal undrstanding: a two tier analysis. School Science Review, 81(2000)296, pp. 41-46.

LEPAITĖ, D.: Kompetencija kaip ugdymo tikslas: pagrindinių skirtumų profesinio vidurinio ir aukštojo mokslo lygmenyse metodologinis pagrindimas. Socialiniai mokslai 2(2001)28, pp. 39-43.

MERKYS, G.: Pedagoginio tyrimo metodologijos pradmenys (paskaitų konspektas). Šiauliai 1995.

NOVAK, J.D./GOWIN, B.D.: Learning how to learn. 1999 (Cambridge University Press).

PIAGET, J.: To understand to invent: The future of education. New York 1972 (Grossman).

WANDERSEE, J.H.: Concept Mapping and the Cartography of Cognition. Journal of Research in Science Teaching 27(1990)10, pp. 923-936.

WEST, D.C./PARK, J.K./POMEROY, J.R./SANDOVAL, J.: Concept mapping assessment in medical education: a comparising of two scoring systems. Medical Education 36(2002)9, pp. 820-826.

Categories	**Subcategories**	**N**	**Examples of notions**
Household	Food	30	Used for preparation of soups, stews, tea, coffee and beverages; canning
	Hygiene	30	Used for person's hygiene; for cleaning of living quarters
	Heating	7	Used for heating of living quarters
	Care of pets and plants	10	For watering of flowers, for keeping and care of pets
Industry	Heavy industry	16	In metallurgy, production of machines
	Light industry	26	In branches of food, chemistry, pharmaceutical and other industry
	Energetics	11	For cooling of reactors, rotation of generators
Agriculture	Cattle-breeding	24	For cattle watering, preparation of fodder, fish raising
	Plant-growing	22	For plants watering
Medicine	Therapeutics	9	Water massage, healing-baths
	Surgery	3	As the foundation for wounds bathe
	Stomatology	1	For rinsing one's mouth
Transport	Water transport	11	For transportation of passengers and freights
Sport	Water sports	14	Yachting, swimming, water polo
	Winter sports	7	Skating, skiing

Recreation	Relaxation	24	For relaxation in the beach, bathing in the sea or lake; water fair, fishing
	Tourism	10	For cruise
Art	Material	6	In sculpture to prepare clay, in painting to paint with water-colors
	Inspiration	6	Repetition of the sea murmur and knocking of rain drops sound in musical compositions
Values	Aesthetic value	12	Fountain, cascade, lake, river, sea, glacier
Religion	Rituals	3	For baptizing, consecration
Traditions	Festivals	2	Sea festival
Problems	Caused by a man	4	Pollution of water reservoirs and their eutrophication
	Caused by nature	3	Heavy showers, floods, storms, hail, drought
Means	Security	2	Used to put out fire, bonfire
Science	Research object	2	Investigated water structure, properties
Emotions	Positive	12	Tea can calm down, cold shower awakens
	Negative	3	It is unpleasant to wear wet clothes, cold rainy autumn weather is nasty and makes a person loose his/her spirit
Physical environment	A part of physical environment	19	Oceans, seas, rivers, lakes, glacier
	Living environment	9	Environment in which living organisms feed, develop and propagate itself
	Cradle of life	8	Life on the earth started in the ocean
	Climate formation	12	Forms weather humidity clouds formation, precipitation
	Relief and landscape formation	7	Weathering of rocks, soil formation
Biotic environment	Component of living organisms	17	Exists in the composition of all living organisms
	Functions carried out inside a living organism	15	Transportation of nutritious matter, participation in all chemical reactions taking part in an organism, etc.

Table 4: Qualitative analysis of categories and subcategories of notions expressing the conception of water.

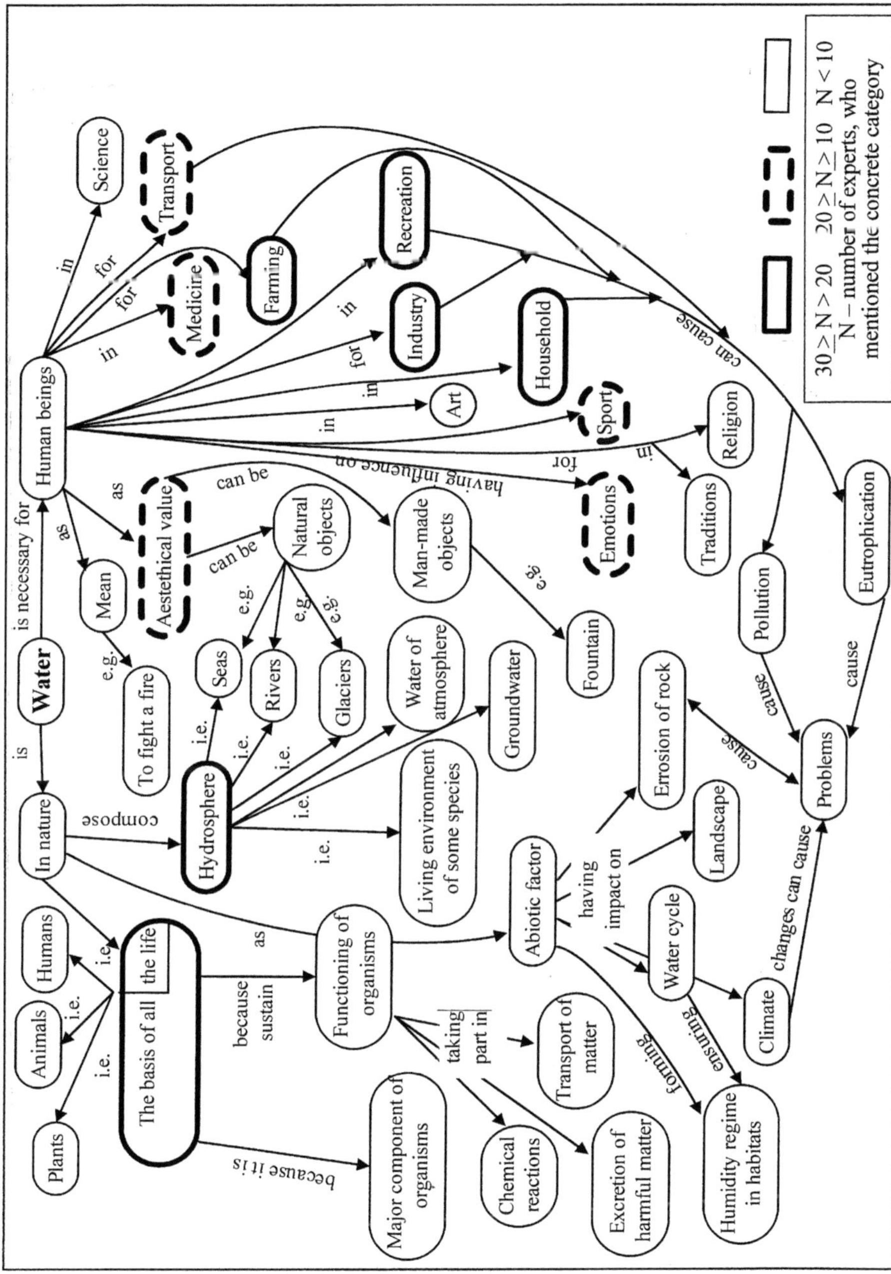

Figure 1: A conzept map prepared for the course „Water"

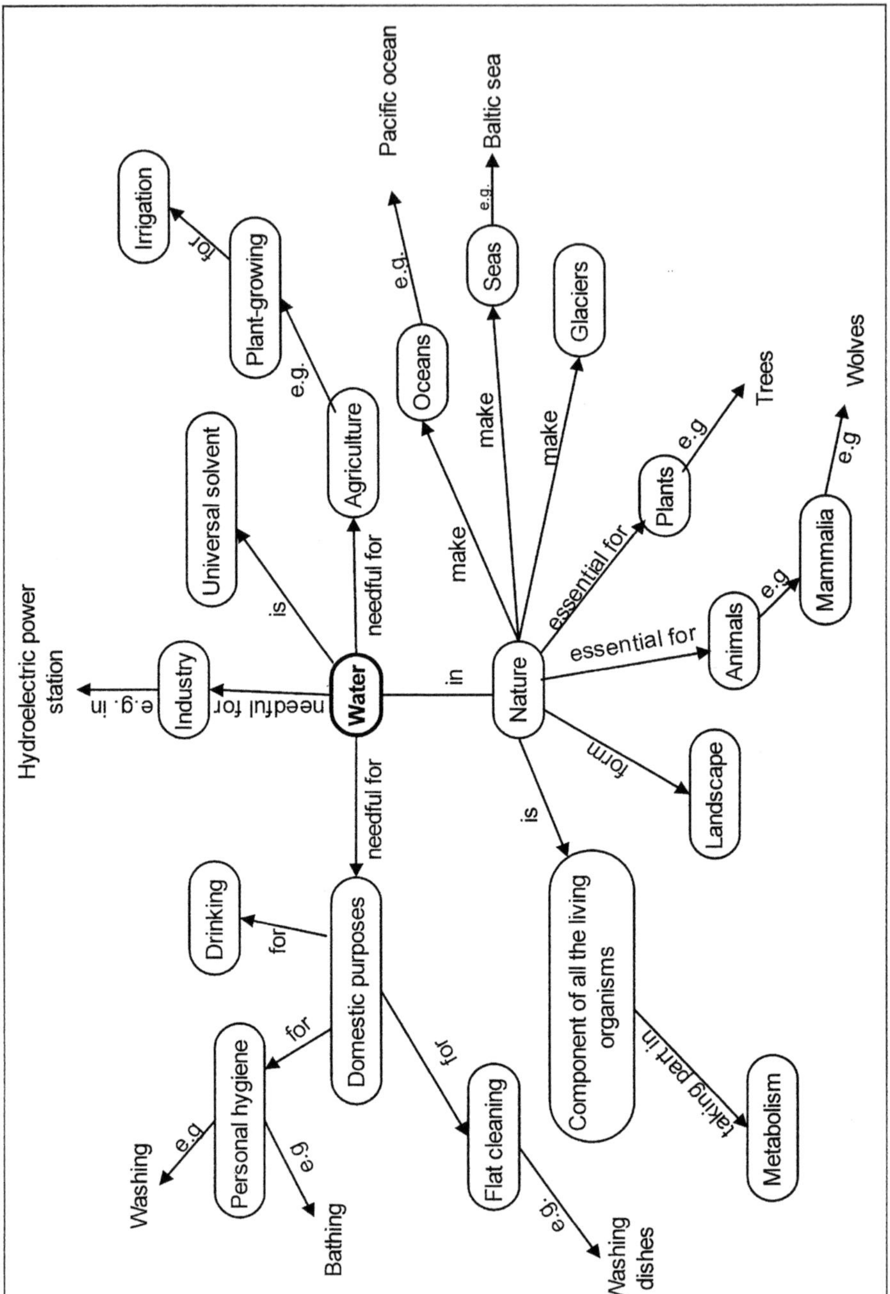

Figure 2: A concept map „Water in my life“ prepared by a twelfth-grade pupil (17 years old)

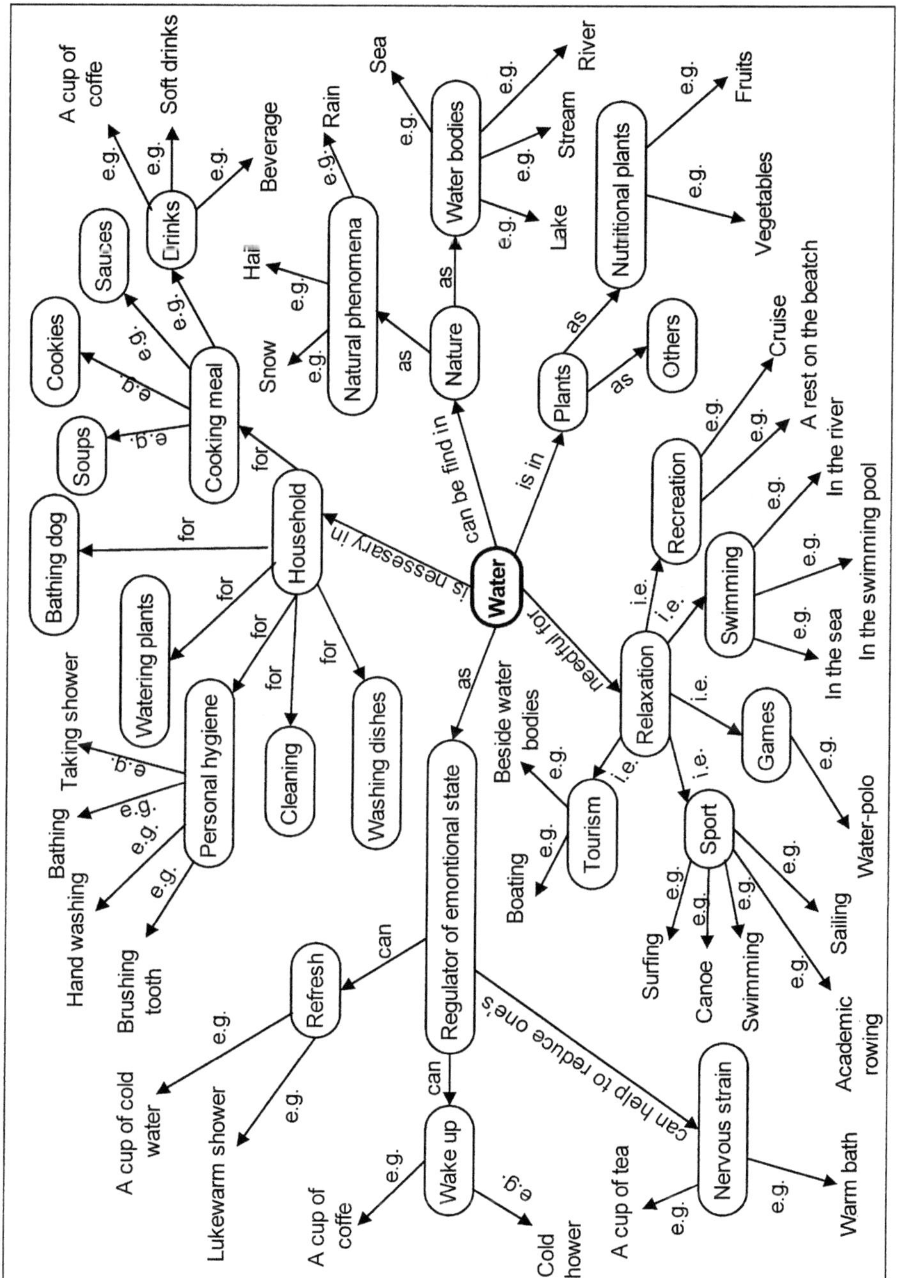

Figure 3: A concept map "Water in my life" prepared by a 2nd year psychology student (19 years old)

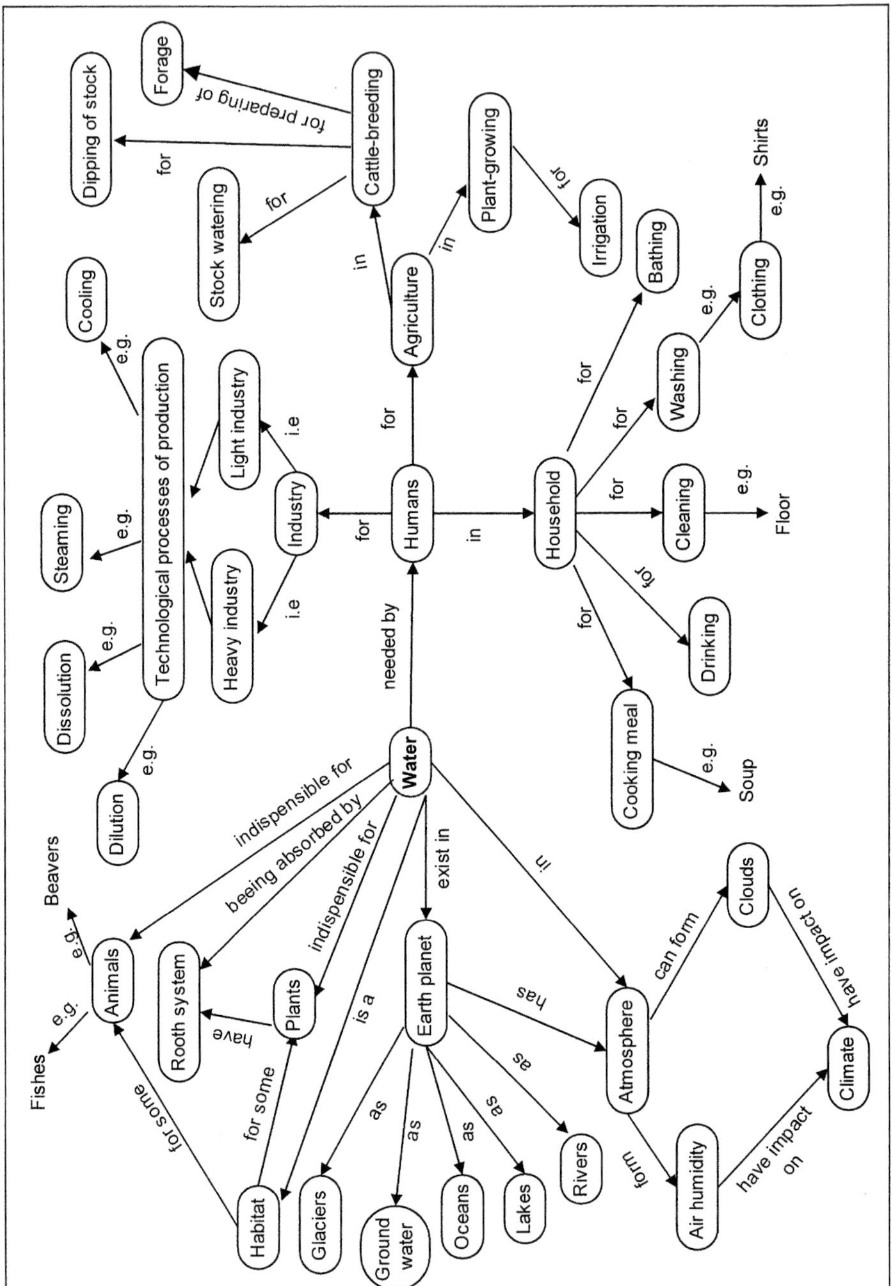

Figure 4: A concept map "Water in my life" prepared by a pharmacist (25 years old)

Vilija Dabrišienė, Ona Visockienė

Manifestation of the Tendencies of Educational Change in Lithuanian Secondary School

Introduction

Change is one of the essential characteristics of the postmodern society. New characteristics of the information and knowledge society are developing during the process of its formation. Knowledge society is the one that creates and maintains knowledge; it is a society of learning individuals and organizations. The concept of knowledge society stresses the relativity of knowledge and emphasizes the necessary abilities: to assimilate information, to apply it and to build new knowledge on the basis of the achieved one. Therefore, one of the most important qualities of such a society is permanent teaching/learning. The other significant tendency named by Drucker (1993) is building of knowledge Knowledge is being created during the process of activeness by reflecting, mediating everything what is done or what is to be done. From the other point of view, the need to balance between intellectual and social development is getting brighter, i.e. the ability to purvey the ideas becomes not sufficient, one feels the need to cooperate with others. These and other tendencies show a direction of changing educational system. The rebuilding of educational system is inevitable and it covers all the elements of pedagogical system: the objective of education, methods of application, the relationship between a pedagogue and a student, their roles etc.

Educational changes are widely analyzed by various foreign and Lithuanian scholars (Stoll (1998), Hopkins, Ainscow, West (1998), Fullan (1998), Hargreaves (1999), Jucevičiene (1998), Janiunaite (1999) and others). However, there are no works, which would investigate the changes in teachers' activities in Lithuanian secondary schools.

Teacher in a traditional school was recognized as a conveyer of information, thus this role has to be changed in the context of the learning society. The teacher becomes a manager of the process of learning: she/he plans, organizes, creates and evaluates conditions, manages the learners and looks after their learning motivation. The learner is a client of a school and his/her needs are the most important, the development of his/her personality needs particular conditions to be built. Consequently, the teacher has to be very flexible, to follow individual needs of a learner attentively, to assess all the new data related to teaching, to learn permanently, to investigate his/her own activity and to develop the quality of actions according to the results of the investigations done. Cooperation that involves all the community of a school, collaborative problem solving, seeking for common goals through internecine support are to be mentioned as very important fields, too. The tendencies of these alternations are highly important and definite. Thus, the issue is focused on the question, whether teachers' activities in the Lithuanian secondary school change under the influence of the world and European educational changes.

For the investigation of this problem the tendencies of educational change, defined by Dabrišiene (2001) were selected and their influence on the change of teachers' activities was analyzed. Those criteria for educational change include: increase in the attention to the individual needs of the learner, changes in the interaction between the teacher and the learner, based on their cooperation, closer cooperation with parents, intensification of teacher's professional activities and role change, closer cooperation between teachers, continuous in-service training and the need to grow, increase in the possibilities of teacher's self-expression, motivation of teachers' individuality and creativity, educational research at school, practice of the teacher as an enquirer.

Thus, the research object is: manifestation of the tendencies of educational changes in Lithuanian secondary schools.

The research aim of this article: to disclose the manifestation of the world tendencies of educational change in the Lithuanian secondary schools.

The aim of the research is subdivided into the following objectives:

1. To analyze and find out whether pedagogue's activities are influenced by the tendencies of educational change.
2. To examine the dependence of the tendencies of educational change on school peculiarities and respondents' demographic characteristics.

The following research methods are employed in the article:

- the questionnaire-based inquiry;
- the statistical data analysis.

The inquiry addressed 500 secondary school teachers from Kaunas, Kaunas County and some major and minor towns in Rokiškis region. The respondents were selected according to the probability choice of a group, the principle of "bouquet" (Kardelis, 1997). Selection of sample size was based on the data of pilot research. The answers to the survey questions were encoded and processed by Statistical Package for the Social Sciences (SPSS) Software Programmes by applying descriptive and non-parameter statistical methods (Bitinas, 1974, Čekanavičius, Murauskas, 2000).

To achieve the research aim, the inquiry addressed teachers with seven questions. The questions were closed, each of them could be answered in four possible ways: “I don’t think so”, “sometimes”, “often”, “constantly”. These answers had to reflect the manifestation intensity of each tendency of educational change. Every answer had been encoded by the numbers (“I don’t think so” – 0, “sometimes” – 1, “often” – 2, “constantly” – 3) before they were processed.

All the obtained data which includes mean scores of attitudes, medians, modes and the characteristics of the distribution of mean scores (standard deviation) and the bias of the mean score is presented in Table 1.

Tendency	**Mean score**	**Median**	**Mode**	**Standard deviation**	**Bias of the mean score**
Closer cooperation with pupils’ parents	1,38	1	2	0,91	0,041
Intensification of teacher’s professional activities	1,52	1	1	0,86	0,038
Closer cooperation between teachers	1,63	2	2	0,75	0,034
The changes in the interaction between the teacher and the learner, based on their cooperation	1,75	2	2	0,79	0,036
Increase in the possibilities of teacher’s self-expression	1,76	2	2	0,84	0,038
Educational research at school	1,61	2	1	0,76	0,034
Increase in attention to the individual needs of the learner	1,75	2	2	0,75	0,034

Table 1: Tendencies of Educational Change: Research Results

We can see above, that the highest evaluation by the teachers is put on the possibilities to increase their self-expression, to pay more attention to individual needs of the learners and on the changed communication between teachers and pupils with a focus on more active mutual cooperation. Less attention is paid to the following tendencies of educational change: closer cooperation between teachers and educational research at school. The tendency in closer relationships with pupils’ parents received the least evaluation by the respondents. The diagram of reliable intervals below (Fig. 1) demonstrates whether the latter tendency of educational change as well as the rest ones has statistically significant differences. Fig. 1, p. 200, shows, that the evaluation of the tendency of closer cooperation with pupils’ parents provides statistically significant difference compared to the evaluation of the rest tendencies. Thus, this tendency of educational change manifests least in the secondary schools.

So, the manifestation of all the tendencies of educational change is observed in Lithuanian secondary schools. The evaluation of their manifestation ranges from “sometimes” to “often”, the statistical differences are not big. The use of mode value (the most frequently repeated meaning that describes typical characteristics of the mean value) allows us to notice, that almost all the tendencies of educational change (except “educational research at school” and “intensification of teacher’s professional activities”) could be described as being “often” manifested. The two latter tendencies could be defined “sometimes” appearing.

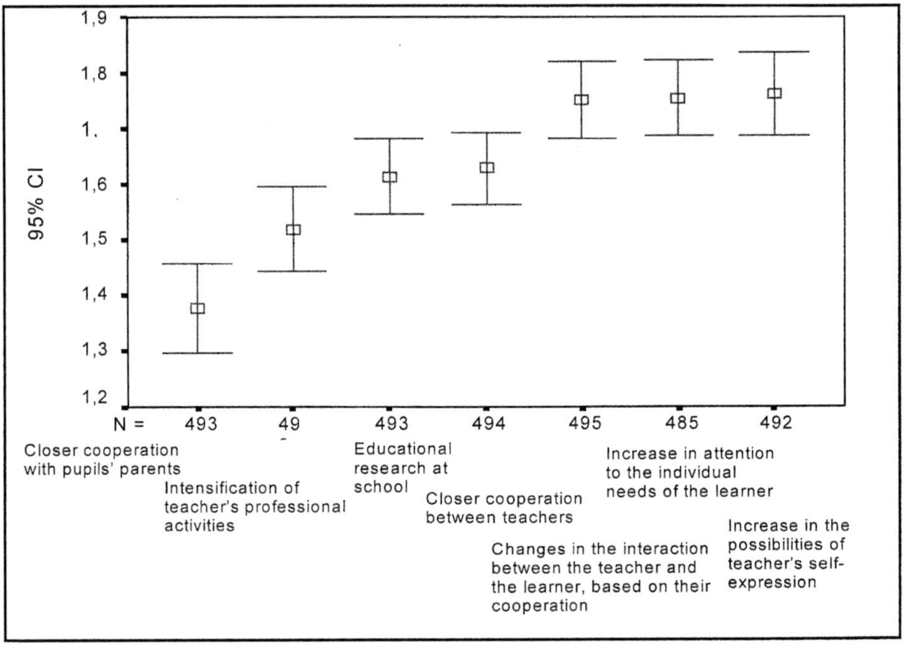

Figure 1: Diagram of Reliable Intervals in the Evaluation of Tendencies of Educational Change

Having a statistically significant difference from the rest ones, the tendency of educational change of “closer cooperation with pupils’ parents” is least manifested. In the city schools such tendencies as “educational research at school” and “intensification of teacher’s professional activities” more often prevail, while in the regional schools the tendencies of “increase in attention to the individual needs of the learner”, “closer cooperation with pupils’ parents” and “the changes in the interaction between the teacher and the learner, based on their cooperation” are evaluated higher.

In summary, we could state, that teacher activities alter in the context of all seven tendencies of educational change.

In order to determine what causes different aspects of evaluation, we performed the investigation into the dependence of tendencies of educational change on qualification category of the respondents, on the subject delivered, on statistical set of samples. For this investigation we employed the methods of non-parameter statistics: Kruskal-Wallis test and $\chi 2$. They allowed defining the dependence (very weak statistically reliable relationship) of the evaluation of some tendencies of educational change on the category of respondents' qualification (see Table 2). The dependence of the evaluation of the tendencies of educational change on the respondents' practical experience and the subject delivered was not observed.

Tendency	Qualification
Intensification of teacher's professional activities	***Mean rank:*** Teacher – 222,86 Senior teacher – 257,17 Learning Resource Manager – 216,76 Expert – 195,50 **$\chi2$=8,970** **α=0,030<0,05**
Increase in the possibilities of teacher's self-expression	***Mean rank:*** Teacher – 234,63 Senior teacher – 238,48 Learning Resource Manager – 272,61 Expert – 363,25 **$\chi2$=7,893** **α=0,048<0,05**

Table 2: Correlation Between the Tendencies of Educational Change and Teachers' Qualification and Practical Experience

Statistically significant dependence of the tendency "increase in the possibilities of teacher self-expression" on teacher's qualification could be explained by the fact that pedagogues have to meet high requirements in order to achieve higher qualification. Usually, teachers who possess higher qualification show their achievements in a variety of activities, even those that are not related to educational process. The research data show, that senior teachers provide the highest evaluation of intensification of teacher's professional activities: the higher qualification category, the lower evaluation. This can be explained by the fact that higher qualification teachers have a tendency to evaluate teacher's activities wider and attribute the functions of non-traditional roles to them.

The research in the dependence of the tendencies of educational change on statistical set of samples was also carried out. Our aim was to find out whether the characteristics that influence teachers' evaluation of the tendencies of educational change come from the inner peculiarities of the school or whether that depends on the geographical position of the statistical set of samples. So, first of all, we used Kruskal-Wallis test and $\chi2$ for the determination of the interdependence between the evaluation of the tendencies of educational change and statistical set of samples. As a result, the evaluation of the two tendencies, namely, "closer cooperation with pupils' parents" and "increase in attention to the individual needs of the learner" was observed to be not dependent on particular school. The evaluation of all the rest tendencies had statistically significant differences. To define how the evaluation of the tendencies of educational change depends on the geographical situation of schools, statistical sets of samples were divided into three groups: Kaunas city schools, Kaunas county regional schools and Rokiškis regional schools. Here some differences in evaluating the tendencies of educational change were observed. We defined whether these differences are statistically significant by using Mann-Whitney test. The research results are presented below.

1. ***Tendency "Increase in the possibilities of teacher's self-expression": evaluation differences:***
   - Kaunas city schools – Kaunas county regional schools: $\alpha = 0,038 < 0,05$
   - Kaunas city schools – Rokiškis regional schools: $\alpha = 0,107$
   - Kaunas county regional schools – Rokiškis regional schools: $\alpha = 0,890$

   Thus, the teachers of Kaunas county regional schools than the teachers of Kaunas city schools evaluate this tendency higher. The evaluation by Rokiškis regional schoolteachers does not show a difference compared to the evaluation of Kaunas city and Kaunas county regional schoolteachers.

2. ***Tendency "Closer cooperation between teachers": evaluation differences***
   - Kaunas city schools – Kaunas county regional schools: $\alpha = 0,119$
   - Kaunas city schools – Rokiškis regional schools: $\alpha = 0,287$
   - Kaunas county regional schools – Rokiškis regional schools: $\alpha = 0,655$

   Thus, the evaluation of this tendency does not show statistically-significant differences.

3. ***Tendency "Increase in attention to the individual needs of the learner": evaluation differences***
   - Kaunas city schools – Kaunas county regional schools: $\alpha = 0,007 < 0,05$
   - Kaunas city schools – Rokiškis regional schools: $\alpha = 0,003 < 0,05$
   - Kaunas county regional schools – Rokiškis regional schools: $\alpha = 0,543$

   Both the teachers in Rokiškis regional schools and Kaunas county regional schools evaluate this tendency equally but higher than the teachers in Kaunas city schools.

4. ***Tendency "Changes in the interaction between the teacher and the learner, based on their cooperation": evaluation differences***
   - Kaunas city schools – Kaunas county regional schools: $\alpha = 0,002 < 0,05$
   - Kaunas city schools – Rokiškis regional schools: $\alpha = 0,578$
   - Kaunas county regional schools – Rokiškis regional schools: $\alpha = 0,035 < 0,05$

   This tendency was evaluated highest by the teachers in Kaunas county regional schools. The average of the evaluation of the teachers in Kaunas city schools was lower.

5. ***Tendency "Educational research at school": evaluation differences***
   - Kaunas city schools – Kaunas county regional schools: $\alpha = 0,010 < 0,05$
   - Kaunas city schools – Rokiškis regional schools: $\alpha = 0,000 < 0,05$
   - Kaunas county regional schools – Rokiškis regional schools: $\alpha = 0,002 < 0,05$

The evaluation of this tendency by the teachers in Rokiškis compared to Kaunas county regional schools and Kaunas city schools present statistically significant difference. This is the only tendency by which the evaluation of Rokiškis regional school teachers differs from that of Kaunas county regional school and Kaunas city school teachers. An assumption could be made here that the reason for this is the availability of higher institutions in Kaunas, which organize Master Studies in Education for many teachers. During this course pedagogues and school principals acquire skills in research, and by practicing investigation in their schools, perceive its meaning and significance for the implementation of innovations. While in Rokiškis regional schools, which are situated far away from higher institutions, the number of such pedagogues is lower. Accordingly, the amount of research as well as the statistically significant evaluation of this tendency is less.

6. ***Tendency "Closer cooperation with pupils' parents": evaluation differences***
   - Kaunas city schools – Kaunas county regional schools: $\alpha = 0{,}000 < 0{,}05$
   - Kaunas city schools – Rokiškis regional schools: $\alpha = 0{,}123$
   - Kaunas county regional schools – Rokiškis regional schools: $\alpha = 0{,}083$

   This tendency has higher statistically significant evaluation by the teachers in Kaunas county regional schools than the teachers in Kaunas city schools.

7. ***Tendency "Intensification of teacher's professional activities": evaluation differences***
   - Kaunas city schools – Kaunas county regional schools: $\alpha= 0{,}000 <0{,}05$
   - Kaunas city schools – Rokiškis regional schools: $\alpha= 0{,}270$
   - Kaunas county regional schools – Rokiškis regional schools: $\alpha= 0{,}011<0{,}05$

   The evaluation of this tendency has statistically significant difference between Kaunas city schools and Kaunas county regional schools. The teachers in Kaunas city schools evaluate this tendency higher.

Thus, we can make a conclusion that the evaluation of the tendencies „increase in the possibilities of teacher self-expression" and „closer cooperation between teachers" depends stronger on the inner characteristics of the statistical set of sample itself. The evaluation of other tendencies of educational change in the city and regional schools differs. City teachers evaluate intensification of teachers' professional activities and educational research at school higher. However, such tendencies as increase in attention to the individual needs of the learner, closer cooperation with pupils' parents, changes in the interaction between the teacher and the learner, based on their cooperation are evaluated higher by regional schools. We suppose, this is influenced by socio-cultural environment of small towns with closer human relationships.

Weak statistically significant interdependence between evaluation of some tendencies of educational change and pedagogue's qualification was observed, however, the research shows, that the characteristics of statistical set of sample (some particular school) plays the most important role in the evaluation of the tendencies. These characteristics involve both school type features (whether it is a city or regional school) and the inner specificities of each school. Since the evaluation by the teachers in Kaunas city and Rokiškis regional schools (though some of them have statistically significant difference) contain an interval (1; 2,1), so we can state, that the tendencies of educational change might be evaluated similarly in the whole of Lithuania. The evaluation of each tendency of educational change mainly depends on the characteristics of statistical set of sample. Here we could define two aspects: first, the evaluation of some tendencies of educational change differs between city and regional schools; second, though statistically significant differences about the school dependence were not observed, but strong differences are defined if we compare separate statistical set of sample. Thus, we can come to the conclusion, that the evaluation of the respondents was mostly influenced by the school (statistical set of sample) inner characteristics. We suggest here that the influence includes school traditions, quality of management, school climate, etc. So, by improving these preconditions and by motivating in-service training of pedagogues and school principals it is possible to achieve higher results in educational changes.

Conclusions

1. Alterations of pedagogues' professional activities in the secondary school manifest themselves in the context of all seven tendencies of educational change. The evaluation of these expressions ranges from "sometimes" to "often". However the intensity of the manifestation of these tendencies is not equal: a statistically significant expression of the tendencies of "closer cooperation with pupils' parents" and "intensification of teacher's professional activities" is weaker.
2. The evaluation shows that the dependence of some tendencies of educational change on the respondents' demographic characteristics is low, while, statistically significant dependence on the characteristics of statistical set of sample is observed. It says that the evaluation differs between the country and city schools. Also, the evaluation highly depends on the inner peculiarities of schools (school traditions, climate, quality of management, etc.).

References

BITINAS, B.: Statistical Methods in Pedagogy and Psychology. Kaunas 1974 (Šviesa).

ČEKANAVIČIUS, V./MURAUSKAS, G.: Statistics and its implementation. Vilnius 2000 (TEV).

DABRIŠIENĖ, V.: Pedagogue's Activities of Individualization Educational Programmes in the Lithuanian Secondary School. Kaunas 2002 (Technologija).

FULLAN, M.: Change Forces. Penetration into the Depths of Educational Reforms. Vilnius 1998 (Švietimo studijų centras).

HARGREAVES, A.: Changing Teachers, Changing Times. Vilnius 1999 (Tyto alba).

JANIŪNAITĖ, B.: Introduction to Comparative Education. Kaunas 1999 (Technologija).

JUCEVIČIENĖ, P.: European Educational Dimensions in Lithuanian Higher Education. Systems and Didactics of Higher Education. 1 book. Kaunas 1998 (Technologija), pp. 3-12.

KARDELIS, K.: Methodology and Methods of Scientific Research. Kaunas 1997 (Technologija).

STOLL, L./FINK, D.: Changing School. Vilnius 1998 (Margi raštai).

Gediminas Merkys, Viktorija Baršauskienė, Dalia Antinienė

The Structure of Social Network: a Diagnostic Study of Lithuanian Students' Population

1. The concept of social network and the rationale behind its empirical investigation

The phenomenon of social network is at the centre of interest of many social and behavioral sciences, such as sociology, psychology, education, cultural anthropology, political science, management, etc. This is natural because the psychosocial development of an individual and the social network are inseparable. The theorists of socialization more or less agree that personality development is a result of two factors: (i) internal mental reality of a person; (2) external social structure. Personality structure reflects the social (micro and macro) structure. L. Vygotsky, a classic of sociocultural psychology, was one of the first who introduced this notion in early 20th century. He argued that any higher (cultural) mental function emerges in human (child's) life at least twice: for the first time, as internal mental reality (through culture, symbols, human relations, historically developed examples of action), and for the second time - as intra-psychic reality which is internalized by individuals as they interact with more experienced society members (Vygotsky, 2000). A few years later a very similar idea was suggested by the author of social learning theory, an American of Romanian origin Bandura (1977): person's socialization is determined by inherent qualities and internal psychological processes, on the one hand, and by social environment, on the other hand. Individuals adopt (learn) social experience when other people, who often belong to the microsocium, demonstrate its examples: parents, friends, teachers, etc. Exceptional role of the cultural and social microenvironment in human socialization was emphasized also by the classic of modern sociology Bourdieu (1977, 1982), who introduced such concepts as 'millieu' and 'habitus'. Finally, a well-known contemporary theorist of socialization Hurrelmann (1993) also argues that socialization is a result of interaction between the internal and external social reality of individual. If we regard the micro and macro levels of social environment as the fundamental preconditions of human development, the concept of social network is one of the most suitable theoretical concepts which can be used for modeling these preconditions deeply and essentially. A heuristic concept of the network is well reflected in the proverb which can be found in many languages: 'tell me who your friends are, and I will tell who you are'.

The rudiments of the concept of social network emerged in 1930, when Moreno, Lewin, Heider and others started analyzing, quantifying and depicting graphically human relations. J. Moreno with his sociometrics could be regarded as one of the key originators of empirical research on social networks. In his book 'Who Shall Survive' (1934), Moreno described the sociometrics as a method for analyzing interpersonal relations in a small group. It was the first impulse for empirical research on social networks. Another big impulse for social network research was Cartwight's and Harary's application of the mathematical graph theory for analysis of group behavior in the 1960s (Scott, 1991).

For a long time, researchers focused their attention on small groups and institutional structures. Their aim was to identify the points of social network (social actors) and to look for connections between them. After the 1950s, the focus changed to analysis of structural characteristics. In the 1970s social network research was strongly influenced by mathematical sociology. A group of sociologists at Harvard University under the supervision of H.White started applying a big innovation of the time – information technology, which in turn opened the new possibilities of using various mathematical models and methods. White initiated the publication of the journal 'Social Network' which featured the articles of scholars from different areas of social sciences. This was a clear indication that the theoretical and empirical studies on social networks got institutionalized as a relatively autonomous interdisciplinary field of social research. Today the development of computer technology again gives a new impulse for social network research, making it possible to statistically model huge amounts of data and analyze the actual complex structures of social human relations (Scott, 1991).

Review of the most recent research on social networks shows that a significant number of both theoretical and empirical studies have been conducted. Some studies of philosophical nature can even be regarded as rather speculative. For example, the advocates of so-called 'biopolitics' attempt to prove the analogy between bio-social structures and society. In their view, the social order of society should imitate the order existing in the nature. They are engaged in the study of the harmony of biological structures and attempt to apply this model to the formation of civic society as a social structure (Oleskin, 1998). Nevertheless, empirical genre is dominating like it was in the beginning of research on social networks. Its object is networks of people of certain age and social status groups: teenagers, youth, adults, senior people (Clark, 2002; Wills, 2002; DiClemente et al., 1996; Marlatt and VandenBos, 1997; Gordon et al., 2002, and others). Some authors analyze the gender issues of social networks (Brass, 1985; Baron and Pfeffer, 1994), others – the structures of various communities, from primitive to contemporary society (Wellman, 1979, 1982; Smith, 1979, and others). A large number of studies are focused on the relationship between social network, physical and mental health (Melchior et al., 2003).

In the context of the abundant and diverse research on social networks, one can argue that a more or less uniform conceptualization of the social network as a theoretical construct is emerging. Scholars of various fields agree that *social network* is a structure of social relations or connections (their characteristics) between people (groups). Social relations within the network can be of very different kinds. Every person (group) is a part of a unique social network.

The concept of social network has been developed to reflect the contexture of social relations and to emphasize that these relations are much weaker than in a typical social group (Matulionis, 2000). Differently from social group, a network is a community, the boundaries of which are not clearly defined and the relations in which are much weaker, sometimes even accidental.

In most cases, researchers distinguish between three levels of analysis of the social structure: macro, mezzo and micro. It implies that networks can be analyzed at the levels of individual, group, organization, community and society. Since recently, priority is given to those methods, which analyze individual's embeddedness in the social structure and the structure's formation based on individual micro relations (Hanneman, 1998). For this reason, in this paper we analyze networks at micro level. Out of two possible types of networks (egocentred and systemic), we have chosen an ego-centered network emphasizing an individual and the type of relationship – 'individual-individual'.

Ego-centered network provides a thorough view of the 'local' network or close relations. It is useful when it is attempted to identify the influence of the social network on individual. Besides, ego-centered networks provide knowledge about certain qualities of the macro structure.

Complexity of interpersonal relations (they can be positive and negative, mutual and one-way, close, intimate and superficial, long-term and short-term) makes research on social networks complex, too. In recent years different research methods were used: survey, observation of interactions, content analysis, etc. Contemporary research methods used in the area of interpersonal relations allow not only to describe, but also to show graphically and to measure their certain parameters. Today all contemporary social research methods are used, or at least could be used in the social network research.

Different researchers of social networks emphasize their different dimensions. Summarizing various approaches, it can be argued that the main characteristics of network are: size (number of the network members), density (extent to which members are related to each other), nature of the relationships (extent to which members can be defined by traditional categories: friend, neighbor, coworker, etc.), and homogeneity (extent to which network members are similar to each other). Relations in social network are characterized by the frequency of contacts, complexity (number of various transactions and support received from the network relations), duration (time period for which individual knows other individuals), and mutuality (extent to which relations are mutual) (Berkman, 2002).

Social network is one of the key factors of person's social integration and social adaptation. It is the milieu in which personal development takes place (Suslavičius, 2000). The key motive behind the formation of the social networks is security. According to Ch. Kadushin, this motive emerges in the early childhood. Pursuit of security is related to dense social networks and social support resulting from them (Kadushin, 2002). An average middle-aged healthy adult has a social network of 55 people (Sullivan, cited in Suslavičius, 2000). Networks of senior people considerably diminish due to many losses (of close people, job, etc.). Normally person's social network consists of individuals of very different competence. Therefore it is very important to which extent an individual through his/ her social relations will be able to use the capital of social network, i.e. resources which are in other members' disposition. The wider the social network and the higher the quality of the relations targeted at a certain aim, the more efficiently the social capital is used. So, a wide and efficient network enables individual's positive social adaptation and integration. One of the key functions of the social network is support, which depends on the size of the network, quality and frequency of relations, etc. The wider the network and the closer the relations, the higher the probability of support is. Greater variety of competences and statuses of the network members produces a potentially greater number of support types. Some relations can provide a person with very diverse support, while others – with highly specialized support of one kind. Social network plays an exceptional role also in the processes of social mobility and career.

Discussions of the concept of social network are often dominated by the emphasis on the socially positive aspects. This is only natural due to highly emancipative attitude, which has always been common in science, and especially social sciences. One of the classics of the structural functionalism Merton (1974) has accurately noted an unconscious inclination of many social researchers to give the priority to the positively functional processes and to relatively ignore the studies of dysfunction. Nevertheless, in certain circumstances, social network can be regarded as a phenomenon 'with a minus sign'. Dysfunctions of social network often turn into a source of many social problems. Here it is worth mentioning a well-known sociologist Durkheim and his classical work 'Suicide', where he suggested a theoretical idea about the impact of social integration and disintegration on the social anomie and suicide statistics. Moreover, this theoretical hypothesis was justified and even relatively confirmed by empirical data, which were very innovative at the time. It is obvious that at least some of the operationalizations of the independent variable used in that study today could be based on the concept of social network and its indicators.

The relationship between the social network and the mental and somatic health has been confirmed by a number of contemporary empirical studies. A number of studies proved that social network impacts mortality, mental disorders, especially depression, and life quality (Achat et al., 1998; Kawachi et al., 1996; Seeman et al., 1993; Vogt, Mullooly, Ernst, Pope, Hollis, 1992; Fratiglioni, Wang, Ericsson, Maytan, Winblad, 2000; Stansfeld, Real, Head, Shipley, Marmot, 1997; Stanfeld et al., 1998; Oxman and Hull, 2001 cited in Melchior et al., 2003). Functional social network and social support performs the role of a 'pillow' for person's health and mental state, when s/he encounters stressful life events. In various crisis situations (unemployment, loss of close people, divorce, serious injury or sickness), the functionality/ dysfunctions of social network often become the key factor, which determines the development of more or less successful, or critical and sometimes hopeless scenario of person's social evolution.

Social network is related to various forms of asocial behavior. According to the theory of social learning, aggressive behavior is mostly learned in the family, to some extent in the peer groups, and to some extent from mass media. Researchers managed to prove some effects of modeling and imitation. For example, it was learned that most of the young prostitutes before establishing their behavior have known at least one professional prostitute. Likewise smoking, drinking and sexually aggressive teenagers have in their social network at least a few friends exhibiting such behavior (Marlatt and VandenBos, 1997). Nevertheless, the role of social network in terms of genesis of individuals' asocial behavior and deviation cannot be fully explained merely by social learning or milieu. As it has already been mentioned, social network as a phenomenon and concept seems to have an even stronger relationship with such constructs as social isolation, social exclusion and social integration. Here again we have in mind both the actual phenomenon and its theoretical conceptualizations. For instance, among the reasons of drug-abuse, alcoholism, prostitution and other forms of asocial behavior, emotional alienation of family members, peers and friends, incomplete families, etc. are often pointed out. So, the 'gaps' of the social network and inadequate quality of the relationships can cause the development of asocial personality. However, contemporary theories of socialization and the practice of social work show the crucial importance of various technologies of social inclusion in the social adaptation and integration of various socially disadvantaged groups of people: prisoners, asocial teenagers, disabled, drug-users, homeless, immigrants, refugees, prostitutes, etc. The central role in these technologies of practical inclusion and integration is in turn played by social worker and client's work with the network and work inside the network. This work includes adjustment and reconstruction of the network, its mobilization and focus, creation of new network formations and maintenance of the existing ones, expansion and diversification of the network, etc.

Despite a large number of network studies, scholars continue to embark on the new research and search for more effective methods. In our view, one of the highly promising fields is diagnostic research of social networks. From the perspective of both theory and practice, it is important to have valid and economically efficient instruments enabling one to 'uncover' and evaluate the social network of the specific person or group. They are important for both socioeducational work and pure theory. Social networks play a crucial role in providing social and psychological helps to individuals, implementing various rehabilitation programmes, etc. On the other hand, it is rather trivial that in most of the empirical studies of socialization, social network normally appears as a highly significant independent variable, as an external predictor explaining functions and dysfunctions of person's psychosocial evolution process.

Even this improvized review of social network research, which could be more precisely defined as an impression rather than full analytical study, enables one to draw one definite conclusion: there exists in the Western world a mass 'research industry' in the area of social networks which is characterized by interdisciplinary direction and diversity of themes. The sate of such research in Lithuania is very different and disappointing. Here no evidence of significant and professional research on social networks exists. This is strange, especially as Lithuania, likewise other post-socialist countries, is witnessing rapid and mass development of the social services sector, which for obvious reasons did not exist in the Soviet era. We have in mind emergence of such professions as social worker and social educator, university departments and journals, initiation of regular academic conferences in this area, etc. All these facts suggest that there exists actual demand in the society for fundamental and applied research in this area. Regretfully, national scholars failed to react to it adequately. Meanwhile the necessity for the national (regional) level research can be justified by a number of important circumstances.

One may ask a question to what extent the conceptualizations of social network developed in Western countries and especially the empirical statistical generalizations can be projected on and applied in the post-soviet cultures? Those acquainted not only with the research methods, but also with the epistemology of social sciences are well familiar with the chronic disease of the social sciences: highly abstract (not to say speculative) doctrines at one extreme versus many super-specialized very context-specific empirical 'mini-theories' at the other. There is a lack of so-called middle range theories, which would have a relatively broad scope of validity and at the same time be empirically tested. All theories of social sciences are limited in time and space, so normally they cannot be formulated using general propositions of the kind 'for all Xs, if a, then b'. Although only relative universality of a theory is possible, it is undeniable that the broader the scope of the theory's validity, the greater number of psychosocial phenomena of a certain kind it explains and the more mature and valuable it is. Consequently, replication in East European cultures of the Western conceptualizations of social network and especially of statistical generalizations enables empirical testing of the limits of theory's validity and universality. On the other hand, different cultures often naturally create quasi-experimental situations, which can be usefully exploited for deeper understanding of the society and individual. For example, theoretical sociology suggests that modernization and such its inevitable satellites as economic welfare and individual autonomy in terms of personal freedoms and rights as well as individual's economic independence are destroying certain traditional social formations – religiousness, communitarianism, stability of traditional family, etc. Thus, one can justifiably assume that social networks in the West and in the Eastern Europe, the culture of which has been relatively less affected by modernization, can have different configurations. More specifically, it is likely that the motives of social and particularly economic security as well as orientation to family and relationships with other close people are more prominent components of the social network in the Eastern Europe, compared to more emancipated and individualized Western society. So, research on social network in a relatively unique national milieu can become a fundamental epistemological problem. In any case, it can become a research problem which is much more important than simply elimination of the gaps in empirical knowledge in a specific small regional research community. The best way of raising and dealing with these problems would be conducting cross-cultural studies of social networks, using both so-called *Etic* and *Emic* methodologies[1] (Matsumoto 2000; Thomas 1993; Berry 1992; Merkys 1999; Triandis & Lambert 1980).

2. Research sample, methods and key results

The results of the social network study presented in this paper are concerned with the student population. More specifically, several researchers of the Faculty of Social Sciences at Kaunas University of Technology are conducting a rather large scope[2] study dealing with the national identity of Lithuanian students,

---

[1] According to the *Etic* methodology, psychosocial phenomena are universal, cross-cultural and generalisable. They can be identified, examined and compared using a uniform methodology. Simply different cultures are characterised by different manifestations of these phenomena, especially quantitative ones, whereas the aim of cross-cultural research is to reveal these differences. The *Emic* methodology is based on the opposite view: psychosocial phenomena are far from universal, can be hardly generalised across cultures and directly compared in terms of statistical indicators, are highly situation-specific, culture-specific, etc.

[2] The overall research sample is 1,002 respondents from various Lithuanian universities and colleges. The dependent variables include national identity, ethnocentrism and emigration orientations, in total 181 items. They were filled by all 1,002 respondents of the sample. Various independent variables were operationalised into 213 items. They were filled by a smaller number of the main sample, which ranged from 245 to 305.

manifestations of ethnocentrism and emigration orientations[3]. The matrix of social network consisting of 35 items was included as one of the independent variables capable of explaining the emigration orientations of the young people (see Table 1). 305 students, i.e. 30.4% of the main sample, filled this matrix[4]. The 'personal' structure of the network was operationalized by the closest social environment – parents, spouses, friends, etc. (7 network subjects in total). The functional structure of the network was defined using such key areas as emotional relations, mutual trust and social-economic security (5 functional areas in total). Every network subject was evaluated in every functional area (7 x 5 =35), using a dichotomous answer format – 'yes' or 'no'. It is interesting to note that the hypothetic expectations of the research group, which regarded the network structure as a factor weakening students' orientation to emigrate, have been hardly confirmed. In the overall 25-item matrix, a statistically significant relationship between the network and emigration orientations was detected only in the population of female students and only in the subscale 'father/mother'. In the light of such relative disappointment and taking into account, on the one hand, the essential role of the social network in person's social development and, on the other, the deficit of empirical studies on social networks in Lithuania, it was decided to analyze the received network data as (a) a closed structure of the attributes and (b) as a psychometric construct, with the view of using this matrix in the future research.

Statistical data received within the 7 x 5 matrix where analyzed on the level of individual items and by creating the subscales. Results by individual network items are shown in Table 1. The dichotomous format of the answers was coded as follows: 0 points ('no') and 1 point ('yes'), so the means can be calculated by dividing percentage estimates shown in the table by 100. It should be noted that the result table (matrix) was produced by ranking. The ranking (from the top to the bottom) begins with the most important network actors (subjects) 'father/mother' and ends with the category of peripheral actors – 'neighbors'. The ranking from the left to the right begins with the network function '*I can borrow money from*', which was selected most often, and ends with the function '*people turn to me for help...*', which was selected least often. To make them more vivid, not only numbers presents the results, but also by color tones: the darker the tone the stronger a certain network dimension by a specific subject and network function is.

---

[3] Emigration of young people, especially students and graduates, to economically developed and attractive Western countries is growing dramatically. This situation is regarded as becoming a significant social, economic and political problem.

[4] The main demographical characteristics of this partial sample: 41.8% male, 58.2% female, age – 18-33 years, Me=20.8, SD=1.90.

	I can borrow money from	I can communicate openly	I can ask for help and delegate important matters to...	If I lost my apartment I could stay with...	People turn to me for help or to discuss their problems with me	MEAN	Cronbach a
Father/mother	96,0	82,5	96,4	99,6	74,5	89,8	0,60
Wife/husband	87,7	93,0	90,4	83,8	87,7	88,5	0,76
Friends	90,5	93,5	86,4	73,0	93,8	87,4	0,72
Brother/sister	81,5	82,0	85,6	84,6	80,5	82,8	0,73
Relatives	66,3	44,6	54,7	80,9	38,0	56,9	0,69
Schoolfriends	61,9	59,4	44,4	27,7	54,2	49,5	0,77
Neighbors	37,8	20,5	17,1	20,5	24,3	24,0	0,78
MEAN	74,5	67,9	67,8	67,2	64,7	68,4	
Cronbach a	0,46	0,47	0,44	0,39	0,44		

up to 30%	Note: the percentage of "YES" answers is shown in the table

Table 1: Ranking of the student's social network (N=305)

It was interesting to perform certain typical psychometric calculations. The internal consistency of scales constructed according to the groups of the network subjects turned to be rather high. For example, the minimal Cronbach-alfa coefficient was 0.60, while its maximum value was 0.78 and its mean by 7 subject scales was 0.72. As it could be expected, when grouping the same 35 items into scales by network functions, the internal consistency was much lower: from 0.39 (minimal value) to 0.47 (maximum value), while the mean was 0.44. This result (low internal consistency by network functions and relatively high internal consistency by network actors) can be explained from theoretical point of view. For example, in case of very close social and psycho-emotional relations with parents (siblings), it can be expected that such 'close' network subjects will provide a multifunctional support: emotional, financial, temporary housing, etc. In other words, in this case it is likely that the network will work positively and will be functional in all life situations. And on the contrary, in the absence of close social relations with network subjects, the functionality of the network is likely to be weak, formal and episodic. One could hypothetically argue that with inclusion of the new items into the matrix, e.g. sincere and systemic help in case of serious illness, disability, becoming a hostage etc., the internal consistency of scales related to network actors would further increase. The fact that the consistency of scales related to network functions is much lower is quite logical. While parents and siblings can help financially or by providing shelter, neighbors and classmates can refuse this kind of help. Thus, it can be hardly expected that some network function will be strong in all rather different groups of the network subjects. It would be difficult to increase the internal consistency of the test even by increasing the number of items. This theoretically meaningful configuration of the internal consistency coefficients can be regarded as an indirect evidence of the construct validity of the test (social network matrix).

Descriptive statistics of the matrix clearly shows that the key figures of the network are parents, siblings and spouses (Table 1). In other words, the closest network exists among people of the immediate environment related either by blood or by marriage. This assumption is confirmed by results of the MDS (Multi-Dimensional Scaling) (Figure 1, p. 215).

It should be noted that the social network actors in the MDS diagram are ranked in the theoretically meaningful way, which is very important when applying this statistical model. The vertical axis reflects the dimension of native social ties versus selected ties. The horizontal axis reflects the dimension of the distance of social ties. In reality, individuals are not able to choose parents and relatives, but they are relatively free to choose spouses and friends (see vertical axis). On the other hand, parents and spouses are closer network subjects than neighbors. The fact that network subjects in the MDS model are ranked in a rather meaningful way again indirectly confirms the construct validity of the network matrix. Also, the MDS model suggests that one can distinguish between the immediate environment network (parents, spouses, siblings, friends), which was mentioned on average in 87.1% of cases, and the further environment network (classmates, relatives, neighbors) – 43.5% of 'yes'.

It should be noted that the key function of the network, marked by a much higher estimate than the others, is economic, i.e. borrowing money (see Table 1). The importance of the financial help function and its distance from other network functions is clearly reflected in he MDS diagram (Figure 1). The vertical axis reflects the dimension of non-financial versus financial help. The horizontal axis can be defined as a dimension of instrumental versus emotional support. For instance, on the left side of the scale (at a distance from 2.4 points from the axes centre), one can see a purely instrumental function 'if I lost my apartment I could stay with', followed by delegating important matters, while on the right side (at the distances of 1.0 and 1.5 points) one case see purely emotional functions: open communication and mutual trust. So, the MSD model is theoretically relevant both in terms of network actor groups and key network functions.

It is worth discussing in more detail the fact that the financial-economic function of the network is obviously dominating. This result can be partly explained by the specific character of the student population. Students are mostly young people without stable and sufficient income ensuring full economic independence. So, students' social role partly explains the situation, when, on the one hand, they lack money and borrow it from time to time, and on the other end, the network subjects of their immediate and further environments respond to these requests for help. One can hypothetically assume that such social perception of the students' role can be a phenomenon which is systematically repeating itself in different cultures: Western and Eastern Europe, Far East, etc.

From a different perspective, this result is in a way unexpected. Likewise other East European states, Lithuania is a country with scarce resources, which is undergoing dramatic social and economic transformation. Most of its citizens have low or medium income[5], still small middle class is developing rather slowly. In this context, 'leadership' of the financial help in the overall structure of the network functions looks really peculiar. The strategic importance of the other measured network functions (getting a shelter, delegating important matters) as well as test instruction[6] suggests that the respondents did not mean borrowing a small amount to survive till next income. Here one means help in case of the trouble. Besides, item formulations reflected not students' expectations and needs form the network, but the actual support received from the network, e.g. 'I can borrow money from'. Thus, in the culture of relatively mass material shortage, the most common function of the network in a paradoxical way becomes economic and financial support.

In this context, it would be interesting to conduct cross-cultural comparison: would the financial-economic function of the social network be at the top in the similar student populations of the more wealthy Western societies? What would be the ranking and the structure of the network actors and network functions? In this context, along with such trivial interpretations and questions it is worth developing the hypotheses based on certain traditional theories and concepts. In this respect, one of the productive concepts could be 'modernization' and Inglehart's (1997) theory of material and post-material values. Perhaps it even makes sense talking about the symbiosis of these two theoretical constructs as an ad hoc hypothesis to interpret the above-mentioned fact. It can be justifiably argued that post-material values (self-expression, interesting occupations, etc.) are more common for modern societies, whereas in the societies less affected by moderni-

[5] At present, the average monthly salary is about 1,000 Litas, i.e. about 330 US dollars or 290 EUR. The major part of the average citizen's income is spent on the basic needs: food, housing, transport, while savings and investment are rather rare.

[6] Test instruction begins with words 'This test will show the circle of your acquaintances and whether you will have someone to turn to in case of a trouble…'

zation material values (economic welfare, security, stability) are likely to be stronger.

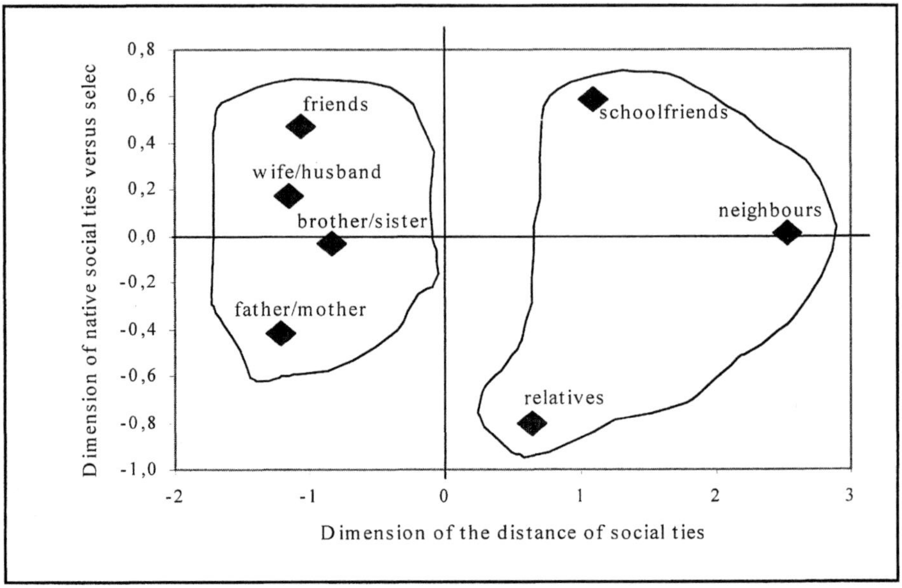

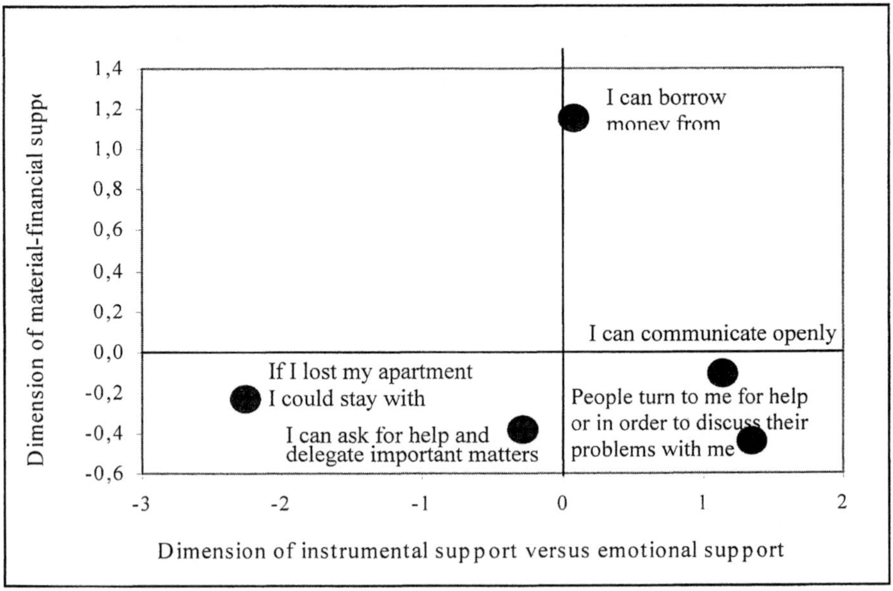

Figure 1: MDS diagram according to the subjects and functions on the network

Further it is worth noting that modernization is, on the one hand, related to (i) deep and very complex organizational 'crystallization' and specialization of the society with well developed services, including social ones; (ii) relatively higher material welfare and related economic and personal autonomy. On the other hand, modernization in part destroys certain traditional structures: traditional family and community based on merely native or blood ties. Besides, it relatively weakens individuals' attachment to the traditional material values. In the light of this argument it is likely that in the modern society, the configuration of network actors and functions can have a completely different ranking than in an East European country or in some region which is even less affected by modernization. In the modern society, the relative 'weight' of the network actors unrelated by kinship or blood ties can be bigger. Further, it can be expected that financial help will not be the first priority in the functional structure of the network. It is more likely that financial aid problems in such societies will be addressed not through 'personal network', but through various modern institutions, such as banks, including student loans, half-time and summer jobs, etc. Due to the influence of post-material values, emotional relations should take the more prominent positions in the network's functional structure. In other words, one can hypothetically assume that in the network structures of Western culture students, the network actors unrelated by blood or marriage will be relatively more prominent. At the same time, the functions unrelated to economic and financial support will be at least relatively more important. In other words, it can be expected that the structure of the network actors and network functions will be more balanced and will not have such prominent asymmetry along the dimensions of native ties versus selected ties and the dimension of financial versus non-financial support.

In any case, this interpretive ad hoc hypothesis could be tested through empirical cross-cultural studies in homogeneous student populations and, if possible, in other populations. These studies should be aimed at comparing networks in at least several Western countries, Eastern Europe and several even more traditional cultures. This effort would enable expanding (or at least testing) the boundaries of the validity of the current empirical network theory. On the other hand, it would allow validating the universality and the prognostic interpretive power of some traditional theories of the higher level of abstraction.

At the same time, it is worth thinking about more local studies, meaning purely practical application of the network matrix as an assessment instrument. Taking into account that (a) a relatively big and rather homogeneous sample was developed; (b) this sample represents clearly identifiable target population – Lithuanian students; (c) good or at least satisfactory indicators of test's psychometric quality were received, it is reasonable to regard the accumulated statistical data of the social network matrix as a test norm. This opens a possibility to make individual assessment decisions about the social network of the specific student. In the view of the network matrix authors, such individual assessment procedure can be useful in various circumstances: first of all, in psychological counseling and psychotherapy of the students on routine basis and in crises; secondly, when applying various rehabilitation programmes after certain traumatic events; thirdly, when providing social support and socio-educational services to students. It should be noted that the assessment procedure is very economical in terms of time and organization. Being rather compact, the social network matrix can be simply attached as an additional module to any package of the diagnostic instruments without considerably expanding it. It is worth assessing the results of the individual respondents at the subscale level (quantitatively) and according to all 35 cells of the table (qualitatively). A particularly useful assessment is according to the network subject scales since they are characterized by a rather high internal consistency. Assessment can be made also by the function scales, which can be regarded as logically valid. However, due to the objective features of the measured phenomenon, research instruments are not very homogeneous. Also, assessment can be made on the basis of the total average test point, which in this study is 0.68 (68%). The main descriptive characteristics of this joint normative indicator are as follows: SD = 0.15, SE = 0.015, Mo = 0.67, MIN = 0.18, MAX = 0.93.

The authors of the described network matrix will be happy to provide interested researchers with all test-related information necessary for cross-cultural studies, from aggregated values – subscales and item estimates, and factorial matrixes to the primary SPSS matrix of statistical data.

Conclusions

1. In the West there exists a massive research industry of social networks. In Lithuania, in spite of the increasing demand of social network research, it is not carried out purposefully.
2. The empirical research of Lithuanian student social network has showed that the answer internal consistence of the "social network matrix" scales comprised according to the network stakeholders is relatively high, whereas the answer consistence according to functions is low. Contiguous social and psycho-emotional relationship is associated with the poly-functional social support of "close" network stakeholders. On the contrary – if there is no close social relationship between network stakeholders, social support will be weakly expressed, formal and episodic.
3. There are two social network structures: a) close environment network that consists of parents, spouse, brothers or sisters, friends; b) further environment network comprising study fellows, neighbors, relatives.
4. In the scale comprised according to network functions, the financial – economical network function predominates. This might be explained by specific student population or

social and economical situation of Lithuania. Presumably, it would be possible to get different configurations of network participants and functions in the modernized Western culture, than in Lithuania.

5. It is reasonable to carry out cross-cultural stages of social network, which would expand the validity limits of the empiric network theory, and would validate the universality and prognostic interpretational capacity of some traditional theories of higher abstraction level.
6. By expanding the normalization basis, the "social network matrix" will be considered as an independent diagnostic instrument or a separate diagnostic module that is to be integrated into other research and instruments.

References

BANDURA, A.: Social learning theory. Englewood Cliffs, NJ 1977 (Prentice-Hall).

BARON, J./PFEFFER, J.: The social psychology of organizations and inequality. Social Psychology Quarterly 57(1994), pp. 190-209.

BERKMAN, L.F.: Social integration, social networks and health. In: International Encyclopaedia of the Social and Behavioral Sciences. 2002.

BERRY, J.W. ET AL.: Cross-cultural psychology: Research and applications. Cambridge 1992 (Cambridge University Press).

BOURDIEU, P./PASSERON, J.C.: Reproduction in education, society and culture. London 1982 (Sage).

BOURDIEU, P.: Die feinen Unterschiede. Kritik der gesellschaftlichen Urteilskraft. Frankfurt 1982.

BRASS, D.J.: Men's and women's networks: A study of interaction patterns and influence in an organization. Academy of Management Journal 28, pp.327-343.

CLARK, M.S.: Social relationship in adulthood. In International Encyclopaedia of the Social and Behavioral Sciences 2002.

DICLEMENTE, R.J./HANSEN, W.B./PONTO, L.E.: Handbook of adolescent health risk behavior. New York 1996 (Plenum).

GORDON, R.A./SAVAGE,C./LAHEY, B.B./GOODMAN, SH.H./JENSEN, P.S./RUBIO-STIPEC, M./ HOVEN, CH.W.: Family and neighbourhood income: Additive and multiplicative associations with youth's well-being. Social Science Research. 2002.

HANNEMAN, R.A.: Introduction to social network methods. Sociology 157(1998).

HURRELMANN, K.: Einführung in die Sozialisationstheorie: über den Zusammenhang von Sozialstruktur und Personlichkeit. Weinheim, Basel 1993 (Beltz Verlag).

INGLEHART, R.: Modernisation and postmodernisation: Cultural, economic and political change in 43 societies. Princeton 1997 (Princeton University Press).

KADUSHIN, CH.: The motivational foundation of social networks. Social Networks, 24(2002)1, pp. 77-91.

MARLATT, G.A./VANDENBOS, G.R.: Addictive behaviors: Readings on etiology, prevention and treatment. Washington, DC. 1997 (American Psychological Association).

MATSUMOTO, D.: Culture and psychology: People around the world. Belmont, CA 2000 (Wadsworth).

MATULIONIS, A.V.: Sociologija. Vilnius 2001 (HOMO LIBER).

MELCHIOR, M./BERKMAN, L.F./NIEDHAMMER, I./CHEA, M./GOLDBERG, M.: Social relations and self-reported health: A prospective analysis of the French Gazel cohort. Social Science & Medicine Vol.56(2003).

MERKYS, G.: Spachliche adaptation und transkulturelle validierung der erhebungsinstrumente. Magister: An International Journal for Psychology and Education Vol.6(1999).

MERTON, R.K.: Funktionale Analyse. In Claus Muehlfeld und Michael Schmid, Soziologische Theorie. Hamburg 1974 (Hoffmann und Campe Verlag).

OLESKIN, A.V.: Setevyje struktury sovremennovo sociuma s tochki zrenija biopolitiki. Politicheskije issledovanija 1. Moskva 1998 (Polis), pp. 68-86.

PATZELT, W.: Socialwissenschaftliche Forschungslogik. Einfuhrung. Munchen, Wien 1986 (Ouldenbourg).

SCOTT, J.: Social network analysis. A handbook. London 1991 (SAGE Publications).

SMITH, R.M.: Kin and neighbours in a thirteenth century Suffolk community. Journal of Family History 4(1979).

SUSLAVIČIUS, A.: Paremiamoji psichologija. Kaunas 2000 (Šviesa).

THOMAS, A.: Kulturvergleichende Psychologie: Eine Einfuhrung. Gottingen 1993 (Hogrefe).

TRIANDIS, H.C./LAMBERT: Handbook of cross-cultural psychology, Vol.1: Perspectives. Boston 1980 (Allyn & Bacon).

VYGOTSKIJ, L.S.: Psichologija. Moskva 2000 (EKSMO Press).

WELLMAN, B.: The community question: The intimate networks of East Yorkers. American Journal of Sociology, 84.

WILLS, T.A.: Adolescent health and health behaviors. International Encyclopaedia of the Social and Behavioral Sciences. 2002.

Rūta Urbanskienė, Virginija Chreptavičienė, Rūta Janulevičienė

Creating Closed Partnership between Business and Science through Social Capital Networks

Introduction

In today's global economy, social capital plays the key role in creating closed partnership networks between business and science. As an extension of the resource-based view, the relational view maintains that competitive advantage derives not from firm-level resources but also from difficult-to-imitate capabilities embedded in network relationships (Dyer & Singh, 1998; Lane & Lubatkin, 1998). By building relation-specific assets, knowledge-sharing routines, and effective relational governance mechanisms into relationships, firms can leverage their relational resources for knowledge acquisition and transaction costs savings. Given that resource limitations of younger firms make them prone to liabilities of newness and adolescence (Amburgey, Kelly & Barnett, 1993; Stinchcomber, 1965), this perspective helps to explain how and why some entrepreneurial firms are able to survive, thrive, and grow despite the lack of significant firm-specific resources.

It is widely believed that social networks exist comprising the fabric of social capital in the community, which have a very desirable quality of life and have a great economic growth potential. Rapidly expanding literature defines social capital as the "...stocks of social trust, norms and networks that people can draw upon in order to solve common problems". These networks involve activities of "civic engagements" such as small business associations, educational institutions, service and charitable clubs, volunteerism and the like.

Another extension of the resource-based view concerns the importance of knowledge as a source of sustainable competitive advantage, as advocated in the knowledge- and learning-based views of the firm (Kogut & Zander, 1992; Grant, 1996; Spender, 1996). Knowledge acquisition is particularly important for technology-based firms: generating knowledge is the core competence of such firms, and survival in high-technology sectors demands that knowledge be continually and rapidly replenished (Lane & Lubatkin, 1998).

As the creation, acquisition, and application of knowledge are predominantly social processes (Kogut & Zander, 1992), social capital may be critical for the long-term success of technology-based firms. Therefore, in the process of the partnership establisment, social capital refers to the network structures of social capital, which relate to and interact with each other in order to solve problems for the common good.

The research problem of the *article* is to investigate theoretically and practically "the best" integrated approach, which arises if organizations act in the same economical conditions but some of them perform better than others and their interest are served better than the interest of others.

The goal of this article is to explore the extent to which technology-based young firms can leverage inter-organizational relationships in order to build competitive advantage. We consider that the degree to which firms can leverage their external relationships for knowledge acquisition and competitive advantage is regulated by the amount of social capital embedded in such relationships. In this context, social capital in a relationship enables the firm to tap into the knowledge resources of its exchange partner. The best practice analysis which is based on the Italian industrial model presented by UNIDO (1997) develops ideas that closed partnership between SMEs grouped in clusters and the intermediate structures e.g. local governance, local and public institutions fosters the system's growth by lowering the barriers of entry for new enterprises. Through close social interaction, the firms are able to increase the depth, breadth, and efficiency of mutual knowledge exchanges. By establishing trust and norms of reciprocity, firms are able to solicit reciprocal disclosure of relevant information, thus increasing the potential for knowledge combination. The breadth of the network of relationships may also affect the opportunities for knowledge acquisition (McEvily & Zaheer, 1999). The knowledge acquired from these sources may then be transformed into competitive advantage via enhanced technological distinctiveness, new product creation, and reduced transaction costs.

The Best practice analysis is a focus on (1) how aspects of social capital (social interaction, relationship quality, and network ties) embedded in relationship of technological firms with key customers increase the firms' knowledge acquisition from these relationships, and (2) how this knowledge acquisition enhances competitive advantage through new product development, enhanced technological distinctiveness, and reduced sales costs.

The article seeks to make a significant contribution to understanding network relationships and their implications for entrepreneurial outcomes. The Best practice analysis acknowledges that the previous work has typically studied the firms' access to tangible external resources, such as manufacturing capacity and distribution channels (Lorenzoni & Ornati, 1988; Shepherd, 1991), whereas we focus on knowledge acquisition from inter-organizational relationships as a source of competitive advantage. By examining social interaction and relationship quality, we also go beyond earlier "relationship-counting" studies, which use the number of partners a firm has as a measure of its use of external relationships (Shan, Walker & Kogut, 1994; Deeds & Hill, 1996). In concludes that earlier work on the firm networking has often focused on a single outcome, such as sales growth (McGee & Dowling, 1994) or innovation (Shan et. al., 1994), while the best practice analysis stressed that closed partnership between the intermediate educational institutions and business associations provides a variety of outcomes, including new product development, technological distinctiveness, and sales cost efficiency. By doing so, this study highlights the important links between social capital, learning, and knowledge-based competitive advantage in business, especially in technology-based firms.

It highlights the importance of social capital networks on the inter-firm cooperation that is exactly those areas with strong intermediate structures, which are the reference points for creation the long-term partnership. This acknowledges the necessity to include a separate course of social capital and networks in management programmes for SMEs executives as the new challenge of globalization.

Methods: The following two methods have been used in the article: the analysis of scientific literature and the analysis of the best practice. Other data have been derived from printed papers, books, and statistics.

Theoretical Framework and Hypothesis

The article builds on knowledge-based view of the firm, which depicts firms as repositories of knowledge and competencies (Kogut & Zander, 1996; Spender, 1996). According to this view, the "organizational advantage" (Ghoshal & Moran, 1996) of firms over markets arises from their superior capability in creating and transferring knowledge. Knowledge creation and innovation result from new combinations of knowledge and other resources (Cohen & Levinthal, 1990; Kogut & Zander, 1992). The accumulation of knowledge through learning constitutes a driving force in the development and growth of young firms (Penrose, 1959; Spender & Grant, 1996), because knowledge acquisition opens new "productive opportunities" (Penrose, 1959) and enhances the firm's ability to exploit these opportunities. The development and growth of technology-based young firms is particularly dependent upon innovatively combining their own firm-specific knowledge with that of external partners because young firms are resource constrained (McDougall, Shane & Oviatt, 1994) and because technology-based firms depend upon knowledge rejuvenation to survive and grow (Autio, Sapienza & Almeida, 2000).

Organizational Learning and Inter-organizational Relationships

Numerous frameworks of organizational learning have been proposed in organizational literature. Many of these conceptualize learning as a process of knowledge acquisition, knowledge assimilation, and knowledge exploitation (Argote, 1999; Cohen & Levinthal, 1990; Huber, 1991). In this study, we follow Huber (1991, p. 89) in assuming that "An organization learns if any of its units acquires knowledge that it recognizes as potentially useful to the organization". Further, we assume that young firms may develop social capital as a strategy to aid in acquiring new knowledge. A competitive advantage results when firm implements a strategy that creates value which other companies cannot efficiently replicate (Hitt, Ireland & Hoskinsson, 1999, p. 5).

Recent studies have proposed that inter-organizational relationships create opportunities for knowledge acquisition and employment (Dyer & Singh, 1998; Eisenhardt & Schoonhoven, 1996; Lane & Lubatkin, 1998; Larsson et. al., 1998). Through interactions with others, firms get access to external knowledge and combine it with existing knowledge. Further, such relationships create a context within what newly created knowledge can be applied and exploited. Several studies have examined how firms pursue learning opportunities in inter-organizational settings, e.g. buyer-seller relationships (von Hippel, 1988) and supplier and customer relationships of entrepreneurial firms (Larson, 1992) and small firms (Uzzi, 1997). Yet, few studies have examined the role of social capital in facilitating learning in inter-organizational relationships.

Inter-organizational relationships include relationships a firm may have with external organizations including customers, suppliers, investors, government institutions, and the like (Dyer & Singh, 1998; Larson, 1992; Ring & Van de Ven 1994). We stressed that the more social capital a technology-based firm develops in a relationship, the more likely it is to acquire new knowledge and exploit it as a basis of competitive advantage. This logic maps closely into Dyer and Singh's (1998) model of relational rents which proposes that the potential of an individual firms has to creating competitive advantage depends not just on its own resources but also on its relationships with other key firms. Their model identifies four potential sources of inter-organizational competitive advantage: relation-specific assets, knowledge-sharing routines, complementary assets, and effective governance.

Therefore, the extent to which technology-based firms can acquire external knowledge from its key customer will depend on the existence of external knowledge, on the ability of the firm to recognize and assess the value of the knowledge, on repeated, intense interaction, and on the willingness of the dyad firms to share information (Cohen & Levinthal, 1990; Dyer & Singh, 1998). We follow Nahapiet & Ghoshal (1998) in arguing that social capital facilitates knowledge acquisition by affecting conditions necessary for the exchange and combination of existing intellectual resources. Lane & Lubatkin (1998) have pointed out that learning relationships in networks involve patterns of interactions that affect the learning of both members; however, as they do, we focus our analysis on an exposition of the knowledge effects on the partnership between business and science through social capital network.

The Concept of Social Capital vs. Social Capital Metaphor

The term of social capital can be traced back to the nineteenth century, where the Italian Risorgimento movement conceptualized *valor social*[1] as an educational feature of growth and practice of self-governing institutions.

The concept of social capital was originally used in community studies to describe relational resources embedded in personal ties within the community. The concept has recently been applied in business, political science, sociology and a wide range of organizational studies dealing with the social context of intra- and inter-organizational relationships (Burt, 1992; Nahapiet & Ghoshal, 1998). An increasing number of research articles about social capital approach the number of different definitions of the concept. Consistent with a relational view of competitive advantage (Dyer & Singh, 1998; Hitt, Dacin, Levitas, Arregle & Borza, 2000; Lane & Lubatkin, 1998) and the network structure that is social capital, the focus of the paper is on how social capital in a key inter-organizational relationship of a firm (i.e. its relationship with public institutions and its key customers) affects the firm's ability to acquire new knowledge and convert it with the aim to enhance new product development, technological distinctness, and reduced sales costs.

But before going into discussion some arguments should be clarified. *The closure argument* is that a network of strongly interconnected elements creates social capital. *The structural hole* argument is that social capital created by a network in which people can broker connections between otherwise disconnected segments. In addition, some specific aspects of structure highlight the point that more than one network mechanism can be cited as responsible for the competitive advantage known as social capital. The two mechanisms cited most often are protection within closed networks and brokerage across structural holes, but there are others around which future work will emerge if social capital continues to be such a popular metaphor.

There is an impressive diversity of empirical evidence showing that social capital is more a function of brokerage across structural holes than closure within a network, while a detailed description is presented below in Figure 1, p. 226, which is an overview of social capital in metaphor and network structure. The Figure 1 is a road map through the next few pages, and a reminder that beneath the general agreement on social capital as a metaphor lie a variety of network mechanisms that make contradictory predictions about social capital.

Many scholars defined social capital as a metaphor about advantage. According to them, society can be viewed as a market in which people exchange all the variety of goods and ideas in pursuit of their interests (Coleman 1990, Bourdieu

[1] Valor sociale is an Italian term roughly translated into social capital.

and Wacquant, 1992, Burt 1992, Putnam 1993). Certain people, or certain groups of people, do better in the sense of receiving higher returns of their efforts. Some enjoy higher incomes. Some lead more important projects. The interests of some are better served than the interests of others. In this case, social capital metaphor means that the people who do better are somehow better connected.

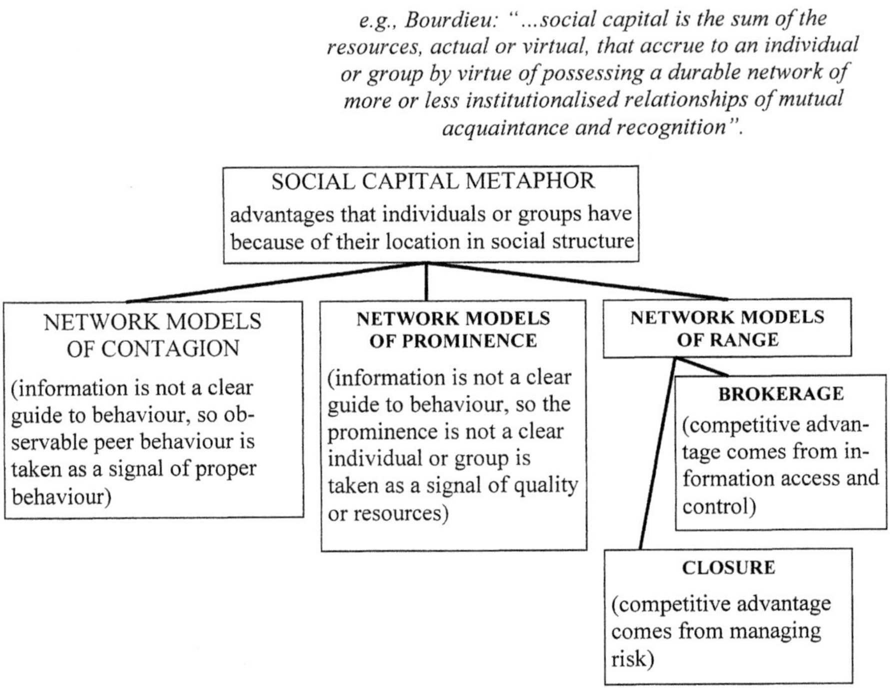

Figure 1: Social Capital, in Methaphor and Network Structure Source: Burt, 2000

Turning to Figure 1 in order to describe the term of social capital, it is reasonable to quote Bourdieu and Wacquant (1992) who define social capital as the resources that result from social structure. According to Coleman's definition (1990, p. 302) social capital is as a function of social structure producing advantage. Putnam (1993, p. 167) grounds his influential work in Coleman's argument, preserving the focus on action facilitated by social structure: "Social capital here refers to features of social organization, such as trust, norms, and networks, that can improve the efficiency of society by facilitating co-ordinated action". Burt (1992, pp. 8-42) agrees with Putnam's definition by beginning his consideration on social capital as the competitive advantage of structural holes. Other descriptions in the literature include social capital as 'the institutions, relationships and norms that shape the quality and quantity of a society's social interactions'

(World Bank)[2]. According to OECD (2001) social capital is networks together with shared norms, values and understandings that facilitate co-operation within or among groups. MacGillvray (2002) argues, that social capital is the networks, norms and relationships that help communities and organizations work more effectively.

Essentially, there are three main types of social capital:

1. *Bonding* social capital-characterized by strong bonds (or "social glue"), e.g. among family members or among members of an ethnic group;
2. *Bridging* social capital – characterized by weaker, less dense but more cross-cutting ties ("social oil") e.g. with business associates, acquaintances, friends from different ethnic groups, friends of friends etc; and
3. *Linking* social capital – characterized by connections between those with differing levels of power or social status e.g. links between the political elite and the general public or between individuals from different social classes. This dimension of social capital, first proposed by Woolcock (2001), is a relatively recent addition to the categorization of different types of social capital (Aldridge et al., 2002, p.12).

The distinction between bonding and bridging capital or between strong and weak ties is crucial, because the structure of prior relations among people and organizations in a market can affect, or replace, information. Replacement happens when market information is so ambiguous that people use network structure as the best available information. Such assumption underlies the network contagion and prominence mechanisms to the left in Figure 1. According to Burt (2000), transactions could be so complex that available information cannot be used to make a clear choice between sellers, or available information could be so ambiguous such that no amount of it can be used to pick the best exchange. White (1981) argues that information is so ambiguous for producers that competition is more accurately modeled as imitation. A market is modeled as a network clique (in other words, a small, cohesive group distinct from an external environment). Producers taking positions relative to other producers on the market schedule determine Price within the clique.

Although contagion and prominence mechanisms can be discussed as social capital, they are more often discussed as other concepts – for example, imitation in institutional theory, or reputation and status in economics and sociology (Burt, 2000), therefore, they will be considered in this description. The other two mechanisms in Figure 1, closure and brokerage do not assume that networks replace information so much as they affect the flow of information and what people can do with it. Both mechanisms begin with the assumption that communication takes time, as prior relationships affect who know whom. A generic research finding showed that information circulates more within than between

[2] http://www.worldbank.org/poverty/scapital

groups – within a work group more than between groups, within a division more than between divisions, within an industry more than between industries (Festinger et. al., 1950).

In this context, knowledge acquisition will depend on three aspects of social capital: (1) the level of social interaction between the firms, (2) the quality of the relationship in terms of goodwill trust and reciprocity, and (3) the level of network ties created through a relationship. Social interaction refers to the extent of social relationships between the focal firm's and the key customer's people (Nahapiet & Ghoshal, 1998; Larson, 1992; Ring & Van de Ven, 1994). Relationship quality refers to the extent that this interaction is marked by the development of goodwill trust and expectations of reciprocity (Dyer & Singh, 1998; Larson, 1992; Ring & Van de Ven, 1994). Customer network ties denote the extent to which the key customer provides the focal firm access or introductions to a broader set of customers (McEvily & Zaheer, 1999; Uzzi, 1997).

So there is a point of general agreement from which to begin a discussion of social capital. The above – cited perspectives on social capital showed diversity in origin and style of accompanying evidence, in which social structure is a kind of capital that can create for certain individuals or groups a competitive advantage in pursuing their ends. Better connected people enjoy higher returns.

Network Closure as Social Capital

Recent work has distinguished three main components (see Table 1) to social capital:

	Networks	**Norms**	**Sanctions**
Function	Network members (access to information, benefits & support)	Rules & understandings (Reciprocity, expectation of co-operation, trust, codified behaviour)	Rewards & punishment for complying/breaking norms
Examples: **Traditional communities**	Neighbors (lending, caring & protection)	Reciprocity, due care of property, challenging strangers	Recognition and respect vs. gossip, social exclusion
New York Diamond Wholesale Market	Diamond merchants	Trustworthy exchange, without payment, of bags of uncut diamonds for examination	Approval, disapproval and exclusion

Table 1: The Three Components Of Social Capital. Source: Aldridge, 2002, p. 11

1. *social networks* – who knows who;
2. *social norms* – the informal and formal "rules" that guide how network members treat each other; and
3. *sanctions* – the processes that help to ensure that network members keep to the rules.

Table 1 illustrates, the view of social capital, which according to Coleman (1990) focuses on the risks, associated with incomplete information. Network with closure, in which everyone is connected in such a way that no one can escape the notice of others, which in operational terms usually means a dense network (Burt, 2000, p. 6). In this case, Burt (2000) argued that closure has to do two things for people in closed networks.

First, it affects access to information in the way it described by Coleman (1990, p. 310):

> "An important form of social capital is the potential for information the inheres in social relations ... a person who is not greatly interested in current events but who is interested in being informed about important developments can save the time required to read a newspaper if he can get the information he wants from a friend who pays attention to such a matter".

Second, and this is the benefit more emphasized by Coleman (1988) who argues that network closure facilitates sanctions that make it less risky for people in the network to trust one another. He illustrates the trust advantage with rotating credit associations (ROSCAs). According to him, without a high degree of trustworthiness among the members of the group, the institution could not exist. When an effective norm does exist, it constitutes a powerful, but sometimes fragile, form of social capital. The consequence of this closure is, as in the case of the wholesale diamond market or in other similar communities, a set of effective sanctions that can monitor and guide behavior. Reputation cannot arise in an open structure, and collective sanctions that would ensure trustworthiness cannot be applied.

Summing up this comprehensive view about social capital, it is important to focus on Coleman's prediction that network closure creates social capital. Therefore, the closure prediction means comparisons between two similar people, one of whom has more social capital. Strong relations among his contacts provide him with more reliable communication channels, and protect him from exploitation because he and his contacts are more able to act unanimously against someone who violates their norms of conduct.

Structural Holes as Social Capital

Participation in, and control of, information diffusion underlies the social capital of structural holes (Burt, 1992). The argument describes social capital as a function of brokerage opportunities, and draws on network concepts that emerged in sociology during the 1970s. Sociological ideas elaborated by

Simmel (1955) and Merton (1968) on the autonomy generated by conflicting affiliations are mixed in the hole argument with traditional economic ideas of monopoly power and oligopoly to produce network models of competitive advantage. In this context, structural holes are thus an opportunity to broker the flow of information between people, and control the projects that bring together people from opposite sides of the hole (Burt, 2000, p. 9).

Therefore, structural holes separate non-redundant sources of information, sources that are more additive than overlapping. According to Burt (2000) there are two network indicators of redundancy: (1) cohesion and (2) equivalence. Cohesive contacts (contacts strongly connected to each other) are likely to have similar information and therefore provide redundant information benefits. Structurally equivalent contacts (contacts who link a manager to the same third parties) have the same sources of information and therefore provide redundant information benefits.

Simmerl and Merton introduced some control benefits from structural holes: The ideal type is the tertius gaudens (literally, "the third who benefits", see Burt, 1992, pp. 30-32). Burt (2000) stressed that structural holes are the setting for tertius strategies, and information is the substance. Accurate, ambiguous, or distorted information is strategically moved between contacts by the tertius. The information and control benefits reinforce one another at any moment in time and cumulate over time.

Thus, individuals with contact networks rich in structural holes are the individuals who know about, have a hand in, and exercise control over, more rewarding opportunities. The behaviors by which they develop the opportunities are many and varied, but the opportunity itself is at all times defined by a hole in social structure. In terms of the argument, networks rich in the entrepreneurial networks and entrepreneurs are people skilled in building the interpersonal bridges that span structural holes. They monitor information more effectively than bureaucratic control. They move information faster, and to more people, than memos. They are more responsive than a bureaucracy, easily shifting network time and energy from one solution to another (Morselli, 2000). Speeding the process toward equilibrium, individuals with networks rich in structural holes operate somewhere between the force of corporate authority and the dexterity of markets, building bridges between disconnected parts of a market where it is valuable to do so.

Summing up, the hole prediction is in comparisons between two similar people, one of whom has more social capital. His network across structural holes gives him broad, early access to, and entrepreneurial control over, information, which creates competitive advantage.

The Social Order of Disequilibrium

The discussion above underlined the differences between brokerage and closure; therefore it is necessary to clarify these differences. Taking closure into consideration, which is about static and about advantages of people in a cohesive group. Strategy guided by the closure argument involves locating a group, and closing ranks with like-minded people.

In contrast, brokerage is about change. The hole argument is about advantages that go to people who build bridges across cohesive groups. Strategy guided by brokerage involves locating a position at the boundary of two groups, and building relations between dissimilar people.

Brokerage must be the more difficult strategy (as indicated by the greater tendency for relations to form within groups, and the faster decay in bridge relations between groups, e.g. Burt, 2000), but the further difference is that brokerage is explicitly about action that cuts across structural holes in the current social structure.

The greater cost of brokerage must be offset by greater gains. The gains can be expected to disappear as more and more people build bridges across the same structural hole. For example, when the first entrepreneurs benefit from synthesizing information across a structural hole, others join them, and the advantages of bringing the hole disappears.

However, Burt (2000) viewed the social capital of structural holes to equilibrium where the short-run advantage of brokerage can become a long-run advantage if social structure is held constant as by culture (e.g. Light and Karageorgis (1994), on socially excluded ethnicity for whom entrepreneurial activities are the route into society), or technology (e.g. Burt (1998) on the industry structure-performance differences in the American economy).

Such situation could arise as follows: An industry of managers and organizations moves towards equilibrium. Managers with more social capital have an advantage in identifying and developing more rewarding opportunities; so the industry begins moving toward a new equilibrium (Burt, 2000, p. 14).

In conclusion, it is important to notice, that the hole argument stands apart from closure both in its empirical predictions and in describing a world of change – a world of discovering and developing opportunities to add value by changing social structure with bridges across holes in the structure.

The argument, describing competitive advantage on the path to equilibrium, is a story about the social order of disequilibrium. Therefore, it is necessary to evaluate the potential beneficial effects of social capital and to discover why social capital is important in creating closed partnership between business and science.

Potential Beneficial Effects of Social Capital

Prior to drawing a picture of potential beneficial effects of social capital it is important to investigate how social capital should be measured, as the later discussion will be based on it. Social capital has been measured in a variety of ways. Putnam (2000) used a composite indicator containing the measures of: the intensity of involvement in community and organizational life; public engagement (e.g. voting); volunteering; informal socializing (e.g. visiting friends); and reported levels of inter-personal trust. Others scholars, like Norris (2000) have questioned Putnam's mixing of associational and trust measures on the grounds that they are in fact quite distinct dimensions of social capital, with associational membership more important for tolerance and other indicators of social cohesion, and trust more important for economic outcomes. Aldridge et. al. (2002, p. 14) prefers to use a simple measure of social capital, e.g. whether people think that other people can, in general, be trusted. Levels of reported social trust vary dramatically across communities, regions and even countries (see Table 2.).

OECD			
Norway	65.3	Italy*	35.3
Sweden	59.7	Belgium*	33.2
Denmark*	57.7	Austria*	31.8
Netherlands*	55.8	United Kingdom	31.0
Canada*	52.4	Korea	30.3
Finland	47.6	Czech Republic*	30.3
Ireland*	47.4	Spain	29.8
Japan	46.0	Mexico	28.1
Iceland*	43.6	Hungary*	24.6
Germany	41.8	France*	22.8
Switzerland	41.0	Portugal*	21.4
Australia	39.9	Turkey	6.5
United States	35.6		
NON – OECD Members			
India	37.9	South Africa	18.2
Chile	21.9	Argentina	17.5
Nigeria	19.2	Brazil	2.8

* 1990-91 data

Note: The question posed in the survey was: "Generally speaking, would you say that most people can be trusted, or that you can't be too careful in dealing with people?"

Table 2: Percentage Of People Saying That Most People Can Be Trusted, Selected Countries, 1995-96. Source: World Values Survey and Knack & Keefer, 1997.

Table 2 indicates, that during the 1990s, around 30 percent of people in Britain said that most people could be trusted. In Scandinavian countries, the figure was around 60 percent, while in Brazil it was 3 percent. It shows that there is increasing evidence that trade at the macro level is influenced by social capital – a common property resource whose value depends on the level of interaction between people.

The literature on social literature suggests that social capital may have a range of potential beneficial economic and social effects on macro and micro levels. Aldridge et. al. (2000, p. 15) highlights the first one:

1. It may facilitate higher levels of, and growth in, GDP.

Increasing evidence is appearing at the macro level which identifies trust, civic norms, and other factors of social capital as a key condition for economic development. Cross-country analysis is pointing to social capital as an important ingredient in economic performance. The World Bank (2002) gives an example that "Developing countries with relatively low social capital seem more likely to have bad policy outcomes, low investment and slow growth. For any two countries with the same level of income, the one with more social capital tends to have more schooling, a more expensive financial system, better fiscal policy and a wider telephone network. Causation could run in both directions, but the results are suggestive". Aldridge et. al. (2000, p. 16) argued that economies with high level of social capital characterized by such features:

- Savings, investment in capital of all kinds, risk-taking and entrepreneurial behavior;
- More effective competition (since, for example, low transaction costs reduce barriers to entry by new suppliers and good information allows consumers to make better comparisons between different suppliers). More effective competition is in turn beneficial for innovation and for the efficient allocation of resources; and
- The organizational forms required by a modern economy. Fukuyama (1995) argued that large private sector corporations depend on high degrees of social trust (if they are not to be encumbered by extensive rules, contracts and bureaucracy). In countries where such trust is lacking (he gives China, France and Italy as examples), the large private corporation is less likely to emerge spontaneously and the State may then have to step in to fill the gap. By contrast, in the US, Japan and Germany (countries, as he argued which have high degrees of social trust), he suggested that there would be less need for the State to intervene in this way.

Moreover, there is evidence of the importance of social capital for economic and business performance at both the national and the sub-national level. At the *national level*, a number of studies by scholars Knack and Keefer (1997); Whiteley (1997) have found a significant correlation between measures of social capital, especially trust in people, and rates of economic growth (controlling for other factors). Indeed the influence of social capital on economic growth appears to be at least as strong as the influence of human capital3. This is particularly true in the case of partnership between business and science, where social capital networks form the linkage among individuals who relate to each other with the aim to solve common problems. Therefore, in countries of transition with weak partnership performance between these institutions as well as in countries with poorly developed legal and financial systems, social capital can be a key facilitator of economic growth.

At the *sub-national level*, there are a number of interesting studies of the contribution of social capital to regional and local economic performance. For example, Helliwell and Put-

---

[3] Human capital include knowledge, skills and competence (Aldridge et. al., 2000, p. 13).

nam (2000) modeled the growth rates of the Italian regions after the reform of Italian regional government in the 1970s. Before the reforms, the higher social capital regions of the North were much wealthier and more productive than the low social capital regions of the South. But the Southern regions were gradually catching up. However, following the reforms, which involved creating a powerful new regional tier of government across Italy's twenty regions, the growth rates of the northern regions with high social capital surged ahead again. It may seem that the high social capital of the Northern regions in some way facilitated these regions' ability to employ the new government to achieve higher growth;

The second potential beneficial effect of social capital presented by Aldrigde (2000, p. 20) is:

2. *It may facilitate the more efficient functioning of labor markets.*

The level and duration of unemployment levels is partly a function of search costs. The networks and contacts that make up social capital can provide highly cost – effective mechanism for facilitating job search. Bridging social capital (networks and contacts with friends of friends and acquaintances) is particularly important since a number of studies by Perri (1997, p.6) have shown that more unemployed people find employment through friends and personal contacts than through any other single route.

And the last one beneficial effect of social capital pointed by Aldridge et. al. (2000, p. 25) is:

3. *It may improve the effectiveness of institutions of government.*

Putnam (1993) has shown that states or regions with higher levels of trust and civic engagement tend to have better quality government, a result that has now been replicated in Germany (Cusack, 1999), Hungary (Schafft and Brown, 2000), and Russia (Petro, 2001). Several pathways are thought to be important:

- Fosters "virtue among the citizenry", making citizens more community-orientated, more law abiding, co-operative with the state and willing to pay taxes. Essentially, it's easy to implement policy in a high social capital context.
- Makes citizens more "sophisticated consumers of politics", offering channels through which they can learn about the wider policy context, mobilize and lobby government to act in ways that benefit the community.
- May make both bureaucrats and politicians better at co-operating, as they carry with them the same skills and social capital resources as the rest of the population.

Evidence for the relationship between high level of social capital and good governance extends from the micro to cross-national examples. For example, at the national level, levels of corruption and bureaucratic quality are significantly associated with general levels of social capital trust, having controlled for GDP per capita (La Porta et. al, 1997). Furthermore, it is important to mention the benefits of social capital to firms. Social capital benefits firms in (1) reducing transaction costs, because its facilitates co-operation and co-ordination. According to World Bank (2002) in the study of flexible production a system has been found that without a co-operative agreement, an automobile company or parts of manufacture might be able to take advantage of the other strategically altering prices. Social capital helps firms protect themselves against such risks in cost-efficient manner. (2) Providing a competitive edge, because it facilitates efficiency gains in time and information, which allows more resources to be allotted to producing and marketing a better product at a higher level.

Therefore, social capital affects firms in four ways:

1. *Within a firm.* Social capital promotes greater co-ordination among individuals and between department. Teamwork can enhance efficiency and quality in small companies as well as multinational co-corporations. For instance, workers and shareholders in a firm are likely to feel an increased sense of pride in their work if they are part of an entity which strives to make a positive contribution to its surrounding communities – communities which are often home to firm staff (World Bank, 2002).
2. *Among firms.* Trust is the foundation for co-operation between enterprises. Businesses banding together in a joint effort are able to establish deeper relationships with one another which can be accessed in the future for other business projects beyond the scope of the original group. For example, research and design costs can be shared when there is a collaborative agreement between otherwise rivalries firms. And this is evident in niche markets, such as software development, where entrepreneurs take advantage of economies of time, by sharing information, and adapting quickly to changing demands. Marketing and lobbying are other opportunities for efficiency and effectiveness gains through co-operation (World bank, 2002).
3. *Across sectors.* Recognizing the importance of cross-sectoral partnership for sustainable business as well as sustainable development, the Prince of Wales formed his Business Leaders Forum to promote synergy, interdependence and social capital across the public, private and civil society sectors. For instance, Indian Business & Community Partnership Trust is a coalition of businesses working across sectors to promote social and economic development in India. Such partnership may translate into better access and more influence during policy debates and increased support from community-based organizations, both are critical components of a successful business initiative (World Bank, 2002).
4. *Within society.* Societal levels of social capital impact businesses positively or negatively. Social capital affects the types of firms, which are successful within a society, and creates either an enabling or disabling environment for private sector development. For example, business endeavors, which have positive spillover for the community, can increase people's level of participation in society and the goodwill they feel towards the public and private sectors. Social cohesion is a key enabling condition for stable politics and profitable business (World Bank, 2002).

	Individual or micro level	**Community or mezzo level**	**National or macro level**
Level of and growth in GDP	Most jobs obtained via social networks. More extensive and diffuse networks are associated with less unemployment, faster career advancement, and higher pay. Longitudinal studies of pathways out of unemployment confirm causal relationship.	Success of many markets rests on social capital, e.g. New York diamond market (Coleman, 1988). Importance of network and knowledge architecture within and between firms for relative success.	National growth rates strongly predicted by social trust in modified neoclassical model – as important than human capital (Whiteley, 1997). Structure of national economies affected by social capital (Fukyama, 1995).

Table 3: The Relationship Between Social Capital And Key Policy Outcomes At Individual, Mezzo And Macro Levels. Source: Inspired by Aldridge et. al. 2000

Conclusion: strongly theoretical and empirical grounds to believe that social capital affects form and level of economic growth. Social capital facilitates the flow of information and lowers transaction costs, notably through increasing trust. Bridging social capital seems especially important, as does the creation of a shared normative framework at a very general level.

Several implications may be drawn from this analysis. As the discussion above showed, different types of social capital are relevant to different social and economic outcomes. Therefore, the beneficial effects of social capital operate at different geographical levels: individual, community and nation-wide. The most important thing is that social capital can contribute to economic growth by improving partnership between business and science. Table 3 below shows some of the evidence on the beneficial effects of social capital on economic growth distinguishing impacts at:

1. *The individual or micro level,* i.e. the effects on individuals of having more or less social capital, such as having more friends or more supportive relationships;
2. *The community or mezzo level*, i.e. social spillover effects from living in a strongly bonded community. The best of these results typically rest on multi-level statistical models that control individual level variables and then calculate any additional effects from the community. Hence, these models can show that even if a given individual does not participate in community life, if those around them do, that individual still benefits.
3. *The national or macro level*, i.e. national or regional impacts (Aldridge et. al. 2000, p.27)

Best Practice in the Perspective of Social Capital and Knowledge Acquisition

The Best practice analysis has been chosen with the aim to reflect the above-mentioned discussion on social capital and to show the extent to which the partnership between SMEs grouped in clusters and the intermediate structures e.g. local governance, local and public institutions can foster the system's growth by lowering barriers of entry for new enterprises and knowledge acquisition through social ties in the economic exchange. It enhances the firm's ability to recognize and evaluate the pertinent external knowledge of the key customers and provides the technology-based firm insight on the specialized systems and structures of the key customer and results in specialized information, language and know-how (Dyer & Singh, 1998; Lane & Lubatkin, 1998). Thus, we hypothesize:

Hypothesis 1: The greater the social interaction between a key customer and a technology-based firm, the greater will be the technology-based firm's knowledge acquisition from that relationship.

The Italian industrial model presented by UNIDO (1997) is known worldwide as a successful example of endogenous development based on SMEs networks and clusters which are strongly rooted in their communities. Italy is impressive, notable in that it is among the largest economies in the world and, 98% of industrial firms have fewer than 100 employees. Almost 90% of firms have fewer than twenty workers. The average size of Italian industrial firms is seven employees. "SME clusters" refer to groups of small and medium sized enterprises located in a relatively delimitated geographic area engaged in the production of the same sort of products. Although there is strong competition among firms, often there is also a high degree of cooperation among firms in that the production processes are divided into distinct phases with separate firms responsible for different phases. The main idea of SME's grouped in clusters is a bottom-up growth approach. *Bottom-up growth* based on small firms that is considered as a main source of increased employment, including self-employment, and a way in incorporating participants, who are traditionally excluded from economic development like women and young people. Bottom growth approach stimulates partnership between local forces and local institutions that increase efficiency and productivity of firms and stabilize the local community in order to establish a foundation for endogenous growth. Therefore, the development of a cluster of SMEs is a long-term process due to the fact that it depends on market perspectives and availability of personal assets. It is, furthermore, facilitated by the formation of local networks and common activities that lower entry barriers and minimize the risk of investment. The need for more sophisticated financial instruments and appropriate policy action for innovation become relevant only when SMEs and the whole cluster need to consolidate.

The competitive advantages of SMEs grouped in clusters are based on *three aspects*: specialization, cooperation and flexibility. A cluster, i.e. firms working together in a productive system characterized by a division of labor among the members is obviously based on a high degree of inter-firm cooperation. Inter-firm cooperation is important, not only in terms of the availability of resources, but also in terms of firm flexibility. Cooperative inter-firm relations help firms to be more flexible in terms of amount of production since firms may outsource more work when there is increased demand, and less when there is a decrease. Cooperation among firms also helps firms be flexible in terms of type of production in that products can be made to order by grouping different contractors together according to the specialties required. The quality of a relationship between the firm and its key customer is reflected in the extent to which the two parties develop common goals, norms, and reciprocal expectations regarding the goodwill trustworthiness of the exchange partners. Goodwill trust and reciprocal obligations have often been seen as alternatives or complements to formal, arms-length, or third party governance mechanisms.

Levels of Local Government:	City Province Region
Local Business Structures:	Business associations & organizations Industry specific service canters Business Innovation Centers and Technology/Industrial Parks Public & Private Firms Credit Institutes
Research & Training Structures:	Local branches of National Research Council (CNR) Local branches of New Technology & Environmental Agencies (ENEA) Technical Schools University (Public) Professional Training Structures Research Consortia
Other Local, Public Organizations:	Chambers of Commerce Convention/Exhibition Centers

Table 4: Italian Intermediate Structure. Source: UNIDO, 1997, p. 10

The costs of sharing know-how in inter-organizational relationships are high, because of the effective governance mechanisms, which allows knowledge sharing and discourage free-riding as (1) the relational governance norms are not time-dependent (i.e., they have no clear end point and may in fact appreciate in value as the relationship progresses), (2) the expectations of reciprocal benefits, actions are more freely undertaken on behalf of the exchange partner and (3) the likelihood of violation is diminished because high-quality relationships are valuable and hard to replicate.

Therefore, Italian SMEs are "governed" by institutions on the level of the European Union, the Italian government (national), regional government and local institutions. Since SMEs tend to be very rooted in their local context, the focus is to show governance on the regional and local levels Table 4. Intermediate governance structures are "intermediate" in that they represent numerous participants that in some way influence economic activity through closed partnership ties. Establishing high levels of mutual expectations should enhance knowledge acquisition because it requires making the systems and cultures compatible.

The closed partnership between the intermediate educational institutions and business associations provides basic business support services and training programmes for SMEs by increasing trust among entrepreneurs and reducing administrative and legal costs.

It is important to take into account that SMEs grouped in clusters may be isolated and may be located within a larger, regional cluster linked together by technological or market similarities in order to increase competitiveness, since it facilitates the diffusion of information regarding markets and technologies, the mobility of human resources between complementary (not only similar) activities and the realization of common initiatives between business and science. Thus, we hypothesize:

Hypothesis 2: The higher the quality of the relationship with the key customer, the greater will be the technology-based young firm's knowledge acquisition from the relationship.

The quality of relationship between business and science positively associated with knowledge acquisition due to the fact that it provides control, increases mutual understanding, quickens exchange processes, and encourages freedom in exchange. SME's grouped in clusters increase the promotion and diffusion of innovation, which occurs through enterprise relationships and interaction, especially if a leader firm or a group of leader firms, exists. An example of the individual success of a leading firm can be powerful stimulant for innovation. Therefore, social capital enhances knowledge acquisition by improving access to external sources of knowledge, by increasing the willingness and ability of exchange partners to identify, exchange, and assimilate knowledge, and by improving the breadth and efficiency of knowledge transfer. "SME's grouped in clusters" acquire knowledge and convert it into competitive advantage in the form of greater new product development, which enhances technological distinctiveness, and reduces sales costs.

The social networks that enhance enterprise performance by allowing information about the technologies and markets to flow between their members have been labeled 'innovation networks', in other words 'brokerage networks'. While those designed to reduce uncertainty by allowing information flow by building trust have been labeled ''solidarity networks', in other words 'closure networks'

	Innovation/Brokerage Networks	Solidarity/Closure Networks
Network function	To enhance enterprise performance	To reduce uncertainty
Type of information flowing through the network	About the new technologies and markets	About the member's conduct, circumstances and intensions
Characteristics of the sets of contacts maintained by network members	Large, diverse, far-reaching networks with infrequent interaction with each other contact	Small, homogeneous, high levels of interaction with each contact which basically based on trust
Overall effect on current enterprise performance	Relatively large	Relatively small
Spill over effects of networking	A low proportion of the overall effect	A high proportion of overall effect
Typical member enterprise	Large enterprises with access to formal institutions	Smaller enterprises with no access to formal institutions

Table 5: Network Function, Structure and Effect on Enterprise Performance. Source: Barr, 1998

Implicitly, two hypotheses suggest that social capital affects competitive advantage outcomes via its effects on knowledge acquisition. In addition, it must be assumed that a district, a cluster, is a grouping of individuals, from whose interaction – both cooperative and competitive – emerges a collective action with evolutionary characteristics. Through interaction, firms establish behavioral norms that influence later behavior. What results is therefore an evolutionary process that is specific (in that each situation is different) and irreversible. It is therefore possible to identify stylized facts that represent the various evolutionary phases of districts, but each specific case must clearly be understood in its complexity. Furthermore, every external instrument designed to favor investments, create common services, promote entrepreneurial aggregation needs to be flexible enough to be adaptable to the specific situation at hand. For this to occur, the first step toward designing policy for the development of districts in an in-depth study of the local level, not only of the history and culture of the area, but also of the existing interaction dynamics among the various social actors. Thus, it is proposed that knowledge acquisition mediates between the three social capital constructs and technological distinctness's, new product development, and sales costs. It highlights the importance of social capital networks on the inter-firm cooperation that are exactly those areas with strong intermediate structures which are the reference points for creation the long-term partnership. This acknowledges the necessity to include the separate course of social capital and networks in management programmes for SMEs executives as the new challenge of the globalization.

Conclusion and Lessons to be Drawn from the Best Practice Analysis

Having completed the discussion and the analysis of literature on social relations and Best Practice, which acknowledge the importance of social capital to business and science. Social capital brings potential effects to the society, social and business groups with the aim to achieve better results by establishing participatory approaches.

1. Social capital is features of social organization in the form of networks based on trust and norms that can improve the efficiency of society by facilitating coordinated action. Social capital networks are an important lubricant of economic system, because it facilitates co-operation and coordination for the mutual benefits of its members. It provides good theoretical bases for further empirical study of cooperative learning. The process of learning in inter-organizational relationships also contribute to closer qualitative and longitudinal examination. A fuller understanding of knowledge as a source of competitive advantage will require more in-depth investigations into the learning process as a whole.
2. The effects of social capital have been analyzed on the example of the Italian Industrial Model presented by UNIDO. The Best practice analysis underlined the significant economic and business development growth through closed partnership networks with intermediate structures, e.g. public institutions.
3. This leads to acknowledge the necessity to include a separate curriculum of social capital in bachelor and master studies for students in Lithuania due to the fact that it is a new field for the research that may reshape economic growth at the individual, community and societal levels.

The Best Practice analysis of the Italian clusters offers two principal methodological lessons to those wishing to act in emerging countries to promote diffused development.

- A strong capacity for territorial and sectoral analysis, combined with the ability to analyze social interaction is required.
- It is necessary to evaluate the results of interpenetration, not only in terms of firm growth, but also in terms of general development of the territory. In this way cumulative effects of changing competencies and know-how available in the area may be fully understood. Methods of monitoring the change process should therefore be built-in to interpenetration so that the local-level actors will be able to effectively observe changes and make future adjustment if necessary.

References

ALDRIDGE, S/HALPERN, D./FITZPATRICK, S.: Social Capital: A Discussion Paper. Performance and Innovation Unit, 2000 April.

AMBURGEY, T.L./KELLY, D./BARNETT, W.P.: Resetting the Clock: The Dynamics of Organizational Change and failure. Administrative Science Quarterly 38(1993)1, pp. 51-73.

ARGOTE, L.: Organizational Learning: Creating, Retaining and Transferring Knowledge. Boston, Dordrecht and London 1999 (Kluwer Academic Publishers).

AUTO, E/SAPIENZA, H.L./ALMEIDA, J.: in press. Effects of Time to Internationalisation, Knowledge Intensity, and Imitability on Growth. Academy of Management Journal, Special Issue on International Entrepreneurship 2000.

BARR, A.M.: Enterprise Performance and the Function Diversity of Social Capital. WPS 1(1998) (Centre for the Study of African Economies; Institute of Economics and Statistics, University of Oxford).

BIANCHI, P./LEE, M.M./BERTINI, S.: The Italian SME Experience And Possible Lessons For Emerging Countries. 1997 (UNIDO).

BOURDIEU, P./WACQUANT, L.J.D.: Invitation to Reflexive Sociology. Chicago, IL 1992 (University of Chicago Press).

BURT, R.S.: Toward a Structural Theory of Action. New York 1982 (Academic Press).

COHEN, W.M./LEVINTHAL, D.A.: Absorptive capacity: A New Perspective on Learning and Innovation. Administrative Science Quarterly, 35(1990), pp. 128-152.

COLEMAN, J.: Foundation of Social Theory. Cambridge, MA 1990 (Harvard University Press).

CUSACK, T.R.: Social Capital, Institutional Structure and Democratic performance: A Comparatitive Study of German Local Governments. European Journal of Political research 1999.

DEEDS, D.L./HILL, C.W.L.: Strategic Alliances and the Rate of New Product Development: An Empirical Study of Entrepreneurial Biotechnology Firms. Journal of Business Venturing, 11(1996), pp. 42-55.

DYER, J.H./SINGH, H.: The Relational View: Cooperative Strategy and Sources of Inter-Organizational Competitive Advantage. Academy of Management Review 23(1998)4, pp. 660-679.

EISENHARDT, K.M.: SCHOONHOVEN, C.: Resource-Based View of Strategic Alliance Formation: Strategic and Social Effects in Entrepreneurial Firms. Organisation Science 7(1996)2, pp. 136-150.

FUKUYAMA, F.: Social Capital and the Global Economy. Foreign Affairs. 1995.

GHOSHAL, S./MORAN, P.: Bad for Practice: A critique of the Transaction Cost Theory. Academy of Management Review 21(1996)1, pp. 13-47.

GRANT, R.M.: Toward a Knowledge-Based Theory of the Firm. Strategic Management Journal 17(1996), pp. 109-122.

HELLIWELL, J.F./PUTNAM, R.: Economic Grown and Social Capital in Italy. In: DASGUPTA, P./SERAGELDIN, I. (eds.): Social Capital: A Multi-faceted Perspective. 2000 (Word Bank).

HIPPEL VON: The Sources of Innovation. N.Y. 1988 (Oxford University Press).

HITT, M.A./IREALAND, R.D./HOSKISSON, R.E.: Strategic Management: Competitiveness and Globalization. Third Edition. Cincinnati, OH 1999 (South-Western College Publishing).

HUBER, G.P.: Organizational Learning: The Contributing Processes and the Literature. Organization Science 2(1991)1, pp. 88-115.

KNACK, S./KEEFER, P.: Does Social Capital Have an Economic Payoff?. A Cross-Country Investigation. Quarterly Journal of Economics 1997/November.

KOGUT, B./ZANDER, U.: Knowledge of the Firm, Combinative Capabilities, and the Replication of Technology. Organization Science 7(1992)5, pp. 502-518.

LA PORTA, R./LOPEZ DE SILANES, F./SHLEIFER, A./VISHNY R.W.: Trust in Large Organizations. American Economic Review 1997.

LANE, P.J./LUBATKIN, M.: Relative Absorptive Capacity and Inter-Organizational Learning. Strategic Management Journal 19(1998), pp. 461-477.

LARSON, A.: Network Dyads in Entrepreneurial Settings: A Study of the Governance of Exchange Relationships. Administrative Science Quarterly, 37(1992/March), pp. 76-104.

LARSSON, R./BENGTSSON, L./HENRIKSSON, K./SPARKS, J.: The Inter-organizational Learning Dilemma: Collective Knowledge Development in Strategic Alliances. Organisation Science, 9(1998)3, pp. 285-305.

LORENZONI, G./ORNATI, O.: Constellations of Firms and New Ventures. Journal of Business Venturing, 3(1988), pp. 41-57.

"Making Trade Work for the World's Poor". Global Economic Prospects and the Developing Countries, Washington, DC 2002 (World Bank).

MAC GILLIVERAY, A.: The Glue factory-Social Capital, Business Innovation and Trust. 2002 (New Economics Foundation for the Association of Chartered Certified Accountants).

MCDOUGALL, P./SHANE, S./OVIATT, B.M.: Explaining the Formation of International New ventures: The Limits of Theories from International Business research. Journal of Business Venturing 9(1994), pp. 469-487.

MCEVILY, B./ZAHEER, A.: Bringing Ties: A Source of Firm Heterogeneity in Competitive Capabilities. Strategic Management Journal, 20(1999), pp. 1133-1156.

MCGEE, J.E./DOWLING, M.J.: Using R&D Cooperative Agreements to leverage Managerial Experience: A Study of Technology-Intensive New Ventures. Journal of Business venturing 9(1994), pp. 33-48.

MERTON, R.K.: Continuities in the Theory of reference Group Behaviour. In: Social Theory and Social Structure. New York 1957, 1968 (Free Press), pp. 335-440.

MORSELL, C.: Structuring Mr. Nice: Entrepreneurial Opportunities and Brokerage Positioning in the Cannabis Trade. Crime, Low and Social Change, 33(2000).

NAHAPIET, J./GHOSHAL, S.: Social Capital, Intellectual Capital and the Organizational Advantage. Academy of Management Review 23(1998)2, pp. 242-266.

PENROSE, E.: The Theory of the Growth of the Firm. Oxford 1959 (Blackwells).

PERRI: Escaping Poverty – from Safety Nets to Network Forms of Organization. Demos 1997/September.

PUTNAM, R.D.: Making Democracy Work. Princeton, N.J. 1993 (Princeton University Press).

RING PS./VAN DE VEN, AH.: Developmental Processes of Cooperative Inter-organisational Relationships. Academy of Management Review, 19(1994)1, pp. 90-118.

SCHAFFT, K.A./BROWN, D.L.: Social Capital and Grasroots development: The Case of Roman self-governance in Hungary. Social problems. 2000.

SHAN, W./WALKER, G./KOGUT, B.: Interfirm Cooperation and Startup Innovation in the Biotechnology Industry. Strategic Management Journal 10(1994), pp. 1150-1174.

SHEPHERD, J.: Entrepreneurial Growth Through Constellations. Journal of Business Venturing 6(1991), pp. 363-373.

SPENDER, J-C.: Making Knowledge the Basis of a dynamic Theory of the Firm. Strategic Management Journal 17(1996), pp. 45-62.

SPENDER, J-C./GRANT, R.M.: Knowledge and the Firm: Overview. Strategic Management Journal, 17(1996), pp. 5-9.

STINCHCOMBER, A.L.: Social Structure in Organizations. In: MARCH, JG (ed.): handbook of Organizations. Rand McNally: (1965)Chicago, pp. 142-193.

UZZI, B.: Social Structure and Competition in Inter-firm Networks: The Paradox of Embeddedness. Administrative Science Quarterly 42 (1997/March), pp. 35-67.

WHITE, H.C.: Where do markets come from? American journal of Sociology 87(1981), pp. 517-547.

WHITE, P.: "Economic Growth and Social capital". Sheffied. (1997)

Antanas Bagdonas, Viktorija Baršauskienė

Congruence of Interests of Different Stakeholders of Secondary School in the Context of Information and Knowledge Society

Introduction

In the contemporary world, where the most advanced countries are concerned about the development of the quality of information and knowledge society, the essential attention is paid to education, and in particular, to the secondary school education, where the possibilities of "equal start" for all citizens should be ensured. More and more opinions of researchers and practical educators emerge, stating, that all the most important decisions on education should be made under the consideration of the interests of all stakeholders in education, who participate in this field. The need to acknowledge contradictory interests of stakeholders in education resides in the very essence of pluralistic liberal democratic society and the philosophical traditions of liberalism. This is characteristic of all countries, which already possess or are in search for the qualities of liberal democratic values and which are developing information and knowledge society.

Abundance and temporality is characteristic of information and knowledge, and due to highly developed information technologies, knowledge becomes available to various stakeholders in education. It is possible to reach all world libraries and other diverse information sources, to use distant education, etc. through the Internet. Intensive flow of information is constantly renewed, and, correspondingly, the models of learning, living and working change rapidly. International mobility among countries and cities increases. Free movement of working force in the countries within the European Union becomes more intensive, and Lithuania is awaiting similar perspectives, too. The conclusions of the European Council, admitted in Lisabon in March, 2000, confirmed that Europe entered the century, of information and knowledge, and this will influence cultural, economical and social development of the countries as well as the lives of concrete stakeholders in education. These issues are actual to education policy makers and practical educators and they cause serious scientific inquiry.

The issues of the development of information and knowledge society as well as the significance of stakeholders in education in the context of lifelong education have been analyzed by Tarasevičienė (2001); Fullan (1998); Hargreaves (1999); Longworth (1999), Jucevičienė (2001); Fryer (1997); Markkula (1998); McLaughlin (1997); Stoll, Fink (1998); OECD experts (2001) and other researchers. They have justified that lifelong education is an inevitable attitude, formally legalized and socially determined in the sense of globalism, spread of scientific knowledge and information technologies, and at the same time – the implementation of these processes in the secondary school is related to the manifestation of the activities of various stakeholders in education, showing incomplete flexibility and increased social differentiation. Irrespective of the common perception about the mission and aims of the secondary school in the society, concrete stakeholders in education and their groups possess different educational interests, and they unequally understand and/or value changes that are in process. The works of the scientists, mentioned-above, help concentrate on the following urgent ***research problem***: *what should be important for all groups of stakeholders in education at the secondary school in order to keep their interests in harmony with the imperative of lifelong learning and to express their decision to act in the context of information and knowledge society.*

The ***aim of the article*** is to identify the congruence (concurrence) among the interests of various stakeholders in education at secondary school in the context of information and knowledge society.

The ***research object*** is the congruence among the interests of various stakeholders in education at school in the aspect of information and knowledge society.

The ***research methods*** of scientific literature analysis and questionnaire survey are implemented in the article.

The article consists of four parts, in which stakeholders in education are presented as clusters of individuals alongside with the congruence of their interests in the context of information and knowledge society.

Stakeholders in education as clusters of individuals at school, functioning in the information and knowledge society

The starting point in the analysis of stakeholders in education as clusters of individuals is the contemplation about the definition of stakeholders. A terminological counterpart of 'stakeholder' in the Lithuanian language is 'subjektas' (lat. subjectum), which may be defined as 'a person, who is involved in cognitive activity and is active (Vaitkevičiūtė, 2001, p. 354). That is the most common definition. Other researchers hold more specific positions as regarded to the definition. Kvietkauskas (1985, p. 469) states, that a stakeholder is considered to be a person, who makes reality the object of his own activity by practically reforming and actively cognizing it. This definition is adequate to social reality, namely, to the context of information and knowledge society. Other authors, while discussing this notion, also do not separate it from practical social reality.

For instance, Plečkaitis (1975, p. 410) defines, that a stakeholder is an actively participating and cognizing person with individual consciousness and will. While psychologists mainly consider a stakeholder as one person and regard him/ her as a personality, researchers in sociology and management refer to a stakeholder not just as one individual, but their group(s) or organization(s). Consequently, *a stakeholder is an individual, a group of individuals or organization, defined by cognitive and other specific activities, conditioned by inner as well as outer characteristics, that disclose interaction of an individual, groups of individuals or organizations with environment.* Thus, one has to admit that in the change of social reality, the definition of a stakeholder may also vary, because the latter is characterized by adequate reaction to the objective reality. With regard to the dominating context of information and knowledge society, such description is actually possible to *a stakeholder in education*.

According to Jovaiša (1993, p. 225), a stakeholder is an individual or social group as an active being of cognitive and practical activity. This is an educator or a pupil, who is active, possesses a specific physical and spiritual structure as well as maintains the system of relationships with self and environment. However, in order to analyze different interests of stakeholders in education at secondary school, functioning in information and knowledge society, it is not sufficient just to provide a general definition of a stakeholder in education. It is necessary to specify what individuals, their groups, organizations might become stakeholders in a specific educational activity, since this is closely related to their interests. Drawing on different research literature (Stoll, Fink, 1998, p.71; Lesin, 1994, p.73; Plečkaitis, 1975, p.410; Simonaitienė, 2001 and others), stakeholders in education, as clusters of individuals, may be grouped according to their following interests:

1. Cluster of educational policy makers and managers of the system of education – individuals who are employed in state and municipal institutions and represent them in the field of education;
2. Cluster of educational service providers at school – teachers, institutional executives, other school staff;
3. Cluster of educational service users – pupils, parents, and society representatives.

Various stakeholders in education possess different realities, conditioned by their individual experience. Therefore, naturally, it is difficult to arrange common activities and perspectives. McLaughlin (1997, p. 140) notes, that in a liberal democratic society it is typical that people possess diametrically different and incongruent attitudes towards life. Therefore, different stakeholders in education might also function in their specific social reality, according to their individual interests. Kuper and Kuper (1992, p. 422) point out, that an interest is related to striving for a certain benefit (not necessarily materialistic). According to specific interests, particular stakeholders' interest groups may be formed; they aim to influence state educational policy to provide better conditions for the development of information and communication technologies, to motivate ideas of lifelong education, to develop a learning society, to update study content, etc. In this context it is urgent to consider if these groups are united and if their interests are congruent. These aspects were analyzed in the research (carried out in 2001) on the interests of different stakeholders in education at school, functioning within the environment of information and knowledge society (doctoral thesis "Evaluation of the Effectiveness of the Secondary School Activity: Peculiarities of Relationship between Manifestation and Congruence of Stakeholders Interests" by A.Bagdonas, 2002). 1415 respondents from 25 educational establishments were surveyed. This article presents the survey of the general scale of stakeholders' interests with the specific attention to the evaluation of different stakeholders in education about how the context of information and knowledge society is significant to them. Thus, out of 106 questions of the survey, given to different stakeholders in education, 10 indicators were selected to reflect their interests at school, functioning in the context of the development of information and knowledge society in Lithuania. These are the questions about *pupils' possibilities to acquire desired education; to choose subjects according to their aptitudes and interests; about the continuous update of learning material, aiming at contemporary requirements; about possibilities to learn the use of computers and their sufficient supply; about the school library reorganization into information center, about supply of textbooks and other resources; about possession of good knowledge of mathematics and foreign languages; well-equipped rooms, laboratories; about partnership networks abroad.* Also, some other relevant questions, related to the interests of stakeholders at school were included in the survey, however, only the above-mentioned statements are selected for the analysis of this article, because they are the best to reflect the readiness of the stakeholders to act adequately in the rapidly changing social reality. On the basis of the analysis results of scientific literature and empirical research, specific interests of stakeholders in education and their significance in the evaluation of the secondary school, functioning in information and knowledge society, are presented.

Interests of stakeholders in the formation of educational policy and its management system at school in the context of information and knowledge society

According to Longworth (1999), many facts of life confirm the link between the prosperity of society and education. Such countries, in which lifelong education has become the strategy of professional activities, flourish considerably in the context of globalization processes. Markkula and Smerla (1998) note, that information and knowledge society is created only through a learning society. Thus, it is not just important for Lithuania to be equipped with computer technologies and to learn how to use them, but also to develop a versatile learning society, which would mainly consist of stakeholders in education, analyzed in this article.

Stoll, Fink (1998, p.214) emphasize, that the context is especially important in the analysis of school and stakeholders, who are acting there. Lithuanian secondary school is functioning in the continuously changing and developing environment. This change has been particularly intensified, when the integration of Lithuania into the European Union started together with the whole world's entrance into the stage of a postmodern society. According to Jucevičienė, Katiliūtė (1998, p.3-12), the characteristics of the 21st century school are the following: a clear school vision to develop pupils' need for lifelong learning; qualitative development of the school to insure high quality education; strong objectives to be equipped with new information and communication technologies alongside with the skills to manage them with regard to continuously changing school environment; professional leadership, that ensures the development of school staff to be able to act in the information and knowledge society. Consequently, one of the most significant priorities for the Lithuanian educational reform is the interest to coordinate the system of education inside and outside school and guarantee the necessary conditions for continuous education. The performed research has also emphasized similar interests of this cluster of stakeholders in education. The strongest interest is the possibility for pupils to choose learning subjects according to their interests and aptitudes. *The cluster of educational policy makers and managers of the system of education rated this interest to be the third out of 106 interests in the general rating scale. As a very important interest at school, (it was rated 5 according to the significance),* was reorganization of the school library into information centre. *Depending on importance, interests ranged in the following order* – preparation of the learning material for pupils in accordance with contemporary requirements (position 14); adequacy of the learning material to the changes in science and industry (position 18); pupils able to use computers (position 29), etc. *The stakeholders of this cluster care the least about* sufficient amount of computers at school (position 69). *The latter reflects the above-mentioned statement that it is important not only to be equipped with technologies, but also to develop skills to use them and implement into educational practice. However, the problem came out with regard to the interest of school relationships with foreign partners (rated position 98 only). This is one of the greatest problems that express the contradiction between information and knowledge society as a global system and the decisions of concrete stakeholders in education to act constructively.*

Indicating that the policy of the state and municipalities and the strategy of its implementation determine the development of contemporary and modern school as well as the perfection of an individual and society can substantiate the received research results. However, as Jucevičius (1998, p.115) points out, in any society (including the modern ones) its institutions, organizations and community should be adequate to the actual times and the actual situation. In a vast diversity of research literature (Zakarevičius, 2000; Arends, 1998; Jucevičius, 1998; Bennett, 2000; Parson, 2001; Jovaiša, 2001; et al.) the current times are characterized as a period of the development into the societies of knowledge and organizations, and, in turn, of the schools, that seek to become knowledge organizations.

Globalization processes will not bypass Lithuania. It is necessary not only to develop new technological culture, but also a learning society. Thus, since the aims of the state structures of Lithuania and municipalities reflect modern nation, open society and democratic country, then in the space of these interests, the perspective to create a learning society will bring democratic principles of education into the Lithuanian school, and this will assist in its entrenching into the system of education in Europe. Consequently, as Bruzgelevičienė (2001, p. 131) notes, the institutions of the state and municipalities should keep to the attitude, that the development of information and knowledge society and lifelong learning are the most significant underlying objectives of education, alongside with other changes in the system of education. However, as Hopkins and others (1998, p. 30) in their analysis about school and the management of its change point out, there is no any greater change, than the change in incredibly increased abilities of stakeholders in education to understand change. The latter is confirmed by the results of the performed research, which reflects the most significant interests of stakeholders in education at school, functioning in the developmental process of information and knowledge society.

Interests of stakeholders as educational service providers at school in the context of information and knowledge society

One of more important interests of school principals and teachers as educational service providers, according to Fullan (1998, p. 99), is the development of schools as learning organizations. Activities of these stakeholders in education are conditioned not just by one interest, but also by the whole group of interests. Consequently, the interests of educational service providers can be referred to as the interests of stakeholders in education to develop a learning organization. They are responsible for the formation of such organizations, where individuals could expand their perception abilities, clarify their vision and develop efficiency of their activities in organization. According to Senge (1990, p.352), leaders and pedagogues of learning organizations are characterized by their abilities to conceptualize strategic, universally perceived insights to make them an open inquiry tat is later critically evaluated and developed. Thus, it is not a chance, that the interests of these stakeholders in education involve promotion of learning among all the staff within educational organizations. Also, educational service users, acting in the reality of information and knowledge society, seek for developmental changes within the group. This is especially relevant nowadays in the context of globalization. The strive of European Union to create a learning society coincides with the Lithuanian need for cultural maturity of democratic society, formulating pedagogues' interest in lifelong learning. The realization of these interests adds to the school development as a learning organization and, consequently, to the creation of information and knowledge society. This is also confirmed by the research results. The cluster of stakeholders as educational service providers (teachers and institution executives) indicated *equal conditions of pupils to acquire desirable education* as the most significant interest. Teachers and school principals rated the latter interest 9 according its importance. It is necessary to note, that school principals indicated the highest rate to the interest of *preparation of learning material that would correspond contemporary requirements at school* functioning in the environment of information and knowledge society. This is the 5th position out of 106 questions in the general rating scale of interests. Teachers indicated this interest at the rate 14, according to its significance. Thus, it is placed in the first quartile of interests as well. Both teachers and school leaders as educational service providers disclosed the interest of *preparation of learning material to be adequate to the changes in science and industry, so that to reconstruct old knowledge* similarly (teachers' rate was 17, while school leaders indicated rate 23). Teachers and school principals as educational service providers expressed similar interests about *the possibility for the pupils to choose subjects according to their aptitudes and interests.* However, in this cluster different interests were reflected, namely, about computerization. Teachers rated it as 36th interest in the general rating scale of interests, while school leaders rated it as 50th. Based on educational practice, it may be assumed that teachers as stakeholders in education are more ready to act in the environment of knowledge society than school principals. This can be related to the implementation of new technologies in educational processes and the availability of skills to use them. Also, one more clear contradiction between the needs of contemporary modern times and the readiness to act in this context in this cluster of stakeholders is observed. The interest to develop school relationships and partnership networks with foreign schools abroad is rated as the last one. Teachers rated 100 according to significance, and school leaders rated this indicator 104 and 106 in the general rating scale. According to Longworth (1999), information society has to be supplemented by a learning society, creative strategic networks of cooperative learning, built on the participation of people from different cultures and races, that would help understand one another wider – internationally, that would teach to collaborate and coordinate different positions and find common goals. As the guidelines of education in Lithuania (2000, p.21) indicate, stakeholders in education will have not only to learn to know each other, but also to act, to learn and to be together. So, the research results confirm, that educational policy of the state in this direction is important and necessary.

Interests of stakeholders as educational service users at school in the context of information and knowledge society

The main stakeholders in education as school and educational service users are pupils and parents. As Coons, Sugarman (1978, p.13) indicate, the rights of children and the most positive interests of parents are expressed by the idea of autonomy. The notion of this ideal is involved in the essence of information and knowledge society, in its principles and values. In the space of parents' interests, in acknowledging these values, the notion 'child's right to have open future' are dominating. This means that parents wish to provide their children with the opportunity of continuing education, with the use of various possibilities, advantages and choices. In this context Stoll, Fink (1998, p. 34) indicate, that the most important interest of pupils at school is to create good conditions for their learning with the help of teachers. While characterizing pupils' interests at school, Edmonds (1976, p.16) mentions the significance of equality in acquiring knowledge. According to Anzenbacher (1992, p. 276), there is no person who could actualize self alone without others. Child's self-realization is concurrent with society's. Bitinas (1996, p. 14) points out, that each society has a functioning system of education. Thus, in the space of pupil's interests, self-realization is very important, and, according to Jucevičienė (2001, p. 17), it is possible only in a learning information and knowledge society.

On the basis on the research results it can also be stated, that in the educational practice one of more important interests of educational service users is *having equal opportunities for pupils to acquire education they desire.* Parents rated its importance as 4th position and pupils rated it as 5th position out of 106 interests in the general rating scale of interests. However, the most significant interest of pupils, reflected in this research, was *the possibility to choose learning subjects, related to their interests and aptitudes.* This was the second position in the general rating scale. Parents rated this interest 17 according to importance. It is worth to note, that almost all the rest interests of this cluster of stakeholders in education coincide. The latter interests include the interest in *teaching material (pupils rated 14, parents rated 10), the interest in computerization, pupils' computer literacy (pupils rated 23, parents rated 25)*, etc. However, discrepancy inside the cluster can be noticed, namely, in *school cooperation and relationships with other schools abroad (pupils rated 45, parents rated 87)*. The latter need of pupils is the strongest as compared to all other stakeholders in education who participated in the research (educational policy makers rated 75, school principals rated 104, pedagogues rated 100). Such conspicuous discrepancy of this interest among educational policy makers, educational service providers and users shows their unequal perception of social reality as the need to communicate and cognize diversity. This result emphasizes the research problem that was presented in the beginning of this article. Considering the fact, that educational guidelines in Lithuania emphasize the importance of partnerships and capabilities to act in the common space of information and knowledge society, a positive solution of this problem is expected in the perspective.

Summarizing the significance of interests of various stakeholders in education at the secondary school, functioning in the context of information and knowledge society, a concurrence (coincidence, congruence) among certain stakeholders can be observed in the first quartile of the general rating scale of interests. Pentagonal (all stakeholders in education) concurrence is disclosed, related to the interest of *at school pupils have equal conditions to acquire education they desire*. This highlights the most necessary precondition for further activities of a person in social reality that is continuously changing. Also, pentagonal concurrence of interests is observed, related to *preparation of learning material with regard to changes in science and industry*. Also, the congruence of all stakeholders in education has been disclosed about the interest *teaching material should correspond to contemporary requirements*. Surely, such concurrence of interests of all stakeholders in education illustrates perception and readiness of stakeholders to act in contemporary society. Also, a triangular (pupils, parents and educational policy makers) concurrence of interests is observed. This is *choosing learning material that corresponds to pupils' aptitudes and interests*. An assumption can be made that the interests of educational service providers (teachers and institutional executives) do not concur in the first quartile, because service providers have not fully cleared out the needs of educational service users, and also, there still exists the tradition of market in education, but not education in market. This is confirmed by clearly unequal significance to the interest *partnerships and relationships with the schools from abroad*. This is why the contradiction is observed between the imperative of lifelong education, the processes of globalization and the decision, capabilities of stakeholders to act in information and knowledge society.

So, the results of the analysis of scientific literature and empirical research allow stating, that the interests of stakeholders in education at school, functioning in the context of information and knowledge society, are diversified. Some generality is detected in this diversity, otherwise, the system of education would be torn apart by essential contradictions, and it would be difficult to expect common agreements, necessary for the system to function and thrive. On the other hand, it is not necessary and even not possible to expect, that all the interests of stakeholders would completely coincide – this does not occur in life. The latter is not even necessary, because then, there would not be any reasons for a discourse, as the main precondition for discussions that determine the development of school as well as the whole system of education.

Conclusions

1. A stakeholder in education is an individual, a social group or an organization, involved in educational activity, possessing specific inner qualities, and correspondingly interacting with the environment. Various classifications of stakeholders in education are possible: according to the degree of their arrangement (individuals, their groups, organizations), according to their position in the system of education in the context of marketing (educational policy makers and managers of the system of education, educational service providers at school, educational service users).
2. The results of the research do not disclose common interests at school among all stakeholders in education – educational service users (pupils and parents), educational service providers (teachers and institutional executives), educational policy makers and managers of the system of education.
3. It is defined, that the three groups of stakeholders in education (educational policy makers and managers of the system of education, educational service providers, educational service users) express the following common interest: pupils should have equal possibilities to acquire education. The following fairly common interests are expressed among pupils, parents and teachers: equal possibilities for pupils to acquire the education, they desire, learning material should be prepared in accordance with changes in science and industry, that is, with contemporary requirements. The first interest can be regarded as the manifestation of democracy, while the second interest discloses innovation.
4. The evaluation of the whole scale of stakeholders in education allows to state, that neither educational service providers nor educational service users express persuasive position about lifelong education.

References

ARENDS, R.J.: Learning to Learn. Vilnius 1998 (Margi raštai). (In Lithuanian).

AZENBACHER, A. Introduction to Philosophy. Vilnius 1992 (Katalikų pasaulis). (In Lithuanian).

BAGDONAS, A.: Evaluation of the Effectiveness of the Secondary School Activity: Peculiarities of Relationship between Manifestation and Congruence of Stakeholders' Interests'. Doctoral thesis. Kaunas 2002 (Kaunas University of Technology). (In Lithuanian).

BENNETT, B.: Cooperative Learning. Vilnius 2000 (Garnelis). (In Lithuanian).

BITINAS, B.: Basics to Philosophy of Education. Vilnius 1996 (VPU). (In Lithuanian).

BORUTA, J./GUDYNAS, P./JACKŪNAS, Ž.: Education Guidlines. Vilnius 2002 (Švietimo kaitos fondas). (In Lithuanian).

BRUZGELEVIČIENĖ, R.: Change of the System of Education in Lithuania. Works on the Educational Reform at the Crossroad of the Millenium: Tendencies, Expectations. Vilnius 2001 (Garnelis). (In Lithuanian).

COONS, J./SUGARMAN, S.D.: Education by Choice: The Case for Family Control. Borkeley 1978 (University of California Press).

EDMONDS, R.R.: Effective Schools for the Urban Poor, Educational Leadership, 37(1976)1: pp. 15-27.

FRYER, R.H.: Learning for the 21Century, First Report of the National Advisory Group for Continuing Education and Lifelong Learning. London 1997 (NAGCELLI).

FULLAN, M.: Change Forces. Vilnius 1998 (Tyto Alba). (In Lithuanian).

HARGREAVES, A.: Teachers in Change, Times in Change. Vilnius 1999(Tyto Alba). (In Lithuanian).

HOPKINS IR KT.: Change and the Development of School. Vilnius 1998 (Tyto Alba). (In Lithuanian).

JOVAIŠA, L.: Introduction to Educology. Kaunas 1993 (Technologija). (In Lithuanian).

JOVAIŠA, L.: Education Science and Practice. Vilnius 2001 (Agora). (In Lithuanian).

JUCEVIČIENĖ, P.: Empowerement of Lifelong Learning: Conception of a Learning City. Kaunas 2001 (ŠUS). (In Lithuanian).

JUCEVIČIENĖ, P./KATILIŪTĖ, E.: European Dimensions of Education Reflected in Higher Education of Lithuania. In: Systems and Didactics of Higher Education. Book One. Kaunas 1998 (Technologija), pp. 3-12. (In Lithuanian).

JUCEVIČIUS, R.: Strategic Development of Organizations. Kaunas 1998 (Pasaulio lietuvių kultūros, mokslo ir švietimo centras). (In Lithuanian).

KUPER, A./KUPER, J.K.: The Social Science Encyclopedia. Second edition. London and New York 1996 (Routledge).

KVIETKAUSKAS, V.: Dictionary of International Terms. Vilnius 1985 (Vyriausioji enciklopedijų redakcija). (In Lithuanian).

LESIN,.B.C.: Instructional Desing Strategies and Tactics. USA 1994 (Educational Technilogy Publikation, Inc.).

LONGWORTH, N.: Making Lifelong Work: Learning Cities for a Learning Century. London 1999 (Kogan Page).

MARKKULA, M./SUURLA, R.: Passion to Learn, Benchmarking Good Lifelong Learning Practice, Report no.9, Helsinki 1998 (IACEE).

MCLAUGHLIN, T.H.: Contemporary Philosophy of Education: Democracy, Values, Diversity. Kaunas 1997 (Technologija). (In Lithuanian).

OECD: Reviews of National Policies for Education – Lithuania: Examiners Report. Vilnius 2001 (Pedagogika, Mokslo darbai).

PARSONS, W.: Public Politics. Vilnius 2001 (Eugrimas). (In Lithuanian).

PLEČKAITIS, R. Dictionary of Philosophy. Vilniuis 1975 (Mintis). (In Lithuanian).

SENGE, P.: The Fifth Discipline. New York 1990 (Doubleday).

SIMONAITIENĖ, B.: The Features' Expression of Learning Organization Activity in Gymnasium. Doctoral thesis. Kaunas 2001 (Kaunas University of Technology). (In Lithuanian).

STOLL, L./FINK, D.: Changing School. Vilnius 1998 (Margi raštai). (in Lithuanian).

TARASEVIČIENĖ, M.: Educational Dimensions of Lifelong Learning. Non-published habilitation thesis. Kaunas 2001 (Vytautas Magnus University). (In Lithuanian).

VAITKEVIČIŪTĖ, V.: Dictionary od International Terms. Vilnius 2001 (Žodynas). (In Lithuanian).

ZAKAREVIČIUS, P.: Paradigm of Management. Kaunas 2000 (VDU leidykla). (In Lithuanian).

Vilma Žydžiūnaitė

The Substantiation of a Junction between Teamwork Competences and Professional Competence

The traditional standpoint, directed to the importance of knowledge acquirement, isn't adequate to the needs of the changing society and work. Teamwork, based on team member's abilities to integrate specialized skills and knowledge isn't the basis of the effective teamwork. Employees should be able to work in complex environments, which are characterized by controversial information, non-formal collaboration, and dynamic processes. The concept of *competences* is related to ability to managing such complex situations and applying knowledge and skills in various levels of activity. That is why this concept has become the research object in social sciences. The concept of competences extended the definition of competence and became the core category in education. It is manifested by the activity, social and personal competences. Competence and competences are in the junction between employees' learning and work. *The purpose of this article is to illuminate the peculiarities of a junction between teamwork competences and professional competence.*

Competence. Stevenson (1996) mentions that change in competences' constructions is related to different contexts. Educational researchers (Jucevičienė & Lepaitė, 2000; Kirschner et al, 1997; Westera, 2001) agree, that personal competence is determined by mastered integrated collection of knowledge, values, skills and standpoints and ability to apply it in different contexts of work. This definition emphasizes on elements of knowledge and skills, which are urgent for every activity.

From the common educational standpoint, knowledge is related to reproduction of facts, procedures, theories and its representation. Different notions define knowledge, e.g. *conceptual knowledge* (Posner & Keele, 1973), *content knowledge* (Gardner, 1975), *declarative knowledge* (Anderson, 1980). In all those concepts knowledge is represented as constant elements and subjects of learning and reminiscence. The important characteristic is knowledge verification and evaluation, but it shouldn't be identified as perception and understanding. Perception and understanding involves intellectual ability to use the information in adequate ways. It is developing, when knowledge is applied in new situations. If knowledge is identified as constant elements that could be "saved", the cognitive skills are related to thinking operations, "producing" knowledge (Anderson, 1980). Though the important role in competence realization is directed to cognitive skills.

Cognitive skills are defined as *procedural knowledge* (Anderson, 1980), *intellectual skills* (White & Mayer, 1980), *mental operations* (Gagné, 1977). These notions include reflection, when the acquired knowledge is used, transformed and supplemented. Cognitive skills are associated with the activities that demand a higher level of thinking (e.g. problem evaluation, decision making, summarizing) and involve processes of analysis, synthesis, and assessment. Evaluation of cognitive skills is complicated. The only way, which helps to evaluate the mastered skills, is based on the employee observation in the context of activity that incorporates concrete skills. If knowledge and skills are not integrated then there is no guarantee, that activity will be effective in complex environments. There is a need to acquire specialized abilities, which include specific context characteristics (Scheeres & Hager, 1997).

Barnett (1994) relates the competence with the behavior in unexpected and undefined situations. According to the author, competence includes skills, activity results, activity assessment/evaluation and the ability to transfer those skills to other contexts. *Ability* becomes the junction between skills and competences (Jucevičienė & Lepaitė, 2000). Skills by themselves can't guarantee effectiveness of work. The abilities, developed in the activity contexts involve realization of skills and are based on the adequate decision-making, but not on the static activity, based on the specialized skills. Ability to manage and apply the complex of skills is central in the competence structure (Westera, 2001). Assessment of skills is grounded by the assessment of procedures or processes. It means that we assess personal incompetence. A competent employee is able to realize the work according its standard, which is related to concrete organization and the concrete activity (Alsop, 2001). Competence is such case includes personal and situational dimensions. Barnett (1994) also actualizes these dimensions: he relates competence with the concrete standards and emphasizes personal competence that confirms adequacy between actions and standard.

Eraut (1994) mentions, that competence should involve positive and negative dimensions, which help to evaluate the activity. Dreyfus & Dreyfus (1986) broaden that idea and include the notions of *expert* and *beginner* into the definition of competence and draw a line between competence and incompetence. That line could mean the levels of competence. The definition of *incompetence* is still complicated. Bowden & Marton (1998) established and grounded the concept of competence levels according to activity levels. This standpoint reflects sociologic and reduction nature of competence. Fish & Twinn (1997) emphasize personal flexibility, related to the changes in activity (it includes ability to adapt and improvize as well).

Competence isn't static: it relates to specific activity context, includes skills, abilities, standpoints, perception *and experience, that are necessary for realiza-*

tion of professional (related to the acquired qualification) and *activity* (related to a concrete activity situation) *roles. These roles should correspond to the standards, urgent for the organization and activity environment. Realization of roles and competence are the teamwork characteristics, and its coordination is the precondition to effective teamwork.*

Competences. Competence and competences relate to skills, personal traits, which are "carried" to work situations by persons, "used" in striving for concrete aims, results and helps to assess/evaluate activity effectiveness. According to Trotter & Ellison (2001), competence is an ability to realize activity aim. Competences relate to personal traits and abilities, which empower the person to act effectively in various situations.

The competence could be characterized in two ways: 1) *as personal ability to realize the objective*, related to professional area and/or specialty, when the precondition of successful activity is the acquired qualification (e.g., competences in psychology, in nursing, in economics); 2) *as personal ability to realize the role/roles in concrete work environment*, when the predisposition of effective activity is personal traits and behavior, which reflect employee's ability to manage social situations.

Jucevičienė & Lepaitė (2000) indicate that competence could be divided into separate parts – competences. According to the authors, in the lowest activity levels (first and second), it is important to apply abilities and skills, and in this context is more correct to talk about competences. Here two important questions arise there for discussion: *1) Are competences analogous to skills? 2) Is competence the structural derivative, which consists of competences?*

1. *Are competences synonymous to skills?*

   Complex and unique situations provoke the premeditated and conscious behavior. Conscious behavior in this case emphasizes personally mastered selection of activity elements, its integration and/or adaptation in new situations. For example, students are taught to develop the effectiveness of activity in complex situations with the predisposition that specialists' thinking, decisions and selection of adequate skills will take a high acceleration in the future, and the result will be a more effective behavior in the activity context. The growing activity speed causes domination of automatism, reduction of thinking and perception components. According to Barnett (1994), the definition of competences can't be based on the ideas of thinking and perceiving, because definitions of those ideas are defined from the operational standpoint. Such standpoint is characterized by the activity observation, assessment of results and technique of the realized skills, but not by the process value. It means that conscious thinking isn't the factor that distinguishes competences and skills.

Activity environments become more complex and can't be explained even by the mastered skills. Here a question arises: "How to define the complexity in order to distinguish competences and skills?" From the theoretical standpoint, comp'lexity of the aim could be explained by various characteristics, e.g. connections between aspects and elements, specificity of processes. Such analysis helps to detail the process, but doesn't define the complexity of the aim. This complexity should be assessed and related to personal characteristics of striving for the aim, e.g. personal qualities, aptitudes, work experience etc. Jucevičienė & Lepaitė (2000) define competences in this way, too and emphasize, that competences include specialized abilities.

Trotter & Ellison (2001) note, that competences are concrete characteristics, which empower a person to realize an activity in many situations and to get perfect results. The authors point out that competences are factors that distinguish the best characteristics from the activity behavior, but competences aren't activity aims and objectives; those competences empower individuals to strive for the aims and objectives. The presented description isn't detailed: authors do not emphasize on the characteristics, which predispose the activity success and do not mention any concrete qualities that help the person to realize his activity role in the best way. Lloyd & Cook (1993) note, that competences are the precondition of successful activity and relate to personal "account" into the activity that includes personal traits, ways of behavior and non-specialized abilities. This definition is related to psychological, holistic basis of competences and illuminates that successful activity depends not only on vocational qualification and specialized competence, but non-professional abilities, too. Trotter & Ellison (2001) mention, that competences are divided into two types: *1) threshold competencies*, which are based on minimal use of abilities and are predisposed by personal motives, values, standpoints, self-confidence that are "carried" into the activity process and direct the mastered skills and knowledge; 2) *distinguishing competences,* which are the key competences in realization of the concrete activity, are related to individual behavior and it is impossible to reproduce the mechanically or imitate; even the perception of the activity aims empowers the individual to use them, because in other situations such competences possibly can't be key competences. The authors do not develop the idea about the two types of competences, but mention that these competences are influenced by individual motivation. It means that competences are related to personal characteristics and could be the precondition of successful activity.

Spencer & Spencer (1993) assert that competences should be concentrated on the work and everyday life. According to the authors, competences, directed to activity, include skills and knowledge; these competences could be developed more easily than competences, related to everyday life, which include personal motives, values, standpoints and self-confidence. Such Spencer & Spencer's (1993) reasoning is synonymous with the remarks of Trotter & Ellison (2001). It grounds the idea that competences could be divided into two types: related to (1) personal characteristics and (2) activity content. Competences shouldn't be treated as skills, because skills are the mastered intellectual or physical actions from the experience, which relate mostly to automatization and demonstrate a person's ability to realize those skills in many ways according to the concrete context (Žydžiūnaitė, 2000). Competences incorporate skills and knowledge. It means that a person should be able to realize adequate activities and to apply the mastered skills.

2. *Is competence a structural derivative, which consists of competences?*

   In order to answer this question, it is important to illuminate whether competences are an autonomous concept, whether this concept is an element of the concept of competence, or whether sub competences and competences are synonymous concepts. Bowden & Marton (1998) note that competences relate to competence, but they are more "narrow" and are directed to work environment and treated as employees' ability to realize the work adequately. According to the authors, that *complex* represents personal competence. Does it mean that competence consists of various competences, and their management and integration represent personal competence? Velde & Svensson (1996; cit. Bowden & Marton, 1998) mention, that the concept of competences should be adequate to the needs of the main activity situation, holistic and contextual. It isn't enough to describe activity objectives according to concrete standards. For the employee, it is important to perceive and understand the integrity of activity characteristics in the context of different objectives. It is the basis of understanding the relation between common and specific parts of competence. The authors emphasize the relation between employee and activity and the competence in such context is a holistic derivative. The context of competence here includes connections with the individual, who realizes concrete objective in the work situation context.

Sandberg (1994) mentions connection between competences and activity content: competence should include individual relation with the personally defined work concept. Spencer & Spencer (1993), Lloyd & Cook (1993), Sandberg (1994), Bowden & Marton (1998), Trotter & Ellison (2001) agree, that competences: acquire the sense in the context of concrete work/activity situation; are in continuing relations with the employees personality; are divided into two types – directed to work/activity objectives and to personality.

Westera (2001) & Alsop (2001) use the notion of *sub competences*, when they define the structure of competence. According to the authors, sub competences also reflect the work functions. The authors do not describe the volume of sub competences. Lloyd & Cook (1993), Trotter & Ellison (2001) do not agree, that competences could be functional and incorporate personal employee characteristics into the definition of competence. Sub competences shouldn't be identified with the competences, directed to work objectives, but the notion of sub competences could be used in the structure of competence. Why?

Competences, directed to activity/work objectives are related to qualification, which is the predisposition of effective activity objective realization. The contemporary work philosophy, according to White (1997): recognizes that the change of qualification becomes an important characteristic of activity; recognizes the change of the activity roles, when they include personal component without the specific education. The structure of personal competences doesn't include qualification, but incorporates personal qualities and abilities; its integration is the basis of effective activity. It is important to assert, that:

- *Competences, directed to work* are called *sub competences* (Westera, 2001) or could be called *functional competences.* They are dependent on activity changes. Because of that a person is empowered to learn continually, if he/she wants to survive in the activity context. From the educational standpoint, these competences could be developed, because they are "external", directly dependent on the activity.
- *Competences, directed to personality* are called *threshold* (Trotter & Ellison, 2001) and in the activity context they realize the functions of "strengthening" and "directing", but their change lasts for a long time, because a person "brings" her/his values, perceptions, convictions, traits, abilities to activity environment; they are influenced by experience and aptitudes (Jucevičienė & Lepaitė, 2000). The internal structure of the personality is characterized by maturity; development and changes of such structure are going more slowly. Competences, directed to personality are "internal" and relate to subcompetencies. Synergy of threshold and work competences in the context of the concrete activity stipulate the realization of *exceptional* (Trotter & Ellison, 2001) competences (e.g., team work, leadership, flexibility and etc.), which are the precondition of effective work (see Figure 1, p. 260):

In the presented model (see Figure 1, p. 260) we can see, that both types of competences – directed to work/sub competences and directed to personality act reciprocally: the junction of crossed lines (marked by dotted lines) reflects openness to environment, and the razed arrows from personal competences demonstrate the "stimulating" influence upon competences, directed to work. Competences, directed to work directly relate to activity aims and objectives. Competences, directed to personality, help the employee to realize the concrete activity roles. Activity aims, objectives and roles interrelate in the work context and change according to the activity needs. Synergy of sub competences and personal competences in the activity environment "gives" the outcome – exceptional competences, which are valued only in the context of concrete activity and realization of those competences is dynamic and dependent on the activity changes. Exceptional competences directly influence realization of employee's roles. When the employee seeks for the activity aims, then the exceptional competences "strengthen" that process and it is a precondition of the activity effectiveness (because of this the arrow in the model is market by a dotted line).

It is important to assert, that: competence is the structural derivative, which incorporates the two main types of competences and those have their specific structure, and their synergy influences the activity effectiveness; the concept of competences is a part of the competence concept, and because of that the concept of competence could be called holistic.

Teamwork competences. Scientists define teamwork competences as: 1) an *autonomous concept:* concept of team competences (Mahmoodi & King, 1992; Salas, 2002); 2) an *element of the other concept: element of professional competence* in the concept of personal soft skills competences (Doe, 2000, 2001); *skill in the concept* of core competences (Management Council Report, University of California, 1995).

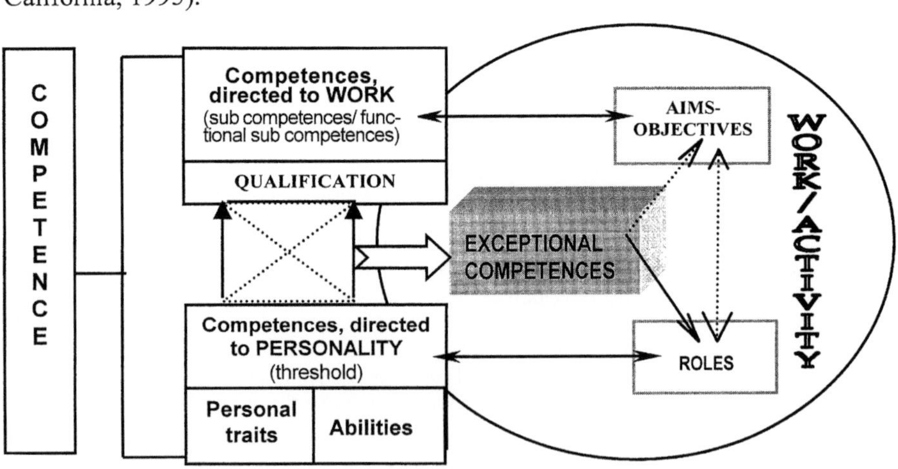

Figure 1: Model of interaction between competence-competences-activity (Žydžiūnaitė, 2003)

Concept of team competences. Mahmoodi & King (1992) investigated library employees and used a questionnaire with closed-ended questions (127 propositions; every proposition was evaluated by three values [(yes, no, don't know]) on Likert scale). Researchers extracted six team competences and characterized their content and their direction (see Table1).

Team competence	Content of teamwork competence	Direction of teamwork competences
Leadership abilities & standpoints	✔ Imitativeness and realization of activity, based on effective collaboration	✔ Formation of future vision ✔ Personnel development
Administration abilities	✔ Structuring personal activities and their combination with activities of other employees ✔ Resource coordination and striving for effectiveness	✔ Personnel development ✔ Activity planning ✔ Development of personal operativeness, coordination
Specific professional skills & knowledge	✔ Applying specific/professional knowledge and skills ✔ Management of information technologies ✔ Negotiations with the representatives of adequate type of organizations	✔ Development of professional knowledge, abilities ✔ Development of negotiation abilities ✔ Collaboration with the community
Cognitive skills & abilities	✔ Management of new information ✔ Continuing learning, based on problems and decisions identification and definition ✔ Application of analytical abilities and it's continuation development	✔ Continuing learning and development
Interpersonal abilities	✔ Interaction with others ✔ Conflict management ✔ Effective negotiation	✔ Creating atmosphere with openness and trust ✔ Communication and collaboration ✔ Development of personal responsibility
Personal qualities	✔ Adaptiveness ✔ Self-control/emotional self-management ✔ Conflict management ✔ Empathy ✔ Self-confidence ✔ Stress management ✔ Ethical behavior ✔ Optimism and enthusiasm	✔ Adaptation to changes ✔ Development of personal self-confidence ✔ Development of personal autonomy

Table 1: Teamwork competences (according to Mahmoodi & King, 1992)

The presented concept of teamwork competences in Table 1 illuminates four characteristics of these competences: 1) *qualification* – team members should acquire specialized professional skills and knowledge; 2) *personal qualities* - influence the effectiveness of collaboration between team members and between team and environment; 3) *management competence as key competence* - its elements are incorporated in every competence; 4) *continuing learning* – the main precondition of team work competences development.

Salas (2002) has investigated more than 4000 various types of teams. The scholar carried out the research in three levels: individual, team, integrated (both – individual and team) and has applied educational interventions (strategies of personal and team learning and team development). The research object was the team competences and learning processes in the activity context. Salas (2002) mentions, that the team work is inseparable from the development of team members' competences and team work competences "stand" on the three columns:

1. *Knowledge:* team mission, aims, norms and resources; strategic evaluation of mistakes; process of dividing the aims and objectives and situational assessment/evaluation; personal obligation, which relates to specific team aims; self-recognition and personal characteristics.
2. *Abilities:* the basic monitoring of activity; flexibility; interpersonal support; team leadership; striving for the aim; conflict management and solving; closed-loop communication.
3. *Standpoints and prejudices:* orientation to team morality; a shared vision; team-members' cohesion; interpersonal trust; perceiving the importance of teamwork; continuing learning and development.

Salas (2002) doesn't mention the value of specific knowledge in the teamwork concept, but emphasizes on teamwork process, where the following aspects are important: 1) Orientation to additional areas, which "stimulates" teamwork: *management* (monitoring of activity, leadership); psychology (self-evaluation, conflict management and solving, interpersonal trust); *education* (continuing learning and development). 2) Orientation to levels: *individual/team member* (personal qualities, personal responsibilities); team (team morality, unity of team members, sharing aims and objectives). Team members have a possibility to understand processes of team work (what is done and why) and communication (giving and receiving feedback) and diagnose the interferences in realization of practical skills, development of abilities and application of team competences because of integration of the three complex elements - knowledge, abilities and standpoints - which are the basis of competences.

The notion of *team competences* is inexact and "narrows" the real meaning of the teamwork concept. The concept of team competences could be interpreted as competences, which are "brought" by team members, but are constant and static without work. Even integrated team members' competences in the environment of teamwork influence the effectiveness of teamwork. Modifications of teamwork competences, given by Mahmoodi & King (1992), Salas (2002), are concentrated on teamwork process. A more exact notion, which reflects interaction between the team and the work, is the concept of *teamwork competences*.

The concept of personal soft skills competences. Doe (2000, 2001) uses a complex notion of soft skills' competences: every soft competence includes concrete *soft* skills. The notion soft is used as common for competences and skills. In that concept soft competences are an inseparable part of professional competence. The notion *personal* helps to understand, that soft competences are the result of personal learning and work experience.

Doe (2000) indicates, that soft competences are equivalent to the technical specialized competences, but they are distanced by more flexible transferability to different work environments. Specialized skills competences are focused on more specific work context. Detterman (1993) emphasizes not the transferability by itself, but asserts, that transferability occurs, if a person is able to use the acquired things in similar, but new situation. Patry (1999) agrees with this idea and indicates, that transferability demonstrates the learned behavior in different situations and it is the main outcome of the learning process. If transferability isn't fulfilled, it means, that the learning process wasn't effective. The characteristic of transferability confirms, that exceptional competences don't appear by it and its acquirement and development are continuing processes. But the origins of exceptional competences are gained in the learning process. It means that an individual, who strives for the first qualification, acquires common elements of competences (knowledge, skills, values and etc.), which could be permanently developed in the activity context. Jungkunz (1999) supports this idea and asserts, that qualification empowers a person to work according to concrete professional demands and even an application of the learned things in work situations (experience) becomes a precondition for competences' development. Personal qualities and abilities, according to Show (1998), become an additional element: it is a "support" for individual and organizational learning and systemic work in groups. It means, that a person should master the basic knowledge, skills, abilities, because without such basis she/he has nothing to transfer.

The concept of *personal soft skills' competences,* presented by Doe (2000, 2001) incorporates 23 soft competences, which are combined with specialized competences in the work context and gives the possibility for the worker to realize professional competence in full value. Soft teamwork competence in such concept is defined as effective and productive work with others and it represents the concept of competence as well. According to Doe (2000, 2001), the structure of teamwork competence consists of the following soft skills: ✔ respecting the team-members and their individual standpoints; ✔ giving the priority to team mission and aims; ✔ acting and making decisions, based on consensus; ✔ striving for the aims and realizing objectives, according the team purpose and obligations; ✔ sharing responsibility with team-members in every situation (successfully/not successfully); ✔ informing team-members about the activity; ✔ supporting the team decisions; ✔ recognizing and valuing the deposit into the activity by team-members; ✔ realizing the work behavior, based on the activity values and mission; ✔ giving the constructive feedback to the whole team and to every team-member; ✔ reacting positively to feedback, given by the team-members; ✔ evaluating teamwork effectiveness, its weaknesses and obstacles and carrying out an open discussion about it.

The presented list of soft teamwork skills reflects the following tendencies: ✔ active interaction between team members and importance of the ethical behavior; ✔ ability of the team to optimize the resources (ability to coordinate the team, team members ability to foresee the needs of one another and compensate one another through allocation of functions); ✔ ability to identify, recognize and regulate the strategy of teamwork and avoid the stress; ✔ team ability to develop collective efficiency. In the structure of that concept the following question arises: What does the structure of teamwork competences include – skills or abilities?

Doe (2000, 2001) presents teamwork competence, which incorporates the complex processes of the activity (interaction, relationships, communication, collaboration, problem solving, non-directional coordination, networks etc.). These aspects point out that teamwork competences are complex and multidisciplinary constructs, which are based on *psychology* (cognitive and communication psychology, group dynamics, organizational psychology), *management* (total quality management, organizational behavior, organizational development, knowledge management), *education* (teamwork competences, competences, directed to work and personality), *sociology* (organizational sociology, work sociology) and it is the research object of social science. It is no faithful to name the elements of teamwork competences as skills. Skills are the smallest elements in the structure of qualification, competence and competences and they are not related to mechanisms of interpersonal interaction. It is more accurate to name the elements of teamwork competences as *abilities. Ability* reflects personal potency, which incorporates personal ability to integrate the mental or physical capacity (Oxford Advanced Learner's Dictionary, 1998, p.2).

The concept of core competences. A group of researchers from California University (Management Council Report, University of California, 1995) present a concept, which consists of ten core competences: facilitation of changes; cooperation; communication; critical thinking; leadership; organizational insight; personnel management; resource management; planning; directing services. Researchers describe core competences as a group of interrelated knowledge, skills and standpoints, which are used together and are called as clusters. The authors of the concept indicate, that each of the ten core competences consists of separate clusters and point out, what kind of processes are incorporated into the core competences. The authors do not explain the content of knowledge, skills, standpoints, which fulfill the content of every core competence. In the presented concept the communication core competence is urgent to teamwork competences concept as well, because it incorporates teamwork, partnership, decision-making, participation process in seeking for the team and organizational aims (these should be above personal interests and aims). Teamwork in the context of core competences is comprehended as a complex element of collaboration and is related to social competence. Teamwork shouldn't be described as an element of collaboration: collaboration is a characteristic and an essential element of teamwork. Is teamwork identical with the social competence?

Lekavičienė (2001) points out, that a socially competent person is able to form groups, to manage those groups, to convince and to win over people. An author presents elements, which characterize the social competence: ability to influence others; personal creativeness; ability to analyze personal aims and plans and to fit to the social environment; striving for changing aims in specific social context; ability to solve problems effectively; managing cognitive, emotional and motor elements of the behavior. The mentioned characteristics of social competence are synonymous to team and teamwork characteristics, but teamwork is a composite construct, which could be treated as environment for realization of social competence. Teamwork shouldn't be identified as social competence, because such comprehension ignores the complexity of teamwork concept and forms the distorted understanding, that teamwork is only a short-term event, not a long-term process of formation and development. Malone (2000) confirms that standpoint: 1) the key characteristics of the effective teams are trust, openness (which includes identification of conflicts and an open resolution of it), cooperation, and cohesion; 2) the central elements of teamwork include problem solving, decision-making, all team members' participation in the work process, engagement and responsibility for the work-related aims and outcomes. Cole (1998), Bern (2001) ascribes such characteristics to social competence as well.

Stewart (2001) states that leadership and membership play essential roles in teamwork. The author indicates, that every team member alternately realizes the roles of the leader and member in work process. Bacal (1998) points out, that for realization of leader's role is important an organizational competence, which incorporates coordination and administration competences. It means, that in the "space" of the teamwork two essential competences are integrated – social and organizational. Those are related to the leader and member's roles in teamwork process. There is no exact answer to the question about the content of teamwork competences: Do teamwork competences include competences, directed to work or are these competences separate?

Mahmoodi & King (1992) indicate, that teamwork competences incorporate specific/specialized competences. Salas (2002) points out, that teamwork competences, directed to work, are separate and this difference influences effectiveness of teams, but researchers don't ground this idea.

In this article it is important to answer the question: Are teamwork competences an element of professional competence?"

The notion *professional* is related to profession (Vaitkevičiūtė, 2001). Profession is inseparable from the specific education and specialized professional training. According to Vaitkevičiūtė (2001), it is the quality of specialized and qualified activity that corresponds to sharing the work and making special requirements. It means that a worker should have adequate qualification and be able to cooperate. Qualification is an element of competences, directed to work (see Figure 1) and cooperation is ability, incorporated in the structure of competences, directed to personality (see Figure 1). It is a key characteristic of teamwork (an element of social teamwork competence). It grounds the idea that elements do not directly relate to specificity of the work (aims and objectives) and are characterized by the intercrossing tendency; qualification and cooperation are urgent for specialized professional activity and teamwork.

The competence includes personal and situational dimensions and realization of roles, according to work aims and standards, which are urgent for concrete organization. Identical characteristics are common for the teamwork concept. The teamwork competences could be treated as an element of professional competence.

The notion *professional competence* in the context of specific activity does not mean only the realization of acquired qualification, and it is interpreted as the employee's or team ability to reach an urgent aim for organization and to realize various roles, which change is dependent on the requirements of the world of work (White, 1997). But teamwork competences are not an obligatory element of professional competence and are urgent for those workers or organizations, where teamwork is an element of organizational development and culture (Stewart, 2001). The model of teamwork competences as an element of professional competence is presented in Figure 2.

The presented model reflects an "entrance" of workers into the team. These workers already have professional competence, which incorporates competences, directed to work and personality. Synergic integration of these competences in "non-team" environment "produces" exceptional competences that are the precondition of work efficiency.

The employee "brings" acquired competences, directed to personality and work to the team environment. These competences help the employee to realize the urgent team roles, which are divided into two main types of leader and member. Team members act reciprocally and in the interaction process develop their exceptional team competences.

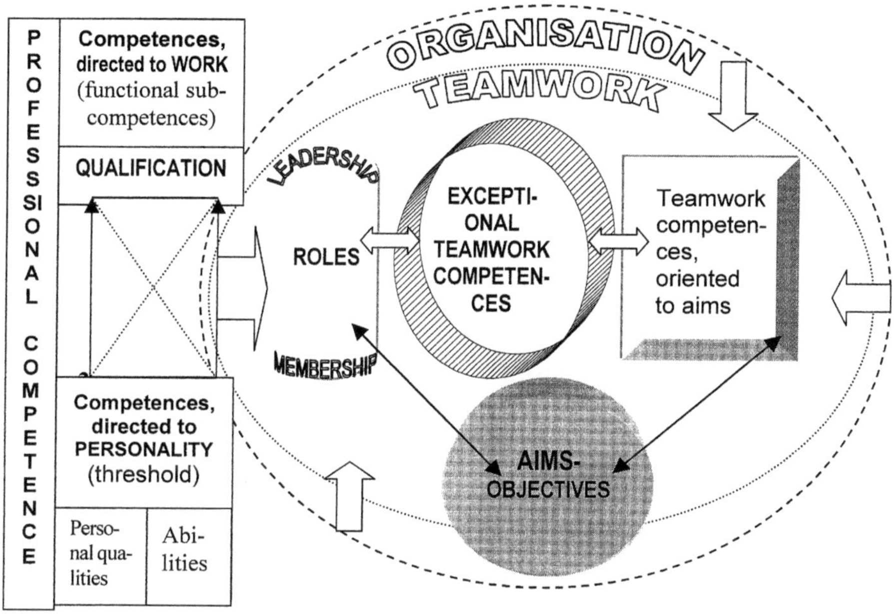

Teamwork competences are directly dependent upon the organizational aims and objectives. Exceptional team competences realize the functions of support and connection and influence coordination between roles and competences, directed to organizational aims. Organizational aims and objectives change and because of that specific team members' roles could change as well. In the process of such case the team will not work successfully before the exceptional team competences are not formed. It means, that in the "space" of teamwork competences, directed to urgent organizational aims and objectives change. Exceptional team competences are "used" and developed by the tendency of permanent change.

Exceptional teamwork competences are the result of interrelation of competences, directed to personality and work. This result is directly dependent on the activity change. Exceptional competences in non-teamwork become elements of short-term activity and these elements could be interpreted as a precondition of effective activity. Permanent change of activity disturbs the development and maturity of exceptional competences, because there is no guarantee that in case of the changed activity aim, the acquired competencies in previous activity will be of value in the context of the present activity. Then the processes of competences' and qualification change become urgent in the non-teamwork.

Teamwork does not ignore the acquired qualification of the employee and urgency of qualification change. Elements of qualification are incorporated into teamwork, which is purposeful, concrete and "produces" estimated outcomes. Application of employee's competence in teamwork means, that teamwork becomes the context of competence (and competences) realization (an element of competence). When work/activity is discussed as the common notion of competence it should be used the notion of professional competence, which incorporates employee's qualification, personal qualities and abilities.

Concluding the model, presented in Figure 2 it is important to point out, that:

- Exceptional teamwork competences are separated from competences, directed to activity aims in the context of teamwork.
- Competences, directed to activity aims, are characterized by change, and exceptional team competences are characterized by the tendency of development and maturity.
- Professional competence is in the reciprocity with teamwork, which includes teamwork competences.
- Development of professional competence and teamwork competences is two separate processes, which are connected by the teamwork context.
- Professional competence could be realized in teamwork and in non-teamwork.
- Teamwork, which includes teamwork competences from the standpoint of: theory, i.e. is an independent construct; activity, i.e. is a type of work; education, i.e. is the context of realization and development.
- The integrated development of professional competence and teamwork competences could be interpreted even in the context of organization, where teamwork is an urgent element of organizational development.

Conclusions

1. Competence, realized in the environment of activity, is an independent construct that includes competences, divided into two types: competences directed to work (subcompetences) and directed to personality.
2. Professional competence is a precondition of effective teamwork. Teamwork is the context of realization of professional competence, where elements of the employee's acquired competence are incorporated into the teamwork roles and acquire a new quality.
3. The collective work of team members influences the origin of competences, directed to activity aims and development of exceptional teamwork competences. Then teamwork

competences could be treated as an element of professional competence, but the main condition of it is the context of teamwork.

References

ALSOP, A.: Continuing Professional Development. A Guide for Therapists. London 2001 (Blackwell Science Ltd).

ANDERSON, J.: Cognitive Psychology and its Implications. New York 1980 (W.H. Freeman).

BACAL, R.: Team Building: The Six Deadly Sins of Team Building/ CSWR Papers. (Center for the Study of Work Teams Newsletter. University of North Texas). Spring. 29 (1998), pp. 2-6. [cit. 25 February 2002]. Available from: <http://www/workteams.unt.edu/ newsletter/Archive/v29.html>

BARNETT, R.: The Limits of Competence: Knowledge, Higher Education and Society. Buckingham, UK 1994 (Open University Press).

BERN, E.: Лидер и группа. Екатеринбург 2000 (Литур).

BOWDEN, J./MARTON, F.: The University of Learning. London, Stirling 1998 (Kogan Page).

COLE, G.: Organizational behavior. London, Great Britain 1998 (Letts Educational).

DETTERMAN, D.: The case for the Prosecution: Transfer as an Epiphenomenon. In: DETTERMAN D./STERNBERG R. (eds.): Transfer on Trial: Intelligence, Cognition and instruction. Norwood, New York 1993 (Ablex), pp. 1 – 24.

DOE, J.: Performance DNA: Personal Soft skills Indicator. USA 2000 (Performance DNA International LTD).

DOE, J.: Performance DNA: Position Analysis Report. – A complete evaluation of the soft skill competencies necessary to achieve superior performance in the position. USA 2001 (Performance DNA International LTD).

DREYFUS, H./DREYFUS, S.: Mind over Machine: The Power of Human Intuition and Expertise in the Era of the Computer. Oxford 1986 (Blackwell Ltd.).

ERAUT, M.: Developing Professional Knowledge and Competence. London 1994 (The Falmer Press).

FISH, D./TWINN, S.: Quality Clinical Supervision in the Health Care Professions: Principled Approaches to Practice. Oxford 1997 (Butterworth Heinemann).

GAGNE, R.: The Conditions of Learning. New York 1977 (Holt, Rinehart & Winston).

GARDNER, P.: The Structure of Science Education. Australia, Hawthorn 1975 (Longman).

JUCEVIČIENĖ, P./LEPAITĖ, D.: Kompetencijos sampratos erdvė. Socialiniai mokslai, 1(2000)22, pp. 44 – 50.

JUNGKUNZ, D.: Išsilavinimas ir kvalifikacija – profesinio rengimo tikslai didaktinių modelių kontekste. Profesinis rengimas. Tyrimai ir realijos 2. Kaunas 1999 (VDU). pp. 22 – 31.

KIRSCHNER, P./VILSTEREN, P./HUMMEL, H./WIGMAN, M.: The design of a study environment for acquiring academic and professional competence. Studies in Higher Education 22(1997)2, pp. 151 – 171.

LEKAVIČIENĖ, R.: Socialinės kompetencijos psichologiniai kriterijai ir vertinimas. Lietuvos studentų tyrimas: monografija. Kaunas 2001 (VDU).

LLOYD, C./COOK, A.: Implementing Standards of Competence. London 1993 (Kogan Page).

MAHMOODI, S./KING, G.: Top Management Team Competencies for Public Libraries. USA Autumn – Winter (1992) 1–2 (Minnesota Libraries XXX), pp. 26 – 32.

MALONE, S.: Learning skills for Managers. Dublin 2000 (Oak Tree Press).

Management Council Report. Process: How did the workgroup arrive as its recommendations? USA 1995 (University of California).

Oxford Advanced Learner's Dictionary. Oxford (1998) Oxford University Press.

PATRY, J.–L.: Perkėlimo įvertinimas mokymo procese: situacijos speficikos įtaka. Profesinis rengimas. Tyrimai ir realijos 2. Kaunas 1999 (VDU), pp. 10-21.

POSNER, M./KEELE, S: Skill learning. In TRAVERS, R. (ed.): 2nd handbook on research on teaching. Chicago 1973 (Rand McNally), pp. 805-831.

SALAS, E.: Team competencies and team training: what we know after 15 years of research and practice. 2002 (University of Central Florida). [Cit. 04 February 2002]. Available from: <http://www.iit.edu/~ciop/news/salas/tsld001.htm.>

SANDBERG, J.: Human Competence at Work: An Interpretative Approach. Göteborg 1994 (BAS).

SCHEERES, H./HAGER, P.: Competences and the curriculum. 1997. [Cit. 05 January 2002]. Available from: <ftp://ftp.swin.edu.au/pub/aare/aare94/conf94/achen94.233.>

SHOW, S.: Defining core skills. 1998.– ETF – AF98 – 010. C:/Vita/Advisory For/SBG – D. DOC

SPENCER, L./SPENCER, S.: Competence at Work: Models for Superior Performance. New York 1993 (Wiley).

STEVENSON, C.: The metamorphosis of the construction of competence. Studies in Continuing Education 18(1996)1, pp. 24 – 42.

STEWART, J.: Managing Change Through Training and Development. 2nd ed. London, Stirling 2001 (Kogan Page).

TROTTER, A./ELLISON, L.: Understanding competence and competency.. In: DAVIES, B./ELLISON, L. (eds.): School Leadership for the 21st Century. A competency and knowledge approach. London & New York 2001 (Routledge Falmer, Taylor & Francis Group), pp. 36-53.

WESTERA, W.: Competences in education: a confusion of tongues. Journal of Curriculum studies, 1(2001)33, pp. 75-88.

WHITE, J.: Education and the end of work: a new philosophy for work and learning. London 1997 (Cassel).

WHITE, R./MAYER, R.: Understanding intellectual skills. Instructional Science 9(1980)1, pp. 101-127.

ŽYDŽIŪNAITĖ, V.: Bendrieji gebėjimai ir jų ugdymas rengiant slaugytojus: magistro tezės. Kaunas 2000 (VDU).

Angelė Kaušylienė

Characteristics of Schools' Network Meant for Educational Stimulation of the Youth

The beginning of educational reform in Lithuania is linked with the European Union convention of educational workers that was arranged in Moscow on December 1988. On the initiative of Lithuanian teachers and pedagogy workers it was emphasized that the educational reform of Lithuania is a little different from conceptual regulations of education of the whole European Union. At first, the development of educational system was implemented in the sections of pre-school and general education. After the restoration of Lithuanian independence a rapid educational reform was started, that influenced rapid changing in the knowledge society. Knowledge information society opens new possibilities in the creation of network organizations operating at the levels of organizations as well as of groups of individuals and pupils. Schools' network organization creates premises for seeking the basic goal – educational stimulation of pupils when an individual has to choose a further way of learning.

Today, when thirteen years of educational reform have passed, all structural elements of our educational system and sub-system set forth in separate directive documents are positively estimated by foreign experts. But a number of problems still remain to be solved: importance of recognition of professional qualification abroad, many children below 16 years old do not attend any school of general education, many students of higher, vocational schools or colleges do not graduate from them or do not study at all, etc. The system of profiled education does not ensure profile variety at each school – not all the pupils are able to choose a way of further learning corresponding to their needs and abilities. Therefore, educational stimulation of the youth related with the inducement of structural elements of educational system for self-development is an urgent matter. The most important function of educational stimulation is to help create harmony between the learning profile chosen by a pupil and his/her further activity.

In the solution of arising problems it is important to strive for coordinated actions of educational stimulation – to unite the institutions of general and further education (vocational school, college, university). Here one of the priority factors could be the creation of network of different types of schools, what is confirmed by different foreign authors (Fend, 1981; Bush, 1995, etc.). The creation of such network is mentioned in the education harmonization directions approved in 1998 by the Ministry of Education and Science of the Republic of

Lithuania, in the Analysis of Permanent Educational System Links published in 2000, in Strategic Regulations of the Development of Lithuanian Education, 2003-2012, etc. The analysis of these and other documents showed that the creation of educational network is not methodologically based in Lithuania.

The objective of this article is to disclose the characteristics of schools' network meant for educational stimulation of the youth.

The objectives are achieved by working on the following tasks:

1. to define the concept of network, its properties and indications;
2. to disclose the general and special characteristics of schools' network meant for educational stimulation of the youth.

Analysis of organization network conception

In XXI cent. it is more and more talked not of individuals, groups or organizations but of the networks that could unite the above mentioned in order they could act much more efficiently. According to Chisholm Elden (1993), the conception of network, as the basic form of organization, was born in the end of XX cent. There is no doubt that in the future the importance of network will increase even more.

Usually, two questions concerning the conception of network are the most important: what is the meaning of this term? What is the aim of networks? (Jay, 1964; Tichy,1981; Bartol, Martin, 1991; Chisholm, Elden, 1993; Chisholm, 1996; Blacher, Crump, McDonald, 2000, etc.). Of course, some more questions also arise: what a exceptional role of networks play? What can a network what cannot be done by the members of organization; how attractive are networks to their participants, etc. In this article the author orientates towards the two above-mentioned questions because in the creation of schools' network meant for educational stimulation of the youth, first of all, it should be understood what a network is and what its aims are.

According to Chisholm (1996) two fields influencing the development and maintenance of network were analyzed. This is a conceptual aspect of network or its two conceptions. The initial position of the first conception is a complicated environment where most organizations are operated (Aldrich, Whetten, 1981; Powell, 1990, etc.). The second conception is nature and indications of network organizations (Jay, 1964; Gerlach, Palmer, 1981, etc.).

The first conception: As networks are developing, the organizations in modern society acquire some new indications that determine the increasing interdependence of organizations, groups and individuals. Chisholm (1996) suggests creation of networks for complicated and interrelated problems. In order to establish them the organizations of different fields (national, private, labor, educational, etc.) and the participants of different levels (local, national, federal, initial, scientific, etc.) are necessary. The composition and structure of an organization must reflect the complication of the environment of its activities because if the network is diverse and able to correspond to the dynamism and complication of the environment, it will help to perceive, control, implement and manage complicated development of situations.

The second conception: A network is defined as organizational and social units linked by specific relations. Jay (1964), Bartol, Martin (1991) supplement this principle. It is emphasized that networks are the structure of collaboration relations with those, whose help is necessary for their efficient functioning. Networks are created in order to exchange information, knowledge and experience fast. By means of networks it is possible to transfer the information on work or inter-support by using informal contacts instead of exchange of formal procedures (Mosley, Megginson, Pietri, 1989). The structures of networks facilitate collaboration among organizations and systems. It is useful for each participant of the network. The basic orientation of an organization is common aims of higher level that unite the organizations. They form the networks seeking to solve "metaproblems" (Chisholm, 1996).

According to Gerlach, Palmer (1981), the aims of networks help to express greater political, social or economical power that helps to create and legalize new social structures. Therefore, it is very important to have a common aim of high aspirations exceeding the aims of an individual organization.

By analyzing the functions of interorganizational network Chisholm (1996) affirms:

- the basic function is embodiment and maintenance of a vision. It concentrates thinking, planning and actions of the participants of the network for common activities of higher level. A unity with such systemic orientation helps to avoid competition among the members of the organization;
- a forum is a role of network to become an arena of discussions. Here complicated matters of development are elucidated, parties concerned are united and their activity is stimulated. Human potential is necessary for that. It would help to link the concepts, to plan, to make action plans, to estimate activities and to make corrections;
- the interchange of the attitude apprehensions, necessary apprehension is a necessity of collaboration among different organizations, renunciation of existing stereotypes of thinking, stimulation of new attitudes to organizations and surrounding world, reorganization of situations. So, the members of each individual organization must realize that their organization belongs to the system of higher level, i.e. network. Therefore, they must estimate all the activities and actions from this point of view;
- communication is an orientation towards the community as a whole where it is important to maintain the openness and regular contacts. Communication is necessary in order to help the members of the network and understand their role in the sys-

tem, link their activities with the achievement of s common aim. Communication enables making influence on each other. All the community must participate in the process of communication development.

In network organization the control of all the members in undertaking the responsibility for the network, determination of mission and tasks, initiation and management of projects and working activities are emphasized. Chrisholm (1996) asserts that the activities of an organization in the implementation of common vision of larger sphere are the way created by its members for the establishment of interrelations in the process of work. Permanent maintenance of network system trend is necessary for the organization to function.

In Table 1 (see pp. 285-286) the definitions of network conception of different authors are given. The definitions, given in the table, show that the network may be described very differently. But all the authors directly or indirectly specify interacting members – participants of the network by emphasizing the structure of their relations.

Furthermore, characteristic features used in the description of network may be perceived:

- aim (purpose) (Hammond, 1983; Hall, 1988; Concise Oxford Dictionary, 1995; Oldroyd, Elsner, Poster, 1996, etc.);
- influence (Adams, 1990; Chisholm, Elden, 1996, etc.);
- information (Concise Oxford Dictionary, 1995; Chisholm, 1996, etc.);
- membership (Roges, Roges, 1976; Gerlach, Palmer, 1981; Handy , 1990, etc.);
- communication and relations (Tichy, 1981; Bartol, Martin, 1991; Garavan, Costine, Heraty, 1995; Oldroyd, Elsner, Poster, 1996; The New Webster's English Dictionary, 1998, etc.);
- points of interaction (Vocabulary in sociology, 1988) and certain relations (Jay,1964; Bartol, Martin, 1991, etc.);
- system (Tichy, 1981; Seed, 1990; Hammond , 1991; Oldroyd, Elsner, Poster, 1996, etc.);
- partnership (Garavan, Costine, Heraty, 1995);
- territorial distribution (Kačerauskienė, 1998; Trist, 1983, etc.);
- decentralization of power (Gerlach, Palmer, 1981; Handy, 1991; Hasting, 1993; Oldroyd, Elsner, Poster, 1996, etc.).

For the position of network analysis and creation the second conception was chosen because in the creation of network it is important to estimate not only the environment becoming more and more complicated and increasing organizations but the nature and indications of network organizations as well.

Network is collaboration among the organizations based on non-hierarchic relations where the aim is educational stimulation of the youth in the network of schools. It is especially important in the creation of the conception of educational stimulation of the youth in the network of schools, because network, as a whole, and each organization, as its component, must be able to stimulate the self-development of the youth.

Characteristics of schools' network

Tichy (1981) researched the properties of network and classified them (see Table 2, pp. 286-287). The properties of the network given in Table 2 show that the author is orientated towards the networks existing inside the organization. But some characteristics may be applied to interorganizational networks. Furthermore, relations differ from each other by the levels of mutuality, complication and intensiveness. Usually, relations remain symmetric but influence is asymmetric. Usually, such are the relations between the manager and his/her subordinate. People play different roles, so the feature of complication shows how two participants communicate. The more relations are between them, the stronger the connection is. Their relations are strengthened by permanent interaction.

The table is orientated towards *the networks inside the organization, but some characteristics may be applied to interorganizational networks, in this case, to the schools' network that is being created.* For example:

1. ***The content of interaction*** consists of four ***exchange measures:***
   - emotional attachment (some organizations are more reliable, it is a greater pleasure to communicate with their representatives in certain activities, etc.);
   - influences (some organizations have greater influence in certain activities, others are weaker);
   - information (organizations differently dispose information, so exchange of information allows creating closer and more objective field of information);
   - commodities and services (commodities and services of organizations usually are not the same).

They may differ from each other by assortment (education provided at educational institutions – higher, college, etc.) and quality (although in a legal sense educational institutions provide adequate education but from the practical point of view their rating is not the same).

2. ***Characteristics of relations:***

Mutual relations may be symmetric and asymmetric. In the case of the schools' network being created, the organizations in the network should collaborate on a parity basis. In the case when relations are equivalent, symmetric relations should predominate, and vice versa.

Other feature of the characteristic of relations is clarity of norms. It is important for interorganizational networks because each organization is a participant and it must be clear how it should behave. Intensiveness and energy are also significant. Sometimes organizations have to fulfill their obligations by sacrificing their interests and the interests of others.

Activities of organizations as participants are usually related with each other in several ways rather than in one. For example, the teachers of general education schools often prepare general programs together with the lecturers of higher schools. And a school of general education trains future students. Moreover, both schools exchange information.

3. ***Structural characteristics:***

It is evident that two structural characteristics are the most important for the schools' network:

- grouping (determines the number of coalitions or other cliques in one network),
- size (shows the number of participants of the network).

Of course, such structural characteristics as visibility would also be significant. It would show if an outsider is able to say who participates in the network. Participants could also set the limits of their network. It is important because the participants of the network and the limits of interaction would be exposed.

The criteria of membership should be characteristic of schools' network (acceptance to membership, hire) as well as openness (relations with other networks, their cosmopolitanism, locality). This would expose the membership and possibilities of participation in the network.

Stability (probable duration of network), copiousness of relations (relations that actually exist among the participants) and reach (desirable number of relations among the participants) should also be an inherent indication of this network.

That is why the network may be described by the density of employment, i.e. what part of professions is involved in the activities of the network. Nowadays the competence of an individual in certain activity is more important. Thus, it seems to be more expedient to emphasize the competence of participants rather than the density of employment.

Vertical density is also quite important. It shows what part of all hierarchic levels is covered by the network (corresponding relations are more or less coordinated by certain organizations). This feature would help to disclose whether the collaboration within the network is carried out on a parity basis.

4. ***Basic participants:***

The property of the network that is under discussion.

Let us suppose that such participant of the network as a recluse could hardly be in an interorganizational network. But it is necessary to determine the number, status and relations of the participants of the network. It is evident that the indication "basic participants" of the network specified by Tichy is important but the classification of participants is not proper. This indication, taking into account the specific character of the network, could be called as a "parity collaboration of the participants".

In the generalization of Table 3, see pp. 288-289, the definition of the first two levels of the network (group inside the network and network of organization) allows stating that they are the most suitable for interorganizational network. The following properties of the network are beyond any doubt:

- content of interaction (emotional attachment, influence, information, commodities and services),
- characteristics of relations (mutuality, clarity of norms, intensiveness, energy, complication),
- structural characteristics (organizational density, grouping, size, visibility, criteria of membership, openness, stability, copiousness of relations, reach, centrality).
- Such criteria like density of employment should not be applied. In an interorganizational network a more important thing is not how many professions are in-

volved in the activities of the network (quantitative index) but what the competence of the participants (qualitative index) is. Thus the indication "density of employment" should be replaced with the "competence of the participants of the network".

Many authors are of the opinion that "partnership" is one of the indications of the network, so "vertical density" should not be characteristic to the network.

Network of organization related with collaboration, mutuality, confidence is of mutual benefit (Garovan, Costine, Heraty, 1995). Here the autonomy of an individual, stimulation of innovations and constant exchange of information and learning are emphasized.

The collaboration is understood as common execution of the activities by at least two persons. This is a bilateral and short-term process based on forms of work, and the work within network covers the concentration of collaboration relations. Such relations are many-layered and supplemented by different forms of long-term work. Therefore, the smallest network must consist of at least three participants. Collaboration agreement is concluded for at least average period of time. In the context of vocational training, networks are described by collaboration of different institutions and coordination of common actions. The criterion of network is in conformity with current situation only when different institutions participate irrespective of specific sphere of activities.

Another very important indication of network is „vision" (Elden, Chisholm, 1993; Chisholm, 1996, etc.). Embodiment and maintenance of a vision is one of the most important functions of network. Broad vision is a powerful factor. It helps changes to arise that convert vision into reality. The participants of the network should act in partnership. Only then the conditions allowing the individuals, groups, organizations to survive in changing environment will be created. Network can unite thinking, planning and actions for common activities of higher level than in an individual organization. Collective activity together with systemic orientation is an indispensable condition for the development of network. Moreover, solidarity helps to avoid competition among the members of organization. Vision, partnership, collective activity and systemic orientation of certain activity are important indications of the network.

The most important aim of schools' network is educational stimulation of the youth, i.e. inducement of schools for self-development of the youth. Participants of the network – schools – must collaborate with each other on a parity basis.

Other partners are also possible. Collaboration of all the members must be equivalent. The vision of network depends on the participants themselves, so each of them may have their own ideas. But they must be as comprehensive and concrete as possible. It can be based on common/aggregate vision of the network: future organization, strategy, necessary resources, abilities (their application), present-future, future situation, etc. (Jucevičius , 1998).

It is clear that the interaction of participants must be united by common content of the network. All four *exchange measures* are important (Tichy, 1981).

Expression of emotional attachment. Among the schools-participants and other organizations-partners the confidence, kind collaboration, etc. must exist. *The influence* of the network schools (general education, vocational, higher, colleges, etc.) *on each other, harmony/unity of the profiles* and their tendencies, etc. are urgent matters. *Interchange of information* is also very important. Changes at one school may have an effect on other participants of the network. Let us suppose that introduction of distant teaching at higher schools may stimulate the schools (general education, vocational, colleges, etc.) to help the pupils/students to prepare for that or at least to perceive the significance of such preparation.

Every school renders certain services. General education, vocational schools help to acquire general education, vocational schools, colleges and, partially, higher schools help to acquire vocational education, create possibilities to raise or change one's qualification. Such "exchange" must also be appraised. So all the *characteristics of relations* that are significant to interorganizational networks are also important to the schools' network:

- mutuality, strong symmetric relations among the members of the network (collaboration on a parity basis);
- clarity of norms (interaction among the members of the network);
- intensiveness, energy (obligations of the members of the network to each other);
- complication (several-type relations of the members of the network).

Not only characteristics of the content of interaction but also *structural characteristics* of the interorganizational networks (organizational density, grouping, size, visibility) may be applied to schools' network. From the practical point of view, the first three characteristics are quantitative (the number of participating organizations, coalitions, cliques, individuals). Therefore, in the network of educational stimulation of the youth it is expedient to specify only the number of schools, groups and youth. Meanwhile the fourth indicator – *visibility* – should be clear to an outsider as well as to a member of the network.

The criteria of partnership are also important to the network. It must be clear to all the members who can be admitted to a membership and on what basis and when it is allowed to leave the schools' network.

Relations with other networks should be established in a cosmopolitical and open way. As here educational stimulation of the youth in the schools' network is analyzed it should be good to orientate towards the locality and collaboration with other networks. For example, if the local schools' networks could not satisfy the needs and interests of an individual, maybe this could be done by labor exchange; higher schools located somewhere else, etc. Significant and expected duration of network existence means its stability and many things depend on that. The schools will try to consolidate short-term collaboration in one manner and long-term collaboration – in absolutely different manner. These indications could be attributed to the characteristic of relations because the relations of the participants of the network are described here.

Centrality is not an exceptionally important indication. But for the purposes of marketing the schools should not involve into the network the collaboration with the "authorities". This could be left to the main participants. The above-mentioned indication, though not necessary, is an important indication to the schools' network. Thus, its participants strengthen the network; increase its prestige, etc.

Furthermore, the analysis of the already created networks shows that, in addition to the above-mentioned indications of the network, it is important to specify the *level of networks* (local, regional, national, international). It is expedient to fix the *date of network establishment* (the beginning of activities) and different status of the members of the network (participants, partners of the network). This could be specified in the section "properties of the network" (structure). Moreover, the *contribution and responsibility of the members* of the network may be fixed in the partnership (concept of the network). From the theoretical point of view an exceptional attention is paid to the analysis of *the structure of network* that is linked with the management (Snow, Miles, Colemen, 1992, etc.).

The participants of network are partners and they have to collaborate on a parity basis, so the hierarchic structure of schools cannot make premises for such collaboration. Horizontal structure is based on a parity basis. Hence, it is acceptable to the schools' networks because it could make premises to allocate the works according to the competence and abilities of the members of the network. Moreover, the workers themselves could become better acquainted with each other and have more possibilities to communicate. However, strongly decentralized structure that cares for the development of innovations is more suitable for an efficient activity.

So, after the evaluation of the data of literature, examples of networks (inter-organizational indications of network), after their modification to the schools' network the matrix of schools' network of educational stimulation of the youth may be worked out by emphasizing the general characteristics of the network and discussing the properties of the network and its indications (see Table 3, see pp. 288-289).

The above general characteristics of the schools' network indicating the properties and indications of the network are a base to disclose special characteristics of educational stimulation of the youth in the schools' network.

Special characteristics of the network of educational stimulation

The characteristics of the schools' network of educational stimulation are linked with the goal of general characteristics of the schools' network and aims to specificity of educational stimulation of the pupils under the conditions of school profiling. Educational stimulation in network organization is disclosed by:

1. *Characteristic of the fullness of selection possibilities:* The network of educational institutions must consist of general education schools and as different as possible post-secondary education institutions – vocational schools, institutions of short vocational courses, colleges, other higher schools. This reflects the feature of diversity that is formed in educational practice of the Western countries and that ensures great selection of educational services.
2. *Characteristic of profiling voluntariness:* In the studied network all profiled and non-profiled yet basic schools of general education should participate, as the possibilities to select desirable profile should be created in the network. This characteristic as though supplements the first one – characteristic of the fullness of selection possibilities.
3. *Characteristic of free migration:* This means that a pupil, who chooses a concrete profile for self-development and thus is attached to his/her school, may study some subjects at other school (of the same network) rather than at his/her school or may move from one profile to another one at the same school.

When a negative attitude of the pupils to their profile occurs in the context of future expectations, an exceptional responsibility falls on the network organization of schools, the aim of which is to help each pupil to solve similar problems and to ensure the flexibility of the educational process:

- to render every kind of advisory support;
- to apply educational technologies to serve a pupil (as a client);
- to familiarize with the system of bridges;
- to help to overcome the difficulties of adaptation, etc.

4. *E-characteristic:* In order to ensure the efficiency of communication of the studied network modern e-communication means (e.g. Internet) are used as well as educational technologies, e.g. distant training, virtual classes, etc.

Computer technology helps to solve the problems of the individualization and differentiation of education that are especially evident in profiling the process of education and when the possibilities to solve them by applying former technologies are rather limited. New technology of education not only rationalizes the activities of an educator but also creates broad possibilities to expand network organizations and to look for new ways of educational stimulation of the youth. The information on schools' networks given in Internet helps to solve one of the greatest problems – lack of information on the profiled teaching and coordination of the admission to colleges and higher schools.

5. *Characteristic of the overlapping of educational content:* At the school of lower level the subjects of higher level of education may be taught (in this case they are taught by the educators of the institutions of higher level of education). Thus, at the school of general education lycée classes, separate subjects or their lessons may appear (e.g. experience of

French lycées). The pupils who complete such subjects of lycée level do not repeat them at a college / university anymore (of course, political and legal resolutions of education are necessary for that). Such classes could be a strong stimulus for pupils to choose an appropriate institution. By assisting the above-mentioned course / subject an educator of the institution of lower level may raise his/her qualification. Such raising of qualification should also be discussed and legalized. The schools' network would help to realize this characteristic.

6. *Characteristic of sharing the good experience:* It means that the educators as well as the pupils constantly and publicly reflect, what the first ones estimate as an influence of educational stimulation, the second ones – as an effect of educational stimulation in this network organization. The good experience discovered is acknowledged and shared in every possible way. No one can realize by himself/herself.
7. *Characteristic of e-form of educational stimulation:* In the website of this network organization, every member of the network, especially the pupils, are given some space. Here everyone can express their ideas, wishes, successes and failures and to receive a response. In case of the need of discussions, consultations are published and "routes" of separate specialties are suggested in the website.
8. *Characteristic of provision of all the information on the profile of school activities and possibility:* All educational institutions within the network provide each pupil with easily available exhaustive information. Thus, objective and successful decisions for further learning are ensured.
9. *Characteristic of desire of geographical compactness:* Educational stimulation is easier executed if the educational institutions operating in the network are closer to each other and the pupils can easier migrate from one institution to another. But this is necessary if e-characteristic is realized in the network organization. Geographical distance then loses its former significance. Although it is desirable that the network of educational stimulation would be in one town (region, county), but it is not necessary, especially when a university (universities) is desired to be included into the network. It would be the most promising to create the network on the basis of one town (region) basing on e-relations and including non-university higher schools and universities situated in other locations.
10. *Characteristic of participation of non-educational institutions enabling/supplementing the network of educational stimulation:* If need be non-educational institutions the activities of which are important from the point of view of educational stimulation may participate in the network. In this sense the institutions of two kinds should be singled out: a) institutions of social work working with teenagers or juveniles who are leaving or have already left schools; b) institutions of career development, professional consulting/orientation that may perform the function of not only educational stimulation but a special function of coordination of the network activities as well. Moreover, the network could be strengthened by the participation of the employers' organizations or their associations in the network activities.
11. *Characteristic of legal recognition and enabling by resources of an educational network as a network organization:* The above-mentioned characteristics can actually operate by grounding the functioning of the network only then, if this organization is recognized by educational administrative structures of higher level that should provide the functioning of the network with necessary resources – organizational, material, financial (the first two being the most important). Organizational resources should ensure the work of coordinating structure of the network and free but not chaotic movement of pupils in the network, material – network computerization – is also very important.

12. *Characteristic of the growth of pedagogical competence of the network of educational stimulation:* The network must always be active. In the development of the society, increase of working requirements the competence of the organizations of the network and their members must also increase all the time. It is especially important to develop pedagogical competence. Network organization has unique possibilities for that – its members (educators) constantly participate in unusual working situations. The system of learning by working activities and recognition of the results of such learning should be established. Such network organization could successfully perform one more function – permanent growth of pedagogical competence – in addition to the function of educational stimulation of the pupils.
13. *Characteristic of holism of the network organization's subject of activities:* The activities of educational stimulation are executed by coordinating the efforts in all three organizational levels: individuals (educators), their groups and their organizations. The most important decisions of the network of educational stimulation are undoubtedly made in the organizational level. But the direct activities of educational stimulation are carried out by the educators and their groups (e.g. educators amalgamated in different aspects: subject, methodical circles, "bridges" helping a pupil to change the chosen profile).

General and special characteristics of the schools' network that were singled out are important in the creation of the network of educational stimulation of the youth that could create the premises for the collaboration of educational institutions of different levels on a parity basis.

Conclusions

1. General and special characteristics of the schools' network are an innovation in the information society of knowledge and they open new possibilities in the creation of the network of educational stimulation of the youth.
2. The missions of the schools' network meant for the educational stimulation of the youth, analysis of its properties and indications (common vision, goals, hierarchy, collaboration on a parity basis, etc.) showed the significance of the properties and indications of the members – organizations of the network and of the network as whole, so the conception was chosen for the creation of this network, the initial position of which is the nature and features of the organizations and the definition of Oldroyd, Elsner, Poster network corresponding to it, after the modification of which to the network meant for educational stimulation of the youth, the conception of the network was defined. It is a network based on non-hierarchic relations, collaboration among the organizations, the goal of which is educational stimulation of the youth in the schools' network. Educational stimulation in the schools' network is a purposeful effect of educational organizations on the pupils helping them to "migrate" expediently and successfully in the educational system.
3. Different analysis of the descriptions of the schools' network made by different authors formed the premises to model general characteristics of the network meant for educational stimulation of the youth: beginning of the activities (date of official registration); legalization (legal registration); goal (distant, medium, close, strategic, tactical, official); vision (future organization, strategy, necessary resources, abilities and their application, the present and the future, future situation); membership (status of basic participants and partners); level (local, regional, national, international); structure (hierarchic, matrix, horizontal); structural characteristic (organizational density, grouping, size, visibility, competence); content of interaction (emotional attachment, influence, interchange of information, services and commodities); responsibility (division of legal and moral responsibil-

ity); management (centralized, decentralized, combination of centralization with decentralization); contribution (investments of the members of the network); characteristics of relations (mutuality, clarity of norms, intensiveness and energy, complication, openness, stability).

4. Special characteristics of the schools' networks consist of:
    - Fullness of selection possibilities showing that the network of educational institutions should be formed of educational institutions of different levels;
    - Profiling voluntariness conditioning the variety of the participants (schools) of the network as a possibility to choose a desirable profile;
    - Free migration enabling the participant (pupil) of the network to choose a desirable profile at any school of the network;
    - Efficiency of e-communication means by applying means modern of e-communication;
    - Overlapping of the content of education by creating premises to teach the subjects of higher level at the school of lower level;
    - Sharing the good experience helping the educators and pupils to reflect systematically and publicly by estimating the influence of educational stimulation and effect of network organization;
    - E-form of educational stimulation by creating a possibility to place the most different information in the website of network organization;
    - Provision of all the information on the profile of school activities and possibility, when the members (organizations) of the network provide each pupil with exhaustive and objective information on the selection of further learning;
    - Desire of geographical compactness, connection by certain territorial principle in order to carry out educational stimulation more efficiently;
    - Participation of non-educational institutions enabling/supplementing the network of educational stimulation that means that the institutions, the activities of which are important from the point of view of educational stimulation, may also participate in the network;
    - Recognition of the educational network as a network organization and its enabling by resources that shows that it can operate only if it is recognized by higher administrative structures and supported by necessary resources;
    - Growth of pedagogical competence of the network of educational stimulation that means that the network must grow by involving other organizations and increasing the competence of the members of the network;
    - Holism of the activities of the subjects of network organization showing that the activities of educational stimulation are carried out by coordinating all three levels of organizations (individuals, their groups and their organizations).

References

ADAMS, J.S.: The structure and dynamics of behavior in organizational boundary roles. In: DUNNETTE, M.D. (ed.): Handbook of Industrial and Organizational Psychology. Chicago: 1976 (Rand McNally), pp. 1175-1199.

ALDRICH, H.E./WHETTEN, D.A.: Managing the Flow of Technology. Cambridge, Mass 1977 (MIT Press).

BARTOL, K.M./MARTIN, D.C.: Management. USA 1991 (McGraw-Hill).

BLACKLER, F./CRUMP, N./MCDONALD, S.: Organizing Processes in Complex Activity Networks. Vol. 7(2000)2, pp. 277-300.

BUSH, T.: Theories of Education Management. London: 1995 (Paul Chapman Publishing Ltd.).

CHISHOLM, R.F.: On the Meaning of Networks. Group and Organizations Management: On International Januar 21(1996)2 June, London (Sage Periodical Press), pp. 211-235.

CHISHOLM, R.F./ELDEN, M.: Features of emerging action research. In: ELDEN, M./CHISHOLM, R.F. (Eds.): Special issue on action research. Human Relations, 46(1993)2, pp. 275-297.

Dictionary of Marketing. English – German. German – English. London 1991 (Peter Collin Publishing).

FEND, H.: Theorie der Schule. München/Wien/Baltimore 1980 (Urban&Schwarzenberg).

GAROVAN, T./COSTINE, C./HERATY, N.: Training and Development in Ireland. Context, Policy and Practice. Dublin 1995 (Oak Tree Press).

GERLACH, L.P./PALMER, G.B.: Adaptation through evolving interdependence. In: NYSTROM, P.D./STARBUCK, W.H. (eds.): Handbook of organization design, Vol. 1. New York 1981 (Oxford University Press, pp. 323-381.

HALL, V.: Networks and networking: a working paper. Internal working paper. Bristol 1988 (National Development Centre for School Management Training).

HAMMOND, S.: Business Studies. Harlow 1993 (Longmann Hause, Burnt Mill).

HANDY, C.B.: The Age of Unreason. London 1991 (Arrow).

JAY, E.J.: The concept of "field" and "network" in anthropological research. Man, 64(1964), pp. 137-139.

KAČERAUSKIENĖ, A.: Kaimo bendrojo lavinimo mokyklų tinklo klausimu. Ugdymo problemos 4(31). Vilnius 1998 (Vilniaus pedagoginis universitetas), pp. 282-289.

MOSLEY, D.C./MEGGISON, L.C./PIETRI, P.H.: Supervisory management. The art of working with and through people. Cincinnati 1989 (South-Western Publishing Co.).

POWELL, W.W.: Neither market nor hierarchy: Network forms of organization. In: STAW, B.M./CUMMINGS, L.L. (eds.): Research in organizational behavior, Vol. 12, Greenwich, CT 1990 (JAI Press), pp. 295 – 336.

RICE, R.E.: New Patterns of Social Structure in an Intermation Society. In: SCHEMENT J.R. (ed.): Competing visions, complex realistics: Social aspects of the Information Society. Norwood 1987 (New Jersey), pp. 107-120.

ROGERS, E.M./ROGERS, R.A.: Communications in Organizations. New York 1976 (Free Press).

SEED, PH.: Introducing Network Analysis in Social Work. London 1990 (Jessica Kingsley Publishers).

SNOW, C.C./MILES, R.E./COLEMAN, H.J.: Managing 21st century network organizations. Organizational Dynamics, Winter. New York 1992 (American Management Association).

The International Encyclopedia of Educational Evolution. HERBERT, J./WALBERG/GENEVA, D. (eds.). Haestel 1990 (Pergamin Press).

The concise Oxford Dictionary of Current English. Oxford 1995 (University of Oxford Press).

The Penguin English Dictionary. London 1992 (The Penguin Books Ltd. Logman Group Ltd.).

TICHY, N.M.: Networks in organizations. NYSTROM, P.C./STARBUCH, W.H. (eds.). Oxford 1981 (University Press).

TRIST, E.L.: Referent organizations and the development of interorganizational domains Human Relations, 36(1983)3, pp. 269-284.

TURBAN, E./MEREDITH, J.R.: Fundamentals of management science. United States of America 1991 (R.R. Donnelley & Sons Company).

WHITE, A./BOORMAN, S./BREIGER, R.L.: Social structure from multiple networks. I Blockmodels of roles and positions. American Journal of Sociology, 81(1967), pp. 730-780.

Author, year	Definition of network
Adams, 1980.	Organizations operating in certain network that has direct or indirect influence on them.
Bartol, Martin, 1991, p. 11.	Relations of communication (structure) with those, whose assistance is necessary for efficient functioning.
Chisholm, 1996, p. 218.	Organizations seeking to overcome out-of-problem situations ("metaproblems") by means of different networks, through which information, influence and emotional attachment are transferred.
Concise Oxford Dictionary, 1995, p. 915	Group of people exchanging information, contacts and experience for professional and social purposes.
Garavan, Costine, Heraty, 1995.	Relations linked by collaboration and mutual confidence leading to common benefit.
Gerlach, Palmer, 1981, p. 323.	Members being able to express greater political, social and economical power and to legalize new social structures.
Hall, 1988.	Voluntary collaboration among the professionals having similar aims and close fields of activities.
Hammond, 1993, p. 344.	System of interrelated events, the element(s) of which moves towards the set aim.
Handy, 1991, p. 90.	„Many-layer" organization. It consists of small groups of professional workers engaged in the most important activities as well as other workers responsible for any part of activities with other organizations.
Hasting, 1993, p. 195.	Implementation of certain processes leading to the decentralization of power and responsibility, abolition of organization boundaries.
Jay, 1964, p. 137	All organizational and social units joined by specific relations.
Kačerauskienė, 1998, p. 282	Network of schools of general education. Distribution of different institutions of these types in certain area.
Vocabulary in sociology, 1995, p. 40	Diagram made of limited number of distribution points where interaction goes on through the joints of those points.
Dictionary of marketing, 1991, p. 123	System combining different points into one.
The New Webster's English Dictionary, 1988. p. 1001-1002	Group made of separate but communicating networks.

Oldroyd, Elsner, Poster, 1996, p. 45.	It is a non-hierarchic organization of people or institutions related and communicating with each other for common aim.
Penguin English Dictionary, 1992, p. 615.	System of intersecting lines or channels. Inter-connected chain, group, system.
Rice, 1987, p. 107	Groups, organizations, societies made of models (elements) of relations.
Rogers, Rogers, 1976, p. 108.	Group of communicating size – from individual to organization.
Seed, 1990, p. 19.	System of certain significance or model of relations among the points; it is usually represented by diagram.
Shement, Lievouw, 1987, p. 107.	Groups, organizations and societies often made of elements called networks – one of the forms of indication of social structure.
Tichy, 1981, p. 245.	Society connected by interrelations as a system of the participants – people, groups, organizations.
Trist, 1983, p. 270.	Interorganizational networks or territories. They are functional social systems taking the position in the space between the public (as the whole) and an organization.
Turban, Meredith, 1991, p. 462.	Logical and chronological sequence of actions and events graphically representing the relations among different actions and events of the project.
Webster's II New Riverside Dictionary, 1984, p. 824.	Open composition, material or structure where strings, wires, threads, etc. are interlaced or woven in certain regular intervals. Network is a group of interrelated stations of transmission.
White, Boorman, Breiger, 1967, p. 730-780.	Organizations related by many networks, through which the commodities, services or emotional attachment flow.

Table 1: Description of network

Properties of network	**Explanations**	**Characteristics of coalitions**	**Characteristics of cliques**
Content of interaction	Four measures of exchange: a) expressions of emotional attachment, b) influence, c) information, d) commodities and services.	Influence Information	Emotional attachment Information
Characteristics of relations: Mutuality	Symmetry of relations comparing to asymmetric relations.	Strong	Strong
Clarity of norms	Are the ways specifying how the participants have to behave by interacting clear?	Very clear	Medium clear
Intensiveness, energy	Is an individual inclined to fulfil his/her obligations by sacrificing his/her own interests?	Poor	Medium
Complication	How one group of participants is related with each other?	By several	By many

Structural characteristics: Organizational density	How many members of organization participate in the network?	Medium	Little or medium
Grouping	How many coalitions or cliques (*in this case the author calls both categories as densely settled region) does one network have?	Medium	One
Size	How many people participate in the network?	Medium number	Several people
Visibility	Can an outsider say who participate in the network? Can the participants set the limits of their network?	Easily	Medium
Criteria of membership	Are the criteria of admission to membership or hire clear?	Very clear	Very clear
Openness	How many relations does this network have with other networks? Are the participants of the network cosmopolitan or do they work locally?	Medium Both variants	Poor Local
Stability	What is a probable duration of network existence?	Short	Long
Copiousness of relations	What part of all possible types of relations does actually exist among the participants?	Medium	Large
Reach	What is an average number of relations separating two participants?	Medium	Poor
Density of employment	What part of professions is involved in the activities of the network?	Large	Medium
Vertical density	What part of all hierarchic levels does the network include?	Medium	Small
Centrality	Do many public "stars" participate in the network?	Centralized	Medium decentralized
Basic participants: Star	Participant having many relations with other participant.		
Ally	Participant who unites two or more groups that in other case would function separately.		
Bridge or link	Participant who belongs to different groups, "clusters".		
Porter	Participant who controls the streams from one side of the network to the other one.		
Recluse	Individual who has no relations with others.		

Table 2: Properties of network (Tichy, 1981, p. 229)

Pos. No.	Property of the network	Indications (criteria)
1.	Goal (common aim, projected result)	* Distant (coinciding with the goals of the participants of the network); * Medium (goals of the near future seeking common aim); * Near (goals of "today"); * Strategic (corresponding to the mission of the participants of the network); * Tactical (showing how strategic goals are implemented); * Official (showing the aim of the network in the correspondence of the mission).
2.	Vision (the whole image of an organization)	* Vision of future organization (image of future network); * Vision of strategy (activity of the network to realize the vision); * Vision of necessary resources, abilities and their application (activities of the participants of the network in the foresight of the growing prospects, resources and abilities and how to provide with them); * Vision of linking the present with the future (abilities of the participants of the networks to obtain information and to see the prospect); * Vision of the future situation (cohesion of the participants of the network to implement efficiently their vision).
3.	Beginning of the activities	Date of establishment (date of unification of the participants of the network or official registration of the network).
4.	Legalization	Legal registration (in national, regional, local level or contract among the participants of the network).
5.	Level	* Local – uniting the organizations of a city, district, village, etc. * Regional – uniting the organizations operating in the region; * National – uniting the organizations operating in the national level; * International – uniting the organizations operating in the international level;
6.	Structure (elements – participants – composing the network and their collaboration)	* Hierarchic (hierarchic relations of power and communication of the members of the network); * Matrix (clear allocation of works and duties of the members of the network; * Horizontal (collaboration of the members of the network on a parity basis).
6.1.	Structural characteristics	* Organizational density (number of the members of the network); * Grouping (grouping of the members of the network into schools, groups); * Size (number of the members of the network); * Visibility (showing, official and non-official registration of the members of the network); * Competence (qualitative indicator of the network members competence of the members of the network).

6.2.	Content of interaction (exchange measures)	* Expressions of emotional attachment (reliability of the participants of the network in communication in certain activity); * Influence (effect of one network on other networks); * Interchange of information (interchange of information of the members of the network in formation of more objective network of services); * Commodities and services (assortment and quality of different services of the members of the network).
7.	Characteristics of relations (relations among the organizations of the network)	* Mutuality (symmetric relations, i.e. partnership among the members of the network); * Clarity of norms (rules and regulations of collaboration of the members of the network); * Intensiveness, energy (obligations of the members of the network by sacrificing their energy); * Complication (activity of organizations as the members of the network related in several ways); * Openness (definition of relations of the members of the network with other network); * Stability (definition of probable duration of common activities of the members of the network).
8.	Memberships (status of elements composing the network)	* Participant (the member of the network participating on a parity basis and providing the basic services of the network); * Partner (participating on a parity basis in the network but not providing the basic services of the network).
9.	Contribution	* Investments of the members of the network (services, human, material, financial-capital).
10.	Responsibility	* Legal or moral responsibility.
11.	Management (what kind of management: aspect of centralization – decentralization)	* Centralized (there are separate participants in the network or institution that concentrate the management); * Decentralized (there are no separate participants in the network and no institution that concentrate the management; the management is distributed among the members of the network that coordinate their activities); Centralization is combined with the decentralization (there is an institution established in the network and the members/ partners of the network entrust some management to it, and other part of management is distributed among the members/ partners of the network).

Table 3: Characteristic of the schools' network

Edita Štuopytė

The Specificity of Socio-Educational Activities of the Community Centers

Introduction

Educational and social problems require new ways of solutions in the changing world, which most researchers (Fullan, 1998, Hargreaves, 1999, Drucker, 1993, etc.) call the post-modern age. In such society knowledge and information as well as abilities/skills to apply them become the basic capital and the premise for the activity. According to Drucker (1993), the person's activity and his/her general skills/abilities become very important in the knowledge society. The need for such qualities as critical thinking, the effective activity, group work, the application of the methods of problem solution, etc. become essential. The acquisition of these new qualities is important both for children, and adults.

Globalization processes disclosed the relevance of the research of community education, community development, and sociality. Lithuanian and foreign researchers analyze the mentioned problems. The problems of sociality were investigated by Karalius (1999), Grigas (1999), Šlapkauskas (1999). The methods of community education as well as other research of the community are presented in publications of Leliūgienė, Baršauskienė (2002), Kuzmickaitė (2003), Kenny (1994) and other researchers.

The development of community centers, the conditions, reasons and further perspectives of their development are analyzed in the works of Durand (1996), Guerrand, Rupp (1978), Kennedy (1986), Passing (1998), Mayo (1998), etc. From the viewpoint of socio-cultural work the community centers are deeper analyzed in the works of Spierts (2003), Baršauskienė and Leliūgienė (2001). The analysis of the socio-educational activity of the community centers working with children has been to analyze recently started in Lithuania, thus this research is relevant and timely.

Referring to Evans (1981), a youngster acquires and accumulates most of the abilities and skills through the unstructured environment. This is also to be said about the assimilation of cultural values, general attitudes, and the behavior models through non-formal institutions (family, various associations, the process of social communication, mass media, museums, etc.). The community centre is one of the institutions, in the activity of which there is a possibility to realize its activity principle: a youngster chooses his/her desired occupation. Where he/she actively participates. In addition, if young person decides to participate in a certain activity, the environment of the institution influences him/her, too. This is important in educating children and young people as the members of the future knowledge society.

Every activity is realized in performing certain functions, consequently in this context the following *problematic question* comes to light: *what socio-educational functions do the community centers perform in the work with children, does the specificity of the socio-educational activity of the community centers reveal itself in performing the functions?*

The purpose of the article is to highlight the specificity of socio-educational activity of the community centre.

The objectives of the article are as follows: to present the research methodology of social educational functions of the community centre and to reveal the peculiarities of social-educational functions of the community centre.

The research methods applied are as follows: the analysis of scientific literature, survey in written, the statistical analysis of the data.

The Research Methodology of the Socio-Educational Functions of the Community Centre

The community centers working with the children perform certain functions. In various European countries the community centers function not for one decade, the work of these centers, however, with the children in the aspect of the socio-educational functions' expression in Europe has not been analyzed yet, therefore the research is significant in the respect of science. Referring to the analysis of the conceptions of community centre and the practical activity, the theoretical model of social-educational functions of the community centre work with the children is created (Štuopytė, 2002).

In this model (Figure 1, see p. 293) theoretical and practical principles of the community centre activity are presented as well as the activity fields and functions. The socio-educational activity is researched through the realization of the functions so that theoretical and practical principles of democracy, partnership, and social education as well as conceptions of people empowerment, attitudes of relationship, communication-in-action conception, etc. manifest themselves through the activity and functions of community centers.

The social educational functions of the community centre, as distinct from the ones of formal education institutions, distinguish by distinctive particularity. *The Empowerment function* manifests itself in the field of *social-care* of the community centre. In fulfilling this function in the community centre the conditions for youngster's self-expression, for the development of his/her activity and action are created. The cozy environment of the community centre, the possibility to express oneself stimulate children to get indulged into the activity of the community centre as well as to get involved into the community life. In this field the *function of care and provision* is also pursued. The workers of community centers take care of youngster's *well feeling* in the centre, *meeting his/her needs* and helping to solve problems.

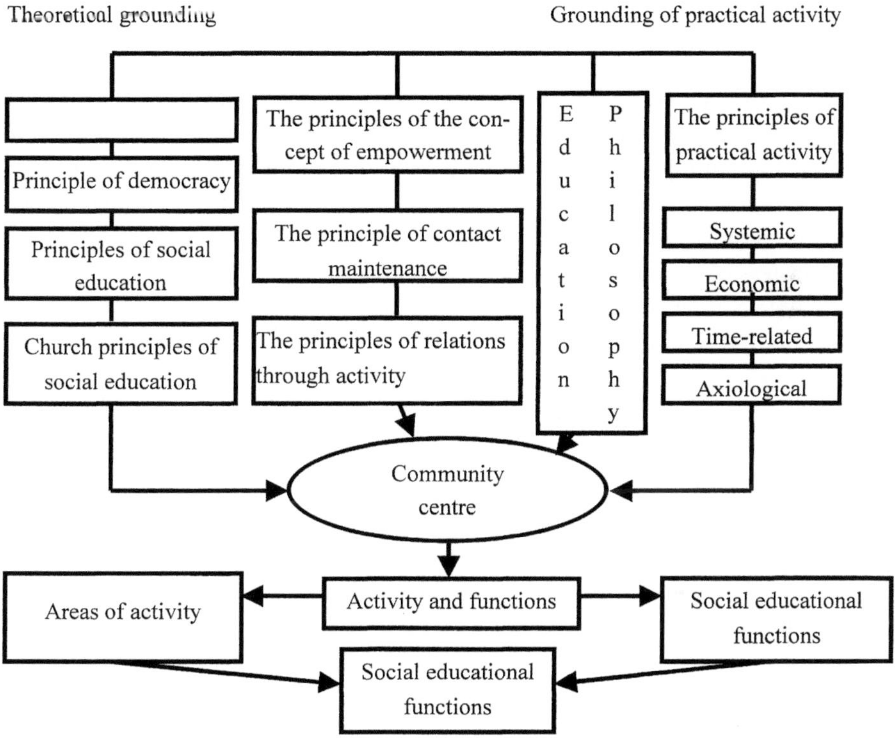

Figure 1: The theoretical model of the socio-educational functions of the community centre (prepared by E. Štuopytė, 2002)

The function of teaching is not so formalized in the centre as at school or other institution. It is realized in two fields. Working in the cognitive technical field the children are helped with their homework, by means of various games their creative powers are disclosed. In the field of social-care, the exceptional attention is focused on the training of youngsters'-volunteers.

The function of developing is fulfilled through the cognitive technical and cultural educative activities. Young people participating in the activity of different teaching subjects as well as optional subjects, not only develop their skills, but competence, too. The development of artistic skills takes place involving the children into performances, organizing different festivals and get-togethers, as well as having the children participated in drama circles (studios).

The function of upbringing manifests itself in ecological exploratory activity, when developing children's relationship with nature. They take care of environment, plant trees, and take part in ecological hikes and other actions. In the field of social-care the exceptional attention is focused on the improvement of youngsters' communication, their caretaking activity in the community.

The function of education is performed in the ecological exploratory activity at the community centre, rendering the youngsters children the knowledge of this activity, as well as the cultural educative activity, when they visit museums, exhibitions and intercept cultural information. It is important that in the activity of the community centre the young people find more information about the history and culture of their town or settlement. In the social field the volunteers of the community centers, fulfill preventive work (they organize activities for the youngsters of social risk, hold seminars on addiction).

At the community centre *the function of training* is performed while developing children's social skills (the social–care field); developing child's physical (the sport field), psychic powers and practical skills (the cultural-educative field). At the centers children play sport games, take part at quizzes, contests, excursions and walking-tours, get involved into the activity of different support and maintenance groups.

While executing *the function of formation* it is most important to stimulate the voluntary activity of the youngster. The young people most frequently work in the ecological, social and sport fields. The centre workers appear as consultants or advisors.

The Organization of the Research

The research of social educational functions is a complex process; thus in order to perceive the researched phenomenon more thoroughly and wider, it is purposeful to apply several research methods. Referring to Kardelis (1997), the principle of triangulation that enables to analyze the phenomenon from several positions was chosen. The research was carried out in all community centers -18 (Figure 2) that functioned in Lithuania at the end of 2000. The research geography involved different towns and regions of Lithuania.

While researching the social educational functions of the work of the community centre with youngster, the research instrument of matrix form was chosen. In the matrix the social educational functions and their characteristics of the community centre, the fields of the centre activity as well as the basic criteria were distinguished (see Table 1, p. 300). During the research, the methodologies of the research of the documents regulating the activity of the community centre, the survey of the community centers' (CC) leaders, the expert evaluation of the CC ergonomic conditions as well as the survey of the CC educators-volunteers and the CC youngsters were prepared. The entire general sample – 34 volunteers, who work with the youngsters, and 116 young people engaged in various programs – were surveyed.

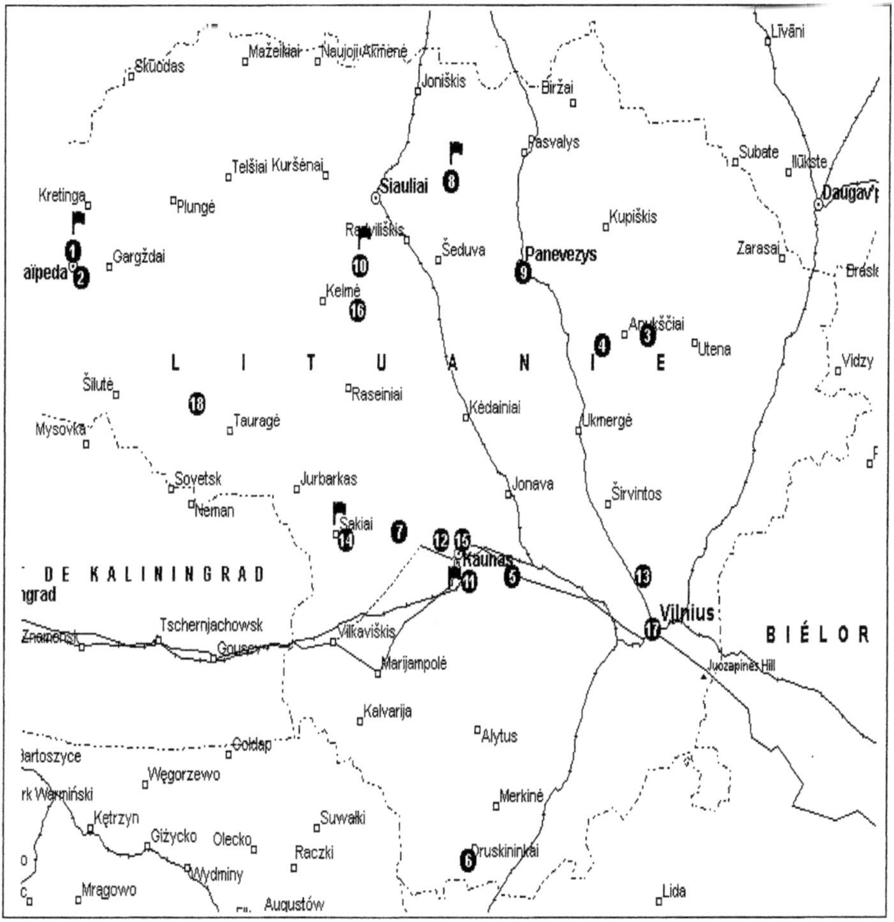

Figure 2: The Community centers (18) in Lithuania at the end of 2000.

The Results of the Research

The analysis of the survey data show that the volunteers working with youngsters fulfill all the social educational functions (except the one of formation). The expression of these functions is not identical. The *function of* empowerment manifests itself very strongly, because more than 80 % of the respondents constantly perform this function. The strength of the expression of the *care/provision, teaching and education functions* is average (more than 50 % of the respondents indicated that they perform these functions). The expression of *upbringing* and *training is strong,* as more than 80 % of the volunteers perform these functions.

Statistically significant differences were noticed between the work of CC volunteers in cultural educative and sport activities. These differences are determined by the oneness of the centers (the microclimate created in the centre; the needs of the children attending the centre; human resources), but not the geographic location of the centre.

The results of the survey of the children participating in various programs of the community show that the function of *developing* manifests itself weakly. There were 12 statements, repeated both in the volunteers' and youngsters' questionnaires.

Applying the methods of mathematical statistics, the statistical significance between the means of volunteers'- and youngsters'-evaluations was investigated. The difference of the evaluation means of the volunteers and youngsters verified by *Student* coefficient enables to state that evaluations of the common actives (homework; playing various games; taking care of the environment; work with computers; competitions, quizzes and tournaments; exercise and sports games; charitable activity etc.) are statistically insignificant.

Analyzing the expression of the empowerment function, the hypothesis whether the youngster's age, gender, his/her parents' family and social status as well as location influences his/her possibilities to unclose at the community centre was checked. The data of the research showed that young men have fewer possibilities in a specific centre working only with young people, than in any other centre (the difference is statistically significant, 95 % probability).

Analyzing the data on the *function of care and provision* the following hypothesis was statistically checked - whether the youngster's state of mind at the community centre depends on the geographical location of the centre, his/her gender and age. The hypothesis is rejected; the reliability level at the evaluations is > 0,05, so it is possible to assume that the above-mentioned factors do not make any influence upon the youngster's state of mind at the community centre. It was statistically verified (95 % reliability) that attendance at the centre in order to meet the need for one's self-expression (Criterion 2.2.) depends on the youngster's gender (statistically significant weak relation in the sense of the monotony/anonymity).

The meeting of the need for security (Criterion 2.2.) depends on the youngster's age. The observed statistically significant weak negative relation in the sense of the monotony allows assuming that the older the youngster's is, the less important the meeting of the need for security at the centre is for him/her. Older youth satisfy their other needs at the community centre. *Analyzing the function of teaching,* the cluster confidence intervals were used in checking the relation between the activity in the centre and the youngsters' gender as well as residence place (region). The statistically significant differences were observed between the following kinds of the activity: preparing homework, learning (Criterion 3.1 Organizing different creative games (Criterion 3.1) the dependence on the youngster's gender and region was disclosed. Analyzing the data on the expression of the *function of upbringing* (Criterion 5.1), the statistically significant difference was identified when the youngster's were looking after the environment at the community centre. The results of significance verification of these activities are presented in Table 2, p. 297. The differences were verified referring to $\chi 2$ and the 95 % probability show the compatibility with the results presented above/earlier.

The data of the youngster's survey disclosed that, while organizing the youngsters' activity at the community centre, all social educational functions presented in the theoretical model (except the one of formation). The functions of *care/provision* and *teaching* manifest themselves most strongly. The expression of the functions of *empowerment, upbringing, education* and *training* is of average strength.

Activity	**Pearson's $\chi 2$**	**Degrees of freedom**	**Level of significance**	**Kendall's τ c**
Preparing homework	34,47	15	<0,05	0,161
Playing various games	40,258	15	<0,05	0,4359
Looking after the environment	25,825	15	<0,05	0,306

Table 2: The Significance of Activities

Analyzing the data of the community centers leaders' survey, it was notice that in most centers there is no with young people's activity because there are no volunteers ready for its activity. Therefore it is possible to draw a conclusion that while working with the youngster's in the community center the issue of human resources is very important. The main way of the work with young people is group work. Only a few centers work with while working with the youngsters individually. The basic goals of the work with children singled out by the leaders of the community centers enable to assume that the centers' most important function in the social field is performing the function of care and provision. The survey results of the volunteers, who work with while working with the youngster's, and young people, who participate in the programs, showed that in their common activity all socio- educational functions were displayed (except the one of formation). Only the strength of the expression of these functions differs.

During the process of educational change non-formal educational institutions – community centers- join the activities of personality development. Every client of community centre can acquire new qualities in the activity of this institution –experience of group work, problem solving. These qualitative changes of human activity are of special significance in creation of the knowledge society.

Conclusions

- The socio-educational activity of the community centre manifests itself through the expression of the social educational functions of this centre. The research methodology of these functions refers to the matrix analysis of the fields and functions of the community centre activity, which enabled to distinguish the research criteria of these functions.
- While working with the youngsters at Lithuanian territorial community centres, all social educational functions defined in the theoretical model are displayed, namely: empowerment, care and provision, teaching, developing, upbringing, education, training and formation.
- The results of the inquiry (interview) of the heads of community centers show that the priorities of the activity of each community center are set according to the needs of the inhabitants of the town, village, or neighborhood. Youngsters are served in all community centers. In the activity of these centers, the functions of development, teaching and upbringing predominate. Manifestation of the function of development is strong. In the cultural-educational activity of the centers, this function is executed by all the studied centers. In the development of youngsters' artistic abilities (criterion 4.2), manifestation of this function exceeds 80%. Manifestation of the function of teaching according to criterion 3.2 (teaching of volunteer children) is medium (over 50%). The investigation of this function showed that in some centers the teaching theory based on double-loop learning is used as the basis for teaching volunteer youngsters. Double-loop learning allows not only changing the strategies of activity, but also gives way to modification of values, attitudes, and norms of the young volunteers.
- The accomplished empirical research disclosed not only the expression of the social educational functions of the community centre, but also the specificity of the socio-educational activity at the community centre. The results of the empirical research showed the irregular expression strength of these functions. The functions of teaching and upbringing are most strongly expressed. The expression of empowerment, care and provision, developing, education and training functions is of medium strength. The function of formation is weakly expressed. The expression of social educational functions mostly depends on the realization of the community centre goals, the understanding/perception of the mission as well as the unique capital of human resources accumulated at every centre.

References

BARŠAUSKIENĖ, V./LELIŪGIENĖ, I.: Sociokultūrinis darbas bendruomenėje. Kaunas 2001 (Technologija).

DRUCKER, P.: Post-capital Society. Oxford: 1993 (Linacre House, Jordan Hill).

DURAND, R.: Histoire des centres sociaux. Du voisinage a la citoyennete. Paris 1996 (Syros).

EVANS, D.K.: The Planning of Non- formal Education. Paris 1981 (Unesco).

FULLAN, M.. Pokyčių jegos. Vilnius 1998 (Tyto Alba).

GRIGAS, R.: Nūdienos pasaulio kontūrai irlietuvių bendruomeniškumom problema. In: Bendruomeniškumas ir savivalda. Kaunas 1999, pp. 4-17.

GUERRAND, R.H./RUPP M.A.: Breve Histoire du service social en France. Paris 1978 (Privat).

HARGREAVES, A. : Keičiasi mokytojai, keičiasi laikai. Vilnius 1999 (Tyto Alba).

KARALIUS, A.: Bendruomenškumo formos ir jų pamatai. In: Bendruomeniškumas ir savivalda. Kaunas 1999, pp. 50-65.

KARDELIS, K.: Mokslinių tyrimų metodologija ir metodai. Kaunas 1997 (Technologija).

KENNY, S.: Developing Communities for the Future: Community Development in Australia. South Melbourne 1994 (Nelson).

KUZMICKAITĖ, D.: Bendruomenės tyrimo metodologijos ypatumai. 2003. [cit. 06 February 2004] Available from: http://www.vdu.lt/seki/

LELIŪGIENĖ, I./BARŠAUSKIENĖ, V.: Bendruomeninio ugdymo metodai ir principai: Užsienio šalių mokslininkų požiūris. Specialusis ugdymas 2(2002)7.

MAYO, M.: Communities and Caring. Basingstoke 1994 (Macmillan).

PASSIG, D.: The Future Community. The Role of Community Center in an Era of Uncertainy and Rapid Change. Jeruzalem 1998.

POURTOISM, J.P./DESMET, H./NIMAL, P.: Quelles competences professionnelles pour les contextes d'abandon? (Non publicated). Belgium 1999 (Universite de Mons – Hainaut)

ŠLAPKAUSKAS, V.: Šiuolaikinės bendruomenės savikūros sąlygos ir veiksniai. Bendruomeniškumas ir savivalda. Kaunas 1999, pp. 30-48.

ŠTUOPYTĖ, E.: Teritorinės bendruomenės centro darbo su vaikais socialinės edukacinės funkcijos ir jų raiškos ypatumai. Daktaro disertacija. Kaunas 2002 (Kauno technologijos universitetas).

	The characteristic of the function	The field of activity	Criteria
1. Empowerment for activity	The creation of conditions for the youngster's natural self-expression and the development of activity	Social-care	1.1. Conditions for a self-youngster's expression 1.2. The development of youngster's activity
2. Care/provision	The safeguarding of youngster's biological and social norms of existence and the meeting their needs	Social-care	2.1. The youngster's environment in the community centre 2.2. Meeting the youngster's needs 2.3. Provision of the youngster
3. Teaching	Support and supervision of the youngsters' learning	Cognitive-technical Social-care	3.1. Assistance in the youngster's learning 3.2. Training of youngster's volunteers
4. Developing	The empowerment of the youngster to develop his/her competence	Cognitive-technical Cultural-educational	4.1. The development of the youngster's skills 4.2. The development of the youngster's artistic abilities
5. Upbringing	The formation of the youngster's relations with the environment, work, and his/her self.	Ecologic-investigative Social-care	5.1. The formation of the youngster's relationships with nature 5.2. The development of youngster's communication 5.3. Youngster's charitable activity
6. Education	Rendering information, and the organization of acceptance and feedback in the youngster's preparation for social life.	Ecologic-investigative Cultural-educational Social-care	6.1. Rendering ecologic knowledge (information) to youngster's 6.2. Rendering cultural information to youngster's 6.3. Preventive work with youngster's
7. Training	The development of the youngster's physical, mental, and practical skills and abilities.	Cultural-educational Sporting Social-care	7.1. The development of the youngster's mental abilities and practical skills 7.2. The development of the youngster's physical abilities 7.3. The development of the youngster's social skills
8. Formation	The youngster's empowerment for activity through which new quality of physical, mental, cultural and spiritual characteristics is acquired	Ecologic-investigative Sporting Social-care	8.1./8.2./8.3 The youngster's volunteer free-time activity

Table 1: The Criteria System of the Study of the Socio-Educational Functions of the Community Center Engagement with Youngster

Auksė Markevičienė

University as a Cultural Phenomenon: the Paradox of Being

Introduction

As early as 1789, Friedrich Schiller defined the ideal of culture (*Bildung*) in the program of the Berlin University among such seminal requirements as, for instance, the unity of research and teaching, as an essential goal of the university. Culture has come to denote a process of development, of the cultivation of character, and the two branches of this process are considered to be research and teaching, specific to the university. Consequently, as long as the university is a center of the educational system where teaching and research are combined, it must also be an institution fostering general culture.

The University of Culture, instituted by W. Frhr. von Humboldt, drew its legitimacy from culture, which had named the synthesis of teaching and research, process and product, the institution and the individual. Thus, the revelation of the idea of culture and the development of the individual is supposed to be one – object and process should unite; the place they unite at is the university (Readings, 1996). However, there are increasing concerns among scholars today that the intensifying market forces characterizing a knowledge-driven economy may sweep universities pushing aside their academic values and traditions and replacing those with the demands of the marketplace. The contemporary university is becoming a bureaucratically organized and relatively autonomous consumer-oriented corporation, with the intellectual activity and the culture it revived being replaced by the pursuit of excellence and performance indicators.

University, traditionally regarded as an agent of culture, has been going through a deep crisis; the idea of a *'University in Ruins'* has been repeated for such a long time that it became a "tiring anachronism" (Derrida, 1994, p.15). The image of the university with two left feet emphasizes the assumption that the university has become an institution, which lost coordination and no longer has a basis for moving forward (Derrida, 1992). Culture used to be such a background; today, however, it is no longer a watchword of the university. Due to the decline of the national cultural mission that has up to now provided its *raison d'etre*, the university has experienced the shift in its role. It has lost its organizing center, the idea of culture as the object, and now is pressed into the service to the state. In short, "the university is becoming a different kind of institution, one that is no longer linked to the destiny of the nation-state by virtue of its role as producer, protector, and inculcator of an idea of national culture" (Readings, 1996, p.3).

Therefore, it is evident that in order for a university not to completely transform itself into a cosmopolitan bureaucratic corporation, it must find ways to sustain the most cherished aspects of their core values, while discovering new ways to respond to the opportunities of a rapidly evolving world.

The aim of this article is to investigate the contemporary university as a cultural phenomenon and to reveal the existing inconsistencies. The speculations about the functions of the university, as a major institution for legislation of culture, will be based on the assumptions of B. Readings, J. Ortega y Gasset, R. Barnett, T. Schuller, and others, with the supporting methodologies of the hermeneutic concept of culture and the Service University conception.

1. Cultural Function of the University

"Finally, there is a function... that is none the less fundamental: the transmission of a common culture [...]. It is a proper function of higher education... to provide in partnership with the family that background of culture and social habit upon which a healthy society depends (Robbins, 1963, cit. Barnett, 1990, p.95).

It is difficult, if at all possible, to provide a coherent picture of the university as a cultural phenomenon. Nevertheless, it would be reasonable to start with the definition of culture in a broad sense and to touch upon its educational potential.

The understanding of culture implies numerous meanings which makes it difficult to imagine culture as a whole. Broadly speaking, culture is a pluralist aggregate which enables the existence of contradicting and even contesting values (Kavolis, 1996). According to Ortega y Gasset, culture is everyone's inevitable urge, the constituent dimension of the human existence (Ortega y Gasset, 1999).

Each culture has a certain core comprised of activity norms, means of behavior, attitudes, and mindsets. Among the fundamental characteristics of culture, singled out by G.P. Murdock, is its instilment which manifests itself by the transfer of skills and knowledge (Murdock, 1993). In the opinion of anthropologists, one of the segments of culture is knowledge (Roser, 1993). Moreover, the "nucleus of culture" is transferred with the help of social institutions, and it could be assumed that one of such institutions, able to echo the perfection and cleansing of culture as an entity, inducing the needs for growth and upbringing (Kavolis, 1996), is the university. Pedagogical action, in the opinion of Salkauskis, is a cultural action. Following his ideas, the role of an institution of higher education could be further interpreted and it could be assumed that the university has to nurture society by fostering relevant cultural frames and dispositions in university students (Salkauskis, 1990).

The cultural function of the university has been emphasized by many contemporary philosophers, and it would be relevant to rely on the ideas of Jose Ortega y Gasset which will contribute to the assessment of the current situation at the university in general and at the Service University in particular.

Higher education provided by a university is comprised of carrying out scientific research, preparing future investigators, and educating intellectual professions (Ortega y Gasset, 1999, p. 112). Knowledge, requisite for the life of a contemporary student, is only a "university minimum". The fact that the university today, at the expense of the extremely expanded professional education and the emphasis on the applied research, has rejected teaching and transmitting culture, is a huge problem. Shouldn't university education be more than just professionalism and research? Can higher education shape a real human being?

Ortega y Gasset claims that a student should be taught not only his/her major, but subjects of the "general culture", i.e., philosophy and arts, and he says, he would consider a course on culture as a nucleus of the whole higher education curriculum. Higher education should, first of all, comprise teaching culture and bear a cultural function. He singles out three functions of teaching at the university, namely, culture transfer, professional education, and scientific research along with preparing researchers (Ortega y Gasset, 1999, p. 120).

Thus, it is important that the university teaches culture. Nevertheless, such an inference does not fully reflect the essence of the university as a cultural phenomenon. A glance from the inside would further a deeper insight into the problem.

The Inner Culture of the University

The cultural function of the university concentrates on the interconnections between higher education and society, and the inner life of higher education institutions is considered to be an essential ingredient in the development of the system (Robbins, 1963, cit. Barnett, 1990). As a matter of fact, close relationship between the general culture of a country and the inner culture of the university has long been the target of philosophers. Moreover, it should be remembered that higher education maintains numerous kinds of culture within itself. Therefore, a close look at the internal culture of higher education would aid in an attempt to describe the university as a cultural phenomenon.

R. Barnett in "The Idea of Higher Education" speaks of two levels of the inner culture of higher education: culture in relation to the academic community, as a form of life with its own values, peculiar to this community, and culture as students' experience. In other words, he specifies the dimensions of culture as a content and culture as a process (Barnett, 1990). Referring to the idea of *Bildung*, we should ask, if it is possible to identify elements of a specific form of life at the university, which would reinforce, in Newman's terms, the cultural '*enlargement*' of an individual student. As Barnett puts it, is it a matter of exposing students to some culturally rich material, that is, of content, or is it more a matter of particular educational experience, or of process? Is cultural development of students relevant only in humanities an social sciences, or does it also have meaning for learners of science; how should the educational process look like for this cultural development to take place (Barnett, 1990)? Do we speak about a unified academic culture, or does it remain a multipartite component of the university?

To answer these questions, culture should be considered in the context of the community of scholars. However, if we look at the university today, signs of internal common culture are not to be seen immediately. On the one hand, the investigation by numerous researchers (Kuhn, 1970, Kerr, 2003, Barnett, 1990, Poskiene, 1998, etc.) has shown that it is very difficult to refer today to a single academic culture; instead, we can definitely speak about numerous disciplinary (sub-)cultures, units based on critical reasoning. On the other hand, it can't be refuted that any academic sub-culture, able of self-examination, or reflection on itself, is a critical enterprise. Consequently, with this distinguishing feature in mind, we can reservedly hypothesize that the academic community has a potential of possessing a single coherent culture.

Under the current changes at the university, this institution more and more comes to represent an educational enterprise, with the staff acting as a corporate body, relying on contracts and managerial performance indicators, with "the governance of academic institutions liable to be reduced to mere technique" (Barnett, 1990, p.104). With this development in mind, it would be reasonable to assume that culture as a process, or cultural experience for students, would be more relevant when facilitating a general enlargement of students' minds as a means to fight arising problems in the commercialized higher education today.

If we presuppose that higher education is, or is supposed to be, a cultural experience, it is relevant to mention that it might take a variety of shapes. First of all, it is the integrating culture, peculiar to subjects and the institution in a general sense, as well as common heritage. Secondly, despite the fact that higher education is currently attempting to move towards the acquisition of competencies and techniques, the educational experience is undoubtedly fostering student's human development, since educational practice has always been based on the background of philosophical values and meanings.

Nevertheless, the above considerations give grounds to the assumption that the two kinds of the university culture (Barnett, 1990) do not bear fundamental differences. That is, the dimensions of the academic community culture and students' academic experience are strongly related, since the quality of student educational practice and their cultural enlargement are evidently influenced by the nature of life in students' academic community and by their interaction with other members of the academic community. This notice is especially valuable if we come to consider the aspect of entrepreneurship development in the academia.

The University and the Culture of Society

It is evident, however, that the inner culture of a higher educational institution is based on broader cultural relationship. Therefore, the investigation of the university as a cultural phenomenon, in addition to singling out the two levels of the internal academic culture, requires highlighting the relations between the university culture and the culture of society.

Heidegger, the creator of the hermeneutic concept of culture, claims that culture is not an activity pointed at discovering pre-existent structures or objective meanings; it does not possess universal standards. Instead, it is a creative process, disclosing new opportunities provided by the past deeds; it is the exploration of "abidance in the world", a humanist ideal and a mystical conception, a root of the thought of humanity. Cultural reality, determined by the "horizon", as *'being there'*, or belonging to the world, is greatly dependent on its interpreter and on the context (Heidegger, 1962).

The university, a societal institution, acts not just as a vehicle for economic reproduction (Barnett, 1990). In the opinion of sociologists, the internal life of academic institutions serves to sustain, reinforce, and reward class-related forms of style, language, and behavior (Giroux, 1983, Apple, 1979). Then, academic institutions help to maintain a flow of the kind of cultural capital on which wider social institutions are based. Ortega y Gasset, Leavis, Bantock, and others have long argued that higher education preserves the cultural fabric of society (Ortega y Gasset, 1999). In other words, the university is the institution to sustain the cultural structure of society and to promulgate certain social ideas.

How could the university input into the culture of society be described? Providing graduates with professional knowledge, skills, and competencies is insufficient; enabling them to critically evaluate, by thought or by action, everything they encounter in the age with no common or unifying culture, empowering them to take a skeptical stance and question societal processes is essential.

In the opinion of J. Ortega y Gasset, only a cultured individual with a contemporary perception of culture and a specialist with the sense of space and time is capable of acting in today's society. To achieve this purpose, it seems reasonable to assume that both the culture of the academic community and the culture of student experience, constituting a general culture of the university with the questioning and a critical element, serve as a "steady injection into the wider culture" (Barnett, 1990).

2. Problem of the Contemporary University

The University in Ruins

The university in the post-modernist and supercomplex world, influenced by globalization and information technologies, is living through the period of radical change. Controversial processes in knowledge-oriented post-industrial societies add new features to higher education. A natural mechanism of the market, created by the economic policy, competition, and other forces, has converted the nature, content, and status of knowledge, has transformed the classical idea of the university, a creator of knowledge and a prophet of culture, and has modified university's intent to advance scholarly inquiry, to cultivate critical thought, and to create fundamental knowledge. Such higher education goals as dedication to the development of intellectual culture and transmission of democratic values have been attenuated with the support for the economy and service to society. J. Duderstadt and a number of other scholars claim that the future belongs to the entrepreneurial (Duderstadt, 2000). As Gould puts it, higher education, becoming more and more like a business, has been "thrown to the market" to solve curricular problems and to foster consumerist culture which drives the economy (Gould, 2003). In the context of on-going financial concerns and growing costs, a certain market-driven flexibility, together with the implementation of corporate culture in managing and generating knowledge, is demanded of the university.

As a result, globalized higher education is today a multi-billion dollar industry (Alderman, 2001). Academia complains that universities have turned into "knowledge factories" where academic ideals are routinely compromised for the sake of money. According to S. Aronowitz, "the learning enterprise has become subject to the growing power of administration, which more and more responds not to faculty and students, but to political and market forces that claim sovereignty over higher education (Aronowitz, 2000, p. 194). To W. Shumar, learning and research have "come to be valued in terms of their ability to be translated into cash or merchandise" (Shumar, 1997, p. 5).

The Service University

With the development of the knowledge society, new tendencies of the interaction between the university and the environment have appeared. As it was implied above, market-oriented higher educational institutions have started reconstructing themselves as enterprises. With the new dimensions added to higher education, such as selling research and providing services to clients both inside and outside the university, a new concept has appeared in higher education – that of the Service University. As a new model of a higher education institution, it does not yet own a complete and universally defined set of criteria; neither does it possess a single and widely accepted shape. Nevertheless, certain features can be further introduced to mark out the Service University.

Thus, with the appearance of this new form of higher education, most values at the university have been substituted by the Service University's financial incentives: non-commercial disciplines have been subject to die out and have been exchanged for trendy programs enabling effective augmentation of skills. To cater to the market, undergraduate and post-graduate study programs have been gradually replaced by extensively marketed short-term professionally oriented courses, sometimes delivered in non-traditional ways. Professors' labor, teaching, and research have been administered by the management through a contract concluded with clients on the purchase of services, either in research, teaching, or consultation (Tjeldvoll, 2000). New processes of learning have appeared, new flexible forms of organization have emerged, and new conditions of employment have come to be relevant, in many cases substituting the traditional scholarship by less paid contracted instructors to administer courses.

As a consequence, certain outcomes might be noted as the costs of the clash between higher education and the market, which reassert the initially mentioned Derrida's idea of the university in ruins, a major challenge for contemporary higher education. Among them, the diminishing academic standards of the university to fit private interests of the donors, academics, and the students-consumers should be mentioned. Furthermore, damaging professional autonomy and fundamental academic values to serve the commercial needs of the clients has become evident. Finally, risking the reputation of the academia as an unbiased body, which has always celebrated academic freedom, must be singled out.

Discussion and Conclusions

How, in a broader sense, is this reality of higher education related to the concept of the university as a cultural phenomenon? Doesn't it directly imply a paradox inherent in the higher education of the 21st century?

In the opinion of Ashby, a university today must be sufficiently stable to sustain the ideal which gave it birth, and sufficiently responsive to remain relevant to the society which supports it (Ashby, 1966; as cited in Altbach, 1999, p.4). Most scholars still agree that the major role of the university is the search for truth and advancement of knowledge, and among the duties of a modern university they see custody and development of the cultural heritage of the state (Readings, 1996; Barnett, 1990, 2000; Schuller, 1991). They claim that the pursuit, promotion, and dissemination of knowledge are important so as to raise the level of intellectual development of an individual (Mosha, 1986) and to develop educated intellectual human talent (McMullen et al., 2000, Kerr, 2001). However, J. Derrida in the famous essay "Mochlos" claims that "there is nothing *in it*" (Derrida, 1992) and affirms that the idea of the university as a solid institution with coherent characteristics and confines, founded on the principles of reason and standing on its original base, is subject to an eternal and inextricable conflict.

Every society needs competent professionals. But it is also vitally important that these professionals are able to live, influence, and make a difference in the world. A human being as a stuffing for the market needs is not valid today; he/she has to be a cultured and "enlarged" person, and educating such a cultured individual remains the essential task of the university. Current developments of higher education and the rise of the Service University do not safeguard the persistence of the academic culture at the university. Will the accelerating entrepreneurial spirit leave space for the independent creation of fundamental knowledge, institutional autonomy, seminal values of higher education, and the academic culture at large? Will a university graduate be able to resist the sweeping chaos, will an educated individual maintain a background enabling him/her to realize that the culture is a "vital system of ideas" in every period (Ortega y Gasset, 1999, p.115)?

Referring to Heidegger's *"being there"*, it can be assumed that the birth of the Service University, as an unavoidable consequence of the changing systems, has been determined by the surrounding world and the "horizon", where it is impossible to appeal to universal values and norms. Thus, there should be no stigma attached this new form of higher education, even if it does not, and will never, resemble von Humboldt's University. For higher education to survive, there is no need to come back directly to the roots of the classical university. To revive the academic culture at the Service University, it rather behooves us to creatively transform and correctly interpret the content of the cultural tradition. Following Heidegger, precisely these new developments constitute *"being there"*. And this is a major paradox of the *being* of the university today.

References

ALDERMAN, G.: The globalization of higher education: some observations regarding the free market and the national interest. Higher education in Europe, XXVI(2001)1, pp. 47-52.

American Higher Education in the Twenty-first Century. In: ALTBACH, P.G/BERDAHL, R.O/ GUMPORT, P.J. (eds.): Social, Political, and Economic Challenges. Baltimore 1999 (The Johns Hopkins University Press).

APPLE, M.: Ideology and Curriculum. London 1979 (Routledge and Kegan Paul).

ARONOWITZ, S.: The Knowledge Factory: Dismantling the Corporate University and Creating True Higher Learning. Boston 2000 (Beacon Press).

BARNETT, R. The Idea of Higher Education. 1990 (The Society for Research into Higher Education and Open University Press).

BARNETT, R.: Realizing the University in an Age of Supercomplexity. 2000 (The Society for Research into Higher Education and Open University Press).

DERRIDA, J.: Mochlos. In: RAND, R. (ed.): Logomachia. The Conflict of the Faculties. Lincoln and London 1992 (University of Nebraska Press), pp. 1-34.

DERRIDA, J.: Specters of Marx: The State of the Debt, the Work of Mourning, and the New International. New York and London 1994 (Routledge).

DUDERSTADT, J.J.: A University for the 21st Century. Michigan 2000 (Ann Arbor).

GIROUX, H.A.: Theory and Resistance in Education: a Pedagogy of the Opposition. London 1983 (Heinemann).

GOULD, E.: The University in a Corporate Culture. New Haven and London 2003 (Yale University Press).

HEIDEGGER, M.: Sein und Zeit. (English translation by J. Macquarrie and E. Robinson). 1962 (Blackwell).

KERR, C.: The Uses of the University (Fifth Edition). Cambridge, MA, London 2001 (Harvard University Press).

KUHN, T.S.: The Structure of Scientific Revolutions. Chicago and London 1972 (University of Chicago Press).

MCMULLEN, M./MAUCH, J.E./DONNORUMMO, B.: The Emerging Markets and Higher Education. New York and London 2000 (RoutledgeFalmer).

MURDOCK, G.P.: Kultūros prigimtis. In: Kultūros prigimtis (Kultūros fenomenas). Vilnius 1993 (Valstybinis leidybos centras), pp. 167-212.

ORTEGA, Y./GASSET, J.: Universiteto misija. In: Mūsų laikų tema ir kitos esė. Vilnius 1999 (ALK/ Vaga), pp. 98-147.

POŠKIENĖ, A.: Universiteto organizacijos kultūra – kompleksinis aukštojo mokslo edukacinis veiksnys. Daktaro disertacija. Kaunas 1998 (Kauno Technologijos universitetas).

READINGS, B.: The University in Ruins. Cambridge, MA 1996 (Harward University Press).

ROSER, P.I./GLAZER, P.M./GLAZER, M.: Sociology: Understanding Society. New Jersey, Englewood Cliffs 1993 (Prentice-Hall).

SCHULLER, T.: The Future of Higher Education. 1991 (The Society for Research into Higher Edu-cation and Open University Press).

SHUMAR, W.: College for Sale. A Critique of the Commodification of Higher Education (Know-ledge, Identity and School Life, Volume 3). New York and London 1997 (RoutledgeFalmer).

ŠALKAUSKIS, S.: Kultūros filosofija. Raštai, 1 t. Vilnius 1990 (Mintis), pp.159-342.

TJELDVOLL, A.: Preparing service university research. In: Studies in Comparative and International Education, 5. Oslo 2000 (University of Oslo, Institute for Educational Research).

Palmira Jucevičiene, Violeta Karenauskaitė

The Role of University in Liberalization of Society in the Countries in-Transition

The spirit of democracy cannot be superimposed from outside. It must come from within.
M.Gandhi

The theory of liberalism claims that every individual should have a right to absolute freedom, which is limited only by freedom of other individuals. In joint activities of individuals, the mind and the consciousness of the moral responsibility have to create conditions for a new form of spiritual, political and economic life. In all historic periods, the ideas of liberalism were popular in many societies and communities; however attempts to combine the approaches 'liberation from what' and 'liberation towards what' inspired discussions and highlighted the problems of liberalism.

In our times, societies in liberal democratic countries face a number of problems which are caused not only by the above-mentioned approaches, but also by the economic and globalization processes, rapid technological development, market liberalization, environmental problems, etc. Along with these processes, changes in the qualities of the contemporary society are taking place: the qualities of the information society and the conflicts related to them prompt the development of the knowledge society, which in turn causes the inevitability of the development of the learning society, since societal changes normally cause educational developments in society. These processes are even more complex and the above-mentioned problems are even deeper in the countries in-transition, which lack democracy, free market and traditions of liberal education.

In liberal democratic countries, the mission of society liberalization is assigned to education, especially to universities, which are relatively independent and emancipate individuals and society. The academic community of universities is at the centre of the intellectual culture and criticism of society. It shapes and promotes the values, which are important for freedom of individuals and society, in particular those related to knowledge, search for truth, rationality, culture of dialogue, positive criticism and continuing personality development. In this way, universities through their educational impact on individuals influence their emancipation and ensure society development and liberalization. Besides, stimulation of the social progress and critical evaluation of the social phenomena makes universities not only a source of intellectual development of the society (a traditional view), but also its engine (a modern perspective).

In the societies in-transition, universities should also perform this liberalizing mission; helping these societies pursue democratic ideals and liberal education of their citizens because, in view of many scholars, intelligent and educated society members have always been the foundation of democracy (R. Barnett, 1990; L. Kraujutaitytė, 1998; et al.). Regrettably, in post-totalitarian countries, higher education institutions, including universities, lack democratic features and values, and emancipation of the society members is a slow and difficult process. Besides, it should not be forgotten that a value set which dominated in the totalitarian regime societies (e.g., collectivism, uniform (often forced) opinions on various issues) does not match the values of democratic society (e.g., freedom of discussion and views, personal autonomy, etc.).

Therefore it is important to try to answer the question, which is of interest for societies in-transition: *how can universities contribute to liberalization of societies in these countries in the complex and problematic context of the world?* The paper is aimed at answering this question.

Universities not only create knowledge, but also educate individuals with broad thinking, who apply this knowledge for the benefit of society, and disseminate this knowledge among its members. Therefore it is important to seek understanding of mutual necessity and trust between universities and society, in this way ensuring their equal partnership and collaboration. Only then universities will have a strong impact on society, promoting dissemination of liberal ideas and democratic values.

Many scholars discussing the issues of democracy in higher education and liberal higher education (R. Barnett, J. Dewey, R. Smith, P. Jucevičienė, L. Kraujutaitytė and others) relate liberal education with the idea of emancipation, arguing that individuals are liberated through learning. It is believed that by getting used to constant observation of the social and psychological barriers restraining their freedom, individuals will be able to actualize their intellectual freedom. In our view, it is purposefully attempted to turn the studies into the emancipation process, which helps individuals to understand their own internal (psychological) and external (social) barriers limiting their intellectual independence. Through dynamic combination of the curriculum, educational forms and methods, universities can empower emancipation of the students (who in the context of the learning paradigm include both members of the academic community and other members of society).

On the other hand, liberation of human thinking requires continuous learning. Universities therefore have to provide opportunities of continuing education for all society members and to take active part in the development of the 'learning society'. Only then they will become a source and an engine of the intellectual development in pursuit of the new qualities of the society.

Agreeing with the view that during transition from authoritarian to democratic social order universities can influence society through empowerment of learning, we will discuss the social and technological aspects of this idea, as well as what and how universities should do seeking the qualities of the learning society.

Since the notion of the 'learning society' has different interpretations, in the first section we will review the development of notion the 'learning society' and its interpretations. In the second section, we will discuss the social and technological aspects of learning. The last section is devoted to discussing the role of social sciences/ humanities and natural/technical sciences in society transition to liberal democracy.

The development of the idea of the learning society

Many authors (R. Barnett, 1990; R. Smith, 2000, L. Kraujutaitytė, 1998, etc.) who analyzed relations between democracy and higher education agree that *the key task of education at any level is the development of society and democracy*, whereas intelligent and educated members of society are a guarantee of democracy. These people, improving their life, improve the life of the entire society. According to P. Freire (2000), education is a key tool enabling the subjects to actively perform in today's world of increasingly rapid and complex changes, where all society members need to be able to solve complex problems. This in turn constantly requires new knowledge and skills. This led to the establishment of the *learning paradigm* which emphasizes permanent learning by society members (M. Ahlberg, 1999). The EU (most of the former countries in-transition, including Lithuania, are seeking the EU membership) emphasize the development of democratic values, multiculturalism, information and communication technologies which help to support life-long learning – the priorities ensuring the development of the 'learning society' (White Paper on Education and Training, 1995; UNESCO, 1998). This development is based on the concept of life-long learning that according to Chapman (1996), encompasses learning not only for personal development and satisfaction, but also for understanding democracy and acting in democratic ways. This is an especially important issue for post-soviet countries, which are still in transition to a democratic learning society and liberal education.

Different authors define such society by emphasizing different aspects. Some of them point out the inevitability of the 'learning society', whereas others criticize this idea.

The theoretical basis of the notion of the 'learning society' has been developed by D. Schon, who called companies, various social movements and management systems 'learning systems' and explained how they can be enhanced (quoted in M.K. Smith, 2002). Later together with C. Argyris, D. Schon developed the idea of organizational learning and suggested the notion of the 'learning organization'. R.M. Hutchins, who related the notion of the 'learning society' to 'life-long learning', pointed out that the speed of changes in society and the increasing leisure time of society members caused the necessity of continuing education. Speaking about the necessity of the learning society, T. Husén integrates these categories emphasizing the 'lifelong learning' process caused by increasing volumes of knowledge and information and the learning of the education system itself using information technologies. Recently the 'learning society' has been interpreted as a virtual global learning community. However, in this case it remains unclear, what citizens of such society have in common, where their roots are. R. Boshier has developed an integrated model of individual's lifelong learning. Also, he has noted the opportunities for democracy in the learning society. R. Edwards (quoted in M.K. Smith, 2002) relates the 'learning society' with liberal democratic society and market economy: 'The normative goals of a liberal democratic society – an educated society – and an economically competitive society – a learning market – are displaced by a conception of participation in learning as an activity in and through which individuals and groups pursue their heterogeneous goals'.

However, other researchers, e.g. Hughes and Tight, Strain and Field (1998; quoted in M.K. Smith, 2002) are critical towards the idea of the 'learning society'. They believe that it is one more contemporary myth of education. Its functions are more related to the interests of capital, power and professions instead of those of the society. According to Strain and Field, 'there is 'out there' a real society in which knowledge and other resources are unequally distributed, to a degree that is not only inimical to the fulfillment of individual capabilities and freedoms, but, arguably, detrimental to the collective survival and development of human society" (quoted in M.K. Smith, 2002). According to R. Smith (2000, p. 188), 'the widespread enthusiasm for the development of a 'learning society' is an especially potent source of confusion here', because the notions of 'lifelong learning' and 'learning society' are frequently confused. Also, this author notes the danger of putting individualism to the first place and incorrect interpretations of this concept as well as the controversy between individualism and 'learning society': 'What could seem more communitarian than a 'learning society'?' I will discuss this problem in more detail in the second section.

S. Ranson, who points out strengths and weaknesses of the idea of the 'learning society' and sees its potential, contradicts these authors:

> 'The notion of the learning society may have some theoretical and analytical potential – but it does require considerable work if that potential is to be realized".

Unfortunately, since recently in the world over, in Lithuania too, a greater emphasis has been placed on the economic motives of learning. However, according to R. Smith (2000, p. 19), we should avoid idealizing the economic purposes of the 'learning society', which are often prioritized in the countries of the European Union (e.g., UK) and the developed countries. This author points out the pressure of business on research (quoted by Coffield who notes that the economic aspects of learning are often given priority over democratic ones, forgetting the genuine aims of education). In this case universities should be careful and critical. In the post-soviet countries, e.g. Lithuania, 'learning society' emerges from education (first of all, universities), contrary to Western countries, where this idea has emerged from the business world. We believe that this has a positive impact on the former countries because in this way universities can influence and stimulate permanent learning in society and at the same time - democratic processes.

To summarize our discussion of the interpretations of learning society, we would like to point certain phenomena which are emphasized in the contemporary research literature and come to dominate in real life. These phenomena mark an essentially new stage in the humankind development, which is related to the necessity and inevitability of such society. Also, we would like to express our own view on the necessity/ inevitability of the learning society.

Firstly, it is related to the *information society* with the following characteristics: the value of information has been understood and increasing; not only the value, but also the volume of information increases, which is related to the development of information technology providing individuals with better and more organized access to information.

However, the 'reverse side of the medal' is that the information overflies people who are capable of working with modern IT can, which is not easy to cope with. This can cause a significant discomfort, which can even be regarded as a conflict brought on by the information society. However, individuals themselves are dealing with this conflict because information society is related to one more phenomenon, which is turning into a strong economic motive for individuals to attempt to cope with huge volumes of information.

The qualities of the information society cause emergence of the new economic phenomenon – knowledge economy, and in turn, *knowledge society*. One of the key features of the knowledge society is that in terms of the key assets, physical and financial capital is replaced by specialized knowledge, which enables to develop increasingly sophisticated products. Therefore professionals possessing specialized competencies are particularly wanted by organizations. In turn, it is in the interest of these professionals to be able to cope with information oversupply and to manage it in order to gain this specialized knowledge and competencies, i.e. to 'take into their hands' (or rather their heads) this key asset of the knowledge economy. Obviously, demand for such competencies requires the knowledge society to have a modern education system.

However, it is not enough because in such a society information is growing at unprecedented speed, which gives continuous opportunities for improving products and their production, at the same time increasing the requirements for human competence. This dynamics again causes tension for individuals: a respected and well paid today's professional tomorrow can realize that his or her competence does not match new requirements. However, continuous demands to develop one's competence can become an unbearable burden to some people. This problem can be seen as a conflict caused by the knowledge society. Can it be solved in the same way as the conflict caused by the information society is solved by the motivation power of the knowledge society?

Let's try to find a logical solution. If professionals working in the knowledge society have to develop their competencies continuously, the only solution is continuous learning. It is not enough to learn within the education system, no matter whether it is formal or informal. What is needed is learning from experience, learning from each other at work and outside work. In this context, efforts of individual people become insufficient – a certain social empowerment is required. In other words, the conflict caused by the knowledge society should be addressed by the new efforts at the society level. We are talking here about the qualities of the learning society, i.e. efforts organized and implemented at the society level to create empowering learning environments and to change the education paradigm from *teaching* to *learning* going on life long and everywhere, in all domains of human life.

The important characteristics of the learning society are individuals capable of learning always and everywhere, learning organizations, and learning-in-partnership networks incorporating both individuals and organizations, such as learning regions, learning cities, learning villages, etc. However, among the many, the two key characteristics are: (i) learning individuals capable of managing their learning, and (ii) learning-in-partnership networks. It is unimaginable that these two phenomena be possible in other than liberal society with the highest degree of democracy. In other words, *development of the learning society is directly related to the liberalization of society.*

As noted by N. Longworth (1999), there will be no other social classes in the future than people who are learning fast enough and those who are not. In this case, the learning society means that everyone or at least the majority is capable of learning fast enough. Does not this mass learning sound like a utopia? It may seem so, however since knowledge economy needs increasing amounts of 'knowledge assets', it can be expected that knowledge organizations will be that economic and social power which will promote and make the development of the learning society real. However, this power alone is not capable of developing the required competencies, it can serve only as a stimulus creating the conditions for other forces with necessary competencies for developing the learning society. Such force is a contemporary university, which is capable of being not only a source, but also an engine of the society development.

One could point out many roles of university, which are important in developing and liberalizing the learning society. However, in this paper we will emphasize one of the key roles of university through which it not only stimulates contemporary human learning and develops a competence of freely creating one's knowledge through interaction with the environment, but also develops individuals' responsibility for their impact on the surrounding environment. We base this argument on the ideas of A. Paškus (1984) and T.H. McLaughlin (1997) that individual's (and, in our view, society's) liberation is a process involving two components – liberation 'from what' and 'towards what'. In case of the learning society, this involves liberation from the necessity to get bound by various legal, economic and other obligations to get access to the educational services: the learning society guarantees them as a matter of fact through empowerment of self-directed learning and learning-in-partnership networks. However, at the same time individuals have to get liberated 'towards something' – towards responsibility for their actions and for acquired and used competence.

Thus, in our paper we will analyze the role performed by university not only in emancipating individuals through empowering them to learn in modern ways and to acquire the competencies required by the environment, but also in developing their responsibility for this environment.

For this reason, we will begin with discussing the social and technological aspects of the learning society, the harmony which is crucial for ensuring continuous learning by the members of the society and their responsibility for their actions.

Social and technological aspects of the learning society

In theoretical interpretations of the notion of the learning society one can clearly distinguish a social aspect (D. Schon, R.M. Hutchins, R. Boshier, R. Edwards, R. Barnet, N. Longworth and others) a technological aspect (T. Husén, A.W. Bate, R. Waterhouse).

The social aspect of the learning society (aimed at ***what***?) comes to light when learning is seen as a process during which people develop their knowledge, understanding, skills, values, attitudes and experiences. This approach emphasizes ability to reflect, listening to other people's opinion, having an independent own opinion, problem solving skills, collaboration, responsibility for one's ideas and opinions, ability to take care of others, active participation in one's community (e.g. academic, urban, etc.), perception of outcomes and results of one's work/study, etc. It should be noted however that some authors *put individual in the first place in the learning process, whereas others emphasize the idea of collective learning.*

According to a number of scholars and education policy-makers, education should be aimed at the development of human potential through the process of continuous support for individual. This process stimulates and empowers an individual through his life to acquire necessary knowledge, skills, values and understanding and to apply them with confidence in all roles of his/ her life. According to Longworth (1999), *individual is 'above everything';* he/she is an active subject whose unfolding requires creating certain conditions. He/she has to know how to change and improve these conditions instead of taking everything passively as a heritage of former generations, to become an emancipated citizen. For this reason, students are at the center of the contemporary education process, while teachers are creators and disseminators of knowledge, change agents in society. As noted by Fullan (1998, p. 30-31), each of them should first of all have a personal goal, only then the way to organizational changes will open. When personal goal disappears, group thinking and superficial, momentous innovations replace it. When many people have personal goals, they gain the capacity to initiate key changes at the collective level. A the same time, the author points out the danger of excessive collaboration (opposing the idea of collective learning) which gradually turns into 'group thinking' – uncritical obedience to the group, unreflective acceptance of the new solutions. He notes that while aspiring for closer cooperation we should not forget the 'good' sides of individualism since the freshest ideas are brought by the group diversity and by those who think in the most original ways (Fullan, p.56-57). This is so because groups much more rapidly than individuals respond to the short-term trends in any area. The tendencies of individualism are reflected in the documents of the European Commission: 'Individuals become the principal constructor of their own abilities' (European Commission, quoted in R. Smith, 2000, p. 192), and UNESCO: '... the right of all people to education and the right of access to higher education based on individual merit and capacity' (UNESCO, p. 28).

However, R. Smith (2000, p. 192) points out the harm of individualization in education, raising the question: 'So this is the meaning of the 'learning society': it is a society in which discrete and disaggregated bits of knowledge and skills are put together by individuals for specific employment purposes? From the just and culturally rich society we move easily to the idea which is based on the atomistic separation of individuals in order to measure and grade them.' He points out the danger of the increasing individualism and of prioritizing the economic aspects for the sophisticated notion of democracy.

However, the aim of higher education is not limited to developing the intellectual abilities; it includes transmission of overall culture and standards to all citizens. This makes the critique of liberalism justified which promotes individualism based on priority of one's own interests over creation of more harmonious life taking into account the others (T. McLaughlin, 1997, p. 45). The 'communitarian' school of liberalism expresses this critique.

For example, the *social theory of learning* postulates human knowledge and understanding as an integral process of the historical, cultural, and social context. This process is continuously nourished by collective spirit, communitarianism and collaboration. Proponents of this theory correctly point out that socialization of individuals is taking place in the multicultural environment, which is characterized by mutual dependence and diversity. Hence, successful learning can be ensured by conscious collective learning and collaboration (Freire), collective reflection. This theory regards *learning as a complex process* related by links between various individuals and socio-cultural forces. In the learning process, it emphasizes construction of collective meaning, interaction between individuals, reflection on the interaction between individual and social interests and other priorities (Harris, 1997). According to these authors, learning takes place through interaction, so they suggest placing greater emphasis on the inevitable education through communication and collaboration to help the students overcome interdisciplinary barriers and to enable them to further develop their knowledge.

Although the social theory of learning emphasizes collective learning, it in no way contrasts it with individual learning suggested by psychologists and proponents of cognitive science. This theory does not attempt to downgrade the importance of individual cognition; it supports a social approach to the 'learning society' and allows looking at this phenomenon from a different perspective.

While Western scholars often criticize individualization of learning and prioritization of economic aspects, individualization and differentiation in education are emphasized in post-soviet countries (more specifically, in Lithuania). This is so because in these countries society is tired of the recently forced collectivism, ideology of one political party, etc. I believe that in such countries as Lithuania, we should realize the weaknesses of these approaches and address them both in the education documents and the academic activity, following Y. Tamir's view: 'no individual can be context-free, however <...> everyone can be free in the context' (T. McLaughlin, 1997).

The technological aspect of learning addresses *how* to realize lifelong learning using modern educational and information technologies, how to support learning and collaboration and relations between students (Nuyen, 1995, cit. Smith, p.193; Waterhouse, 2000; White Paper on Education and Training, 1995, UNESCO, 1998). The technological approach places a special emphasis on the future of the virtual learning society and the necessity of educational and information technologies which would encourage students to look at the more distant environment and to find the ways of integrating into democratic society. It is frequently argued that EIT can be more effective than the traditional methods of knowledge transfer. This is an arguable proposition, which we will not address in this essay.

Modern EIT can help to develop the key skills of permanent, autonomous learning and to reach various sources of information. They can encourage communication and collaboration, expand the learning space outside university and involve new people into educational process. According to McCormick (1992), ability to use modern information and communication technologies and to get, process and disseminate information will determine citizens' possibilities to participate in the activities of social and political communities of various countries, to influence society development and to understand democratic principles and values.

However, in the post-totalitarian countries and those in-transition, where information society is in its first stages, there is a gap between the information-rich and information-poor individuals (or societies) and the priority is often given to 'universality' of IT instead of their 'accessibility'. Different possibilities of students to use IT and to acquire information skills may become a reason of social inequality and social polarization, and development of the 'learning society' should reduce these differences (White Paper on Education and Training, 1995, p. 30).

Another problem is that the existing information and communication technologies (ICT) in many cases have not been developed for educational and collaboration purposes. Besides, there is a lack of special ICT tools for high quality learning/ teaching (Bates, 1997). In most cases, all tools are developed for private business to meet the needs of global markets instead of supporting and facilitating a democratic dialogue in the educational process. These technologies are not ideal for citizens who are developing democratic society. Hence, members of the university academic community (those with relevant competence, both computer science specialists and people from other disciplines) could implement the technological aspect of the 'learning society' by contributing to the development of special educational programs and tools. Besides, to make educational and information technologies effectively serve the needs of all citizens, it is important to assess the quality of already existing technologies, to identify the gaps related to their use in educational processes and to encourage research in this area (Bates, 1997).

Summarizing it can be concluded that in the countries in-transition, including Lithuania, having realized the weaknesses of both individualistic and collectivist approaches, these approaches should be combined in the academic context, following Y. Tamir's view that no individual can be free of context, but ... everyone can be free within the context (T. McLaughlin, 1997). Besides, overemphasizing its certain aspects, namely, by separating the social and technical aspects, which would raise controversies and tension; therefore the elements uniting these aspects, would impede successful development of the learning society and reflecting the idea of learning should be sought. Also, ways should be searched for in which universities could ensure the liberalization of single individuals and at the same time of the entire society. We will discuss this question in the third section of this paper.

The role and problems of humanities/ social sciences and natural/technical sciences in relation to liberalization of society

It is obvious that the social aspect of the 'learning society' can be realized through humanities/ social sciences, whereas the technological aspect can be implemented through natural/ technical sciences. Frequently these two groups of sciences and their role in society are regarded from opposite perspectives (I. Weijers, 2000, p.121), arguing which sciences play greater role in society liberalization.

Democratic education is aimed at educating independent, constructively critical, active citizens who are well familiar with democratic principles and practices (L. Kraujutaitytė, 1998, p. 82). In liberal democratic societies, particular attention is devoted to liberal education as a factor ensuring the development of qualities, which enable individuals to better adapt to the changing environment.

On the contrary, in authoritarian societies the key subjects of the curriculum were related with the single-minded ideology. Disciplines that had the greatest impact on the development of students' worldview (philosophy, history, literature, etc.) were based on the dialectic materialism and any criticism of this ideology was suppressed. Such system was based on the clearly controversial theory of good and was contrary to the education systems of democratic countries (illustrated by T. McLauhglin, 2000, p. 14). Obviously, universities and in turn humanities and social sciences at that time could not serve as sources of democratic values and liberal ideas.

Traditionally, in liberal democratic societies, *universities influence societies and democratic values are disseminated primarily through humanities and social sciences.*

According to Gudaityte (2001), humanization of higher education is important as a precondition for the development of democratic society. The ideal of democracy is a free individual in a free society. Only in this case, optimal interaction between individual, society and its certain groups is possible. Humanities constitute the core of the curriculum of liberal higher education and have the biggest impact on individual's emancipation in the process of study. This idea is supported by K. Williams (2000, p. 64):

> 'Literature, however, is perhaps the most potent avenue to a liberal civic education. More than any other area of the curriculum, literature can help us to learn that 'every other person is basically you''.

Social sciences in turn enable one to deepen knowledge about social processes and their changes. An exceptional role in the integration of disciplines is played by philosophy and sociology, as these disciplines can effectively reveal the general meanings of other disciplines: philosophy constantly reminds about the meaning of existence and cognition, whereas sociology is an important tool for learning about social interests and ideologies (Barnett, 1990). Besides, these sciences can be used to teach students democracy at theoretical level, as noted by L. Kraujutaitytė (1998, p. 82):

> 'The will for communication reflected in the humanities and social sciences shapes the direction in which communication requires to emphasize not as much the results of the activity, as collaboration, tolerance, mutual understanding, care and multisided and equivalent impact on decisions, when the ideals and principles of collective life are purposefully coordinated, not imposed'.

In social sciences and humanities, these twelve years of Independence, which followed after fifty ideologized years, are a period of great transformation, liberation from the centralized monolithic system, integration into the global academic community and education of the young researchers free from the Soviet ideology. Hence, it is important what approach is dominating towards the role of social sciences and humanities as compared to other sciences, and whether it is recognized in post-totalitarian societies that they should be given bigger attention. Positive answers to these and similar questions would serve as evidence that society with no democratic traditions is really facing the democratization process: first of all, the trends of social evolution are analyzed and the opportunities of optimal changes are searched for (Gudaitytė, 2001, p. 30). Failure to admit the priority of humanities and social sciences in higher education leads to depriving it of opportunity of liberalizing the society and ensuring its intellectual emancipation.

Can the natural/technical sciences play a role in liberalization of the society? As it run paradoxically, exactly the key features of liberal education (rationality, critical thinking, analytic skills, independence) at universities were realized in the curriculum of natural and technical sciences. This was so because these sciences had relatively close relations in their areas with the academic community of the rest of the world. Besides, these sciences by their nature are further from ideology, so transmitting some democratic values within their disciplines was much easier than in the ideologized humanitarian and social sciences of that period. Therefore it can be argued that *in the totalitarian system natural sciences in a way served as disseminators of the ideas of liberal education and as an engine of society development*. So in the end these sciences had limited possibilities for liberal education, and they depended on teachers' flexibility and ideological position.

Today in Lithuania and in other countries in-transition, natural and technical sciences are more dependent on research infrastructure, equipment and materials, so since lately research in this area has been focused on the market. While seeking greater funding, the social purposes of education are often forgotten. From the technological perspective, the role of these sciences in the society development cannot be denied; they can ensure the realization of the technological aspect of the 'learning society'.

Knowledge in the field of natural and technical sciences is related to modern technologies, health care, ecology and other areas important for society. These sciences are crucial for understanding the surrounding world; they enable technological progress and the development of other sciences as well as their integration for economic development. They help to develop research skills and thinking, i.e. make significant contribution to the development of information and knowledge society. The significance of natural and technical sciences and the ability of human beings to use them to affect the environment is increasing, in turn increasing the threat to the environment and to human beings themselves. Therefore, these developments should be accompanied by the increasing responsibility of people using them. How can this be achieved?

University has a mission to develop and realize its curriculum in a certain way determined by the philosophical approach to knowledge and educational theories of the natural sciences.

This *philosophical approach* invites to look at the knowledge of natural and technical sciences from the social perspective, because these sciences must help humankind lead a better life instead of leading to self-destruction. Thus, contemporary university should not only provide knowledge of the technological disciplines, but also show the importance of knowledge in the context of interaction 'science – technology – environment – society' (Zoller, 2000). Only then students will understand their relationship with the environment and will seek harmony between economic, technological, ecological and social development in their activities. In this way, integration of knowledge and values in pursuit of valuable knowing will be achieved (M. Ahlberg, P. Dilon, 1999). Forgetting these goals may lead to educating the individuals with no sense of responsibility for just and safe use of knowledge in their activities.

An educational theory based on constructivism argues that learning is a constructive activity pursued by an individual; it cannot be transmitted. Every individual who is involved in learning, by continuously rethinking his/ her knowledge already possessed at the theoretical level (not passively or actively absorbed) and testing it in practice, constructs his/her own theory and develops his/her competence, integrating various concepts into a whole which is significant for him or her. This theory emphasizes developing not only knowledge, but also general and specialized skills, and relates learning to values.

Problems caused today by science and technology development are related to *environment* (climate changes, ozone holes, diminishing resources of fresh water, etc.) and *quality* (caused by global competition). This causes increasing demand for universities to combine in various disciplines intellectual development (education leading to understanding) and social development (learning through collaboration, learning to work in groups, etc.) because today's world needs professionals with educated mind and transferable skills, which enable to solve these complex problems (I. Weijers, 2000, p. 125). At the same time, today one of the key tasks of university is to bring learning closer to the real life situations and to apply the principle of program integration, enabling to cross the boundaries of the specific context and time period (A. Markevičienė, 2001).

So, how can university realize the above-mentioned mission, address the problems and meet the new challenges in terms of both social and technical aspects of the learning society?

In our view, this can be done through socialization of the curriculum of natural and technological sciences (not undermining the importance of the discipline-specific knowledge) and by empowering learning of the academic community through creation of educational learning-intensive environments at university.

By socialization of the curriculum and teaching/ learning methods of the natural and technical sciences we mean emphasizing the topics and social aspects of the discipline which are important for society development; showing the applied side of these sciences in the specific area of study; emphasizing the teaching/ learning methods (interdisciplinary projects, problem-based methods, etc.) which help to develop the skills of constructive learning through collaboration. *In this way, natural/ technological sciences can help to realize the social aspect of the learning society* (***aimed at what***).

For example, medical students should be familiarized not only with the use of ionizing radiation for biovisualisation, but also with its impact on human body, possible harm, etc. To explain the students the problems of pollution, they should get involved into interdisciplinary projects integrating knowledge of various sciences (physics, chemistry, biology, etc.) and stimulating learning through collaboration. These are only a few examples of many possible approaches.

Empowerment of learning by members of the academic community (which constitutes not an insignificant part of society) means providing support in terms of curriculum, methods and *rich educational environments* in order to ensure the maximum recognition of such environments and their transformation into student's individual learning environment. A broad range of knowledge, skills and talents of the university academic community should be mobilized to serve the students, at the same time showing the value and attractiveness of learning. In pursuit of these goals, university itself has to become a learning organization, capable of supporting learning at all organizational levels and among all its members, independently of their position and educational level. Only then university will be able to develop students' skills and competencies crucial in the learning society: meta-learning, communication, information management, problem-solving, etc., and enable the students to transfer them to the rest of the society.

Rich and effective learning environments can be created by using information and communication technologies, i.e. by realizing the technical aspect of the learning society (how?), of course, keeping in mind the problems discussed in the second section of this paper and looking for their solution.

In our view, socialization of natural/technical sciences and creation of effective educational environments in this area stimulates a motivated purposeful learning and development of the learning skills, which leads to individuals' emancipation and at the same time to liberalization of the society.

Summarizing it can be argued, that the social aspect of the 'learning society' can be realized not only through humanities/social sciences, but through natural/technical sciences too.

So, the creation of the 'learning society' is necessitated seeking tolerance and respect for different views and thinking in the academic community and to transfer these values to society, for representatives of these sciences to engage in an open and impartial discussion instead of confronting each other. Both humanities and social sciences have their purpose, therefore the role of neither of them in the academic community and social development should be overemphasized. I agree with P. Smeyers (2000, p. 116) that *university education should first of all raise questions what is going wrong in society, while* ***inside every discipline it should bring clarity about what is the most valuable***. Societies on the way to liberal democracy should note a suggestion by Barnett (1990) and McLaughlin (1997) not to limit oneself with narrow overspecialized curriculum and textbooks, which provide a one-sided view of various controversial issues. Secondly, since open and impartial discussion is one of the key democratic values, it should be used in different forms and methods of study (and in various disciplines) because only in this way students will be encouraged to admit the complexity of diversity and its continuously controversial character. According to Barnett (1990), university programs should encourage independent study and problem solving, stimulate students to formulate and emphasize the reasons of their actions, give them possibilities to work in groups and involve them in such activities in which socially significant problems of morality and competence are solved.

Also, universities could enhance their role in society through liberalization/ democratization of the study process: enabling students to make free choices, equality-based relations between members of the academic community, use of active teaching methods, and competence-based study programs.

Conclusions:

- The learning society is an inevitable necessity, a realistic prospect of society development, and a condition of successful operation of the knowledge economy. It is caused by the internal conflicts of the information and knowledge societies and by society's constant pursuit of new qualities.
- Development of the learning society is directly related to liberalization, because the key characteristics of the learning society can be realized only in the liberal society. These characteristics are: learning individuals capable of managing their learning and learning-in-partnership networks.
- University is a source and an engine of the social development and a force enabling the development of the learning society in the countries in-transition. It can perform this role by emancipating individuals through enabling them to learn in modern ways and developing their responsibility for the surrounding environment.
- The technical aspect of the learning society (how?) can be realized through natural/ technical sciences, by using information and communication technologies and by creating rich and effective learning environments at university.
- The social aspect of the learning society (aimed at what?) can be realized through humanities/ social sciences, educational technologies and socialization of the curriculum and methods of the natural/technological sciences, the latter (socialization) being the precondition of the quality of life in the knowledge economy.
- In pursuit of the qualities of the learning society, universities have to combine the social and technical aspects of the learning society because their separation causes contradictions, problems and tension. Implementation of these aspects, emancipation of the academic community and liberalization of the society should be based on the harmonious balance between social sciences/ humanities and natural/ technical sciences.

Reference

Ahlberg, M./Dillon, P.: Materials for Constructive Environmental education, mainly to be used in pre-service and in-service teacher education: An integrating approach to environmental learning. In: De Paz, M./Pilo, M. (eds.): European Project for Environmental Education. Genova 1999, pp. 3-33.

Barnett, R.: The Idea of Higher Education. London 1990 (SRHE and OU Press).

Bates, A.W.: Restructuring the University for technological Change. London 1997 (The University of British Columbia). [cit. 19 January 2004]Available from: <http://bates.cstudies.ubc.ca/carnegie/carnegie.html>.

Freire, P.: Education for Critical Consciousness. Vilnius 2000 (Tyto alba). (In Lithuanian).

Fullan, M.: Change Forces. Vilnius 1998 (Tyto alba). (in Lithuanian).

Gudaitytė, D.: Methodological Aspects of Research on the Transition from the Elite to Mass Higher Education. Journal Social sciences 2(2001)28, pp. 28-38. (in Lithuanian).

Harris, E.: Community learning: Not relevant to research and teaching? Canadian Journal for the Study of Adult Education, 11(1997)1.

Higher Education in the Twenty-first Century. Vision and Action. Paris 1998 (Unesco).

Jucevicienė, P.: Higher Education in Lithuania. Quis es? Quo vadis? In: Studies in Higher School. Kaunas 1997 (Technologija), pp. 9-31. (in Lithuanian).

Kraujutaitytė, L.: Democracy of Higher Education: the worldwide experience and situation in Lithuania. Doctoral thesis. Kaunas 1998 (VDU). (in Lithuanian).

McCormick, R.: Curriculum Development and New Information Technology. Journal of Information Technology for Teacher Education 1(1992)1, pp. 23-49.

McLaughlin, T.H.: Contemporary Philosophy of Education: Democracy, Values, Diversity. Kaunas 1997 (Technologija). (in Lithuanian).

McLaughlin T.H.: The European Dimension of Higher Education: neglected claims and concepts. In: Crawley, F. and others (eds.): Universities Remembering Europe. New York, Oxford 2000 (Berghahn Books).

Paškus, A.: Individual and liberty. Chicago 1984 (Ateitis). (in Lithuanian).

Smith, M.K.: The Theory and Rhetoric of the Learning Society. (2002) http://www.infed.org/lifelonglearning/b-lrsoc.ht

Smith, R.: Democratic Education and the learning Society. In: Crawley, F. and others (Eds.): Universities Remembering Europe. New York, Oxford 2000 (Berghahn Books).

Waterhouse, R.: The Distributed University. In: Scott P. (ed.): Higher Education Re-formed. London, New York 2000 (Falmer Press).

Weijers, I.: Bildung Versus Benefit? In: Crawley, F. and others (eds. Universities Remembering Europe. New York, Oxford 2000 (Berghahn Books).

White paper on Education and Training. Teaching and Learning: Towards the Learning Society. 1996

Williams, K.: Realism, Rationalism, and the European Project. In: Crawley, F. and others (eds. Universities Remembering Europe. New York, Oxford 2000 (Berghahn Books).

WILLIAMS, S./GRAY, P.: A Comparative Analysis of Education Reform in Lithuania. 27th conference "Learner-Centered Universities for New Millennium: Education Reform and Teacher Training". Vilnius 2002 (VPU), pp. 474-479.

ZOLLER, U.: Interdisciplinary Systemic HOCS Devlopment – The Key for Meaningful STES Oriented Chemical Education. Chemistry Education: research and practice in Europe, 1(2000)2, pp. 189-200.

Nijolė Zinkevičienė, Brigita Janiūnaitė

Conservatism and Innovativeness as University Values in the Aspect of the Knowledge Society

The world nowadays is constantly changing because of the technical and social progress; increasing competition, strive for democracy and freedom. Change has become a usual phenomenon in most life spheres, but its speed is so rapid that it stresses not only an individual, but also organizations and even systems.

Toffler (1972) says that in order to avoid a shock of future, a person should become more flexible that earlier. He will have to find completely new ways of preservation of vitality, as former religious, national, family, etc. background is no longer reliable to outstay the constantly increasing speed of change. Not long ago stability and durability were the ideal qualities. Whatever a person created – a pair of shoes or a church – he tried to make his product last as long as possible. Now our life is based on the motto "use – throw away". The reality of life encourages us to hurry, to keep up with the others, to look for our own place and to fight for it. Sometimes these aims upstage the spiritual spring of an individual. In such a life-time hurry there is no time to realize whether that for what we are making so hard physical and intellectual efforts, is moral and worthy.

Education as a system faces the same problems of permanent changes. But the relation between education and change is even more complicated because *education is one of the most conservative social institutions.* It is conservative by its own origin (Simon, 1989; Fullan, 1993). According to relationship with traditions education is sometimes seen as similar to the Church (Grambs and McLure, 1964). Though, as Handy (1994) states, in this age of paradoxes even these institutions must change, as they are affected by strong power.

Universities, as specific organizations with their unique heritage (features of the Antique and Middle Ages university, the works and thoughts by the great scholars) and former most stable institutions also meet the challenges of the knowledge society. In their context a controversial problem appears – to change or to keep status quo. The answer is not to be easily given, as the tendency of the massive higher education causes a conflict of values – a confrontation of the new and the old, the conservative and the innovative. It is no longer possible for a modern university to remain traditional. It is influenced by lots of institutions, global processes, and constantly changing information. The postmodern vision – service university – is trying to subjugate this educational institution to the common free market regulations.

Universities, according to Kumar (1997), having kept a distance from the society for long ages, nowadays are thrown into the fuss of every-day political and economical life. Academic people have become the governmental and business advisers, scholars and researchers have joined large projects. Universities especially by their scientific research functions have become the major figures vouchsafing the further success of developed industrial societies

In this context there arises a problematic question - *Is conservatism a totally negative phenomenon in higher education? Is the model of the traditional university to be considered as old-fashioned and should it make way to the idea of a service university? What is the balance between conservatism and innovativeness in higher education, if possible?*

The aim of this article is to analyze conservatism and innovativeness as university values, the peculiarities and problems of their interface in the conditions of constant change.

The article consists of four sections. The first section deals with the semantic meaning of the relation between conservatism and innovativeness. In the second section conservatism and innovativeness are discussed as the values of higher education. The third part analyzes the peculiarities and problems of the interface of these two features in higher education in the conditions of constant changes. The fourth part touches upon how it is possible to obtain the balance between conservatism and innovativeness in higher education.

1. The relationship between the concepts of conservatism and innovativeness

Conservatism, according to Webster's Third New International Dictionary (1993), is a political ideology based on the sense of strong tradition and social stability, emphasizing the importance of the established institution (religion, property, family or class structure) and giving priority to gradual development and preserving the best elements from the past against unexpected change; it is a tendency to accept an existing fact, order, situation or phenomenon and to be careful towards change; strong resistance to innovation.

The term "conservatism" has derived from the Latin word *conservare,* which means "to preserve", "to save". It is widely used nowadays to define one of the political ideologies, which is the background of many political parties of the modern European countries. Conservatism in its primary sense has been known from old days as a negative attitude towards any change. But it became a political ideology only in the modern times.

According to the conservatives, society is more than a usual collection of casual people, their arithmetic sum. Society is rather a union of people's power, as while acting together and in an organized way, they can achieve more and better results that by individual activity. That is why the main value for conservatives is preserving stable human relationship and nurturance of stability and order. The stable order in the society might be changed very carefully and deliberately, as it might be easily damaged and would be hard to restore.

The conservatives declare that people are much happier when they feel not only free as personalities, but first of all the members of their family, religious community, their nation and state. A person can realize his real abilities, his relationship with other people and his own role only in a stable society. Thus the stable order that has been built for ages goes together with the caste distribution of people and a heritage of monarchal government, and it is not a misunderstanding or relic of the past. (G. Vitkus, 2003).

Traditionalism is a concept close to conservatism. According to the Dictionary of International Words, (1986), it is keeping to traditions, relying on them, conservatism. In this context certain connotations of conservatism – positive and negative – can be distinguished.

It is positive by its *succession* – a desire to carry out the dateless university mission, the essence of which is to transfer people‘s experience gained through the ages, to cherish the national culture, customs, traditions, heritage and be rational towards change. But in the conditions of the rapid changes of the modern world a negative connotation of conservatism can be distinguished. It is denial of rapid changes, negative attitude towards any new thing or phenomenon, morbid nostalgia of the past. From this point of view conservatism is related to depression, stagnation, insularity and dissociation.

Innovativeness in its common sense is a feature associated with "possession" of innovation, as an idea or a process, renewal.

The origin of the word *innovation* is Latin and means "renewal", "change" (The Dictionary of International Words, 1985, p. 344). The Oxford Dictionary of Advanced Learners (1995, p. 614) describes innovation as a process of changes, implementing various new ideas, methods etc. The International Encyclopedia of Social Sciences (1972, p. 339) describes innovation as a process during which new products or technologies are implemented into economic system. According to these definitions, it can be said that innovativeness can be understood as innovation – implementation of new ideas and methods. According to Toffler (1997), it is an ambition to create a certain deviation from the present situation by joining various different features in order to create an innovation. Barney and Griffin (1992) state that innovativeness is organizational efforts to create and develop a new product, service or possibilities of using present products. Vasiliauskas (2002), according to Lynch (1997), treats innovativeness as one of the organizational resources – ability to generate and implement ideas. It is usually related to big risk, as not all changes are successful. Podlasyj (1999) is under the opinion that in educational context innovativeness is not an invention, but a useful application of known ideas, implementation of progressive novelties, that enrich pedagogical experience.

Summarizing the review of the concepts of conservatism and innovativeness ir becomes evident that they are of oposite meanings. Conservatism in its general sense emphasizes keeping to traditions, experience and status quo, whereas innovativeness means changes, receptivity and efforts to change, cretae, developa and enrich.

2. Conservatism and innovativeness in the context of university values

The term "value" may have a very wide and very narrow meaning: a wish, pleasure, temptation, moral obligation, enjoyment, duty, ideal and many other options when something is preferred.

M.S.Bassis (1988, p. 70) states that one and the same value can be the background for several norms, and the same norm can be based on several values. A value becomes the standard of the desired and aspired, and this does not depend on the situation. These are wide, abstract and prevailing standards defining what is good, right and suitable. The normative meaning of values is clearly disclosed in J.P. Shaver and W. Strong's definition, where values are standards and principles determining worth , that is, they are criteria by which we evaluate things (people, objects, ideas, actions, situations), whether they are good, worth, desired, or – on the contrary- bad, undesirable, hated (1976, p. 15). This position allows to consider a value as a specific standard, which is the background for action or evaluation, but these authors do not indicate prevalence of values and how the standard is chosen.

In this article I keep to M. Rokeach's conception of values, presented by M.A. Smith. It is wider and more comprehensive.

A value is a long-term ***successive*** *belief that a particular principle of activity and existential ideal both in personal and public sense has a priority from the point of view of certain alternative and opposite activity principles and existential ideals* (M.A.Smith, 1977. p. 224).

Philosophical theory of liberalism distinguishes two other levels of values: *public and personal*, based on the "thin" and "thick" theories of good. Public values are considered to be fundamental that help people to live together in the society and reach their prosperity. (T.H. McLaughlin, 1997, p. 24). Public values are incorporated in legislation and cover the main democratic principles and ideals. These values are related to the "thin" theory of good that does not depend on different individual opinions and contradictions and are compulsory to all members of the society.

The other level of values – personal – is more complicated and diverse. These values are based on the "thick" theory of good, that cannot be applied in the whole society, as they are different and abundant, individual and group values, ideals and believes, by which personal lives are created.

The "thick" theory of good, however reasonable and comprehensive it might be, cannot determine public values.

T.H. McLaughlin (1997) thinks that sometimes it is difficult to outline the limits between public and personal values. In the conditions of such diversity, it becomes hard for public values to remain the fundamental background of the society, but it is thought that the more individual and group values cover each other the thicker public values become, as they are determined by the thickness of cover.

Some researchers (J.K. Schermerhom, J.G. Hunt, R.N. Osborn, 1994; W.G. Ouchi 1982), analyzing organizations, point out a specific sub-level of values, typical of different organizations and social groups. These values are identified with beliefs, norms and rules, followed by the members of an organization inside and outside and supported by the organization, in reaching its aims. Every organization may have its primary and secondary values, which determine its successful activity.

In this paper I follow the philosophical tradition of liberalism that distinguishes fundamental, public and personal values that are based on the "thin" and "thick" theories of good. All the three levels affect each other, as the center of everything is a human being.

Higher education as an educational phenomenon, affecting society and reacting to external changes, also has its values, which are important in analyzing any institution of higher education. The system of values is the background of any individual or organization, thus its determination and evaluation may allow better understanding of the processes and individual or organizational behavior in different situations (Jucevičienė, 2000).

Values in an institution of higher education are the keystone of its culture. There are different positions in trying to define the features and values of higher education. Some opinions are even cardinally opposite. J.M. Halstead (1995) represents one of the attitudes. He declares that educational system must remain neutral, that is, in fulfilling its mission and functions it must develop the values of pupils, students and not enforce those of its own. On the other hand, educational institutions and teachers working there influence young people by transferring knowledge, by involving them into common activity, by being popular. Society is that concerned subject which wishes educational system (making influence on an individual) to transfer public values. It is considered that higher education cannot remain neutral, where personal values of individuals would be developed elementally. Educational system and higher education in particular are closely connected with public values (T.H. McLaughlin, 1995).

R. Barnett (1990) describes values in the context of the features of higher education. Some features of higher education – scientific activity, study activity, parietal relationship, critical thinking, autonomy, academic freedom, openness to the society, scientific and qualification degrees – coincide with values. That is so because the idea of higher education (from the theoretical point of view) and the activity of the higher education community (from the practical point of view) are based on the same values.

In general, the background of higher education values is a set of aims, values and common ideas that is traditionally related to higher education and determined by the majority of the academic community. (R. Barnett, 1990). Researchers (T.H. Mc Laughlin, J.M. Halstead, R. Barnett, A. Gutmann and others) discuss the tangle of personal, public and specific values of higher education, that forms an entire background for education of an individual, i.e. higher education is as if obliged to an individual to educate him an integral personality (through the effective educational process at an institution of higher education and society) with the hope that the process of cherishment and transference of certain public values to other generations will be carried out in series, without any confrontation with the society. Specific values of higher education are treated as organizational values that distinguish higher education institutions among others, join the academic society members for creation of scientific knowledge and studies and have specific worth and the limits of validity. This set of values is not a set of rules – it is a certain philosophy, which is a fundamental background in reaching a convention in an institution of higher education.

Universities have always been the treasurers of the intellectual culture of society. It is a place where the young generation is obtaining the cultural heritage and traditional values. One of the main aims of the university has always been integration of an individual into the national culture, evoking the feeling of belonging to a certain social group and society. This aim is rather conservative by its content.

For a long time there was a chain-like relationship between higher education and society (Barnet, 1994), where higher education was in the position of an educator of the society elite and was not related to every-day needs of the society. In this relationship society is passive and does not influence the process of higher education knowledge creation. From this point of view the university is considered to be the source of the society development and the creator of the knowledge that is not to be discussed, absolutely correct and pressed on the society. This chain relationship reflects a classical condition of the higher education in the society when the knowledge created in universities was the directing post of the society, and the society was obediently following that direction (P. Jucevičienė, 1997).

But today university is no longer the same as it was many years ago. The interface of the new and the old is evident in many activity fields of the university.

While the traditional science and study activity of the classical universities was orientated towards the fundamental objectives – to create scientific knowledge and on its basis to educate the elite of the society, the mass higher education considers the application of this knowledge equally important by bringing the whole social environment together. The position of the higher education is illustrated in the best way by Barnett (1994) "triangle", the essence of which is that the knowledge created in higher education is checked in the real life and comes back to universities in the form of new curricula or courses. The relationship between higher education and society is interactive and its essence is that society needs high level professionals prepared by universities, whereas higher education influences the development of the society not only by ensuring this preparation of high level professionals, but giving them overall education.

In the modern culture (educational culture included) there exists a clearly expressed layer of innovations, which breaks and reconstructs cultural traditions. Innovativeness, as a university value, is significant by its positive attitude towards change, new traditions, and appearance of new forms. This encourages the progress of the university, its development, strive for new qualities. But breaking long traditions and norms follows the intensity of socio-cultural life and its complexity. Sudden spread of the mass culture influences higher schools too, and determines the increasing rate of the modern personality crisis, as here public values often contradict the personal ones.

3. Peculiarities of interface between conservatism and innovativeness at university in the context of constant change

The word "university" readily evokes an image of changeless tradition: ancient buildings and unworldly intellectuals surrounded by students. Today universities, teachers and students are quite different from earlier stereotypes. As Smith & Webster (1997) say, the contrast between image and reality alerts us to an important feature of universities – they have never been a fixed entity, frozen in form. These authors think differently from the others that universities have been ever changing, always adapting to any new circumstances.

Kumar (1997) states that the function of universities is not to swim with the tide but to go against it. It is this conviction that underlies the many accounts of the university that stress its maintenance and enrichment of a certain high culture against the encroachments of business, politics and daily life. This was the influential and attractive vision for the universities of John Henry Newman, Mattew Arnold, F.R. Leavis. Universities are, in this view, islands of culture and high-minded thought in a sea of commerce and banality. This line of defense is no longer tenable. The concept is perfectly acceptable in a society, when universities are primarily concerned with educating cultural and political elites. Such is no longer the case in most contemporary societies. Bauman (1997) is right to say that there is simply no longer a consensus on what constitutes the high culture, nor agreement that this would be a desirable aim even if were possible.

Just imagine that universities might be able to say to society, "Go your own postmodern way. Let commerce and culture be one. Unleash the forces of cultural anarchy. We, however, will continue to insist that there is a high culture; we will cherish the "great tradition", and seek to transmit it to succeeding generations of students. If they are more interested in electronic games than great works of literature, this is regrettable but perhaps not fatal. We, the scholars, will cultivate the great legacy of the past, and preserve it as best we can until society comes to its senses".

Even if this position could be politically or economically approvable, it is no longer a real option. The postmodern turn has affected the universities even more than the wider society. It is within the academy that postmodern theory was first formulated, and here it has been most vigorously elaborated and promoted. But the paradox is that postmodernism destroys everything related to customs and traditions.

The increasing influence of market is typical of the higher education of the end of the 20th century. It stimulates higher education institutions to meet the requirements of business and industry. As Jucevičienė (1997) says, the conceptions of critical thinking, life-long learning being the expression of public qualitative needs and a condition for further development, determine big changes in the conceptual and practical levels of the educational phenomenon, educational system and in higher education.

Higher education has already lost its elite position with the aim of educating the elite of the society. Universities are no longer the only creators of knowledge. They compete with various firms, organizations, and other alternative educational organizations.

In order to win this competition, it is necessary to be able to qualitatively react towards the changes in the society and its needs. In order to win this competition, it is necessary to meet the changes in the society and its needs in the most qualitative ways. The demand for higher education is increasing. This determined the transformation of higher education into mass higher education in the end of the 20th century. The interface of conservatism and innovativeness is especially significant in this process.

Massive higher education is a positive tendency, as it leads to a highly intellectual society, to the competency of the university as a service provider and society as a service client.

But the massive higher education has brought other problems, related to the university preparation for qualitative changes. As Smith and Webster (1997) say, it is hard to predict what the transition from elite to massive higher education will bring. There is a big diversity of trends in higher education: a huge leap in student numbers coinciding with unceasing reductions in funding; a visibly declining university infrastructure which is in urgent need of renovation and which is, in many cases, architecturally inappropriate for handling the large numbers of students that today are routine; a significant diminution of the standing and status of academics; an emphasis on the vocational dimensions of university education; a culture of audit and assessment, all in the name of greater accountability to the public paymaster; a decisive shift away from collegial forms of governance towards distinctively managerial methods; a bewildering number of new arrangements introduced to facilitate and encourage working with outside agencies; an astonishing growth of new subject areas and associated knowledge; an almost universal conversion to modularization of undergraduate programs…

These and other changes pressurize the university system very much. As a result, university life has changed a lot. Most academics feel not being able to control their activity, as most initiatives come from the central management group and a group of politicians, having no relations to the university. Universities become "work places" but not researchers' communities, though the latter was the reason why most of the academic people had chosen their careers.

Smith and Webster (1997) say that students nowadays are different, too. Expanding numbers so rapidly while simultaneously cheapening the unit of resource has not only crowded campuses with new constituencies, but also reduced the individual attention given to those students, making learning more anonymous and self-determined. With this goes a decline in the intimacy between students and their teachers that has been for many a defining feature of higher education in the past. And with the loss of intimacy goes a decline in satisfaction and purpose.

For many scholars and the author of this paper it is tempting, confronting the transformation from elite to mass higher education and all the attendant pressures of squeezed resources and greater external direction, to give up on it all. The appeal of a Golden Age, one of small classes, high standards, well-prepared students, and generously funded and autonomous institutions, is compelling.

Whatever its appeal, nostalgia must be tempered with a full appreciation of some major achievements of the move to mass higher education. The doors have been opened for so many who would formerly have been excluded from academe. The easing of entry barriers, enabling many mature students to attempt higher education, while it has placed some additional strains on the university, has undoubtedly been an advance, introducing into the university a sizeable proportion of people whose experience and orientations can deepen the education of the rest while at the same time giving those who as school leavers missed out on university a chance to achieve at degree level.

But together with more opportunities for university studies, today the qualifications obtained at university do not guarantee a middle-class occupation. Job patterns are more and more insecure. There is a problem that students' expectations do not match the sorts of career that will be open to them. Many graduates of today's universities are to meet with disappointment in their working lives.

Diverse scholars (Smith and Webster 1997, Zygmund Bauman, Peter Scott 1997) suggest that higher education is not simply adapting, but is rather transmuting, into a radically new phenomenon, the postmodern university. It means not only teaching more students on fewer resources, or paying more attention to alternative sources of university income, or trying to maintain standards in a tumultuous time. The changes are altogether more fundamental, such that the postmodern university bears little resemblance to what went before. This new university is heterogeneous, so much so that the title "university" indicates little if anything by way of common institutional features.

Such thinkers have conceived this as a transformation of knowledge, a move away from what may be termed Mode 1 knowledge which is homogeneous, rooted in strong disciplines which are hierarchical, and transmitted to novitiates in an apprentice-master relationship, to Mode 2 knowledge's which are non-hierarchical, pluralistic, transdisciplinary, fast changing, and socially responsive to a diversity of needs such as students' dispositions and industrial priorities (Gibbons *et al.*, 1994). This plurality of knowledge announces an end to establish and common purposes of the university; they're being no identifiable unity and no possibility of agreement on goals or even methods of work.

Instead there is a multiplicity of differences: different academics pursuing different knowledge's, different teams of researchers combining and recombining to investigate shifting topics, different sorts of students following different courses, with different modes of study and different concerns among themselves, different employment arrangements for different types of staff – difference everywhere in this the postmodern, flexible, accommodating university.

There has been a good deal of comment on the decline, even the death of the university as institution. Alternative sources of knowledge emerge and undetermined the university's monopolistic position. The development of Internet, the spread of multimedia technologies, cable TV is frequently highlighted in this regard. Beyond the walls of established higher education institutions, floating in cyberspace, one may find discussion groups, research networks, bulletin boards, and associated forms of the emergent "virtual university".

Teaching and learning have changed too. Why visit the university to learn when one may connect, at one's own convenience, to on-line databases from office or home for much richer sources of instruction? This raises the wider issue of the spread of distance learning, a movement which overcomes the limits of space imposed on gaining a traditional university education, while potentially allowing any student to get on-line access to the leading figures in any given field. But can all this be considered as better ways of the learning process in comparison with the direct communication and cooperation while learning between a teacher and a student?

Some years ago there was an experiment carried out by teachers of foreign languages at KTU.

As many students expressed an opinion that learning foreign languages in classrooms with the help of a teacher was boring and inefficient, they were offered a chance to use computer programs and not to attend classes. If I remember rightly, it took a week until the students said, "Thank you. We want our teacher back. It is boring to learn by computer." This is only one example of how innovative methods and new technologies should not be overestimated. Interpersonal, direct communication, an open discussion and a simple talk will always remain the main way of learning as long we are a society of human beings. Other ways, however efficient and attractive they might seem, may be only supplementary.

Kumar (1997) emphasizes the informal side of university life, not as a residual but a central feature of universities. He says that the university aura, the feeling of being together, is much more appealing that the advantages of the virtual university: "I want to see universities as bright and energetic students of all ages have experienced them at all times: as places to explore themselves with others, in speaking, writing, performing, playing, imagining, stretching themselves in mind and body. Nowhere else, and at no other time in their lives, irrespective of age, will students encounter each other with so much time and so many resources to do so much, unconstrained by the requirements of job or family. The university is indeed a place, a physical space with buildings and grounds that exist to facilitate the pursuits of students and teachers.

According to A. Smith, F. Webster (1997), mass higher education is determined by changes and transformations all over the world, thus it is hardly predictable. Though new consolidations (joining, alignment) have normalized around 1990, but the elite university culture has been irremeably affected by the open mass culture. The academic youth is no longer interested in the unifying symbols; traditions have degenerated because of a commercial shade.

All this impoverish students' lives, the former romance is missing. Could the postmodernists answer the rhetorical question – can the virtual or service university satisfy students as personalities, depriving them of constant eye-to eye cooperation and communication?

The changing environment – new technologies, abundance of information, the spreading international market, etc. – in one way or another affect all universities and make them similar and exceptional. As S. Duke (1997, p.7) says, the spreading environment is usually culturally alien to university traditionalism, but it should not lead it to crisis, but open new possibilities for development, diversification and assessment of university values and priorities. Thus the question arises, how deeply the changing environment affects university, and whether we should worry about the university crisis.

Today's practice shows that however impressive the quantitative expansion of massive higher education, it has not been accompanied by a qualitative transformation of fundamental academic values or institutional instinct.

It is the most important sphere that needs special attention while discussing universities. Values and aims reflect the changes in the university priorities. They point to the directions where the institution moves. Efficiency is a business category defining the amount of the produced product in comparison to the input. In universities it is related to the knowledge that is reflected in the numbers of graduates, research, publications, projects, conferences, etc. Knowledge efficiency at university is determined by adequacy of aims and quality that do not allow this process to be elemental and characterizes university knowledge. The concept *continuous learning* discloses one of the functions of contemporary university, which provides it with the exceptional role in the society – to be its *intellectual motor.*

K. Kumar (1997) is of the opinion that diverse changes have affected universities more than the wide society. They are more than acceptance of new norms and standards, nor distribution of new academic roles. These are qualitative changes of the university as institution and university idea as such. These changes influence the role of the university in the society and its attitude towards the university. But this author emphasizes, that university efforts to imitate market and commercial structures are harmful if not damned, as universities cannot compete with better equipped and more experienced private commercial organizations because of lack of resources and personnel. Could anybody deny these arguments?

I would certainly agree with the author. More over, I would try to find more sound arguments against too big commercialization in higher education.

4. The balance between conservatism and innovativeness in higher education

According to R. Grigas (2001), people's traditional life, based on various ethno-cultural symbols, customary norms of behavior, proves that there existed rather stable social structures, their "specialization" and interaction between those structures. There was no sharp contradiction among the natural or traditional and the "innovative", more formalized social organizations. The level of innovation did not exceed the capacity and receptivity of social organization, thus there was no confusion. The same situation can be noticed in the activity of the traditional university over ages.

Civilization and modernization (as an attribute of civilization) always disorder and change the traditional social organization. Modernization usually raises concrete aims that demand reformation of the old structures and creating the new ones. This way an artificial, virtual social environment is created and it starts confronting and conflicting with the traditional natural, spontaneously formed social order.

Whether or not we choose to call our times "postmodern", it is clear that universities now exist in an environment vastly different from that of the 19th century, when the basic idea of the modern university's function was established. Departing in many ways from earlier, medieval ideas, the nineteenth century's view was formed as a reaction to industrialization and modernization. Universities would not oppose these developments, but they would maintain a critical distance from them. It would be their function to preserve the high culture against the onslaughts of modernity – not as an escape from modernity but as a critical engagement with it. (Kumar, 1997) Why not follow this example in the postmodern times? Today, very enthusiastically we are trying to destroy everything that is traditional without thinking whether the new structures match in the traditional context. These structures become rather aggressive in such cases when they are attempting to imitate those of other cultures, formed by other historical conditions. Imitation, voluntaryistic order forms the fields of social tension, where contrary forces are evoked. Confrontation of contrary forces makes empowers more active modernization.

The reality of life proves that mechanical imitation of other experience, gained in other culture, other social space and time in a certain degree may only tangle the problems related to the practices of social and political organization.

Optimal innovation, i.e. functioning of the structure when introduction of an innovation does not infringe the structure of the general construction, the origin of its continuity, its teleological singleness, is positive and it is called *a dynamic balance*. It is such a state of a social structure when stability and continuity is preserved together with rather intensive changes.

This is necessary in our turbulent times, when education has lost its stability in the international level. Wise people all over the world have always resisted misunderstood and unreasonable changes in the continuity of the university. These factors determined the success of the university in the modern era.

Because of commercialized influence of the Western culture, more and more attention is given to short-term surrogates of culture, not the real values. After Lithuania regained its independence, many stereotype opinions and illusions appeared concerning the situation of the modern Western culture, arts, education, etc. Undeniably, there are many positive aspects in the Western culture, education and other spheres, but together with features of humanism, there are many evidences of decadent culture (unsound self-reflection, escapism, deformed cult of sex) (Andrijauskas, 1995).

The background of the whole our reformed educational system is a human being and his culture. (The Directions of the Lithuanian Educational Reform, 1993).

A human being is the highest property. It is necessary to create all possible conditions for his individuality to prosper in the context of cultural values. A human being and culture are very closely connected: conditions for maturity and self-realization of an individual are created through culture, whereas an individual continues and cherishes the culture of his nation and mankind. Thus the object of education is undeniably in the cultural context. It is not possible to educate an individual in an isolated way, not considering the needs of a certain nation and culture.

The author of this article agrees with Lukšienė (1993), that the situation in Lithuania is not normal because we are having a human being who has been injured by the communist regime. This is one of the main reasons of today's problems not only in higher education, but in other spheres of life, too.

Universities must meet the requirements of the society. Traditional cultural models are no longer suitable. Whether the university culture is strong, depends on its members. As Stoll, Fink (1998) write, the culture of school as organization can be a product and a process. As a product, it embodies the wisdom of those who lived before. As a process it is constantly renewed and recreated, because new members are taught of old rules. Its participants create culture, and when they change, the culture changes too.

Universities as organizations have to be flexible in the conditions of constant changes. In this context I would emphasize the significance of university leaders, academic personnel. It is necessary for university people to be creative in the unlimited and often ambiguous environment.

Universities today need charismatic personalities, trying to break the routine structures. Their activity has to be innovative, creative. But one of the main requirements is identification with the common culture. It is no longer enough to gain certain qualifications and to demonstrate technical competence. For the managerial work at universities a candidate is looked upon as the whole personality, a certain "personality package" or "competence portfolio" that consists of professional qualifications and charismatic features.

In general, it is possible to state that in today's global world universities' role is exceptionally important. They have to realize the contemporary mission of higher education – to be the source and the motor of the intellectual development of the society. This mission is based on the traditional (to be the source of intellectual development) and the modern (the motor) origins of higher education. So the mission of contemporary universities in its general sense is to match the two – traditional and modern – aspects of higher education in order to guarantee fast intellectual development of the society. All this should be done trying to avoid any (or at least big) conflicts in matching the qualities of the knowledge and information society.

Conclusions

1. Conservatism in higher education has two connotations. It is positive as preservation of the best experience, cultural traditions, heritage, and rational attitude towards change. It is negative as strong resistance to change, innovations, and unconditional affection to old order, inflexibility, and stagnation.
2. "The university is, has been and can only be a place where the teaching is part of the unending dialogism of the outer society, where thought takes place beside thought. There must be a future for the university in its work of thinking, which goes on outside the instruction package of corporate excellence, one that has survived the attractions and repulsions of the nostalgic and the romantic (Smith and Webster, 1997).
3. In order to reduce the ideological flatness of higher education and trying to expand the limits of rationality of this public sphere, the relationship between conservatism and innovativeness in higher education should be based on academic rationality, the essence of which is that there should be a dynamic balance between these two categories in the higher education – a state of a university structure and culture when stability and continuity is preserved together with rather intensive changes. It can be achieved through open and objective discussions, critical thinking, intellectual and academic freedom. The results will depend upon how strong the university culture is, how wise and far-sighted its leaders, the central managerial group and the academic community are.

References

ANDRIJAUSKAS, A.: Postmodernizmo ištakos ir neklasikinių diskursų erdvė. Miestelėnai. Miestas ir postmodernioji kultūra. Vilnius 1995 (Taura).

BARNETT, R.: The Idea of Higher Education. Buckingham 1990 (SRHE, Open University Press).

BASSIS, M.S.: Sociology. New York 1988 (Random House).

BAUMAN, Z.: 2 Universities: Old, New and Different. The Postmodern University? Milton Keynes 1997 (Society for Research into Higher Education & Open University Press).

BROWN, P./SCASE, R.: Universities and Employers: Rhetoric and Reality. The Postmodern University? Milton Keynes 1997 (Society for Research into Higher Education & Open University Press).

GIBBONS, A.: Innovation and the Developing System of knowledge Production. Brighton 1997 (University of Sussex).

GRIGAS, R.: Sociologinė savivoka. Specifika, metodai, lituanizacija. Vilnius 2001 (Rosma).

GRIGAS, R.: Socialinių įtampų Lietuvoje laukai. Vilnius 1998 (VPU leidykla).

GUDAITYTĖ, D.: Aukštojo mokslo tapimo masiniu reiškiniu ypatumai: pasaulinės tendencijos ir jų projekcija į Lietuvą. Daktaro disertacija. Kaunas 2000 (KTU).

HALL, E.: Beyond Culture. New York 1989 (Anchor Books).

HALSTEAD, J.M.: Values in Education and Education in Values. London 1996 (The Falmer Press).

JUCEVIČIENĖ, P. Lietuvos aukštasis mokslas Quis es? Quo vadis? Studijos aukštojoje mokykloje. Konferencijos pranešimų medžiaga. Kaunas 1997 (Technologija).

JUCEVIČIENĖ, P. IR KT.: Universiteto kultūra ir jos tyrimas. Kaunas 2000 (Technologija).

KUMAR, K.: The Need for Place. The Postmodern University? 1997 (Society for Research into Higher Education & Open University Press).

MCLAUGHLIN, T.H.: Šiuolaikinė ugdymo filosofija: demokratiškumas, vertybės, įvairovė. Kaunas 1997 (Technologija).

MELODY, W.: Universities and Public Policy. In SMITH, T./WEBSTER F. (eds): The Postmodern University? Contested Visions of Higher Education in Society 1997 (Milton Keynes: Open University Press), pp. 72-84.

PODLASYJ, I.P.: Pedagogika. Kniga 1. Moskva 1999 (Vlados).

SHAVER, J.P./STRONG, W.: Facing Value Decisions: rational building for Teachers. Belmont, CA 1976 (Wadsworth).

SMITH, A./WEBSTER, F.: Changing Ideas of the University. The Postmodern University? 1997 (Society for Research into Higher Education & Open University Press).

SMITH, A./WEBSTER, F.: Conclusion: An Affirming Flame. The Postmodern University? 1997 (Society for Research into Higher Education & Open University Press).

STOLL, L./FINK, D.: Keičiame mokyklą. Vilnius 1998 (Margi raštai).

TOFFLER, A.: Valdžių kaita. Vilnius 2001 (Tyto Alba).

VITKUS, G.: Europos Sąjunga ir ne tik ji: 99 terminai. 2003 (Išleista Europos komisijos Delegacijos lėšomis).

ŽELVYS, R.: Švietimo politika. Švietimo studijos. Vilnius 2003 (Garnelis).

Vida Stasiulionienė

Deep Learning at University: Does It Always Mean Quest for the Truth?

Contemporary university is inevitably affected by the changes in society which shape the mission of this higher education institution. This mission includes creating, maintaining and developing the intellectual potential of the society, which enables turning knowledge and science into the key welfare factor of the society/ state. In recent academic discussions (Drucker, 1993; Jucevičienė, 2001; Flinders, 1998; Longworth, 2000), it is frequently emphasized that in the information and knowledge (society people feel constant need of developing and renewing their personal competence, i.e. the increasing demand for higher education. According to Bowden and Marton (1998), therefore, university can no longer be defined as merely a higher education institution accumulating teaching and research. In addition, it has to be seen as a learning university, recognizing the fact that I learning with great variety of its forms involves not only all students, but also teaching and research staff, i.e. the whole academic community. This is the reality of the post-modern university (Hargreaves, 1999), which stimulates more and more discussions about the quality of learning, the so-called deep learning. From philosophical perspective, any kind of learning (the cognitive process) is normally related to searching for the truth. This issue has always interested the thinkers of different schools. For example, the founder of the idealistic philosophy Plato believed that human cognition is not capable of discovering the truth in the material world because this world is constantly changing. For this reason, people should strive to understand the world of ideas, seek philosophical wisdom. Plato's followers Augustine, Hegel and others admitted that not the *truth per se* is the most important, but *searching for it,* because every idea can be contradicted. Similar views were held by the realists Aristotle, Tom Aquinas and others who argued that while *closely exploring the world* one can get to know God because He created it so that human beings have a way to get to know it (Anzenbacher, 1992). Hence, looking from philosophical perspective and taking into account the contemporary requirements to individuals and their learning, the university studies should be primarily focused not on acquiring new knowledge about the reality but on learning the ways of getting to know that reality. This would enable one to understand the objects and phenomena of the objective reality in a more adequate way.

These arguments suggest that both the category of truth and the term of 'deep learning' in search for the truth raise a lot of questions. *What is the truth and how one can discover it? What is the essence of deep learning, especially when adopting different methodological positions? Is deep learning always good and does it always lead to the actual truth,* even if individuals learn and get to know the world in rather different ways? *How to avoid the contradictions related to deep learning and to discover the real truth, or at least to get closer to it?* These questions shape an important ***research problem:*** *can it be the case that, taking into account the limited possibilities of getting to know the objective reality as well as the differences in students' perceptions, thinking and learning, deep learning at university may lead to a distorted view of the world, i.e. it does not always lead to the real truth?*

The latter question is the focus of the discussion presented below.

The article consists of three sections. In the first section, the category of truth is discussed by introducing its different interpretations and attempting to answer the question about what the truth is and how to discover it. In the second section, taking into account different philosophical perspectives and not attempting to give a single answer, the notion of deep learning at university is discussed. The third section is devoted to highlighting some contradictions and controversies related to deep learning and to suggesting some ways of avoiding them on the way to the actual truth. The article ends with conclusions summarizing the key propositions of the discussion.

What is the truth and how to discover it?

Analysis of the deep learning as a phenomenon, which is inseparable from searching for the truth at the individual level, calls for discussion of the notion of the truth itself.

For many years a number of philosophical schools attempted to answer the questions 'What is the truth? How can it be discovered? Are human beings able to get to know the world and to discover the actual truth?' According to Aristotle, *the truth is correspondence between the object of cognition and its reflection in the cognizing subject's consciousness* (Barrow and Woods, 1988). This notion of the truth is quite common. However, according to Bitinas (2000), it would not raise any doubts if the subject did not have to express in some way the *results of cognition,* which we *normally understand as knowledge about the reality.* Representatives of some philosophical schools argue that knowledge is the effect of the sensory perception of reality, or the result of intuition, contemplation, and in some cases of insight. We can gain knowledge from many sources: books, teachers, research, etc., and its value to every cognizing (learning) individual is different. An important fact is that knowledge is acquired through searching for the truth in the cognitive process, when an individual (a student) penetrates into the object of cognition and reflects it in the form of his own images. It should be noted that although the cognizing individual, according to Bitinas (2000), normally does not identify oneself with these reflection effects (images), he actively attempts to create them. So, these images/knowledge is nothing else but *personal constructs.*

A question arises *how to determine whether this expression of the cognitive results (knowledge), which is shaped as a unique construct of the cognizing individual, is correct and reflects the essential and true qualities of the cognition object? Does this cognitive act produce the actual truth, reliable knowledge? Is human cognition at all capable to discover the actual truth and to get to know the reality?*

The author of this article agrees with Bitinas (2000), who argues that answering the question about whether the expression of the cognitive results is adequate to the object, i.e. correct, one has to understand some controversies related to this adequacy: every partial truth specified by an individual in a traditional way involves an element of error. It is caused by the limitation of the cognitive act itself because, on the one hand, it is difficult to specify the essential features of the cognition object, and on the other hand, what may seem essential in this object to one person, may appear secondary to another. It should also be noted that various philosophical schools differently answer the above question, emphasizing different problems related to the cognitive process and its results.

For example, the follower of the classical agnosticism Hume argues that *individual in principle is not capable of crossing the boundaries of his experiences.* It is therefore *impossible to learn what the relationship between the objective reality* and human experience is, i.e. to what extent the truth V. discovered through experience is adequate to the object of cognition. According to Anzenbacher (1992), modern agnostics note that the substance of some truths (concepts, definitions, axioms, hypotheses, etc.) discovered by humans is not discussed in terms of its corre*spondence to reality; it is the matter of the rational agreement between the scholars because no alternative solution has been found so far.* The adherents of the epistemological dogmatism believe that *there are truths (e.g. religious dogmas), which are true independently on human experience and even on whether they are accepted by human mind.* Some Antique thinkers (Plato, Aristotle) believed that *in human cognition it is important to distinguish between what is knowledge about reality and what is only a judgment about it.* Aristotle argued that search for the actual truth and its results should be first of all based on the necessary assumptions and categories, while judgment requires merely probabilistic assumptions. Hence, judgment should be seen as knowing which errors and lie may involve, while knowledge acquired through learning is knowing which is much closer to the truth and more adequately reflects the reality. Bacon, one of the adherents of realism, argued that *adequate reflection of the real* *world is negatively influenced by certain distortions (or attitudes) of cognition* which he called the *'idols of cognition': idols of species,* i.e. characteristic to the whole humankind which make people believe what they like; *idols of the cave,* i.e. distortions of the individual cognition, when people see the reality as if from the depth of a cave, that is they see only its shadow; *idols of the field* related to imperfections of language, e.g. when the reality is described by inappropriate attributes; and *idols of the theatre* which emerge as a result of the uncritical judgment of other people's views, blind imitation of the authorities (based on Bitinas, 1996).

Thus, the arguments of different thinkers representing various philosophical schools suggest that it is not easy to specify how one can determine whether the results of cognition (especially of individual cognition) are adequate to the object of cognition. There are some factors, which should be taken into account before answering the above question.

Further analysis of the category of the truth leads us to the concept of *coherence.* Kant, one of the developers of this concept, and some adherents of the neo-positivist philosophy argued that *knowledge is correct if it is a component of the coherent system.* However, this concept is also I limited, according to Ozmon and Craver (1996). It is unclear to what extent knowledge corresponds to reality when a certain inner inconsistency of knowledge may reflect the difficulties of the cognitive process. Another problem/ limitation is that verifying the inner consistency of knowledge is possible only in rather simple cases. Consistency of complex knowledge systems cannot be verified, and it is often an immanent quality.

One more rather controversial description of the truth was offered by the pragmatists (Bacon, Locke, Comte, etc.) who argued that the *effectiveness and usefulness of certain knowledge about the world does not imply that this knowledge corresponds to reality, i.e. is true.* As pointed out by Bitinas (2000), according to pragmatists, this is so because the reality is not accessible since human beings and reality are separated by intermediary, i.e. certain activities. Summarizing the above discussed notions, it can be concluded that ***a quest for the truth is a controversial process bounded by continuous attempts to defeat error, while knowledge is true to some extent if the truth itself is cognizable.*** Hence, our knowledge can be and is incomplete, one-sided, partial, etc. It is true if it reflects the reality. In other words, human beings are capable of cognizing the reality but *their knowledge is never fully reliable and deep* (the more we 1earn, the clearer we perceive the limitations of cognition), *i.e. the truth they discover is always relative, and the completeness and deepness of the acquired knowledge is determined by the circumstances in which cognition takes place.* So, *it is not the functioning of the truth but the quest for it, which is important.* In the educational context, it implies that the students will verify the truths discovered during the learning process throughout their entire life. They will change their attitudes towards various propositions and reality not once. Hence, it is important to teach and to accustom them to search for the truth. As noted by Diesterweg, a bad teacher supplies the *truth, while* a *good* one teaches how to discover it.

The above discussion suggests that individual cognitive process repeats social cognition Andrijauskas, 1995), i.e. that which was pursued by other individuals attempting to discover the reality. According to Bitinas (1996), this repetition is relative and covers only the results of cognition and not the cognitive process, which involves a lot of individual and collective quests, errors and deviations. So, there are differences in the social and individual cognitive processes (not the results), which should be highlighted in the context of this article revealing certain *characteristics of the individual cognitive process (quest for the truth).*

One of the key characteristics of the individual cognitive process is the *difficulty related to assimilating the cognitive content itself* as knowledge accumulated by humankind for a long time, according to Bitinas (1996), is converted into a personal construct of the learning individual. This idea is in a way contradicted by the idealistic philosophy, which argues that individual cognition only recreates, and explicates the ideas, which inherently reside in the individual's spiritual sphere. Consequently the knowledge acquired by the cognizing individual is not dependent on his individual experience; he only seems to assimilate the generalized human experience, hi reality, he only recognizes and recreates what has been inherently hidden at his spiritual level. Meanwhile realists support the idea of the *tabula rasa* (Locke), arguing that no *a priori* knowledge exists before individuals 'meet' knowledge, which is new for them, and before it turns into their mental structure. Likewise pragmatists, realists believe that individual cognition is a certain contract between the cognizing subject and his environment admitting that the truth is always relative and is important only from the perspective of its practical application. An interesting perspective on the cognitive content is offered by the existentialism: individuals themselves choose the truths and knowledge, first of all focusing on self-knowledge.

The above discussion suggests that the cognitive content covering a very broad range of phenomena and objects of the reality determines the individual cognitive process and its results (a rather syncretic perception of the reality), at the same time raising a number of questions for discussion.

One more characteristic of the individual cognitive process, according to Ozmon and Craver (1996) and Barrow and Woods (1988), is the fact that *systematic learning/ search for the truth by the cognising individual is determined by actions of the educator.* On the one hand, whatever teaching methods the educator uses, they all are an imitation of the social cognitive process because he normally already knows the solution of the problem given to the students. On the other hand, to be (able to adjust certain methods of cognition to the students, the educator has to know them well. In this sense, in the individual cognitive process the students are not only the subjects, but also the objects of cognition: the educator constantly adjusts their cognitive activities and quest for the truth. It should be noted that *the latter idea is important even in the context of the newly emerged learning paradigm because the role of educator is not eliminated from the learning process, only the character of this role changes.*

In the process of individual cognition/ search for the truth, individuals get prepared for life. Their personal cognition is induced by the desire for relevant education, competence, etc. In this context, one more characteristic of the individual cognitive process comes to light: *knowledge acquired by the individual is valuable only to him/her. It enables him/her to reorganize to some extent the surrounding reality.* The latter quality is particularly important in the context of deep learning since a rational way to analyze this phenomenon is by focusing on the individual cognitive space.

What is deep learning at university?

To answer this question, first of all the phenomenon of *learning* should be discussed. In the literature one often can find a rather general definition: learning is a continuous process of quest and discovery, which enables human beings to change their views and behavior and to renew themselves (Ozmon and Craver, 1996). However, from a philosophical perspective it is important to discuss the nature of this phenomenon in the university context.

An interesting perspective on learning at university was offered by Bowden and Marton (1998) who argue that it always embodies perception and study of various situations, i.e. the objects and phenomena of the objective reality, when every phenomenon or situation accumulates/ involves a certain number of critical aspects, which can be recognized and evaluated as a whole. The authors note that there exist different ways of perceiving an object or phenomenon which depend on the critical aspects a student is able to discern and evaluate as a whole. Besides, this ability to recognize and discern emerges only after experiencing so-called variation, i.e. certain changes occurring during the learning process which involve the ways of individual's perception and experience.

Learning in the university context is related to the problem of *transferring the acquired knowledge into the real life situations.* It could be argued that the concept of transfer or generalization is meaningless because when we learn something we have to use it in different situations, and transfer, according to Smedslund (1953), is taking place at every single moment of learning. Hence, transfer is nothing else but learning. However, transfer of knowledge acquired at university into real situations only at first sight seems straightforward, hi fact, **a** very important factor should be taken into account: students learn referring to certain situations, which they normally never encounter outside the university. Thus, there exist a lot of unpredictable, unknown differences in these situations. According to Bowden and Marton (1998), everything that a student can use in this case is his knowledge about the objective reality, what he knows and what is known. So, to prepare for the future, which is unknown, student's learning has to be deep.

What kind of learning is deep?

This is a philosophical question because it has no single answer, as the phenomenon of learning itself (as every other phenomenon) can be analyzed from different methodological perspectives.

Placing the above suggested interpretation of the learning phenomenon and the concept of the knowledge transfer into the context of deep learning, it could be argued that deep learning at university on the way to quest for the truth has the following characteristics:

- *Students are able/ learn to discern (perceive) as many as possible key critical aspects of a certain phenomenon* (because their perception allows to understand the essence of the phenomenon itself), experiencing different dimensions of its variation/ change;
- Students learn/ are able to transfer the acquired knowledge into different situations, i.e. *their learning crosses the boundaries of the learning situations;*
- Students use different methods of perceiving/ understanding the objective reality, taking into account the limited possibilities of cognizing this reality.

As it has been noted, in various literature (Spielberg, 1982; Marton and Wenestan, 1988), the notion of *deep learning* is related to perceiving the dimension of the present and the past, hi this context, very relevant is Reardon's (2000) idea that the best way of 'joining' the past and present is a continuous reflection by students of themselves and their surrounding environment. This approach is somewhat similar to the idea of the existentialism (Sartre, Heidegger, Husserl) that learning can be regarded as good (and deep in the sense analyzed here) when it helps individuals to better see themselves both in good and in bad sense. As a thinking method enabling a human being to critically analyze and to realize oneself (Jovaisa, 1993), to learn from one's own others' experience and to create new knowledge (Cochran-Smith and Lytle, 1990; Argyris et al., 1985), reflection can be related to deep learning at university. As a result, one more answer to the question what deep learning is can be the following: this the kind of learning when students, being well aware of the importance to get prepared for the future, learn to *look at different learning situations from perspectives of the past and future referring to their accumulated experience and the continuous reflection on it.*

Emphasis on the meaning of reflection in the context of deep learning in general raises some questions. If students' experience can be regarded as one of the dimensions of deep learning, *shouldn't the idea of the deep learning at university be related to the 'Socratic' approach to phenomena and situations,* when questions (e.g. raised to reveal the critical aspects of these phenomena) are more important than answers? (McLaughlin, 1997). On the other hand, *isn't deep earning related to the popular notion of critical thinking,* since it is critical perception and evaluation of various situations that according to Bowden and Marton (1998), allows understanding their nature?

Of course, these questions do not have a single correct answer. However, taking into account the origins of the university learning (the Socratic method of questions and answers and critical evaluation of things and phenomena by means of research remain the basis of the university studies), it can be argued that the very idea of university emphasizes primarily 'deep' and not 'superficial' learning. *Both the Socratic approach and critical thinking therefore can be related to deep learning at university.*

The latter argument has prompted the author of this article to engage in discussion on one more question: what is the relationship of deep learning at university with such important factors as *students' attitudes, views and learning motivation?* Bowden and Marton (1998) offered an interesting example relevant to this discussion: when a student follows the 'superficial' *approach,* it is important for him to get the instructions on what and how to learn, to repeat the text which he has read, to answer the specified questions, etc. Meanwhile a student following the *deep learning approach* is motivated by interest in learning something new about what is already known.

This example, on the one hand, emphasizes the relationship between individual's (student's) attitudes, motivation and learning, and on the others hand, the complexity of this relationship because, according to Stewart (1996), an individual does not always realize if he or she has certain attitudes, whereas once they are realized learning they can be changed through if there is strong inner motivation. Thus, it can be assumed that *deep learning at university is related to strong inner motivation and deep approach to learning situations.* However, at the same time it should be taken into account that deep learning is influenced by the above discussed cognitive distortions-certain preconceptions that limit the adequate reflection of the objective reality and make search for the truth more difficult.

What are the controversies related to deep learning at university and how to avoid them and to get closer to the actual truth?

One of the important controversies related to deep learning is related to the concept of transfer during the learning process, which has already been discussed in the second section of this article. The essence of this controversy is that a university student, teacher or other individual engaged in learning, according to Bowden and Marton (1998), will never find oneself in the real life situation, which is identical to the learning situation. On the other hand, knowledge acquired in learning situations quickly becomes obsolete. Besides, it is not easy to determine whether it reflects the actual reality, i.e. is true. Thus, individuals *learn for the unknown, uncertain future, applying what is known and what they know, being unable to predict the future situations.*

This problem has been discussed by the thinkers of idealistic philosophy (Augustine, Hegel, Kant, etc.) who believed that individuals are not able to cope with the current problems not due to the lack of certain facts or knowledge, but due to inability to relate them to the 'great' ideas and to use this (Ozmon and Craver, 1996).

This controversy suggests that deep learning at university should be also related to teaching the students to critically assess what has been learned, i.e. to reflect on their experiences and knowledge. Besides, as correctly noted by Bowden and Marton (1998), to avoid this controversy, it is important to learn to experience the variations of the critical aspects of the analyzed phenomena in the present and in the past, or according to Dewey, to learn from one's own and others' experience and through it. University teaching and learning in this sense are inseparable from application of the relevant variation models, creation of learning environments facilitating deep learning and getting closer to the truth.

Following the dual ontological approach that argues that human perception (and learning) should be treated as the relationship between the perceived and real objects or phenomena, i.e. as the relationship between a part and the whole, one more controversy related to deep learning can be noticed. *No matter how deep learning is, every human being perceives the world differently, and these perceptions never coincide with the way the world really is.* Consequently, the truth discovered by an individual is relative, whereas the deepness and completeness of the acquired knowledge is determined by circumstances in which the cognitive process is taking place. The controversy lies in the limitations of human perception/learning, as every phenomenon, according to Bjorklund (1990), is inexhaustible and can be perceived in different ways. A similar controversy is created by the phenomenological perspective, which argues that *the world and its phenomena, which we experience and perceive, are as they are and that the others see them in exactly the same way.* However, following the idea of the deep learning, as it has been mentioned, it is obvious that the way individuals experience and perceive the world changes in the process of learning. So, different people cannot perceive and experience the world in the same way. Therefore, in case of deep learning in general, one probably should not attempt to get to know as many as possible aspects of the surrounding world or to discover the ways of perception which would allow to see the world as it really is. Perhaps more important than that is to learn a variety of ways of cognising and perceiving the world.

Bowden and Marton (1998) note that in this sense one should learn to understand perceptions of other people, in this way enriching one's outlook on the world, since by participating in each other's perception process, individuals are able to better perceive the world, to create collective consciousness (e.g., at university). Besides, it should be remembered that an important way of learning at the university is research (in its relative sense). However, even when conducting research on the same object, due to the individual cognitive differences, different people do not explore identical problems: they raise additional questions related to different aspects of the research object. Hence, as noted by Axelrod (1984) and Gambetter (1988), when research is focused on the problems, which have not been addressed before, it is important to discern the variations of the different critical aspects of the problem and to share them publicly, expressing them in the collective context. This allows avoiding one of the problems related to deep learning: to disseminate incomplete, untested ideas, which can be adopted by other people.

Referring to the importance of the variety of the objective knowledge/perception of the world in the context of deep learning, it should be noted that this variety, the differences in perceptions of students and teachers and of researchers exploring different research topics is often undervalued at university. In this sense, there is a lack of academic discourse. Besides, in the context of the questions raised by the above controversy related to deep learning, it should be remembered, according to Bowden and Marton (1998), that the development of the new ways of perception is an important, but not the only, dimension of deep learning.

According to Stewart (1996), one of the controversies related to deep learning can be caused by the *preconceptions and views of the students* (Bacon called them cognitive distortions, or idols), *when they come to believe that the world is seen as it is.* Bowden and Marton (1998) believe that this is *an important dilemma for university,* especially in case of the creation of knowledge that is new for an individual (during the teaching and learning process) and knowledge, which is new in the absolute sense (e.g. created as a result of certain research).

It is interesting to note that, despite a rather clear understanding of the influence of individuals' attitudes on the learning process which has been proved by research, some scholars (McLaughlin, 1997) discuss whether university should shape students' attitudes or to restrain from such influence encouraging them instead to think about it and to come to their own conclusions. In this context, Booth's (1992) idea is important that deep learning is inseparable from such fundamental attitude, as *'I understand that I can be wrong, so I respect others' views and attempt to look at problems from different perspectives'*. Besides, adopting this attitude allows understanding that the perception, which is different from the common truth, is not absolutely wrong.

According to Bowden and Marton (1998), this, as well as other attitudes, should be 'brought' to university from secondary school, i.e. the latter should build the foundations of deep learning.

Conclusions

The category of truth and especially the problems related to the quest for it remains at the centre of the philosophical discourse. The truth is often regarded as the correspondence between the object of cognition and its reflection in the consciousness of the cognizing subject. Although it can be admitted that this interpretation of the truth is rather objective, it cannot be accepted without discussion and evaluation of other interpretations, since the formation of this reflection (knowledge as the result of cognition) is influenced by different factors: sources of knowledge, value of knowledge to the cognizing subject, some problems related to the content of knowledge, certain attitudes of the individual (cognitive distortions), etc. In the context of the above discussion, the following ideas can be highlighted:

- *The truth is a controversial process bounded by the continuous attempts to defeat error, whereas knowledge as the result of human cognitive process is true if the truth itself is open to cognition;*
- *The truth discovered in the process of individual cognition should be regarded as relative, and the deepness and completeness of the acquired knowledge is determined by the circumstances in which the cognitive process is taking place. Therefore,* ***not the functioning of*** *the* ***truth, but*** *the quest* ***for*** *it is important;*
- The phenomenon of the search for the truth (cognitive process) taking place at the individual level is characterized by the difficulties related to assimilation of the cognitive content and by certain problems. The educator plays an important role in this process, whereas the value of the knowledge acquired by own space, when this knowledge enables him to some extent to reorganize the surrounding objective reality.

The idea of deep learning has been an object of interest to researchers for a long time. Thinkers representing different philosophical and methodological schools have different perspectives on this idea and its implementation at university, offering unique interpretations, critique and inviting for discussion. In the context of the above discussion, the following key aspects outlining the nature of the idea of deep learning can be highlighted:

- *Deep learning is an integral dimension of the university learning because university is the place where people learn to act, live and resolve future problems;*
- *Deep learning can be described as learning by using various ways of cognising/ perceiving the objective reality, reflection and critical thinking, following deep attitudes and strong inner motivation to get new knowledge. In this process, the critical aspects of the certain phenomena of the objective reality are identified perceiving different moments of variation of these phenomena and transferring the acquired knowledge into the situations, which are not adequate to the learning situations. Deep learning is inseparable from understanding the cognitive limitations and difficulties.*

The controversies highlighted in the discussion offered in this article indicate that such a complex phenomenon as deep learning calls for complex analysis. Being undoubtedly positive, at the same time it is related to certain problems:

- *The learning situation is never identical to the future situations, meanwhile it teaches to act in them applying what is known and understood, taking into account that what is known (knowledge) quickly gets obsolete and does not necessarily reflect the reality;*
- *Independently of the deepness of learning and principal knowledge of the world, every individual (student) perceives it differently due to his attitudes, values and motivations, and their perceptions cannot be identical to the way the world really is. However, these perceptions, which are different from the common truth, are not absolutely wrong.*

To avoid the controversies of deep learning at university and to get closer to the actual truth, it is important to encourage and teach students self-reflection, to emphasize use of different ways of perceiving the world in the process of learning, and to show the importance of learning to understand other people's perceptions.

References

ANDRIJAUSKAS, A.: Origins of Post-modernism and the space of non-classical' discussions. In: ALISANKA, E. (ed.): Townspeople: Town and Modern Culture. Vilnius 1995 (Taura), pp. 9-33 (in Lithuanian).

ANZENBACHER, A.: Introduction to Philosophy. Vilnius1992 (Kataliku Pasaulis). (in Lithuanian).

ARGYRIS, C./PUTNAM, R./MCLAIN, S.D.: Action Science. San Francisco 1985 (Jossey-Bass Publishers).

AXELROD, R.: The Evolution of Cooperation. New York 1984 (Basic Books).

BARROW, R./WOODS, R.: An Introduction to Philosophy of Education, 3rd e d. London 1988 (New York).

BITINAS, B.: Fundamentals of Philosophy of Education. Vilnius 1996 (VPU). (in Lithuanian).

BITINAS, B.: Philosophy of Education. Vilnius 2000 (Enciklopedija). (in Lithuanian).

BJORKLUND, S.: The Research Connection. Studies of Higher Education and Research. Stokholm 1990 (National Board of Universities and Colleges).

BOWDEN, J./MARTON, F.: The University of Learning. London 1998 (Kogan Page).

COHRAN-SMITH, M./LYTLE, S.L.: Inside/Outside: Teacher's Research and Knowledge. New York 1990 (Teacher's College Press).

DRUCKER, P.: Post-capitalist Society. Oxford 1993 (Linacre House, Jordan Hill).

FLINDERS, D.J.: Teacher's Isolation and the New Reform. Journal of Curriculum and Supervision, 4(1988)1, pp. 17-29.

GAMBETTER, D.: Can we Trust? In D. Gambetter (Ed.), Trust, Making and Breaking Relations. New York 1988 (Basic Blackwell).

HARGREAVES, A.: Teachers Change, Times Change. Vilnius 1999 (Tyto Alba). (in Lithuanian).

JUCEVICIENE, P.: An Integrated Approach to the Theory and Practice of Social Work: Response to the challenges of the 21st century. Ada Paedagogica Vilnensia, 8(2001), pp. 189-198.

LONGWORTH, N.: Making Lifelong Work: Learning Cities for a Learning Century. London 2000 (Kogan Page).

MARTON, F./WENESTAM, C.G.: (1988). Qualitative differences in retention when a text is read several times. In: GRUNEBERG, M.M./MORIS, P.E./SYKES, R.N. (eds.): Practical aspects of memory: Current research and issues 2, pp. 370-376.

MCLAUGHLIN, T.H.: Contemporary Philosophy of Education: Democracy, Values, Diversity. Kaunas 1997 (Technology). (in Lithuanian).

OZMON, H.A./CRAVER, S.M.: Philosophical Foundations of Education. Vilnius 1996 (Leidybos Centras). (in Lithuanian).

REARDON, J.: Development of Critical Thinking. Theory and practice. Vilnius 2001 (Girnelis). (in Lithuanian).

SMEDSLUND, J.: The problem of 'what is learned?'. Psychological Review, 60 (1953), pp. 157-158.

SPIEGELBERG, H.: The Phenomenological Movement: A Historical Introduction. Hague 1992 (Martinus Nijhoff).

STEWART, J.: Managing Change through Training and Development. London 1996 (Kogan Page).

Rūta Klimašauskienė

Knowledge Mobility between Higher Education and Educational Practice: Pursuing Qualities of the Learning Society

Introduction

Information and knowledge role is becoming more substantial in the contemporary society. The transition from industrial to knowledge-based (or in other words – learning) economy involves a complexity of economic and social processes. On the one hand, it holds the promise of increased productivity and an improved standard of living. But on the other hand, it simultaneously implies that individuals and organizations face major challenges in adjusting to new circumstances (OECD, 2001, p. 7). The emerging new forms of economic activity affect the characteristic nature of work and the types and levels of skills required in the economy. In this context there is a widespread agreement that the production and distribution of knowledge are increasingly significant processes in determination of economic development and competitiveness. The privilege of production and distribution of knowledge havs been assigned to institutions of higher education, universities in particular. However, the situation in the contemporary society is changing and universities are losing their privilege of new knowledge production, since the number of institutions, which are willing and are able to manage knowledge as main strategic source of competitive advantage, is growing (Jarvis, 1998). Snitka (2002, p. 26) points out, that new knowledge produced "is not valuable if it is not applied for practical and useful purposes". Different researchers highlight the dissemination aspect of knowledge, produced by higher education. The modern conception of higher education, which treats higher education as a mover of the society, stresses the importance of its activity in disseminating new knowledge. Higher education is transforming into a radically new phenomenon of a post-modern university rather than adapting itself in a changing environment (Smith, Webster, 1997, p. 104). This process of transformation first of all changes the system of knowledge production when a discipline-based (Gibbons (1998) refers to it as Mode 1), homogenous, and clearly hierarchical system is replaced by a system, transgressing the boundaries of a single discipline and responsive to a variety of students' needs and industrial priorities. Successful dissemination of new knowledge is a precondition of innovation stimulation, which is one of the most important conditions of development on regional, national and international levels. Higher education directly influences the growth of economy through the education of competent specialists, production of new knowledge, accumulation of knowledge, improvement of knowledge accessibility, and knowledge application in local contexts. The growth of contemporary economy is based on the accumulation of knowledge rather than capital (The World Bank, 2002). Traditional approach to knowledge production and application should be replaced by a systemic approach to innovative processes (EC, 2000). Therefore it is very important to apply the knowledge produced in higher education for the development of new products and value. Here one can deal with a relevant point of *searching for the ways, which could promote the mobility of knowledge produced in higher education and enable practice to absorb that knowledge. How does the knowledge distributed in these ways could pursue the qualities of the learning society?* This article is concentrated on how should the knowledge mobility between higher education and educational practice occur in order to increase the practical value of knowledge produced in higher education and thus pursue the development of the learning society. The methodology of this paper is based on the modes of knowledge production (Gibbons et al, 1994; Gibbons, 1997, 1998), the ways of knowledge dissemination, the knowledge creation theory (Nonaka, 1994; Nonaka and Takeuchi, 1995), which emphasizes the importance of tacit knowledge in the processes of knowledge creation. The changes of contemporary society are analyzed in the first part of this article, and possible ways of knowledge mobility and their influence to formation of qualities of the learning society are investigated in the second one.

The context of changes in society

Information processes, occurring since the middle of the past century are radically changing the way of mankind's thinking, lifestyle, and all society in general. These changes are entitled in different ways by various researchers (Toffler, Masuda, Castells, Habermas, Baudrillard, Giddens et al): postindustrial society, information society, postmodern society, knowledge society, etc. Every of these descriptions involve one common aspect: they are used to describe the main feature of the contemporary society – the usage of information processing and communication technologies in the everyday activity (VU Methodical Material, 2000). That means that society has found itself on a threshold when everyday activity is transforming into the activity of a new character and new quality. As Juceviciene (2000, p. 2) points out, in spite of different aspects separated by the above mentioned researchers, one can find the intrinsic characteristics of this society: the intensively growing rates of information, high-standard and constantly developing information and other related technologies. Bell (in VU Methodical Material, 2000) notices that an increased demand of theoretical knowledge and its usage is one of the features of such society. The reason is in the changing nature of the work – it becomes more intellectual and requires high-level specialized knowledge (Jucevičienė, 2000). Consequently, knowledge but not material or financial resources, becomes the main asset of the contemporary society.

The necessity of knowledge brushup or even replace inevitable because of constant intersification of information flows that influence the requirements for a workplace. Thus everyone can face the need for specialized knowledge in information society by deepening his/her knowledge, i.e. by learning and improving competence. Intensive growth of information flows requires the establishment of certain centers of information accumulation and analysis (Castells, in VU Methodical Material, 2000). But in such cities of information one can see the development of two-ply class system, depending on possibilities to manage information. The first (upper) class is composed of those whose activity enables to dispose information, whereas the second one (lower) consists of those who are busy with ordinary, simple work. Castells names the first class professional providers of information. They cover a relatively small part of the society, but dominate in culture, represent the elite and play the main role in the city of information. This class manages the flow of information, has possibilities to communicate with various people from all over the world and is cosmopolitan. Whereas the second class communicates only in constricted, insular circle and is localized in a certain place, has a relatively low level of enterprising. The key difference between those classes is unequal opportunities to reach the information. If the first one knows about the existence of various ways how to reach information and uses all of them, the second class knows about them but due to particular reasons does not use or can't access them. As Juceviciene (2000) claims, in pursuance to obviate so-called class of 'offcasts' (i.e. of those who do not catch the time to renew their knowing) in the information and, later – in the knowledge society, it's necessary to set another goal – the aspiration of the learning society. This could enable everyone to maintain his/her knowledge and competence in a proper level and to change them radically by necessity. The search of information and access to it determine knowledge possesion. An important point here is that the search for information as a process, having a certain finite result, is being replaced with the act of searching as a continuous, endless process, which becomes a lifestyle in the contemporary society. If an individual wants to keep up with changes, one of the main forms of life becomes innovative activity. Heylighen (2002) notices that rapid development of information technologies and the growth of information flows shows a paradox – he calls it information overload. This brings to that "people find it ever more difficult to cope with all the new information they receive, constant change in the organizations and technologies they use and increasingly complex and unpredictable side effects of their actions". That is because knowledge mobility between higher education and educational practice (when thinking about it not only in terms of formal educational system, in the aspect of learning process) is crucial. Definitely knowledge produced in higher education is of great value as having idiosyncratic potential to changes and development. That is why it is important to have this knowledge reachable to everyone who needs it. However, according to Melnikas, Jakubavicius, Strazdas (2000, p. 14), only a quarter of those ideas that are at the basis of innovation find their way to research structures. The contemporary society sets the demands for higher education not only to produce new knowledge, but to build the capacity to access the existing stores of global knowledge and to adapt it to local use (The World Bank, 2002). Knowledge per se is not valuable. It is valuable only when is applied for practical and useful purposes (Snitka, 2002, p. 26). The knowledge mobility has to become the main premise expanding value of knowledge and affecting innovations. So the second part of this paper is investigating the ways of knowledge mobility between higher education and educational practice, which are creating preconditions to enhance innovations, which in turn is on the basic elements of the learning society.

Higher education produced knowledge mobility as the main element of the learning society

Contemporary higher education doesn't limit itself with the functions of knowledge production and storing. Growing demands of the society add one more function – the function of knowledge dissemination. As changes in society influence higher education, the processes of knowledge production and knowledge dissemination are changing as well.

Let's look at the traditional higher education at first. The traditional higher education is characterized by a disciplinary structure. This implies that university research is carried out in accordance with scientific significance of its results. This conventional practice provides researchers with some guidelines about the importance of problems, the ways of solving them, the agents of this process, and a valuable input in a certain field. This structure is referred to as disciplinary-based knowledge production mode (Gibbons et al, 1994), Gibbons, 1997, 1998). In other words, disciplinary-based structure defines the concept of 'good science' and indicates what a student has to know in order to become a good researcher. Gibbons refers to this mode of knowledge production as Mode 1. Because of their institutionalized disciplinary-based structure, universities were the pioneers in validating this system of knowledge production. This structure has been the main mode of organising and managing the university. It has provided a rationale for study programmes and teaching; it has been the main element that integrates teaching and research, as well as grounds the importance of their integration.

Knowledge dissemination in traditional universities may take place following the linear model, which may be summarized as a premise that consumers adopt innovation as soon as they learn about it and that there is no difference between information reception and innovation adoption (Kriaucioniene, 2002, p. 15; OECD, 2000, p. 39). According to this model dissemination of knowledge is successive; it takes place from fundamental science to marketplace (Lithuanian White Paper on Research and Technology, 2001, p. 60): fundamental research, applied research, technological development, market research, product development, production, and marketing. Thus knowledge produced in Mode 1 is prescribed to society as if knowledge is able to automatically stimulate innovation, as well as society development. This model is based on the traditional approach to dissemination: knowledge has to be accessible to those who are interested in it (EC, 2000 b, p. 28).

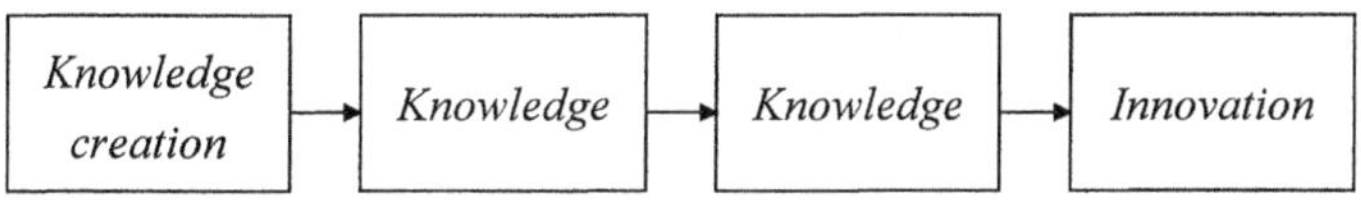

Figure 1: The linear model of knowledge dissemination (adapted after OECD, 2000, p. 39)

The linear model of knowledge production is related to two fundamental problems (OECD, 2000). First of all, the model comprises at least seven complex processes (knowledge production, verification, comparison with existing knowledge, dissemination, adoption, implementation, and institutionalization), which also involve a number of factors, which may determine a failure in knowledge transmission. The second problem is that these components are conceptualized as certain stages. However, not all of them have to be present in every case. Moreover, in practice these stages are interrelated with each other by feedback what the linear model does not reflect. One could find even more problematic aspects when talking about the linear model of knowledge dissemination. Another aspect is that this model ignores tacit knowledge of individuals, and the context to which knowledge is prescribed is considered as completely blank – some kind of *'tabula rasa'* – and without any prior experience. Regrettably, practice shows that processes of knowledge dissemination are occurring in a different way. The so-called agents play important role in adoption and dissemination of new knowledge. Their tacit knowledge is central in the process of absoption (EC, 2000b, p. 2).

The linear model is closely related to traditional conception of higher education, when it was usual that higher education, being in the 'top' of the society, influenced its development by production of new fundamental knowledge (Jucevičienė, 1997). This concept emphasizes the exceptional position of higher education in the society, when knowledge produced in the university was 'prescribed' to the society and became an unquestionable landmark of development. Society depended on higher education, but had no influence on it. Universities had their privilege and autonomy. Knowledge produced by universities was the source of social development and was accepted without no doubts, questions or discussions about their truthfulness or objectivity. Thus one more problematic aspect emerges: knowledge disseminated in accordance to the linear model is entirely of traditional nature: it has disciplinary nature and is designed to solve problems derived from concrete disciplines. Such traditional understanding doesn't meet the demands of the modern society. It needs knowledge suitable for practical purposes and use, and scientific knowledge, employed to gain such practical knowledge is required to be integrated from different disciplines.

One could say that pecularities of knowledge production in Mode 1 dictate static and unflexible character of knowledge dissemination. One more weakness of the linear model is that because of the absence of feedback it is difficult to disseminate new knowledge exactly where it is most of all needed. The linear model could be considered as suitable for the coming of new technologies or technics, but it is not for ensuring the competetiveness of these technologies as well as their viability in the global market (Lithuanian White Paper on Research and Technology, 2001, p. 61). But the main point is that in the contemporary society a number of institutions emerge that are able to produce knowledge. This knowledge is actively discussed in society and there is no more unquestionable truth. Thus we should talk about the process of a different – interactive kind.

Gibbons et al (1994), Gibbons (1997, 1998) measuring changes in society, points out that evidence of a new system of knowledge production have been emmerging. He refers to the new mode of knowledge production – Mode 2. This way of knowledge production differs from Mode 1 firs of allt in the context of knowledge production. Mode 2 is employed for problem solving in a broader context of application comparing with Mode 1, intended to be useful more generally to someone in industry, government or society. This implies that application is not related to improvement or development of some specific product for a specific branch of industry. Application is conceptualised in a much broader way that it normally is when we refer to implementation of ideas into a certain practice. The new mode of knowledge production does not limit itself with single disciplines in the production of new knowledge also because in the knowledge creation process specialists from different fields of disciplines are employed. The solution is bound to transgress disciplinary boundaries and become multidisciplinary, even transdisciplinary. Multidisciplinarity boundaries mean collaboration of different scientific fields: using different methods for some time. Transdisciplinarity boundaries imply that this kind of collaboration of different scientific fields and social sectors is constant and systematic (Mittelstrass, 2000; Klein, 2001). Transdisciplinarity boundaries are expressed not only on the level of application, but also in the inner cognitive structure of science (VU methodical material, 2002). This point is important in pursuance of successful activity in the learning society, as the learning society is requiring knowledge of different kind. Besides one may assume that the new transdisciplinary knowledge will be easier to transfer and adapt in different contexts.

Another inescapable impulse for emergence of the new mode of knowledge production is skill profile needed for succesful problem solving. The new form of knowledge production requires comprehensive skills and experience; this requirement is constantly changing, says Gibbons (1998). Individuals, who possess very different skills, abilities and experience, produce knowledge. In contrast to Mode 1, it is often very difficult to predict problems in Mode 2. In addition, a vast number of other potential knowledge producers emerge in non-university institutions; communication networks (electronic, organizational, informal, social, etc.) can help them to become a single entity. Differentiation of specializations in these networks forms preconditions for new forms of useful knowledge.

Solving problems of such kind requires different organizational characteristics. Flexibility and response time are the major factors in Mode 2 (Gibbons, 1997). For this reason research groups are not fully institutionalized in this mode: they are formed only in order to solve a certain problem. Problems are temporary, so research groups are not established for a long time. Matrix and network structures normally dominate in these organisations; research groups are formed to solve specific problems. Competence gained in this process is easily transferred to other contexts, which makes it more valuable. When specialists of different competences and activity areas return to their familiar environment, they have already changed their system of knowledge and expanded their understanding. They disseminate new knowledge to their peers; thus it reaches society easier and in a shorter period of time. According to Snitka (2002, p. 29), this kind of direct or indirect dissemination of knowledge stimulates the development of new technologies, processes or products in companies, and even may provide some economic profit.

Perception of potential influence of research and technology development on social development encourages the formation of different interest groups that seek to influence research results. This is reflected even in the composition of research groups, as researchers and practitioners of different fields often work together (this is required by the nature of research). Social accountability is recognizable in the entire process of knowledge production not only when results are interpreted or disseminated, but also when problems are pointed out and research priorities are set (Gibbons et al., 1994). The contemporary society is no longer a passive viewer of the process of knowledge production – it is an active partner in this process. Researchers have to encode dissemination parameters in the solution of a problem; this is related to values and attitudes held by different interest groups that have been kept away from the field of research before. As problems are solved in the context of application, scientists and researchers surely become more reflective: they are active participants not only in defining and solving problems, but also in evaluating research. This is how social accountability increases. In Mode 2 those that are important for the context of application supplement the criteria of intellectual interests: social, economical, political, etc. The degree of competitiveness of the solution, its costs, social acceptability and other aspects are taken into consideration. Thus the system of quality control becomes more flexible and socially variegated. This happens because it is increasingly difficult to define a concept of a 'good science'. Gibbons (1998, p. 10) points out a question of whether this kind of control negatively affects the quality of results. But a broader system of control does not imply lower quality of results. They will only be different from those obtained in a disciplinary-based system, i.e. of multidimensional nature.

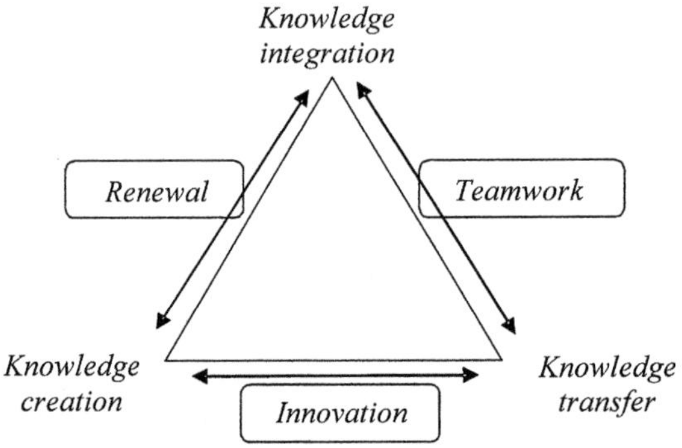

Figure 2: Interactive model of knowledge dissemination (OECD, 2000, p. 73)

One could find each of the mentioned aspects in the contemporary concept of higher education (Barnett, 1994, p. 12), noticing that interrelation of higher education, society and knowledge takes a new shape. The process of new knowledge production and introduction to society now is changing from delivering of unquestionable truth to a discussion and critical evaluation of new knowledge. In this context, when society becomes an active partner in a knowledge creation process, the processes of knowledge dissemination and its absorbtion take different – interactive – character. It shows the complexity of relations between institutions of higher education and educational practice.

According to this model, knowledge is produced using human resources (through teamwork). It differs from any other capital good in its flexibility: reallocation of human resources can result in new teams that are able to produce new forms of specialized knowledge (Gibbons, 1998). Ability to do this is treated as a major precondition of success. Innovation is considered as an inseparable part of this model. It is reached when new knowledge is implemented in practice and realized. Human beings are important "carriers" of knowledge (EC, 2000 b). Knowledge is inside them: it is not codified or transformed into information, but is essential for a successful functioning of both organization and society. Thus tacit knowledge plays substantial role in the process of knowledge production and dissemination, especially when practitioners, not only members of a certain academic community take part in this process (OECD, 2000). Tacit knowledge is the knowledge that is referred to as 'vitally important for innovative process' (EC, 2000b, p. 15). But the problem is that scientists mostly use explicit and codified knowledge, whereas practitioners rely on tacit knowledge. Researchers simply cannot evaluate practitioners' tacit knowledge, as it is different with every individual; researchers also have a system of individual tacit knowledge. It is possible to solve this problem of knowledge dissemination if we treat it from the point of view of Mode 2. As in Mode 2 knowledge is produced by diversified interest groups rather than individuals, they are compelled to share not only their explicit, but also tacit knowledge. The collaboration between researchers and practitians is important in order to reduce a gap between scientific knowledge and knowledge required for practice. Mode 2 and interactive model of knowledge dissemination precondition closer cooperation between practitioners and researchers. When participating in the process of knowledge production practitioners raise problems, which are relevant for practice; hence the knowledge gained in such process will bring more practical value and thus reduce the gap between scientific and practical knowledge. The gained benefit could be mutual for science (eg. in the context of deper understanding of a certain object) a well as for practice (enhancing innovations through the usage of new knowledge).

As the process of knowledge production incorporates not only higher education but also other social establishments, and the process is becoming more diversified with respect to profile and locality, another model of knowledge dissemination, network model, is also applied (EC, 2000b, p. 20). Knowledge production and application takes place in industry, services, and other fields. Knowledge is produced not only in scientific, but also in other organizations. Complex network relationships of activity, institutions and agents of dissemination interconnect it. Knowledge networks are conceptualized "as social networks between knowledge actors, in order to allow the creation and transfer of knowledge on an individual, group, organization and inter hierarchical level" (Seufert, von Krogh, Bach, 1999, p. 186). The network model, according to the above mentioned researchers, enables interconnections between 1) different levels and areas of knowledge: individual knowledge types (e.g., explicit and tacit), different knowledge levels (e.g., individuals, group, organization) and different areas of knowledge (e.g., customer knowledge, R&D knowledge); 2) knowledge work processes and knowledge network architectures: different physical places (real, virtual or mental), different timezones, etc.; 3) knowledge work processes and facilitating conditions.

Implementation of the network model is also important because it enables to temper the differences between the so called 'oases' and 'deserts', i.e. regions with high level of innovation implementation and those with low level of knowledge production, dissemination and application. When we point out the importance of absorbing and integrating new knowledge and innovations rather than merely accepting them, we also have to note the importance of effective communication networks (EC, 2000b, p. 24). A significant role in this process of course is played by information technologies, which facilitate the access to information and knowledge (The World Bank, 2002). Even here, however, the processes of globalization reveal a paradox that 90% of knowledge is produced not in the place where it is most necessary (Gibbons, 1998). Moreover, revolution of information technologies brings another danger: the differences in the development rate of information technologies in different regions are growing. Therefore, organizations involved in knowledge production face a challenge of transferring knowledge to those places where it can be most effectively applied. Networking can help organizations to find knowledge that is necessary but cannot be produced inside the organizations, as well as apply it for successful innovation implementation. Networking of different sectors can also help to reduce competition of different regions; development of complementary networks for medium-sized sectors can lead to a synergy effect (EC, 2000b, p. 81).

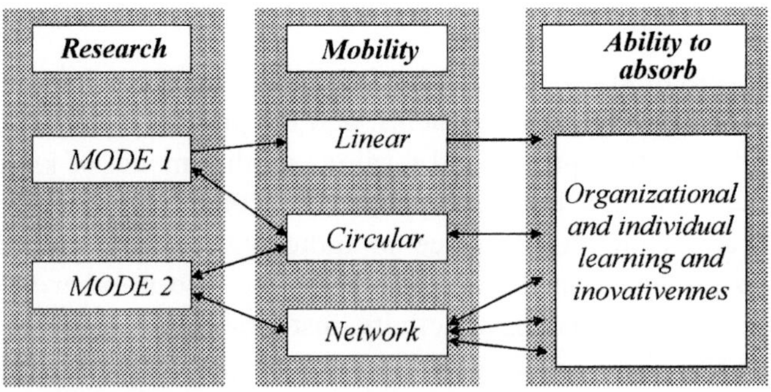

Figure 3: The relationship between knowledge production, dissemination and ability to absorb

But the main point is that networking allows increasing the mobility of existing knowledge. Mobility of knowledge in this structure is directly related to the flow of human capital, as knowledge can be (and is) transmitted not only by some medium of codified knowledge (books, journals, electronic databases), but also by people who contain both codified knowledge and the 'invisible side' of knowledge, i.e. tacit knowledge.

Not only mobility of knowledge is important when talking about the creation of the learning society. Ability to absorb new knowledge is relevant as well. One could find this point in The First Action Plan for Innovation in Europe (1997, p. 10), which says that "the crucial factor for innovation is the link between research (the production of knowledge), training, mobility, interaction (the dissemination of knowledge) and the ability of firms to absorb new technologies and know-how". Definitely the ability to absorb new knowledge and to aplly it afterwards is an important factor of successful activity in the learning society. The ability of individuals or organizations to absorb new knowledge and transform it into a certain form of activity directly influences their potential to innovations. Accordingly, knowledge in the form of innovation could be regarded as an output of the learning activity (OECD, 2001, p. 12). Learning in this case covers both levels – the individual (when information, knowledge, understanding or skills are aquired by individual people in some form of education – formal or informal) as well as the organizational (which is constructed on the basis of individual knowledge) (OECD, 2001, p. 14). A resumptive scheme of the changes in the society is drawn in the Fig. 3.

One could find a substantial body of research, which shows that the linear model of knowledge produced in the traditional higher education dissemination is an exception rather than a rule (OECD, 2001). The lack of feedback and the prescribtion of the 'truth' to society are no longer sufficient. The new ways of knowledge production require absolutely new ways of knowledge mobility. The present mode of knowledge production could be called transitional from Mode 1 to Mode 2. The traditional, disciplinary science still is the base of the higher education, but a lot of elements of the new knowledge production system could be found in universities. Of course, knowledge produced in this system is disseminated under the circular model, which displays the complexity of knowledge transfer and absorbtion. Whereas network model refers to dissemination of knowledge produced not only in higher education, but in other social institutions as well; the network nature of this process means that the elements of production, transfer and absorbtion are closely interrelated in the aspects of participating institutions, individuals, intercourse between them likewise multiple natures of the processes. Contrary the traditional approach emphasizing the spreading of information to a large audience, nowadays in the process of dissemination the interest is more and more directed to disseminating the results in a way that could affect the development of individuals and communities (IACEE, 1999). Possibility to collaborate in the knowledge production and dissemination processes stimulates individuals and organizations to be active because this is the way they may influence the production of the knowledge that is the most useful for them at the moment. And this is how the circular as well as the network model enable to avoid the class of "offcasts" in the learning society. Participating in these processes enriches the existing knowledge and so contributes to the development and learning. It also promotes the innovative potential thus retaining high level of competetiveness on the one hand and creating more comfortable workplace on the other one.

But concerning the problemic aspects of the learning society, information overload should be mentioned. According to Heylighen (2002), this means that because of the existing information abundance, a big part of it may not be fully assimilated, another part could be ignored or forgotten, stay overlooked or lost in any other way. And yet the most highly educated people seem to sufer most acutely from information overload. While they may be objectively most competent to make decisions, they are subjectively most aware of the gap between the information that is out there and the one they can effectively assimilate. Enhancing knowledge mobility and collaboration between practitioners and scientists may presuppose that thus produced and transferred knowledge will be of a major practical value in a certain sphere and there will be a smaller probability for this knowledge to be forgotten.

Conclusions

The processes of transformation in society reveal the necessity to talk about new relationships between knowledge produced in higher education, their mobility and society. A new infrastructure of knowledge production with higher education and other social institutions involved, influences changes in knowledge production in higher education. It gradually changes from disciplinary - based, hierarchical, and exclussively academic, into transdisciplinary, application context - oriented, which is of a "flatter" hierarchic structure. These changes precondition the participation of 'practitioners' in knowledge production; this is how the differences between pure and applied knowledge are tempered. The traditional linear model of knowledge mobility unfortunately does not consider the importance of individual tacit knowledge in knowledge dissemination and innovative processes and is not suitable when striving for the qualities of the learning society. The changing forms of knowledge production require changes of knowledge mobility modes. The non-linear (interactive, circular) and network models are used to disseminate knowledge produced in the contemporary higher education.

They are flexible and give opportunities for both researchers and practitioners to take part in knowledge production, dissemination and innovative processes. They also point out the importance of tacit knowledge because new knowledge can become innovation only when it becomes tacit knowledge – only when an individual is interested in it, is able to know about its existence, find and assimilate it too. The network model not only interconnects different types of knowledge in dissemination, but also highlights the importance of incorporating knowledge produced in different time and place; it also preconditions knowledge dissemination in those places where it is the most necessary and the most useful. Consequently, pursuing the qualities of the learning society, reducing information overload and creating premises for everybody to learn in time, the main attention should be drawn to the knowledge produced in higher education mobility according to the circular and network models, equally to processes of learning, which enables individuals and organizations expand the abilities to absorb new knowledge.

References

BARNETT, R.: The Limits of Competence. Knowledge, Higher Education and Society. Buckingham 1994 (SRHE and Open University Press).

GIBBONS, M./LIMOGES, C./NOWOTNY, H./SCHWARTZMAN, S./SCOTT, P./TROW, M.: The New Production of Knowledge. London 1994 (Sage Publications).

GIBBONS, M.: Innovation and the Developing System of Knowledge Production. Brighton 1997 (University of Sussex).

GIBBONS, M.: Higher Education Relevance in the 21st Century. In: KRIŠČIŪNAS, K./RINKEVIČIUS, L. (eds.): Aukštojo mokslo, technologijų plėtros ir inovacijos politika. Compendium. Kaunas 1998 (KTU).

EUROPEAN COMMISSION: The First Action Plan for Innovation in Europe. Buletin of European Union, Supplement 3(1997).

EUROPEAN COMMISSION: Innovation in a Knowledge-driven Economy. No. 567 final. The Commission of the European Communities. Brussels 2000.

EUROPEAN COMMISSION: Innovation Policy in a Knowledge-based Economy. No. EUR 17023, The Commission of the European Communities, Directorate General, Luxembourg 2000.

HEYLIGHEN, F. (2002). Complexity and Information Overload Society: why Increasing Efficiency Leads to Decreasing Control. Draft Paper. Available from: <http://pcp.vub.ac.be/HEYL.html>.

IACEE: Effective Dissemination: A Guide on the Dissemination of the Results of International Educational Projects. Helsinki 1999.

JARVIS, P.: Besimokanti visuomenė – šiuolaikinio pasaulio reiškinys. Mokykla 1-2(1998), pp. 1-6.

JUCEVIČIENĖ, P.: Lietuvos aukštasis mokslas. Quis es? Quo vadis? Studijos aukštojoje mokykloje. Konferencijos pranešimų medžiaga. 1 knyga. Kaunas 1997 (Technologija).

JUCEVIČIENĖ, P.: Ar įmanoma veiksminga švietimo reforma be esminės požiūrių į mokslą kaitos? Mokslas – studijos – mokykla: VII tarptautinė mokslinė konferencija: plenariniai pranešimai. Vilnius 2000 (LR Švietimo ir mokslo ministerija, Vilniaus pedagoginis universitetas, Egerio Karoly Esztierhazy mokytojų rengimo kolegija), pp. 17-24.

KLEIN, J.T.: Unity of Knowledge and Transdisciplinarity: Contexts of Definition, Theory and New Discourse of Problem Solving. 2001. (<http://www.mines.edu/newdirections/essay2.htm>)

KRIAUČIONIENĖ, M.: Nacionalinės inovacijų sistemos vystymo metodologija. Daktaro disertacija. Kaunas 2002 (KTU).

Lietuvos mokslo ir technologijų baltoji knyga. Vilnius 2001 (Mokslo ir studijų departamentas prie švietimo ir mokslo ministerijos).

MITTELSTRASS, J.: Transdisciplinarity – New Structures in Science. 2000. Available from: <http://www.mpiwg-berlin.mpg.de/ringberg/Talks/mittels%20-%20CHECKOUT/Mittelstrass.html>

NONAKA, I.: A Dynamic Theory of Organizational Knowledge Creation. Organization Science, Vol. 5, No. 1(1994), pp. 14-37.

OECD: The Knowledge-based Economy. Paris 1996.

OECD: Knowledge Management in the Learning Society. Paris 2000.

OECD: Cities and Regions in the New Learning Economy. Paris 2001.

SEUFERT, A./KROGH, G. VON/BACH, A.: Towards Knowledge Networking. Journal of Knowledge Management, Vol. 3, No. 3(1999), pp. 180-190.

SMITH, A./WEBSTER, F.: The Postmodern University. 1997 (SRHE and Open University Press).

SNITKA, V.: Mokslinių tyrimų, technologijų, inovacijų politika ir žinių ekonomikos plėtra. Kaunas 2002.

THE WORLD BANK: Constructing Knowledge Societies: New Chalenges for Tertiary Education. 2002

VU: Metodinė medžiaga: informacijos visuomenė ir jos vadyba. Vilnius 2002.

Dalija Gudaitytė, Vida Vilkienė

Towards the Knowledge Society: Co-operative Learning Encouraged by European Mobility Projects in Teacher Training Higher Non-University Educational Institution

Introduction

The permanently changing world is in demand of new abilities, skills and competencies; human resources is the fatal factor in determining progress of teachers' traditional roles to deliver information, to organize and manage the teaching process, and acquire the new quality. Modern teacher as an agent of moral and social changes has to assist in the change of regaining the moral aim of teaching that is necessary for further knowledge development and teaching as well as learning.

The experience gained in European foreign language teacher training mobility projects as well as its dissemination in the context of cooperative learning enriches teaching and learning. Therefore, it is essential to ascertain the importance of the impact, influence and dissemination at College through the implementation of the foreign language teacher-training programme. Mobility is beneficial not only to teacher trainers or students but also for communication and discussions about problematic situations, comparison of culture as it creates favorable learning atmosphere. It is necessary for the parity of communication; it also enables both teachers and students to learn overcoming culture barriers and country borders from each other.

Different institutions in different countries implement European educational mobility projects; it is a way to ascertain teaching content, the interdisciplinary and communication aims. The quality of studies at teacher training institutions, colleges included, is in dependence with skills and possibilities to perceive now concepts of learning.

The global and European education dimensions require a new competence. In the context of the dimensions mentioned above foreign language teacher training occurs in the transfer of the traditional teaching paradigm, namely, from teaching to life-long- learning. Therefore much importance is attached by Western researchers – Bourner; Katz; Watson (2000), Fry; Ketteridge; Marshall (2000), Toohey (1999),Coll (2002), Reece; Walker (2000), Fullan (1998), Jensen (2001), Ramsden (2000), Ulriksen (1998), Gibbons (1998), Zay (1998), Ball (1999), Grenfel (2002), Hunt (1996) and others. The teacher training problem has been considered as well as analyzed by Lithuanian scientists Jucevičienė (1998, 2000, 2001, 2002), Adaškevičienė; Janiūnaitė (2001), Baranauskienė (2002), Dabrišienė; Jotautienė; Tidikis (2001); Cesevičiūtė; Minkutė-

Henrickson (2001), Gudaitytė (2000), Tumėnienė (2001), Rudaitienė; Vyšniauskienė (2002), Lipinskienė (2001), Kavaliauskienė (2001), Saulėnienė, etc.

The article deals with the research problem of the experience gained in European mobility projects and its dissemination at College as a new structural subdivision of the Lithuanian system of education as enabling to seek for modern learning goals at modern higher education institution; it also discusses if the dissemination of the experience can become an organic part of the programme implementation process and assist in improving the quality of studies.

The aim of the article is to analyze the experience gained in European mobility projects and its dissemination at College as a new sub-division of the system of education in Lithuania. The possibilities of achieving the aims of higher education are discussed seeking to improve the quality of studies by disseminating the experience gained as well as to integrate it into the modern foreign language teacher training programme.

The research methodology used in the article is based on:

- The European dimension in education;
- The philosophy of establishing a learning environment for teachers and students' learning in partnership enabling cooperative learning (Wenger, 1998); Juceviciene, 2003);
- Modern as well as classical principles of higher education underlying the parity of students and teachers in communication and cooperation within the study progress (Barnett, 1990; 1994).

The methods used scientific literature analysis, questionnaire, descriptive statistics.

The article consists of three parts: the first part deals with the analysis of the peculiarities of implementing foreign language teacher training programmes in the context of the European dimensions at college – a higher education institution. The second part focuses on the learning possibilities provided by European mobility projects through the implementation of foreign language teacher training programme. The third part discloses the analysis of the results of the research on the dissemination of the experience gained in European mobility projects in the context of cooperative learning at College.

1. The peculiarities of implementing modern foreign language teacher training programmes in the context of the European dimensions at college, a higher non-university teacher training institution

The conclusions of the analysis by Adaskevicienė and Janiunaitė (2001) on the European dimensions in the Western countries state that the object of the European dimensions – from the conception to values – reveal the problematicity and the multi-segmentation of the concept. From the point of view of the implementation it should be stated, that the process is possible through the educational policy in the contents as well as in everyday activities of the teacher and the learner. From the point of view of the segmentation in the process of implementation the development of the European dimensions is directed to teachers and learners' attitudes, approaches as well as values and citizenship.

The European dimensions are to be perceived as the inter-subject integration of the teaching contents on grounds of cooperation, information technologies and group work. The major part in the process of the European dimensions implementation is of *projects* developed by educational institutions in different countries as one of the ways to set common purposes for communication, teaching contents and integration of study subjects. A teacher who is directly involved in projects at the institution implementing the European dimensions through teaching contents is in need of new experiences. This is in inevitable connection with teacher training as one of the segments of the European dimensions. The experience and qualifications gained by teachers in European projects impact on the process of their implementation, teaching staff mobility and cooperation. The skills in European languages and information technologies are indispensable. Thus, the global and European dimensions are in the background of modern teacher's space; the traditional role of the teacher changes and a demand of new competences appear: a teacher is to be a promoter, a researcher and a catalyst, a mediator of changes, a colleague in cooperation and an innovator. The new roles require a qualitatively new approach to teaching as well as an ability to take the major responsibility in mediating changes.

In the context of the European dimensions and in the course of the change of the traditional vocational training dimension into the modern one an educational institution aims at training not just a specialist to work in a certain field (namely, a foreign language teacher) but at enabling to acquire new skills, abilities and knowledge to be applied in today's world and the further developing of competences (Jotautiene, Tidikis, 2001; Lepaite, 2001). The foreign language teaching qualification is in accordance with vocational competence; they contravene in competence levels. Furthermore, the vocational qualification gained is a further stimulus to develop teaching competences.

In summary, the training of future foreign language teachers is in the background of the change of traditional paradigm into life-long learning and comprises applying the communicative approach in foreign language teaching, intercultural competence as an absolutely important necessity for future foreign language teacher mobility and inter-cultural flexibility, development and transition from behaviorist as well applied science model to holistic-interactive-reflexive model in higher education. The confrontation of the mentioned aspects with the social background, the system of values, attitudes to culture, politics and the heritage of historical and ideological traditions of the country call forth the problems of foreign language teacher training programmes. The new skills as well as the appearing new roles of life-long-learning teachers involved in the changes influence them in the course of the teacher's career and are closely interrelated with competence enhancement and reflection. Future teachers are oriented to acquire the mentioned attitudes in the course of their studies.

2. The peculiarities of foreign language teacher training programme at college – a higher non-university institution

The binary system of higher education in Lithuania was legitimated in March, 2000 seeking to solve the problems of higher education (for instance, insufficient influence of social partners on the processes of the higher education system, the non-existence of non-university sector, poor regulation of qualifications offered, etc.). College is characterized as a non-university institution in which non-university study programmes prevail and majority of students studying within them. Thus, two ways of acquiring profession are available – by choosing university study programmes or gaining a concrete qualification by unity of studies and work. Moreover, a right to choose is implemented in the democratic society of today's Lithuania.

The training of future foreign language teachers as well as teacher training in general are closely linked with the aims of education, namely, humanism, democracy, nationalism, innovation, accessibility, contextuality, effectiveness, continuity and diversity. The modern teacher training programme assists in seeking for goals and aims of education development – effectively and accordingly providing accessibility, social justice and continuity with quality assurance and expanding the interrelationship of education and life practice and meeting the challenges of the 21st century. Modern teacher's profession is permanently challenged by new social, educational as well as training requirements, therefore, the quality of education depends and will always depend on the quality of teacher training as well as ceaseless improvement of teaching methods and regulations of higher study programmes. Qualitative teacher training enabling the teacher to represent wholesome life quality in education impacts and influences the welfare and future of the society.

As Watson (2000) states, higher education has always been associated with the acquisition of profession and the exceptional feature of higher education institutions has always been to unite academic values linked together with study subjects and vocational values related to practice and profession. Therefore, the urgency of vocational higher education enhancing partnership and cooperation within as well as outside institution is maintained in study programmes, teaching contents and its presentation. On one hand the writing and development of study programmes based on partnership combine with the responsibility for their quality assurance and the wholesome assessment of institution. On the other hand, the cooperation of the institution with vocational and statute structures encourage the improvement of vocational programmes. On the basis of knowledge and practice 'the surplus value' of higher education is created as well the tradition of vocational higher education is enhanced. More flexible structure of study programmes and the ability to work independently in order to gain wider than specialized qualifications and skills are challenges to be met in the process of study programme implementation and seeking the aims oriented to learners (Ulriksen, 1998).

Development of higher education curriculum is based on five philosophical approaches to curriculum, as Toohey (1999) states, namely, traditional or discipline-based, systems-based, cognitive, experiential and socially critical. The ideal study programme comprising the best features is not always achievable and it is of benefit to students to experience the diversity of approaches. The ever-lasting higher education curriculum is non-existent (Ulriksen, 1998); it has to be reflexive and its thoroughness and consistency as well as incontestability are to be ceaselessly discussed and interpreted. It is obvious that the goals of curriculum are out of negotiation. It also means that teachers as well as students have to be aware of the totality they are seeking in the processes of teaching and learning. Therefore, the reflective approach in common with an active interest becomes the most important requirement for teaching improvement in order to enable the quality of studies. The quality assurance is one of the essential priorities of education development. The quality assurance has to be in accordance with a human's labor market - oriented needs and life in the open civilized society, assisting students in adapting to reality and assessing it critically, as well as maintaining the identity, independence and seeking for self-development and self-expression.

Therefore it is of importance to form both general and vocational as well as public abilities and skills. Meeting the challenges of the change from learning to teaching is expected from teachers at higher education institution, university or college. They must to have excellent teaching skills, be flexible in professionally changing teaching methods while delivering subject courses in order to meet the diverse students' needs.

According to Jucevičienė (2001), a higher education institution seeking to enable students to deal with problems in the future has to create a system of teaching and learning on grounds of deep student's attitude towards learning as well as to establish an atmosphere favorable for studies. Learning environment with a high responsibility as well as a possibility of autonomy make learners self-confident and encourage their motivation. Enabling learning is an educational value and an assistance in acquiring a life-long-learning competence.

Teacher training is a learning continuum. A learning society cannot be created without a teacher involved in life-long-learning. Teacher as an agent of moral and social changes has to promote moral goals of learning in order to further develop knowledge and skills that are necessary for an institution to change as well as to meet today's challenges. A modern foreign language teacher is obligated to acquire general as well as specialized subject knowledge; teacher has to become a moderator of independent learning processes and be enabled to work on expanding the functions of educational institution; to cooperate with colleagues, learners and their parents as well as the society; to project and continuously improve the activities of the educational institution. In this world of today with its permanent changes and the development of human resources being of vital importance, new abilities, skills and competences in higher education are set to be the holistic competence enhancing the learner to seek for changes in action (Lepaitė, 2001). Therefore, modern foreign language teacher training programme has to be created on grounds of holistic competence as well as is to be oriented to competence development. In other words, the implementation of the competence-oriented modern foreign language teacher training programme is called forth to gaining qualifications and developing the ability of action based on knowledge, skills, attitudes, personal qualities and values.

As Jucevičienė (2001) reveals the concept of European dimension developed in accordance with the functions of educational activities, it can be stated that a modern foreign language teacher training programme is oriented to the acquisition of skills in at least one (recently emphasized-two) European languages, modern information and communication technologies, implementation of educational innovation (European mobility projects are considered to be one of the recent possibilities for innovation) as well as in European thinking competence. The latter in the context of modern foreign language teacher training programme implementation gives way to European thinking and enables both a future foreign language teacher and a teacher trainer in promoting educational innovation in the form of the European dimension as well as the awareness of the education systems of other countries. Modern foreign language teacher training programme seeks to develop general competences, namely, of communication, information selection, analytical, critical thinking and problem solving, case and action research, team work, personal knowledge and improvement, learning, projecting, action in organization as well as openness to society and innovation. In the course of teacher training programme implementation the future foreign language teacher is enabled to transfer general skills into teaching process.

Summing up, a modern foreign language teacher training programme to be implemented at higher non-university educational institution has to enable future teachers to actively participate in public life, to meet the learners' needs and the demand, to develop professional skills and to interrelate their learners' attitudes, experiences and conceptions with the changes of the conception of today's world, to apply deep approach to learning in accordance with the conception of life-long-learning and continuous professional development. To achieve the aim, a higher non-university institution has to search for new methods and ways in order to assure the quality of studies.

Collaboration in the context of cooperative learning

The demand for cooperation is enhanced by the need to improve teaching and learning. If teachers are involved in achieving the aims of the teacher-training programme and are eager to cooperate with their colleagues, to meet the challenges of the permanent change as well as to enable learners in cooperating effectively, the success of cooperation is guaranteed. Cooperative learning in the group develops partnership skills in listening, receiving and sharing information, managing and challenging as well as assists in dealing with problem solution.

As Zay (1998) points out, the concept of partnership links with that of 'reflective practice' and in the context of cooperative learning the activities of a learning community are based on reflective thinking-in-action. Jucevičienė (2001) points out the criteria of successful cooperation: benefit for all partners (the more numerous their participation, the more successful the partnership), the responsibility for the success of the cooperation, systematic meeting of collaborating teachers, pro-active management that enables the partnership freedom as well as right interests.

Reflection is considered to be of the essential importance in transforming and integrating new experiences and conceptions with the former experience. Reflection acquires major value while learning from experience. The situations invented in the teaching process enable students to reflect; this is achieved by applying appropriate teaching strategies in order to empower learning, change the former knowledge and form new skills in reflection seeking for life-long-learning. Zay (1998), Schon (1983) refer to John Dewey first formulated the notions of 'reflective thinking' that links education, experience and life. Schon (1983) adds that 'reflection-in-action' or 'in the course of action' gives rise to a restructuring of the practice. According to Schon, 'reflection- on – action' is an orderly application of logic to the known data on order to resolve a problem. 'Reflection – for-action' implies a programming into the future. 'Reflection –about – action' occurs when wider contexts, namely, the social, economical and political ones are taken into account. It is often accompanied by the wish to improve activities.

The essential aspect of reflection is time, present and past. Major share of the best practice is on the grounds of thinking before starting an action. 'Reflection-in-action' represents the partners' manipulations of time between the interrelated actions in order to re-think, reflect and to evaluate. Zay (1998) and Day (1995) think that all forms of reflection are of equivalent value and neither of them is more superior than the others, thus, teachers should apply it in their life-long-career. Different kinds of reflection can be perceived as reflective practice continuum. A necessity of teachers' (involved in cooperative learning) reflection can cause changes in their attitudes and practical skills that might impact on the institution as well as on training or vocational training programme.

Summing up, learning and activities of teacher trainers involved in cooperation is based on reflection; furthermore, reflective didactics of higher education empowers the development of democracy as one of the main principles of higher education. Teachers who learn in partnership with their colleagues and reflect on their experiences and activities enhance their professional development, self-development and self-expression.

3. European mobility projects: possibilities of learning and implementing a foreign language teacher-training programme

European mobility and cooperation projects are implemented in partnership. Partners have to work together in order to improve the training of future teachers and build up a strategy in order to assist in improving the quality of teaching and learning. Projects and project activities are the means to achieve the results foreseen in accordance with the demand of the study programme, course, methods or teaching strategy on the grounds of practical experience; The European dimension in teaching should be highlighted in general as well as in specialized study subjects delivered by newly trained future teachers.

European mobility projects can assist in identifying general needs of learning groups as well as improving the present or newly developed study programmes for teacher training. Especially important elements are defining and implementing the teaching contents based on the European dimension, the continuum and impact of the project. Because of the diversity of initial teacher training in Europe, it is frequently rather difficult for those future foreign language teachers who take part in mobility projects and activities to make the most use of the project benefit at the same time mobility being one of the most potential ways of disseminating the experience of the European dimension in teaching (Comenius: school education, 2001). Therefore, participation in mobility projects and activities might and should improve the quality of teacher training, to orient to practical activities and observation instead of acquiring the theoretical knowledge instead.

Criteria	Statements in the questionnaire	Answer average	Mode
Interrelation of knowledge	**Participation in European mobility project assists in integrating the newly acquired knowledge with the former one**	**5**	**5**
Impact of experience on personal experience	Participation in European mobility project assists in integrating the new material into everyday personal experiences	**4,6**	**5**
Diverse learning possibilities	Participation in the project provides more various learning possibilities	**4,6**	**5**
Encouraging self-expression	Experience gained in the project encourages self-expression	**4,5**	**5**
Impact of dissemination of experience on learning	Sharing experience assists in generalizing student's own learning experience	**4,5**	**5**
Need for communication with other students	It is necessary to communicate with other students in order to generalize the experience gained in mobility programme	**4,6**	**5**
Impact on the relations with teachers	Cooperation sharing experience assists in positive cooperation with teachers	**4,2**	**4**
Impact on relations	Cooperation makes communication and sharing	**4,3**	**5**
New learning possibilities	Cooperation provides more learning possibilities	**4,3**	**5**
Interrelation between cooperation and communication	Cooperation while sharing experience has strong influence on communication with other people	**4**	**4**
Perceiving possibilities of cooperation	I learned a lot about cooperation and its possibilities	**3,9**	**4**
Intensiveness of contacts with teachers	Cooperation strengthens my communication with students and teachers	**4**	**5**
Activity	Students and teachers actively cooperate learning from each other through the implementation of the foreign language teacher training programme	**4**	**4**
Openness	Teachers are open in discussions with students	**4**	**4**
Perception	**New ideas appear in discussions with students**	**5**	**5**
Interrelation of learning and cooperation	Learning and cooperation with teachers supplement one another	**4**	**4**
Improvement of learning	Cooperation with other students assists in learning	**4**	**4**
Impact of cooperation on learning and teaching	It is necessary to cooperate with students and teachers in order to perceive the essence of learning and teaching	**4**	**4**

Table 1: Mean and Mode of Students' answers

Criteria	Statements in the questionnaire	Answer average	Mode
Courage and self-confidence	Participation in European mobility project enables to gain courage and self-confidence at work	**4,7**	5
Respect	It is noticeable that respect at work increased after having participated in the mobility project	**3,8**	4
Responsibility	New and more responsible tasks are given after having participated in the mobility project	**3,9**	4
Meaningfulness	Participation in European mobility project assists in perceiving a new meaning of teaching and work experience	**4,6**	5
Knowledge satisfaction	**It is pleasant to perceive that it is still possible to learn new things**	**4,8**	5
Impact on professional con-tacts	**International experience expands personal as well as professional contacts**	**4,7**	5
Life and work activity	Cooperation with teachers make life and work more active	**4,6**	5
It is easier to solve problems	It is easier to solve problematic situations and to find solutions	**4**	4
Impact on professionalism	**Cooperation enhances professionalism**	**4,7**	5
Impact on the quality of communication with colleagues	Cooperation impacts on the quality of communication with colleagues and assists in establishing contacts with new people	**4,4**	5
Impact on self-evaluation	Stable personal contacts assist in adequate self-estimation	**4**	4
Innovation and flexibility	Cooperation encourages to try new things and to change flexibly	**4,3**	5
Activity	Teachers and students cooperate actively by learning from each other in implementing this foreign language teacher training programme	**4,1**	5
Impact on teaching and learning	Cooperation in the group promotes developing teachers‘ teaching and learning skills success-fully	**4,4**	4
Interaction of coop-eration and im-provement	Partnership supports and assits in teacher‘s professional improvement	**4,5**	5
Impact of interaction on students‘ learning results	Interaction between teachers and students en-ables successful learning	**4,5**	5
Impact on self-evaluation	Cooperative learning assists in seeking for mutual learning that is indispensable or teaching	**4,1**	4
Impact on the effec-tiveness of mutual teaching and learning	Cooperative teachers and students‘ learning impacts on the effectiveness of mutual work and studies	**4,1**	4

Table 2: Mean and Mode of Teachers' answers

Project work and project education oriented to the problem being applied in the teaching process within teacher training programmes link to European mobility projects and activities. Eventually, the project organization, concentration on the problem, the purposefulness of the participants, interdisciplinary aspect appears to be the fundamentals of the experience gained in European mobility projects.

Project education is a challenge to project-oriented teacher training programme (Ulriksen, 1998); it is one of the ways to acquire knowledge (Gibbons, 1998) that promotes self-development, improvement and reflective thinking as well as acquiring new skills and expansion of experience.

Projects is effective means to learn research and form the researcher's attitude towards learning (Marshal, 1999), this attitude enables developing skills in analyzing, synthesizing and assessing as well as promotes transferred skills and enhances the possibilities of future teachers on the labour market and strengthens motivation.

Taking part in European foreign language teacher training mobility projects assists students in perceiving the essence of European learning; while teachers – in teaching; both teachers and students develop competences for the reason that European mobility projects is one of the most innovative possibilities for implementing a foreign language teacher training programme in the context of the European dimensions. They stimulate cooperation, improve the quality of teaching and research, enhance creativity, initiative and responsibility, providing an opportunity for every teacher or student to acquire theoretical, practical as well as behavioral knowledge. Discussions, the analysis of problematic situations, comparison of cultural experiences, sharing experience and particular learning environment empower cooperative learning, with the overcome of culture barrier and crossing the country boundaries.

According to Tumenienė, Janiunaitė (2000), in the context of the global and European dimensions the teacher's as agent's role links interest, professionalism and cooperation with innovative educational collaborative environment, continuous responsibilities of organizing teaching and learning from others' experience.

After having participated in international mobility programmes teachers are of opinion that they are of vital importance for their teaching practice as they provide a number of possibilities: to try new subjects, new teaching frames, to acquire knowledge and experience that are missing at their institution.

The frequencies of teachers and students' answers as well as the percentage distribution have been calculated. On the base of the data of questionnaires, it can be stated that students mostly assessed the importance of the experience gained by participating in European mobility projects. Students think that participation in an international project assists in integrating newly learned study subjects to the experience possessed; they also consider that it is necessary to share the experience gained as well as to generalize the learning experience.

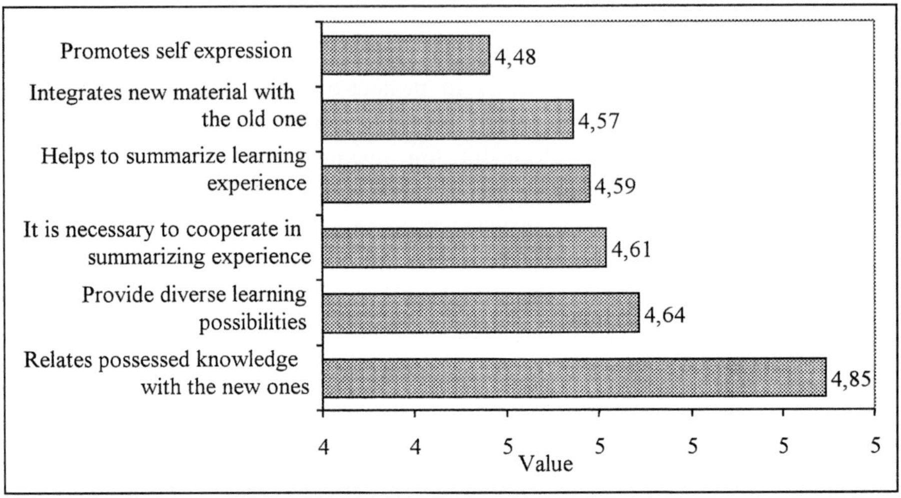

Picture 1: Rating the categories of future teachers' approach to the experience gained in European mobility projects

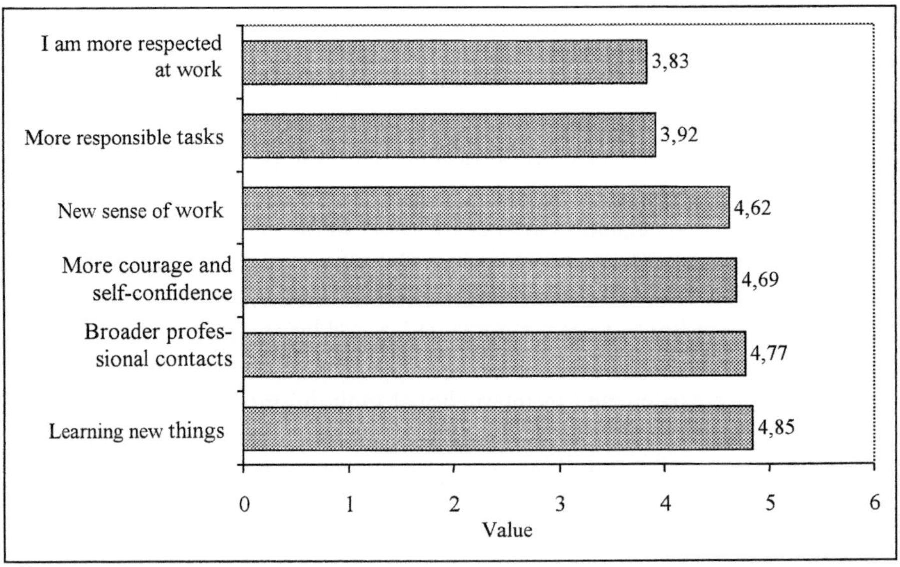

Picture 2: Rating the teachers' approach to the experience gained in European mobility projects

By expressing their approach to cooperation on the grounds of the experience gained by participating in European mobility projects, students (34 respondents) emphasize the role of cooperation in strengthening their relations with students and teachers.

The students' approach to the cooperative learning is illustrated by their attitude towards the importance of cooperation with teachers and students in order to perceive the essentials of teaching and learning. The students' assessed approach to the experience gained through participating in European mobility projects is illustrated in Picture 1, p. 384.

The research data show that students-respondents while expressing their attitude towards the experience gained in European mobility projects, mostly emphasize its impact on the inter-action between the newly acquired and the possessed knowledge (mean - 4,85).

Rating the teachers' approach to the experience gained in European mobility projects is presented in picture 2, p. 384.

The data of the answer frequencies and percentage show that teachers-respondents mostly emphasize the impact of the experience gained in European mobility projects on their satisfaction at the knowledge gained, professionalism and the work meaningfulness. By expressing their approach to the experience gained in European mobility projects, teachers mostly point out the possibility to learn new things (frequency of answers 4,85), thus revealing their seeking for knowledge; to encounter new meaning of work (frequency of answers 4,62); to expand professional contacts (frequency of answers 4,77); to gain more courage and self-confidence (frequency of answers 4,69).

Summing up the students future teachers' approach to the experience gained in European mobility projects, it can be stated that they positively estimate the impact of the experience, interrelating the newly acquired knowledge with the possessed one, ascertain it as more diverse than other learning possibilities, namely, sharing experience in cooperation, personal contacts, self-expression and learning. Summing up the results of the research on the students future teachers' approach to the experience gained in European mobility projects as well as their approach to cooperative learning in the process of implementing foreign language teacher training programme, it can be stated that students acknowledge the positive role of students and teachers' cooperation as well as communication with other people. Thus, the necessity of cooperative learning is ascertained on the grounds of openness and open discussions with teachers, learning improvement and activeness, namely, active students and teachers' cooperation, the inter-action between learning and cooperation as well as the perception of the essentials of learning and teaching.

The analysis of the research results illustrating the teachers-respondents' approach towards the experience gained in European mobility project substantially disclose that the emphasis has been laid on the teachers' learning and self-satisfaction, development of professional contacts and links as well as on teachers' emotional self-assessment, courage, confidence and the new perception of teachers' work and teaching experience. The results of the research on the teacher trainers' approach to the experience gained in European mobility projects illustrate that cooperation activates teachers' life and work, impacts on their professionalism and the quality of their communication with the colleagues. Cooperation assists in finding the best solution of problematic situations, dealing with innovative teaching and learning methods flexibly, while consistent personal contacts add to adequate self-estimation. Teachers-respondents absolutely agree with the statements of the questionnaire and the values implied, thus confirming the necessity, need, importance, impact and influence of the cooperation while sharing experience.

To identify the interrelationship of the questions and the answer groups the Spearman's non-parameter correlation measure has been applied. It presented the strength of the relation from the point of view of monotony. The fixed strength of the relations among the three groups has been identified – statistically it differs from zero and is positive. It shows that there is a relation among the groups: the increase in the first group of answers is in accordance with the increase in the second and the third answer groups. The correlations of the students' answers are presented in table 3.

	L1 Average	**L2 Average**	**L3 Average**
L1 Average Pearson correlation Meaning N	1,000 56	,453** ,000 56	,200 ,140 56
L2 Average Pearson correlation Meaning. N	,453** ,000 56	1,000, 56	,405** ,002 56
L3 Average Pearson correlation Meaning. N	,200 ,140 56	,405** ,002 56	1,000 56

**Correlation is important at 0, 01 level.

Table 3: The correlations of the students' answers

As it can be seen from the data, the students' approach to the experience gained in European mobility project that is presented by the first group of answers, closely links with the students' approach to cooperation on the grounds of the experience gained; the latter is inter-actively related with the students' approach to cooperative learning in the process of implementing the foreign language teacher training programme.

The correlations of teacher's questionnaire answers are presented in table 4.

	L1 Average	L2V Average	L3 Average
L1 Averago Pearson's correlation Meaning N	1,00 13	,806** ,001 13	,929** ,000 13
L2 Average Pearson's correlation Meaning N	,806 ,001 13	1,000 13	,723 ,005 13
L3 Average Pearson's correlation Meaning N	,929** ,000 13	,726** ,005 13	1,000 , 13

* Correlation is important at 1,1 level

Table 4: The correlations of teacher's questionnaire answers

	L1 Average	L2V Average	L3 Average
Spearman's L1VID correlation Meaning. N	1,000 , 13	,621* ,023 13	,884** 000 13
L2 Average Pearson correlation Meaning. N	,621 .023 13	1,000 , 13	,601* ,030 13
L3 Average Pearson correlation Meaning. N	,884** ,000 13	,601* ,030 13	1,00, 13

*Correlation is important at 0, 05 level
**Correlation is important at 0, 1 level

Table 5: Non-parameter correlations of teachers' questionnaire answers

Non-parameter correlations of teachers' questionnaire answers are presented in Table 5. As it can be seen from the data table the teachers' approach to the experience gained in the European mobility project expressed by the first group of statements closely relates with the teachers' approach to cooperation based on the experience gained; the latter is appropriately associated with the teachers' approach to cooperative learning, participating in the European mobility project through the implementation of foreign language teacher training programme.

The conclusion can be drawn, that the experience gained in European mobility project enables students and teachers in cooperating and sharing the gained experience as well as in learning from it in the context of cooperative learning.

To conclude on the results of the research, it can be stated that:

- The teachers and students' approach to the experience gained in the European mobility project reveals the impact of the experience gained on teachers' learning and self-satisfaction; the development of realations and contacts as well as on the teachers' emotional self-estimation; courage, confidence, assessment of teaching; on students' learning and the interrelationship of their former knowledge with the newly acquired, more versatile possibilities of learning, sharing experience in partnership, personal experiences and self-expression.
- Teacher trainers and future teachers' approach to cooperation based on the experience gained in the European mobility project discloses its impact on teachers' professionalism, the quality of communication with colleagues and other people, on dealing with problematic situations, adequate self-assessment and innovation; students and teachers' cooperation, communication with other people, students' cooperative learning, openness in discussions with teachers, learning improvement and activity.
- Teachers and students' approach to cooperative learning through the implementation of the foreign language teacher training programme reveals the importance of cooperative learning for the interactive learning between teacher and student; development of skills and interaction, professional development, effectiveness of studies and work, interaction between learning and cooperation and the perception of the teaching and learning essentials.
- Teachers and students' approach to foreign language teaching and information technologies ascertain the attitude towards the inseparable links between them, thus underlying the teachers and students' innovation in the implementation of the foreign language teacher training programme.
- The distinguished approaches to the experience gained in European mobility project, to the cooperation based on the experience gained in European mobility project, to cooperative learning through the implementation of the foreign language teacher training programme are interactively interrelated and influence, impact and affect one another on the way to dissemination.

Conclusions

- The modern foreign language teacher training programme should enable future-teachers to actively participate in public life, meet the learners' needs and demand, to develop vocational skills, to relate the learners' attitude, experience and concepts to the changes of the global concept, to apply deep learning approach in the background of life-long learning and continuous professional development.
- In the context of changes and the implementation of foreign language teacher training programme is a problematic process; there is no ideal study programme; it has to be flexible, dynamic, various, reflective and oriented to holistic learner's competence. Experience gained by teachers in international programmes justifies internationalization as the principal goal of higher education policy as educational paradigm of the twentieth century, when teachers and learners' cooperative learning is of essential importance while based on reflection; all forms of reflection, namely, reflection-in-action, reflection-on-action, reflection – for-action are of equal importance; reflective didactics of higher education enables development of democracy as one of the principles of higher education.

- European mobility projects as one of most innovative possibilities to implement foreign language teacher training programme assists teachers in perceiving the teaching dimension, while students in perceiving the learning dimension; both teachers and students can improve and develop competences; thus, cooperative learning empowers teachers and students to teach and study in the full meaning of the word.
- Teachers as foreign language teacher trainers and students as future foreign language teachers' approach to cooperation based on the experience gained in the European mobility project discloses their impact on teachers' professionalism, the quality of the communication with colleagues and other people, solving problematic situations, adequate self-assessment and innovation; students and teachers' cooperation and communication with other people, cooperative learning, openness in discussions with teachers, learning improvement and activities.
- The experience gained in European mobility projects and its dissemination at College as a new structural division of the educational system in Lithuania enables to achieve aims of modern learning as well as of implementing higher education study process; the experience dissemination can become an organic part of the implementation process of modern foreign language teacher training programme as a means of assuring the quality of study improvement.

References

ADAŠKEVIČIENĖ, V./JANIŪNAITĖ, B.: Europinių švietimo dimensijų kilmės ir sampratos erdvės problemos. Socialiniai mokslai 3(2001)29, pp. 17-26.

BRAZDEIKIS, V.: Bendrosios programos ir informacinės technologijos.Vilnius 1999.

BUTLER, J.A.: Progressing from Didactic 'Expert' to Partner in Learning. Capability 'Beyond Competence' Conference. November 1995.

COLE, A.L.: Impediments to Reflective Practice: toward a new agenda for research on teaching. Teachers and Teaching: theory and practice vol.3, no.1 (March 1997), pp. 7-27.

Comenius: mokyklinis ugdymas EK. SOCRATES Europos Bendrijos švietimo programa 2000-2006. Kandidatų vadovas. Kaunas 2001 (Šviesa).

DANIELS ET AL: New Directions in Professional Higher Education. Philadelphia 2000 (SRHE and Open University Press).

Europos matmuo švietime. Green Paper in Education. 1993 (EC).

Evaco Comenius 2.1 Project. Manual of EU Project Evaluator Competence Development. Kaunas 2002 (KTU).

FRY, H./KETTERIDGE, S./MARSHALL, S.: A Handbook for Teaching and Learning in Higher Education. Enhancing Academic Practice. London 2000 (Kogan page).

FULLAN, M.: Pokyčių jėgos. Vilnius 1998 (Tyto Alba).

GIBBS, G.: Improving Student learning Project: Briefing Paper for Participants in the Project. Oxford 1990 (The Oxford Centre for Staff development).

GUDAITYTĖ, D./JUCEVIČIENĖ, P.: Elitinio aukštojo mokslo tapimo masiniu proceso esmė: paradigma ir charakteristikos. Socialiniai mokslai 3(2000)24, pp. 112-119.

JOTAUTIENĖ, M./TIDIKIS, R.: Profesinio rengimo ir aukštojo mokslo problemos. Profesinio rengimo sistemos kaita: kai kurių kontroversinių aspektų analizė. Socialiniai mokslai 3(2001)29, pp. 53-62.

JUCEVIČIENĖ, P.: Europinės švietimo dimensijos Lietuvos aukštajame moksle. Aukštojo mokslo sistemos ir didaktika. I knyga. Kaunas 1998 (Technologija).

JUCEVIČIENĖ, P.: Mokymasis žinių ir informacinėje visuomenėje. Verbalinis pranešimas. Kaunas 2001 (KTU).

JUCEVIČIENĖ, P.: Intensyvus transnacionalinis kooperatyvus mokymasis. EMISTE MODULE seminaras. Insbrukas, 2003 02 18-23.

LEPAITĖ, D.: Kompetencija kaip ugdymo tikslas:pagrindinių skirtumų profesinio vidurinio ir aukštojo mokslo lygmenyse metodologinis pagrindimas. Socialiniai mokslai 2(2001)28, pp. 39-42.

Lietuvos švietimo koncepcija. 1992

KAVALIAUSKIENĖ, V.: Teacher Factor as an Educational Concern. Socialiniai mokslai 3(2001)29, pp. 27-32.

RAMSDEN, P.: Kaip mokyti aukštojoje mokykloje. Aidai 2000 (MM).

RIKOWSKI, G.: Globalisation and Education. Paper prepared for the House of Lords Select Committee on Economic Affairs, Inquiry into the Global Economy. [Cit. 2002 01 22]. Available from:
www.ieps.org.uk.cwc.net/rikowski2002d.pdf

SCHON, D.: The Reflective Practitioner. New-York 1983 (Basic Books).

TERESEVIČIENĖ, M.: Mokymasis bendradarbiaujant. Vilnius 1999 (Garnelis).

TOOHEY, S.: Designing Courses for Higher Education. 1999 (SRHE&Open University Press).

TUMĖNIENĖ, V./JANIŪNAITĖ, B.: Pedagogo veiklos pokyčiai pasaulinių ir europinių švietimo dimensijų kontekste. Socialiniai mokslai 2(2000)23, pp. 24-32.

ULRIKSEN, L.: Project - work and curriculum – a Danish experience. Conference on Higher Education at University of Central Lancashire. Preston 1998 07 06-08.

WHITEHEAD, J./TAYLOR, A.: Transcultural Competence:Foreign Native Speakers Training as Modern Foreign Language Teachers in England.Paper presented at the British Educational Research Association Annual Conference, University of Sussex at Brighton, 2-5 09 1999.

ZAY, D.: Reflection-Practice-Partnership.Questions and Points of Tension in Conceptualizing European Perspectives.Paper presented at the European Conference on Educational Research Ljublljana, Slovenia 1998 09 17-20.

ŽELVYS, R.: Švietimo reformos. Švietimo studijos 4.Vilnius 1998 (Švietimo studijų centras). Available from: <http://www.sln.org.uk/mfl/p4d.htm>.

Loreta Rudaitienė

Self-Directed Development of Emotional Intelligence in Professional Performance

Since the new educational paradigm is oriented towards personal initiative, educational tendency of an individual to construct knowledge by self is more and more vivid" (Selden, 1998). In the context of change inner personal factors condition individual and organizational success, therefore, special attention to educational possibilities of emotional intelligence is gradually increasing. Penetrative awareness of emotional process and skills, its effective management by rational correction of perception and by informing conscience, turns out to be one of the main factors for maturity processes of individuals, organizations and societies. Theories and models of emotional intelligence by Bar-On (1997), and Salovej and Mayer (1990), the concepts of emotional intelligence and emotional competence by Goleman (1995) formulate new professional conception in various professional and scientific fields. Development of emotional self-awareness, management of emotions, motivation, empathy and handling relationships (Goleman, 1995) foster emotional literacy – a variety of skills, strategies, maps and tools that enable emotional potential in communication and, consequently, through emotional fluency stimulate individual and professional ethical decision making. Unique mission of emotional intelligence to support self-regard and self-confidence, to develop self-efficacy and unlock empathy, to cultivate adaptive skills, service orientation and to strengthen influence could help in solving problems and reinforcing effective professional performance. This could be done by reflecting not just only stressful events and tension in professional performance or some wrong actions of the staff, but on the whole by evaluating effectiveness of professional behavior from the point of view of personality, so that to foster professional quality and energy.

Qualitatively new character of contemporary professional action causes new dimensions of ethical judgments, which were not emphasized by traditional ethics. The system of market objectively forces a person to be responsible and accountable for the consequences of his/her own actions. Implementation of tentatively solved ethical decisions, based on humanistic ideas and pragmatism through constructivism would assist any practical performance. By relating this context to educational psychology, it would be significant to analyze dynamic relationship between emotional intelligence and emotional competence as transformational aspect of a person when he/she experiences emotional tension or encounters an emotional situation. Here the following questions emerge: are moral meanings cared in professional performance? Are emotionally problematic situations observed and disclosed in work situations? Are emotional and ethical problems solved and how? Are ethical dimensions of professional action considered instrumentally? Is it taught how to sense and perceive ethical problems? We should acknowledge, that the process, in which ethical and educational power of emotional intelligence develops into emotional competence is not yet considerably analyzed (an exception is altruism-empathy and ego identity, Goleman, 1998).

The article presents the *research problem: What is the concept of self-directed development of emotional intelligence in professional activity, that is, how can emotional intelligence foster professional performance?* The *aim* of this article is to justify theoretically self-directed development of emotional intelligence in professional activity with the view to develop effective level of emotional competence. The *research method*, used in this article, is the analysis of scientific literature.

1. Ethical meaning of the relationship between emotional intelligence (EI) and emotional competence (EIC)

The emphasis should be made on the fact that so far the researchers are still holding discussions about the potential differences between Emotional Quotient (EQ) and Intelligence Quotient (IQ), about different definitions of the notion of emotional intelligence, about its relationship to emotional competence and their measurement. Based on the analysis of the main controversies about different theories of emotional intelligence, we could present the following three trends:

1. The representatives of the *first view* (Mayer and Salovej, 1990, 1997; Mayer, Salovej, Caruso, 2000 and others) integrate emotional intelligence into the intelligence model as it relates to processing information. In 1990 Salovej and Mayer were the first to present the term emotional intelligence (EI). They formulated its concept and defined emotional intelligence as the *'ability to monitor one's own and other's feelings and emotions, to discriminate among them, and to use this information to guide one's thinking and action'* (p. 187). We should notice, that these scholars distinguished EI abilities from social traits or talents and created the theory with a cognitive emphasis. Their model views emotional intelligence as integrating both the cognitive and emotional systems in process and it is sub divisible into four branches (Mayer, Salovej, 1997). The branches are as follows:

    - *emotional perception:* this branch of EI begins with the capacity to perceive and to express emotions and feelings.
    - *emotional assimilation-identification:* this branch of EI deals with emotional facilitation of thought. Emotions enter the cognitive system both as cognized feelings and as altered cognitions. The emotional facilitation of thought focuses on how emotion enters the cognitive system and alters cognition to assist thought.
    - *emotional understanding:* this branch of EI involves understanding and reasoning with emotions.

- *emotional management:* this branch of EI shows emotional management. Only with such understanding a person can have the necessary knowledge to fully manage and cope with his/her feelings.

These scholars argue that EI should be distinguished from personality traits. Since in this model EI is defined as an ability to perceive and understand emotional information (Mayer et al., 2000), then according to this approach EI could be developed as an *ability*.

2. The representatives of the *second point of view* (Goleman, 1998; Boyatzis, 2001; Cherniss@ Goleman, 2001; Bocchino, 1999 and others) formulate EI in the context of performance theory. Initially Goleman (1995) attempted "to capture in the EI notion almost everything but the IQ: this framework expanded the definition of EI beyond its limits and involved the factors of an individual, e.g. personality traits, motivation, etc., that is, he stressed non-cognitive aspect of emotional intelligence". There seems to be some progress in clarifying Goleman's (2001) concept of EI recently as the author himself made a distinction between EI and emotional competence. He defined EI as 'our potential for learning practical skills that underlie the four EI components – *self-awareness and self-management (Self) and social awareness and relationship management (Other)'*. According to the author, 'EI provides the main principles to develop a whole number of competencies as learned capabilities that result outstanding performance at work". The latter form *emotional competence* (EIC) and it '*shows how much of EI potential an individual has realized by learning and mastering skills and translating intelligence into on-the-job capabilities'*. According to the author, EIC includes:

   - *self-awareness:* knowing one's internal states, preferences, resources and intuitions. It involves the skills of emotional self-awareness, accurate self-assessment and self-confidence;
   - *self-management:* managing one's internal states, impulses and resources. It involves the skills of emotional self-control, trustworthiness, conscientiousness, adaptability, achievement drive, initiative;
   - *social awareness:* handling relationships. It involves the skills of empathy, service orientation and organizational awareness;
   - *relationship management:* inspiring and guiding groups of people. It involves the skills of developing others, influence, communication, conflict management, visionary leadership, catalyzing change, building bonds, teamwork and collaboration.

To generalize this, we should notice that, according to the second approach, EI is defined as a background for the development of EI-based competences, thus it could be developed as *EI-competencies*.

3. The representatives of the *third point of view* (Bar-On, 1997 and others) integrate EI into the context of personality theory in process. Mayer et al (2000) referred to models such as Bar-On's (1997) as *mixed models* because they combine skills that can be characterized as mental abilities (such as emotional awareness or problem solving) and personality traits (such as self-actualizations or optimism). The term Emotional Quotient ('EQ') was first introduced by Bar-On in 1988 as a counterpart to the intelligence quotient (IQ), in the attempt to assess EI as a measure of well being (Bar-On, 1997). He attributed EI to such personality factors as warmth and communication and as an important factor in determining one's ability to succeed in life. Now Bar-On (2000) defines EI as '*an array of emotional and social knowledge and abilities that influence human's overall ability to*

cope with environmental demands effectively'. This *whole* includes: 1) the ability to be aware of, to understand, and to express oneself; 2) the ability to be aware of, to understand, and to relate to others; 3) the ability to deal with strong emotions and control one's impulses; and 4) the ability to adapt to change and to solve problems of a personal or a social nature. The main domains in his model include:

- *intrapersonal skills.* They consist of *self-regard* that is the ability to respect and accept oneself as basically good; *emotional self-awareness* that is the ability to recognize one's feelings; *assertiveness* that is the ability to express feelings, beliefs, and thoughts and defend one's rights in a non-destructive manner; *independence* that is the ability to be self-directed and self-controlled in one's thinking and actions and to be free of emotional dependency; *self-actualization* that is the ability to realize one's potential capacities.
- *interpersonal skills.* They consist of *empathy* that is the ability to be aware of, to understand, and to appreciate the feelings of others; *social responsibility* that is the ability to demonstrate oneself as a cooperative, contributing, and constructive member of one's social group; *interpersonal relationship skills*, that is the ability to establish and maintain mutually satisfying relationships that are characterized by intimacy and by giving and receiving affection.
- *adaptability.* It consists of *reality testing* that is the ability to assess the correspondence between what is experienced and what objectively exists; *flexibility* that is the ability to adjust one's emotions, thoughts, and behavior to changing situations and conditions; *problem solving* that is the ability to identify and define problems as well as to generate and implement potentially effective solutions.
- *stress management.* It consists of *stress tolerance* that is the ability to withstand adverse events and stressful situations without "falling apart" by actively and positively coping with stress; *impulse control* that is the ability to resist or delay an impulse, drive, or temptation to act.
- *general mood.* It consists of *optimism* that is the ability to look at the brighter side of life and to maintain a positive attitude, even in the face of adversity; *happiness* that is the ability to feel satisfied with one's life, to enjoy oneself and others, and to have fun.

We should state that according to the third approach EI presents cognitive and personality aspects, it joins the first two groups of EI theories, presented above. Thus according to Bar-On (1997, 2000) model, EI could be educated through *cognitive and personality characteristics.*

For the analysis of the relationship between EI and EIC Saarni's (2000) idea that 'emotional competence is the demonstration of *self-efficacy* in emotion-eliciting transactions' is valuable, because it confirms, that the desired outcome of emotionally competent person is deeply integrated with his/her moral commitments. According to the author, this is because self-efficacy is led by the belief that one has the capacity and skills to achieve the desired outcome, which reflects cultural values, beliefs and norms, by transforming them into personal meanings. In the context of this article we would like to emphasize, that the potential of EI, which matures EIC, shows, that person's emotional responses are deeply influenced by *moral character and ethical values* in ways that promote and support 'personal integrity; thus, matured emotional competence should reflect this wisdom. Saarni (2000) claims that personal integrity is observed, when a person lives in harmony with his/her inner moral perception (or moral disposition), which means that one's life is balanced emotionally and a person is functioning adaptively and competitively – on sympathy, self-control, fairness and a sense of obligation (Campbell and Christopher, 1996; Flangan, 1991; J.Q.Wilson, 1993)'. Consequently, a balanced life, characterized by personal integrity, reflects mature EIC, and emotionally competent people are committed to justice, they are open and express care, they are flexible and sensitive.

Arnold and Gasson (1954) note, that 'emotional process involves not just psychical functions, but the whole personality, and regulation of action is conducted not through consciousness, but through personality consciousness; consequently, the regulating influence of emotions is determined by the whole set of individual and personal characteristics'. Therefore, it is better to refer not that much to emotion, influencing action, as to an individual, experiencing emotion alongside with peculiarities of the action. Here we come upon such concepts as personality self-development, self-discipline, self-analysis, self-expression, dignity and self-regard, perception of one's mission, self-realization, culture of emotions and feelings, etc. In this context, the third approach to EI, the *mixed models* of emotional intelligence, relating to both - cognitive and personality contexts - are more attractive. Reflections, presented above, state that emotional and rational abilities, integrated in EI, and used to perceive, identify, understand and manage emotions in social contexts condition moral disposition, self-expression, culture. Meanwhile, Mayer and Salovej's (1997) concept of EI, that defines 'emotional intelligence as an ability to recognize the meanings of emotions and to use that knowledge to reason and stimulate emotional and intellectual growth of a person' is rather narrow in the context of ethics, because here the desired outcome of an individual is not indicated.

In conclusion we should state, that *emotional intelligence is inner potential of an individual to perceive, identify, understand and manage emotions and by moral disposition to influence the development of emotional competence as a competence of personality.*

2. Ethical decision making as the main characteristic of emotional intelligence

In the aspect of value perception and value change through ethical decision making such notions as *sympathy*, the importance of which in moral responses was emphasized by the ethical theory of David Hume, and *consistency,* an idea, which was derived from Immanuel Kant, play a significant role. According to Griseri (1998), both of these aspects reflect emotional and intellectual elements in ethical dispositions of a person. The author claims, that sympathy, that is an emotional element, works in relation to the disposition to self and how it relates to others, directly and affectively influencing behavior and communication. Consistency, that is a cognitive element, relates to self-concept and its influence on others, disclosing itself in thinking and imagination. According to the scholar (see Fig. 1), ethical behavior (Griseri, 1998) involves relationship of the following three different elements of an individual – self-concept (consistency), self-disposition and disposition-to-others (sympathy). The first and the second both lead to a closer examination of one's self-disposition through disposition to others. In their techniques, one disturbs stability and the other blocks inertia in the system of values, thus creating *tension*, which manifests itself as *doubt*. A person may continue to keep to his/her old point of view, but the feeling of doubt brings a pressure to change and/ or a need to rethink one's own attitudes and opinions. The capacity for doubt is a core competence for both a person and a professional, because it reflects an important aspect of *how people may be able to develop themselves ethically*. The tension between consistency and sympathy emerges inside a person and puts pressure on his/ her self-concept through disposition to self and others.

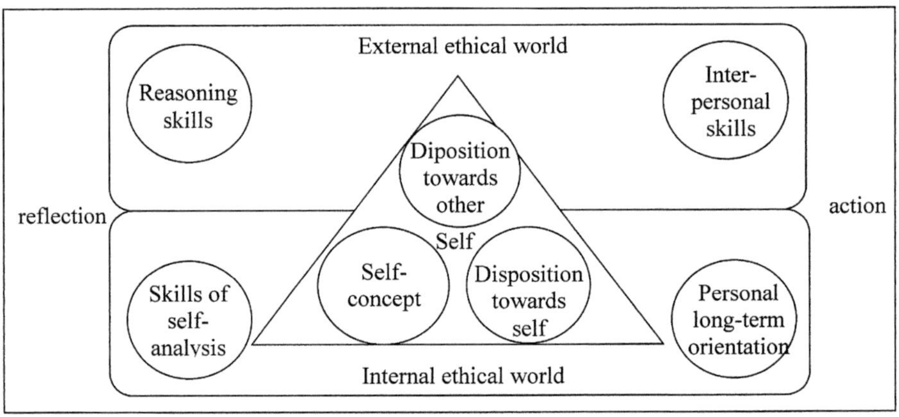

Figure 1: Personal skills and abilities: context and interaction in ethical

Emotional intelligence is integration of sympathy and consistency, and this notably happens in ethical decision making. The area of EI is the notion of self-disposition through one's disposition to others. Self-disposition is what a person *feels emotionally* about himself/herself. It includes (see Part One) *emotional awareness*, which is also one's opinion about personal abilities, strengths and weaknesses that creates *self-regard, assertiveness and independence*. EI also involves disposition to others that expresses knowledge how to understand and evaluate emotions and feelings of others, that is, *empathy*, it also includes *social responsibility*. When an individual feels sympathy and is eager to help others, he/she compares this orientation with his/her self-disposition ("if I do this, will I feel bad or will I respect myself and my feelings? Or: "Why do I refuse this feeling about the other?"). A person usually seeks to equalize discrepancies between self and others. Since self-disposition *includes self-respect, dignity and self-pride* and represents itself as a natural human need, a person strives to possess such characteristics and qualities, which could earn him/her public respect and self-satisfaction. Due to this, an individual constructs his/her behavior to be congruent with social values, and that, in turn, generates not only external but internal orientations: *individual perception of responsibility, conscience and the parameters of dignity*. Here conscience plays the most significant role, because it urges a person to perform his/her own spiritual self-control that happens through *doubt*, originated from the tension between affective and rational aspects and it urges an individual to evaluate his/her mistakes and in this way harmonize life's reality. According to Vasiljevienė (2000), nowadays, rational approach to conscience is increasingly being influenced by the pragmatic context of the contemporary world. This means that as a person is naturally prone to hedonism (search for pleasure and avoidance of torture), then in order to escape torture, he/she should make every attempt not to fault conscience. The idea of self-concept implies that values may change via cognitive means, while self-disposition implies that change is possible on an *emotional base*. If doubts or ethical tensions are not considered seriously, if a person does not strive to perceive, identify and understand ethical problems and avoids making ethical decisions, he/she is doomed to return to *self-deception*. In this case double morality informs a person about itself by emotional expressions. Actually, such truths like honesty, equality, respect and human worth are clear per se and do not have to be discussed, however the reality uncovers quite different phenomena: dishonesty, selfishness and the cult of materialism.

Ethical decision making is also supported by other elements of EI, namely, *impulse control, stress tolerance, reality testing, flexibility, optimism, interpersonal relationship skills, happiness, self-actualization* and others (see Part One). In the confluence of emotional experiences and analytical thinking, the existing reality is approaching ethical ideality by joining emotional expressions (conscience, the feelings of guilt, shame, penance, responsibility) with values, ethical principles and norms. Under these considerations we could state that the power of EI presents the possibility to tune instinctive impulses with moral norms and this motivates *ethical self-development through self-reflection, self-criticism and ethical decision making*. In summary of the above-presented analysis, we could claim that *a disturbed balance of the value system of a person implies his/her perception, that certain action is not congruent with the demands of ideal principles; this incongruity is experienced more or less emotionally, so a person tries to soften emotions, by making ethical decision, based on the overall power of EI.*

3. Significance of inner-ethical decision making for a deeper understanding of professional role and for the nurture of personal responsibility in professional performance

In this part of the article our analysis deals with professional aspect of human performance, thus it is necessary to emphasize, that according to individual ethics, a person is free up to the boundaries of freedom of other people; while in professional ethics a person inevitably encounters the rules that are imposed by society and organizations. Therefore, on individual level a person cannot avoid tension between ethical reality and ethical necessity, whereas as a representative of some professional action or as a performer of some function, he/she has to conform to the imperatives of professional obligation. Codes of ethics, as professional functional obligation, are the main institutional means of ethics in culture, profession, scientific community and organization. They help to analyze ethical dilemmas, seek to harmonize functioning and humanity, motivate formation of *moral competence*. However, we would like to notice, that the problem of stagnation of ethical codes is getting more and more vivid and non-reflective implementation of standards causes learning ethical rules by heart. This phenomenon should be critically reflected so that not to allow codes to change into the means of artificial regulation of behavior and control, that enables manipulation, alienation of social relationships, inconsistency between personal and professional interests.

Consequently, we would like to emphasize, that, though the norms of ethical codes are objective, they become effective and functioning only when they are *personal*. This can be achieved if an individual performs inner reflection, which involves his/her professional mission, and role, responsibilities and duties, searches for motivational possibilities, makes true and responsible decisions. This means, that a professional in the contemporary world should become a *moral subject*, the one who regulates self and feels personally responsible and accountable for the laws and codes of the organizational structure, conditioned by its aims of an organization. When a person assimilates order psychologically, when one perceives, identifies and understands who he/she is and what his/her aims are, when one takes responsibility and quits being a function or role, one becomes free.

Hereby, today, *intensive, sophisticated professional life, that demands concrete responsibilities, forces a person to eliminate tension between 'who he/she is' and 'who he/she should be' and helps to establish order. In other words personal interests of self-realization enable individuals to link professional reality to professional necessity.*

4. Self-development of emotional intelligence as a potential for fostering emotional competence and a premise for nurture of professional role and personal responsibility in professional performance

Educational aspect of EI in professional performance becomes evident in making endeavor to depolarize ethical reality (intentions) and ethical necessity (expectations), that is, trying to perceive, understand and open oneself and diminish distance between Real-Self (that is, 'who and what I am') and Ideal-Self (that is, 'who and what I would like to be'). This can be achieved by clarifying the significance and substantiality of professional ethics, professional role and responsibility in one's life and profession. According to Jucevičienė and Lepaitė (2000), 'value background of an individual and personal characteristics alongside with professional knowledge, skills and abilities in action, the latter providing a person with formal professional qualification, accommodate a person with the possibility to become competent'. This relates to the congruence among – the attitude of an individual to self as a professional, his/her ability to perform a professional role and professional identity with the reality of professional functioning. According to the authors, the mentioned adequacy 'presents complexity of competence in the highest (holistic) level'.

We could state that self-development of EI provides a possibility for an individual to foster personal competence in professional performance. This educational process is observed during the breaks of his/her reflection, based on EI, that overlap with the stages of ethical decision making. The breaks are accompanied by a sense of surprise, the unexpected or urgency, they offer certain learning proposals that are accepted by an individual if he/she wishes to change.

This self-learning process is not always fluent or linear; it might be instant or very long and repetitive. Such inner reflection in its nature matches ethical as well as rational contemplation, because in this process a person formulates a problem, searches for alternatives, argues and counterarguments, finds ethically rational way and methods to solve the problem, and makes an independent, free and voluntary individual, yet, subjective decision. The problem still remains, however, because it demands objectivity, that could be reached by justifying the accepted position, by holding academic arguments and by developing descriptive ethical discourse, in other words, by opening oneself and implementing one's change externally.

The article presents the model 'Self-Development of Emotional Intelligence through Ethical Decision Making (context of professional action)' (see Fig. 2, p. 400), based on the transformation of the paradigm between ethical reality and ethical necessity, the conception of value change (Griseri, 1998), the proaction model (Covey, 1990), theories of emotional intelligence (Bar-On, 1997, 2000; Mayer ir Salovej, 1990, 1997; Goleman, 1998, 2001), notion of emotional competence (Goleman, 2001; Saarni, 2000), self-directed change (Boaytzis, 2001), various theories of ethical decision making.

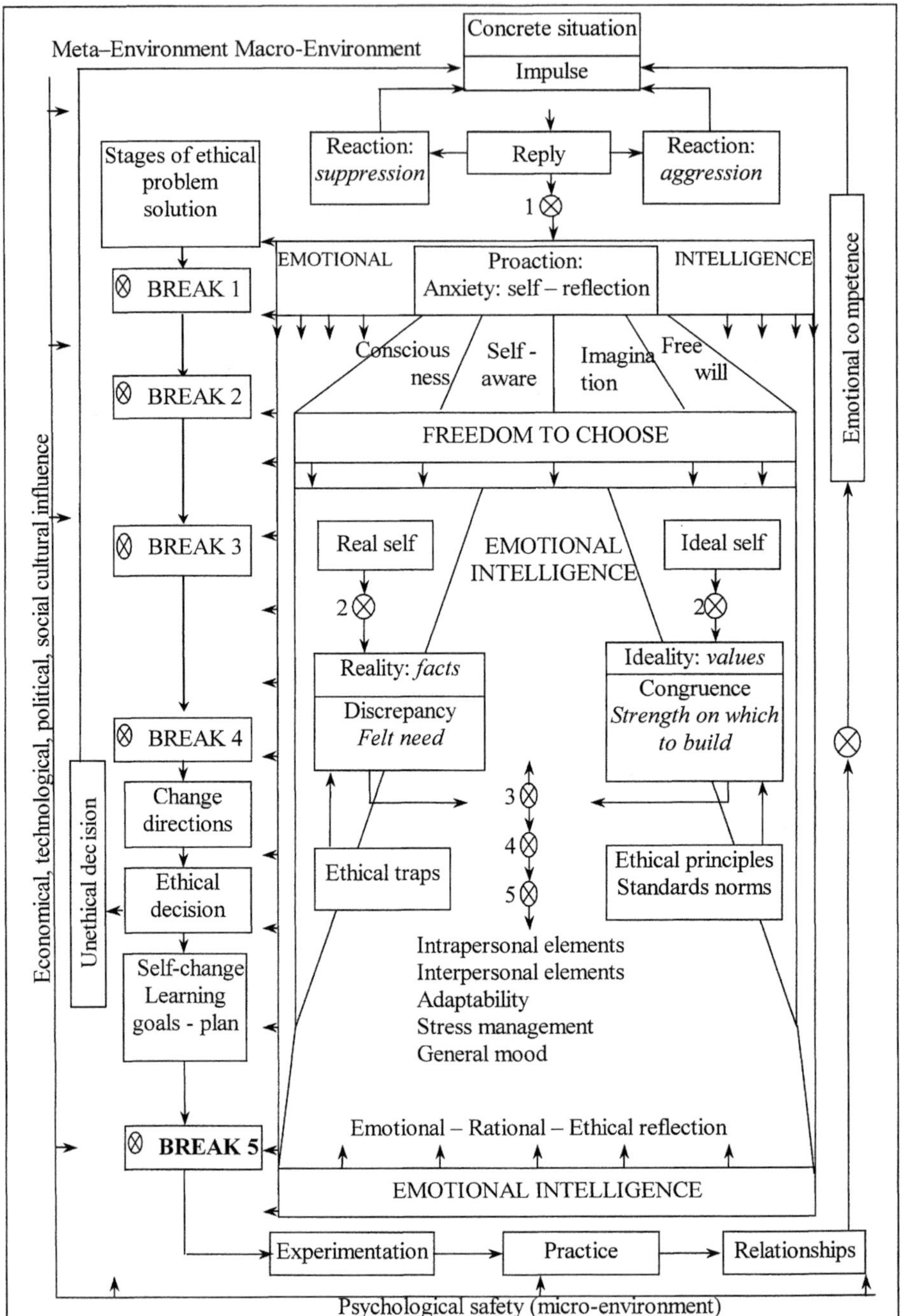

Concrete situation
Impulse
Meta–Environment Macro-Environment
Reaction: suppression
Reply
Reaction: aggression
Stages of ethical problem solution
1
EMOTIONAL
Proaction: Anxiety: self – reflection
INTELLIGENCE
Emotional competence
BREAK 1
Consciousness
Self - aware
Imagination
Free will
BREAK 2
FREEDOM TO CHOOSE
BREAK 3
Real self
EMOTIONAL INTELLIGENCE
Ideal self
2
2
Reality: facts
Discrepancy Felt need
Ideality: values
Congruence Strength on which to build
BREAK 4
Economical, technological, political, social cultural influence
Unethical decision
Change directions
3
4
5
Ethical traps
Ethical principles Standards norms
Ethical decision
Self-change Learning goals - plan
Intrapersonal elements
Interpersonal elements
Adaptability
Stress management
General mood
Emotional – Rational – Ethical reflection
BREAK 5
EMOTIONAL INTELLIGENCE
Experimentation
Practice
Relationships
Psychological safety (micro-environment)

Figure 2

The dynamic environment involves the two following modes of emotional behavior: reaction and proaction (Covey, 1990). Reaction is a stage, when an impulse of *self-defense* is expressed. According to Hans Selye (Furst, 1999), during reaction the following adaptation phases are observed: 1) exhaustion, when adaptation mechanisms are not able to function; 2) resistance; 3) anxiety, during which a person searches for ways to adapt oneself. It is important here to emphasize the fact that during the latter phase, reaction changes into *proaction* Covey (1990), if an individual opens his(her) emotions, feelings and moods inside self and reflects on them (see Fig. 2, p. 400, Break One).

Here through *self-awareness, imagination, conscience and independent will* a person makes his(her) *decision freely*. Here, the mode of impulsive reaction changes, depending on the human psyche, and an impulsive reaction may turn into a motive. The presented model (see Fig. 2) shows that during proaction an individual goes through the following main stages of ethical decision making, where he/she experiences breaks, based on EI:

- ***Break 1:*** Perception and identification of ethical problem (impulse control);
- ***Break 2:*** Understanding, formulation and correction of ethical problem (setting criteria for decision);
- ***Break 3:*** Process of ethical decision making (creating alternatives and their evaluation);
- ***Break 4:*** Taking final ethical decision (predicting consequences, short-term and long-term goals);
- ***Break 5:*** Implementation and consequences of the final decision (experimentation, practice, relationships).

We have already mentioned (Part One), that EIC is 'a developed capacity based on emotional intelligence, that results in outstanding performance at work' (Goleman, 1998, 2001). However, according to the author, 'having social awareness, the skill at managing relationship, does not guarantee that an individual has mastered the additional learning required to handle the other person adeptly or to resolve a conflict – it means just that a person has a potential to become skilled at these competencies'. The model, we present here, demonstrates, that emotional competencies could be and should be developed as professional abilities and skills through ethical decision making, based on EI. Here individual voluntarily-evaluative Self emerges and strives to enlarge one's value by emotionally integrated cognitive processes of imagining and self-criticizing one's own character features, abilities, appearance, social significance, moral disposition, etc. While developing EI inside oneself, a person might change his/her own attitude towards self and self-concept. This process reminds of action science theories (Argyris, Putnam, Smith, 1985). Argyris and his colleagues called these theories Model I Theory-in-Use and Model II Theory-in-Use and presented their relationship as a distinction between defensive and open social systems. Model II is engaged in inquiry rather than in inhibition, and stimulates double-loop learning, when the governing variables change themselves due to the emphasis on open inquiry instead of suppression of a conflict. Open confrontation on difficult issues, minimally defensive interpersonal relations and collaboration make quality of life more positive than negative, stimulate the effectiveness of problem solving and, increases long-run success. Similarly, according to Griseri (1998), in the area of value change, there exist three levels of change: *alpha, beta* and *gamma change*: 1) *alpha change* is similar to what Argyris et al (1985) calls single loop learning – a simple amendment of attitude or belief, without any structural implications for the concept involved; 2) *beta change* is where an individual re-evaluates the parameters of a concept; 3) *gamma change* is where the concept itself changes. Consequently, in our context, those who avoid acknowledging their mistakes and are prone to repeat wrong actions without efforts to understand them, demonstrate low level of EI, such people are passive and inert in their action.

Break 1: Perception and Identification of Ethical Problem

Impulse-Response: reaction or proaction

If emotions are suppressed and a person does not open them to self and is not eager to understand them, then proaction may not happen. As a consequence, problems are not solved and defensive actions emerge with the above-mentioned characteristics of exhaustion (suppression) or resistance (aggression), followed, in the first case, by the elements of lack of trust, self-protection, physical and spiritual destruction or, in the second case, violence, attack, control.

During this stage emotions direct attention to pressing concerns of action and through experienced *doubt or tension* signal what should be the focus of immediate consideration. Additionally, a person can identify an experienced emotion as irrelevant to a decision, so, he (she) can take steps to discount and manage it so, that it will not be a source of error in decision making. The following elements of EI – *emotional awareness, impulse control, reality testing*, stress tolerance and others (see Part One) – help to perceive and identify ethical problems.

Break 2: Understanding, Formulation and Correction of Ethical Problem

Real Self - Ideal Self: setting criteria for decision

According to Boyatzis (2001), self-directed change matures when an individual contemplates in the aspect of who he/she is (that is, Real Self) or who he/she wants to be (that is, Ideal Self). Many psychological theories state, that the balance between personal Self-Image depends on the fact, whether an individual is open to oneself or closed, whether he/she accepts unfavourable information about oneself. Consequently, Self-Image can be labile (changing) and rigid (stagnant). The more a person is open to information about self (either positive or negative), the more possibilities of change he/she possesses. This is especially important in order to adapt to new, unusual situations that challenge people with new demands. Meanwhile, a person with stiff Self-Image limits oneself by strong psychological defensive actions, during which new information is reconstructed in order not to damage one's stagnant self-image. This hinders a person to adapt to new environment and situations, to transform the attitude towards self.

During Break 2, when a person seeks to discover who he/she is and who he(she) should be, the tension between reality (facts) and ideality (ethical principles and professional ethics) is perceived and contemplated. Personal aspirations, dreams, and desired states are shaped by values, philosophy, life and career stages, professional roles and motives. We should notice here, that values are not considered to be achieved – everyone possesses them – the aim of a person here is to reflect, evaluate, change, refuse or take them. During this stage an individual, who accurately perceives his/her emotions and can determine their causes, can effectively use emotional input and get important information about ethical problems, opportunities and recourses to solve them (Schwarz, 1990). In the context of professional performance, a person has to adapt him/herself to the values of the organization where he works, and not the opposite. Standards and norms of professional ethics, professional duties and individual responsibility have to be transformed into the manifestation of subjective will, sincere intensions and the voice of conscience (see Part Three).

EI helps a person step aside at a distance, appear in the position of the other or watch and evaluate one' own behavior and communication with others or through the eyes of others. Such ability to observe oneself like an object is one of specific human features. As we have already mentioned (see Part Two), self-disposition (what a person *feels emotionally* about oneself), integrated with self-concept, (that helps build *emotional self-awareness*), strengthens *self-regard, assertiveness and independence*. Certainly, awareness and exploration of oneself exists in the context of awareness and exploration of the other - disposition to others (see Part Two). Interrelationship of these categories is especially vivid when decision is related to long-term strategies and their consequences. EI demands new cycles of re-evaluation if final decisions are not ethical. Besides, EI serves as the means of defining *realistic goals*, based on criteria for self-evaluation, because emotions urge feedback and reconstruction of the defined goals in one's professional performance. The elements of EI - *self-regard, assertiveness, independence, empathy, social responsibility, self-awareness, reality testing* and others (see Part one) - provide insights into the ethical depths of emotional situation, assist in formulating and clarifying the problem on individual and professional level, help to define goals and criteria for the final decision.

Break 3: Process of Ethical Decision Making

Creating alternatives and their evaluation

As a person already knows his/her strengths and weaknesses and admits his/her values, he/she determines the balance between these inner aspects he(she) wishes to preserve, keep and relish and the aspects he/she wants to change, stimulate to grow, or adapt to his(her) environment and situation. During this break emotions are used to facilitate cognitive processes intuitively and through meta-mood regulation, people may realize that different moods and emotions cause them to view issues differently, consider and choose among several different types of options or alternatives (Damasio, 1999). Positive moods can facilitate creativity, integrative thinking, inductive reasoning, while negative moods can strengthen attention to detail, detection of errors and problems and careful information processing (Isen etal., 1987; Salovej et al., 1993). Shifts in emotions can lead to more flexible planning and a broadened perspective on problems (Salovej and Mayer, 1990). We should also emphasize, that increased flexibility deriving from EI, may help individuals identify relationships among many issues they are confronted, enabling them to respond to multiple issues simultaneously. Majority of the elements of EI – *empathy, social responsibility, problem solving, flexibility, optimism, emotional awareness, reality testing, stress tolerance, impulse control* and others (see Part One) – help create and evaluate alternatives.

Break 4: Taking Final Ethical Decision

Predicting consequences, short-term and long-term goals

Here final decision is taken and from contemplation an individual moves to the preparation for change. This stage is as much emotional and ethical as intellectual, because a person, knowing one's strengths and weaknesses, discrepancies and congruencies between Real-Self and Ideal-Self, and what he/she wants to preserve and what to adapt, predicts consequences, short-term and long-term goals. Alongside with that he/she determines directions and goals, and they are articulated. Boyatzis (2001) emphasizes that 'setting of goals and creating of plans to achieve those goals is an integral part of various models and theories of change processes, and in particular self-directed change processes, for several centuries (Kolb and Boyatzis, 1970)'. The process of taking final decision should involve deep reflection about its consequences, thus it becomes clear, that a person has to possess all the information – factual and rational as well as emotional and intuitive. During this stage if a person accurately perceives emotions and their causes, he/she can manage emotions which might interfere with effective decision making or he/she also can use certain emotions as proposing courses of action. EI facilitates seeing connections among divergent information, and thus may help understand how issues are interrelated. In the context of integration between EI and ethical decision making Forgas (1995) affect infusion model (AIM) should be analyzed as well, because it helps understand how affect influences cognition, judgment, opinion and decision making. Affect priming is an important mechanism through which affect infuses judgments, selects and encodes information and various opinions *congruent with individual's current affective state*. This model is especially 'valuable when decision makers are faced with complex tasks that need extensive and constructive information processing, when ambiguity and uncertainty exist, new information needs to be assimilated, and decision makers desire to make accurate judgments and good decisions' (Fiedler, 1991; Forgas, 1994, 1995). All the elements of EI (see Part One), especially *empathy* and *social responsibility* as well *as self-regard, assertiveness, emotional awareness and independence, problem solution, flexibility, reality testing, optimism* and others help a person penetrate deeper inside, reflect his/her environment, imagine and define the consequences of the final decision, determine short-term goals (reactions) and long-term goals (social and professional responsibility), make an ethically true decision and stimulate person's inner change. This is how personal responsibility is being formed, since ethical decision making is free and voluntary.

Break 5: Implementation and Consequences of the Final Decision

Internal and external change: experimentation, practice, relationships

Not always EI is characterized by kindness and affability, it is closer to openness, sincerity, persistence, and, if necessary, it might be distincted by strict decisions; however, the latter are usually made emphatically and reasonably. If the final decision is wrong, an individual perceives negative consequences through his/her personal emotional experience, which at any price, urges the other cycle of emotional reflection. The cyclic return shows that ethical knowledge, principles of justice and moral norms insist to be perceived. They should be evaluated as necessary, universal and true. Only then psychological basis to realistic formation of introvert moral orientations is achieved, and the process of self-learning and self-regulation of a *moral subject* happens. Here morality gets the possibility to become functional. During Break 5, the desired changes are experimented and practiced. Acting on the plan and toward the goals involves numerous activities with new behavior. Educational content of this model involves the process of personal reconstruction, carried out by imagination in proactive thinking, when person's acquired EI skills help to develop new EI abilities and skills and, consequently, new actions are mastered easier and quicker. In this context ethical truths become a habit inside a person, they lose abstractiveness, they enable an individual to be rather an active performer than a passive viewer. According to Covey (1990), value orientations influence formation of effective habits, that condition long-term success. An emotionally matured person cares, observes and reflects self and others, he/she is prone to search and find new activities, to feel increased competence, self-confidence and novelty.

In ***summary*** we should state, that *self-directed learning of emotional intelligence in professional activity is the formation process of emotional competence as the competence of personality of a professional, where an individual fosters his/her abilities of emotional intelligence(perception, identification, understanding and managing of one's and other's emotions), through ethical decision making in professional action, consequently, deepening personal comprehension of one's professional role and professional responsibility.*

We should emphasize that inner readiness of an individual and the quality of one's self-determination to perceive and solve problems, based on EI, might show the results only in supportive micro-environment, because *without promotion of emotional culture*, even emotionally intelligent people might not demonstrate their emotional competencies, instead they might suppress these talents inside themselves.

Conclusions

1. Short analysis of the main theories of emotional intelligence (Bar-On, 1997, 2000; Salovej and Mayer, 1990, 1997; Goleman, 1995, 1998, 2001) revealed *cognitive, personality* and *mixed* aspects of the concept of EI. *Firstly,* emotional intelligence can be developed as *an ability; secondly,* emotional intelligence can be developed as *EI competencies*; and *thirdly* it can be developed through both *cognitive and personality characteristics.* The distinction between emotional intelligence and emotional competence (Goleman, 1998, 2001) helped penetrate *ethical and educational* potential of EI to be developed into emotional competence in the context of action theory.
2. Inner power of an individual, while experiencing *doubt or tension* between affective and rational activities in human consciousness, brings a person to ethical decision making. Here a person demonstrates abilities that help link instinctive impulses, life reality and pragmatic attitudes to ideality, eternal values, ethical principles and moral disposition. These abilities represent emotional intelligence in its overall power, thus, ethical decision making is justified as the manifestation of overall power of EI and the main characteristic of EI in the context of action theory.
3. Professional performance presents a unique opportunity for ethical reality to blend with ethical ideality. Self-actualization and success demand an individual to adapt oneself as a personality to specific job requirements and responsibilities. Development of individual reflection and ethical decision making is a significant factor for the formation of deeper understanding of professional role, for fostering personal responsibility in organizations as a premise for better functioning of ethical codes and effective performance of organizations.
4. The presented model 'Self-Development of Emotional Intelligence through Ethical Decision Making (context of professional action)' demonstrates evaluating and educational power of EI, that enables an individual and a professional performer to conduct one's inner change through the process of ethical decision making, that consists of experiential breaks of the reflection, based on EI. The power of EI urges a person to perceive and solve ethical problems in his/her professional context through emotions: their perception, identification, understanding and managing. Self-directed development of EI in professional performance happens when emotions, generated from human reactions (impulses), change

into the feelings of *self-regard, dignity and responsibility* alongside with self-control and effective behavior. A person is mature and respectable just only when he/she perceives, that there is a close connection between reason and consequence, when he/she is capable to be individually accountable for one's behavior and who balances one's actions in the way, that others feel well too. Emotionally competent person makes ethical decisions freely, inspired by his/her inner responsibility, that encourages creation of alternatives against status quo. The analysis, performed in this article, enables *emotional competence* to be confirmed as a *personality competence*, a significant and integrating part in the holistic concept of professional competence. In this context EI is a potential to bring new quality to teaching and learning.

References

ARGYRIS, CH./PUTNAM, R./SMITH, D.: Action Research. USA, San Francisco 1985 (Jossey-Bass Publisher).

Feelings and emotions as dynamic factors in personality integration. In: ARNOLD, M.B./GASSON, J. (eds.): The Human Person. New York 1954 (Ronald), pp. 294-313.

BAR-ON, R.: Emotional Quotient Inventory (EQ-I): Technical Manual. Toronto, Canada 1997 (Multi-Health Systems).

BAR-ON, R.: Emotional Social Intelligence: Insights from the Emotional Quotient Inventory. In: BAR-ON, R./PARKER, J. (eds.): Handbook of Emotional Intelligence: Theory, Development, Assessment, and Application at Home, School and in the Workplace. San Francisco 2000 (Jossey-Bass, Wiley), pp. 363-389.

BOCCHINO, R.: Emotional Literacy: To Be a Different Kind of Smart. USA 1999 (Corwin Press, Inc.).

BOYATZIS, R.E.: How and Why Individuals Are Able to Develop Emotional Intelligence In: CHERNISS, C./GOLEMAN, D. (eds.): The Emotionally Intelligent Workplace: How to select for, Measure, and Improve Emotional Intelligence in Individuals, Groups, and Organizations. San Francisco 2001 (Jossey Bass).

CHERNISS, C./GOLEMAN, D.: Training for Emotional Intelligence: A Model. In: CHERNISS, C./GOLEMAN, D. (eds.): The Emotionally Intelligent Workplace: How to select for, Measure, and Improve Emotional Intelligence in Individuals, Groups, and Organizations. San Francisco 2001 (Jossey Bass).

COVEY, S.R.: The Seven Habits of Highly Effective People: Restoring Character Ethic Australia 1990 (Mc Pherson's Printing Group), pp. 20-45, 85-161.

DAMASIO, A.: The Feeling of What Happens Body, Emotion and the Making of Consciousness. London 1999 (Vintage), pp. 35-82.

FIEDLER, K.: On the task, the measures and the mood in research on affect and social cognition. In: FORGAS J.P. (ed.): Emotion and social judgments. Elmsford, NY 1991 (Pergamon Press), pp.83-104.

FORGAS, J.P.: Sad and guilty? Affective influences on the explanation of conflict episodes. Journal of Personality and Social Psychology 66(1994), pp. 56-68.

FORGAS, J.P.: Mood and judgment: The affect infusion model. Psychological Bulletin 117(1995), pp. 39-66.

FURST, M.: Psychology. Vilnius 1999 (Lumen leidykla). (In Lithuanian).

GOLEMAN, D.: Emotional Intelligence. Why It Can Matter More Than I.Q. New York 1995 (Bloomsburry).

GOLEMAN, D.: Working With Emotional Intelligence. New York 1998 (Bantam Books).

GOLEMAN, D.: An EI-based Theory of Performance. In: CHERNISS, C./GOLEMAN, D. (eds.): The Emotionally Intelligent Workplace: How to select for, Measure, and Improve Emotional Intelligence in Individuals, Groups, and Organizations. San Francisco 2001 (Jossey-Bass).

GRISERI, P.: Managing Values. Ethical Change in Organizations. Great Britain 1998 (MacMillan Business), pp. 89-161.

ISEN, A.M.: Positive affect, cognitive processes and social behavior. In: BERKOWITZ, L. (ed.): Advances in experimental social psychology, vol.20. San Diego, CA 1987 (Academic Press), pp. 203-253.

JUCEVIČIENĖ, P./LEPAITĖ, D.: The Concept of Competence. Socialiniai mokslai 1(2000)22, pp. 44-50. (In Lithuanian).

MAYER, J.D./SALOVEJ, P.: What is Emotional Intelligence? In: SALOVEJ, P./SLUYTER, D.J. (eds.): Emotional Development and Emotional Intelligence. New York 1997 (Basic Books).

MAYER, J./CARUSO, D./SALOVEJ, P.: Selecting a Measure of Emotional Intelligence. The Case for Ability Scales. In: BAR-ON, R./PARKER, J. (eds.): Handbook of Emotional Intelligence: Theory, Development, Assessment, and Application at Home, School and in the Workplace. San Francisco 2000 (Jossey-Bass, Wiley), pp. 320-340.

SAARNI, C.: (2000). Emotional Competence: A Developmental Perspective. In: BAR-ON, R./PARKER, J. (eds.): Handbook of Emotional Intelligence: Theory, Development, Assessment, and Application at Home, School and in the Workplace. San Francisco 2000 (Jossey-Bass, Wiley), pp. 68-92.

SALOVEJ, P. ET AL: Emotional intelligence and the self-regulation of affect. In: WEGNER, D.M./PENNEBAKER, J.W. (eds.): Handbook of mental control. Englewood Cliffs, NJ 1993 (Prentice Hall), pp. 258-277.

SALOVEJ, P./MAYER, J.D.: Emotional Intelligence. Imagination, Cognition and Personality, 9(1990)3, pp. 185-211.

SCHWARZ, N.: Feelings as information: Informational and motivational functions of affective states. In: HIGGINS, E.T./SORRENTINO, R.M. (eds): Handbook of motivation and cognition: Foundations of social behavior, Vol.2. New York 1990 (Guilford Press), pp. 527-561.

SELDEN, A: Constructivism. GOUEVA, F.Q. (ed.). MAA Online. USA 1998 (The Mathematical Association of America)

VASILJEVIENĖ, N.: Business Ethics and Codes of Conduct. Kaunas 2000. (In Lithuanian).

Aldona Augustinienė, Nijolė Čiučiulkienė

The Educating of Students' Decision Making Competence in the Problem Based Learning Context: Facing the Perspective of the Contemporary Knowledge Society

Introduction

In the context of the post-modern information and knowledge society change one of the most urgent priorities is the quality of the personal decision making. It is a very complex social, physiological, educational and philosophical phenomenon, the concept of which can be revealed with the help of multidisciplinary aspectual approach.

Analysing the way of life considerations, choices and decision making situations it is evident that people face necessity to accept responsibility for their actions, purposefully act in the changing conditions, when they have to contemplate about the projection of their needs, desires and actions in the future. In the traditional society the decision used to be taken once and for all times, but the changing work environment causes transformations in the way of life choice situation. The greatest achievement would be a new system of creating our well-being – based on the power of mind, rather than muscles (A. Toffler, 2001, p. 26). A personal success in life would become no longer just an individual value, but society value as well.

Therefore special skills are necessary, the complexity of which can be called decision making competence. Decision and choice making give rises to various character stresses for a person. According to Z. Bauman (1993) stresses can be classified into the dichotomies under the following pattern: between lure and compulsion, between choice and lack of it, between possibility of self-creation and absence of such possibility, between the personal independently percept decision making and on the other hand, forced categorical evaluations experienced as a constraint and weakness.[1] *The stress raises importance of capacity to make competent decisions.*

Therefore, the ***research question*** arises: who has to determine the individual choice regarding the way of life: what is the roll of education in forming decision making competence? *Thus, the aim of this article is to offer the philosophical critical exploration of the decision making competence and the ways of its formation, problem-based learning being among them.*

[1] Bauman, Z. (1993). A Sociological Theory of Post-modernity. – In: Intimations of Post modernity. – London ant New York. Tonehstone. Cited in R. Grigas, 2001, p. 144.

The exploration is performed according to the following methodological guidelines:

- J. Raz (1986) concept of personal autonomy;
- P.J. Hartung and D.L. Blustein (2002) interpretation of Parsons's Career Decision Making Model;
- P. Thagard's (2001) ideas on emotional coherence based theory of decision making and Informed Intuition model;
- M. Savin-Baden (2002) research on problem based learning experience domains.

The article consists of two parts. The first part, while being orientated into different philosophical, psychological perspectives contains the discussion of the following issue:

"The Understanding of the Decision-Making Competence Concept and Who is Going to Predetermine Personal Choice and Decision-Making".

The second part deals with the exploration of some discrepancies in decision making and educational means to form decision making competence.

The article is generalized while presenting the conclusions with the main discussion principles summarized.

1. What is the decision making competence and what problems accompany the choice of a life path?

As the main message of this article is to analyse the decision making competence and the educational possibilities of its formation, the exploration of the decision making[2] competence will be interpreted with the starting point of educational understanding.

V.Žydžiūnaitė (2002)[3] states that characteristics of the "competencies" concept from the standpoints of education are: the elements (knowledge, skills, attitudes, standpoints, values); components (behavioural and mental); complexity of the activities; transferring and evaluation of competencies; dimensions (personality, expertise of the activity). The notion of competence reminds successful activity, effective use of recourses and adequate choices.

---

[2] ***Decision*** – a choice that you make after thinking carefully: the process of making a choice after thinking carefully, the ability to make choices quickly, confidently, and effectively. ***Decision making*** – the process of deciding what to do about something (*Macmillan English dictionary*, 2002, p. 359).

Competence – the ability to do something in a satisfactory or effective way: a person's range of skills or knowledge; a particular skill; ***competent*** – capable of doing something in a satisfactory or effective way (*Macmillan English dictionary*, 2002, p. 280)

[3] According to research by Barnett (1994), Kirschner (1997), Westera (2001) and P. Jucevičienė, D. Lepaitė (2000), D. Lepaitė (2001).

According to P.J. Hartung and D.L. Blustein (2002), the crucial component of the competence concept is the third element – adequate choice that might be interpreted as a wise choice, especially stressed in the classical F. Parsons model:

In the wise choice of a vocation there are three broad factors: (1) a clear understanding of yourself, your aptitudes, abilities, interests, ambitions, resources, limitations, and their causes; (2) a knowledge of the requirements and conditions of success, advantages and disadvantages, compensation, opportunities, and prospects in different lines of work; (3) true reasoning on the relations of these two groups of facts. (Parsons, Chosing a Vocation, 1909, p. 5, quoted in P.J. Hartung and D L.Blustein (2002).

Still, the precise meaning of the "wise choice" component of Parsons's model remains unclear even today (Spokane & Glickman, 1994, quoted in P. J. Hartung and D. L. Blustein, 2002). Did he mean that wise vocational choices are only rationally, cognitively, independently, systematically, deliberately, and methodically made? Or, as K. E. Mitchell, Levin, and Krumboltz (1999, ibid) have recently argued, is there any wisdom in making choices intuitively, emotionally, collaboratively, spontaneously, and even haphazardly? For most of these past 100 years since Parsons's original work, counsellors, theorists, and researchers have replied "yes" to the former question and "no" to the latter question. In the manner of René Descartes, "true reasoning" has been relentlessly interpreted to mean logic and rationality devoid of context. Career development and counselling professionals may have, as Krumboltz (1998, p. 391, ibid.) recently put it, brought into the notion of true reasoning "hook, line, and sinker". P.J. Hartung and D. L. Blustein (2002) conclude that Parsons may not have intended such a literal meaning for the phrase that is gradually expanding to include both rational and nonrational meanings.

Recent work by Phillips (1994, 1997) suggests that *"true reasoning" may be best interpreted as a part of two basic types of* career decision making models: (a) rational–choice models and (b) alternative–to–rational–choice models. Rational–choice models traced directly to Parsons and long prized in research and practice as the most accurate portrayal of career decision making, value *reason, logic, objectivity, and independence*. These models view the wise decision maker as an *"objective scientist" who is "methodical, systematic, independent, and unimpulsive ... and maintains as an ultimate goal the maximization of personal gain"* (Phillips, 1997, p. 276). As F.Parsons stresses a single person's profession choice, it is important to analyze the works of E. Ginsburg, S. Axelrod, D. Herma, and D.E. Super. A special attention is paid to the D.E. Super's career conception, which includes human roles, working place, and position. D.E. Super treats vocational orientation as a process paying a lot of attention to ensuring the competence, necessary for decision making and choices, avoid hasty and irrational decision making and its consequences. According to D.E. Super psychological-pedagogical measures may prevent these career choice difficulties, such as insufficient maturity, internal conflicts, lack of self-confidence, information or problem solving skills. J.L. Holand in his profession choice theory (1973) reveals the importance of correct information about oneself and considers a personality the main factor of profession choice as the result of genetic potencies and interaction with environment.

By contrast, alternative or "other–than–rational" models *value intuition, emotion, subjectivity, and interdependence.* Such models point out that the decision–making process is replenished with ambiguity and uncertainty that frequently activates the process. Rational models emphasize *the individual decision maker* (i.e., the person). Alternative models emphasize the circumstances surrounding the decision–making process (i.e. the environment or context). In contrast to rational models that focus on maximizing personal gain, alternative models incorporate meaningful contextual and circumstantial elements of person's lives. Some scientists point out, that human rationality has its own limits. In most cases people fail to get the most important information, they tend to pick out data rather subjectively. Thus, a lot of data remain beyond the individual control.

Hartung and D. L. Blustein's (2002) position is that a systematic integration of rational and alternative–to–rational approaches is both timely and theoretically defensible. The timeliness is based on the reality that the intellectual climate and supporting research and theory are now moving to a more pluralistic phase. Consistent with this pluralistic movement is the awareness that cultural and individual differences are to be valued, understood, and respected.

Applying Savickas's (1993, ibid), post-modern discussion of career counselling to this analysis, rational models of career decision making reflect the modern era's emphasis on logical positivism and objectivistic science. Alternative models align more with the post-modern emphasis on "interpretivism," meaning making, relationship, agency, and community.

Like Krumboltz (1998) and others, ibid) believe, however, that a post-modern perspective interprets true reasoning to mean different things for different people depending on their worldviews, decision making styles, cultural value orientations, and life circumstances. Rather than emphasizing rationality and reason devoid of context it is better to recognize the complementarity of these perspectives rather than viewing them as conflictual. [4]

P. Thagard (2001) also offers a synthesis and partial reconciliation of intuition and calculation models of decision, using a recently developed theory of emotional

[4] As evidence of this thesis, one of the guidance materials produced in the Vocation Bureau, and titled "Plan of Life" presaged an alternative perspective on career decision involving narrative and constructivism.

coherence. This theory builds on a previous coherence-based theory of decision-making.

Understanding decision making in terms of emotional coherence enables persons to appreciate the merits of both intuition and calculation as contributors to ***effective practical reasoning***. Experts on decision making recommend a more systematic and calculating approach.

Decision-making is a kind of inference. Many philosophers have advocated coherence theories of inference but have remained rather vague how to maximize coherence (e.g. Harman 1986, Brink 1989, and Hurley 1989, ibid).

The theory of *Informed Intuition* [5] combines the strengths and avoids the weaknesses of the intuition and calculation models of decision making. Like the intuition model, it recognizes that decision making is an unconscious process that involves emotions. Like the calculation model, it aims to avoid decision errors caused by unsystematic and unexamined intuitions.

Scientific consensus concerning competing scientific theories can emerge from a process of individual coherence and interpersonal communication (Thagard 1999, ch. 7, ibid.), but conflict resolution concerning what to do require a more complex process of comparing and communicating the diverse goals driving the various decision makers. A crucial part of this process is becoming aware of the emotional states of others, which may benefit as much from face-to-face interactions involving perception of people's physical as from purely verbal communication. This does not mean that practical and theoretical reasoning should be sneered at. Making reasoning explicit in decisions helps to communicate to all the people involved what the relevant goals, actions, and facilitation relations might be. *If communication is effective, then the desired result will be that each decision maker will make a better-informed intuitive decision about what to do.*

Improving inference is both a matter of recognizing good inference procedures such as *Informed Intuition* and watching out for errors that people commonly make. A full theory of decision making would have to include an account of where human goals come from and how they can be evaluated. People who

[5] The theory of *Informed Intuition* includes the following steps: 1. Set up the decision problem carefully. This requires identifying the goals to be accomplished by your decision and specifying the broad range of possible actions that might accomplish those goals. 2. Reflect on the importance of the different goals. Such reflection will be more emotional and intuitive than just putting a numerical weight on them, but should help you to be more aware of what you care about in the current decision situation. Identify goals whose importance may be exaggerated because of jonesing or other emotional distortions. 3. Examine beliefs about the extent to which various actions would facilitate the different goals. Are these beliefs based on good evidence? If not, revise them. 4. Make your intuitive judgment about the best action to perform, monitoring your emotional reaction to different options. Run your decision past other people to see if it seems reasonable to them.

base their decisions only on the goals of sex, drugs, and rock and roll may achieve local coherence, but they have much to learn about the full range of pursuits that enrich human lives.

Wisdom of decision making regarding a life is connected with the person's none/maturity and available in/competence. People's misconception and reality indetermination mark their decisions with the sign of *inferiority* (R. Grigas, 2001, p. 36). The behaviour's inferiority of the social process participants increases proportionally to the increase in the complexity of a social object or expression and its changeability rates. This is the objective part of the decision inferiority. The subjective one depends on the subjects' readiness for accumulation of information and disposition to follow it. Various environments and causes often determine not only lateness of cognition and practical actions, the degree of their inferiority but also the correction of everything.

Decision making is a complex process. The fact that it encompasses irrational elements raises a question whether preservation of human independence is possible if the person takes irrational decisions and whether the development of his decision making competence is possible.

Thus, it is important to discuss what can decrease the misconception.

2. The ways of decision making competence development: problem based learning (PBL) phenomenon

In this part some controversies are highlighted and stressed on the kind of decision making ways offered, in attempts to get closer to the right choice of the way of life while investigating PBL educational possibilities.

The theory of continuous professional improvement (Heathcote, 1997) is based on the conceptual idea of personal identity and autonomy development and preservation of throughout the career. In this respect personal identity and autonomy as the expression of combined value is one of the preconditions of the career success. In fact the decision-making mandate often falls on to different variables (accordingly to N.A. Fouad, 1995; D.E. Super, B. Sverko, 1995, F.A. Ibrahim, H. Ohnishi, R.P. Wilson, 1994; quoted in D. Brown, 2002) such as values of social relations and family interacting with sex, socio-economic status, abilities and etc instead of the people themselves. This proves that the choice makers are taking the "decision burden" in a rather passive way.

Personal autonomy can only be achieved when individuals have the basic resources enabling them to act consistently with the tasks and expectations imposed upon them by the society. Some degree of resource equality is a prerequisite for achieving autonomy and sharing the society benefits and burdens. Some minority groups may have no opportunities for choosing profession due to the lack of information and therefore (D. Brown, 2002), they also have less possibilities for autonomy. To be autonomous and to have an autonomous life, a person must have options which enable him to sustain throughout his life activities which, taken together, exercise all the capacities human beings have an innate drive to exercise, as well as to decline to develop any of them, so not the number of options available but the variety matters. The autonomous person is the one who worships his personal identity, "*makes his own life and may choose the path of self-realization or reject it*" (J. Raz, 1986, p. 376).

Based on the concept of liberal education, personal independence and autonomy is one of the social morality values which instilment is a part of the moral task of education, raised by the secondary school. The achievement of personal thinking autonomy and independent thinking of the students is the main focus. It is not simple for the students, therefore it is doubted that those, who haven't made choices regarding a certain morality and, based on the topic of this paper, regarding the issues of choosing a way of life, are not improperly influenced or indoctrinated. Therefore "critically thought over decisions" of the individuals are important not only for the "personal morals" (T. McLaughlin, 1997, p. 24). For instance Swann in his report of the Education Research Committee regarding the education of children and ethnic minorities groups (Great Britain Parliament, House of Commons, 1985, ibid.) stresses the development of children's independence and autonomy, as revelation of their individualities, so that they become free from beforehand and forced stereotypes, because that would mean the limitation of "his or her freedom of self decision making regarding the choice of further way of life. This requirement to grant "the individuals a free choice that everybody could act and educate themselves in the pluralistic society structure as they please" at least theoretically does not allow the communities deciding and controlling adolescents' beliefs and lives. The internal group and society member's limitations seeking cultural purity and solidarity may initiate the coercion of an individual, therefore it is a significant line not supposed to be crossed – asserts W. Kymlicka (1995, ibid.).

From the psychological point of view adolescence and young adults gain emotional independence from their parents, whose authority starts competing with peer groups. This is the age when people contemplate about the future profession, when existential issues becoming very important, while lacking experience and conscientiously perceived criteria of values for addressing them. Young people do not make final decisions even after choosing further studies as P. Johnson, C.N. Nichols Jr., W.C. Buboltz and B. Riedesel (2002) summarize.

On the other hand, in the traditional culture the relationship between the older and younger generation was certainly predetermined by the tradition of authority. But this tradition became devaluated along with diminishing process of all traditions. The younger generation is noticing that "any novelty which can be offered by the elder is certainly older them". According to H. Arendt (1996, p. 197) it is not adequate to the changed world and not suitable for the youngsters, while all educational and other influence is concentrated in the hands of adults. There is only one way out – the younger generation should be provided with as much self-governance as possible with adults only helping. The positions of post-modern state scholars differ due to the moral responsibility of the acting subject. But this responsibility is not left to one person, stressing the importance of the social institutions, such as vocational orientation or specialists' obligation *"to develop the ability of the individuals to choose profession (career) in a free and informative way"*.

Nevertheless, it is usual for the future plans to be developed in the family first. It is based on parental birthrights and responsibility, assigned to the parents by society. Therefore a controversy reveals itself – even though it is accepted in the concept of the liberal society that the decision necessarily lies upon the child, the parents influence his/her decisions regarding the profession, choice of school etc.

Counsellors make another group of people, participating in choice and decision-making regarding the profession and the way of life. According to R. Tatsuno (2002) career guidance in schools is affected by the changes in companies' hiring and management systems. Those changes that are taking place in industries will accelerate changes in career guidance. The focus of career guidance in schools will shift from "which company they choose" to "how students should develop their own career". It is not guidance from a short-term perspective that is connected to school education. Hartung and D.L. Blustein (2002) argue that counsellors can once again be inspired to give back to their communities by helping all citizens find satisfaction in their lives.

Nowadays, the vocational and career guidance theorists and practitioners are urged to rethink paradigms and concepts, come up with new approaches, theories and techniques helping the individuals to acquire knowledge, attitudes and skills, to self-manage their educational and work paths. Counsellors are urged to turn to the role of a coach or a consultant who works together with the client in order to help clarifying his/her difficulties and setting individual strategies and plans for life and career. More than decidedness, counsellors should help clients – students or workers – to improve their decision-making skills and ability to cope with change, they have to include activities to promote the development of self-esteem and ego-strength, purposefulness, self-management and life-career planning skills, job search skills, and skills to cope with career transitions and unemployment (Krieshok 1998, Harris-Bowlsbey 1996, quoted in Tractenberg, ibid. 2000). Career guidance to be effective in the 21st century can no longer focus on career decisions alone, but has to broaden its horizons to a more comprehensive approach in order to help the individuals to think strategically and holistically about work, education, leisure and other aspects of their life.

The third group of people, participating in choice and decision-making regarding the profession and the way of life are teachers who can foster student's autonomy. It is not meant to reduce it to a set of skills that need to be acquired. The teacher and the learner can rather work towards personal independence and autonomy by creating a friendly atmosphere characterized by "*low threat, unconditional positive regard, honest and open feedback, respect for the ideas and opinions of others, approval of self-improvement as a goal*" (Candy, 1991, quoted in D. Thanasoulas, 2000). Promoting learner's independence and autonomy does not mean leaving learners to their own devices or learning in isolation. There has to be a teacher who will adapt resources, materials, and methods to the learners' needs and even abandon all this if it is necessary.

> But even if learner's independence and autonomy is amenable to educational interventions, it should be recognised that it "*takes a long time to develop, and simply removing the barriers to a person's ability to think and behave in certain ways may not allow him or her to break away from old habits or old ways of thinking*" (Candy, 1991, ibid).

Nonetheless, as R.N.Evans, G.McCloskey (2001, ibid.) assert, many aspects of the current school program actually discourage *decision-making* by students. Most young people make tentative occupational choices several times before entering higher school. If a child aged 7 or 17 announces that he wants to be a lawyer or a truck driver, we may be reasonably sure of three things:

> (a) this tentative decision is made on the basis of inadequate knowledge of his own characteristics and of the demands of the job; (b) the school has done little to provide either type of knowledge; and (c) the school will say, in effect, *"You are too young to concern yourself with such things. They should be decided later"*.

Some educators seem to have an almost irrational fear of teaching decision-making in relationship to the world of work. They seem to feel that such instruction will lead to early, irrevocable occupational decisions, which will minimize future student options. Thus we may argue that students' autonomy problem is closely related to teachers' autonomy.

While summing up what was said before it is necessary to stress the importance of educating in decision making. Such educating should have in mind the personal identity and autonomy, the level of personal experience, which dominates in decision making process; it should also create the enabling environment helping the students to match rational and emotional levels of decision making, supporting self-confidence and exercising informational, problem-solving skills, providing the effective communication the result of which would be "a better informed decision about what to do". Problem based learning might be one of the solutions, because as one of the problem based learning researchers Maggi Savin-Baden (2002, p. 3) states:

"(P)roblem based learning can offer students opportunities to engage with complexity and help them both to see: ambiguity and learn to manage the ambiguities that prevail in professional life".

The latter approach stresses the importance of personal experience, which is closely linked, with the concept of personal stance identity and autonomy. *According to M.Savin-Baden (2002, p. 67) learner's experiences and stances* can be presented in the figures 1 and 2, p. 420.

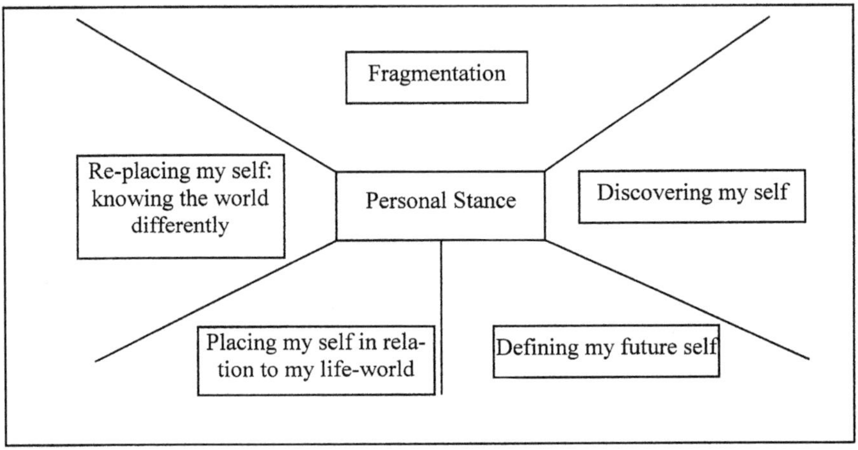

Figure 1: Domains within personal stance (M.Savin-Baden, 2002)

The term "personal stance" was used by M. Savin-Baden (2002) with the purpose to depict the way in which staff and students see themselves in relation to the learning context and give their own distinctive meaning to their experience of that context. Personal stance encompasses the means of which they discover, differ and place themselves within the problem-based learning environment and express the interplay between what they bring to, and take from their learning experiences. The ways in which people speak about themselves, view their profession, their peers, the facilitator and the institution, their decision making strategies are explored within the conceptual framework of personal stance.

Personal stance by its way is always transient, inspiring the context of problem solving and decision making. Personal stance starts with fragmentation domain, which is so characteristic for the initial stage of decision making, because a person has to evaluate rational and emotional aspects of their problem solving and decision making. In "fragmentation" domain students experience challenges to their values and beliefs, which appear to be at risk or are threatened through this challenge and reluctant uncertainty. Educating through problem based learning may challenge students' current sense of themselves and their way of both seeing the world and acting within it:

"This is because of the way in which problem based learning encourages students to assemble their own body of knowledge and to make decision about what counts as knowledge. This leads to a sense of fragmentation" (M. Savin-Baden, 2002, p.6 0).

Fragmentation inspires another domain – "*discovering my self*". In "*discovering my self*" domain students' experiences of self-validation (i.e. the self-re-examination) are the means by which self-discovery occurs. Self-discovery arises through forms of problem-based learning that promote a reflexive search for self-know ledge and self-improvement within students. Alternatively problem based learning may cause conflict which forces a re-appraisal of personal values and re-evaluation of learner's identity, at the same time forming the context for building the personal autonomy, the kernel aspect of personal improvement.

One more personal stance is "*defining my future self*". In this domain students position themselves in the learning context in terms of their view of themselves as future professionals.

They seek to understand themselves and the learning they are undertaking in terms of a perceived future role "*and they see the material being learned in relation to their future self. The way students place themselves in the learning environment as a future professional governs the possibilities and limits of engagement with the material, the context and other people*".

The next stance is connected with future perspectives and may be called "*placing myself in relation to my life-world*". As M. Savin-Baden (2002, p. 61) states, students gain a heightened understanding of their world and the world of higher and professional education and all that entails. Learning in a problem based way challenges students to confront the relationship between the previous experiences of their life-world (Habermas, 1989, ibid.) and their new experiences emerging from interaction with the objective world. This can lead to new understanding of their own reality, based on problem solving strategies".

The last personal stance is "*re-placing my self: knowing the world differently*". It captures the idea that students are able to frame their learning experiences for themselves, to take up alternative perspectives in order to challenge both themselves and the knowledge society. The analysis of alternative perspectives involves the student into rational communication with the purpose to make decision. In this stage the personal stance intermingles with interactional stance.

Interactional stance encompasses the way in which students interpret the way they see themselves as individuals, and others with whom they learn, construct meaning in relation to one another. Interactional stance comprises the domains shown in Figure 2, p. 420

Domain "*the ethic of individualism*" is characterized by the individual placing himself at the centre of the value system and therefore learning within the group is an activity that is only valuable in terms of personal gain for the individual, i.e. for desired competence formation: in our case decision-making competence formation.

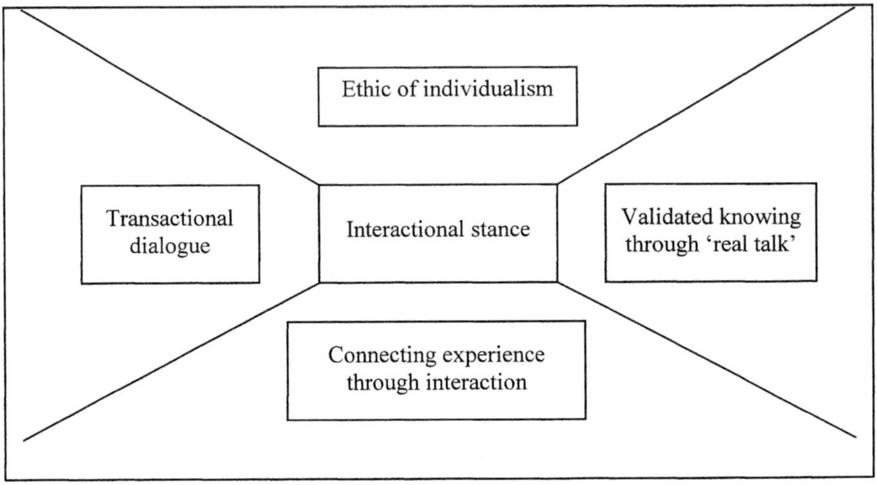

Figure 2: Domains within interactional stance (M. Savin-Baden, 2002)

"*Reality talking*" requires careful listening; it implies a mutually shared agreement that together you are creating the optimum setting so that superficial ideas can grow. According to M. Savin-Baden (2002), for students in this domain learning in groups is often accompanied by a boost of self-esteem and confidence and, as a consequence, a renewed exploration and facilitation of individual learning aspirations. Being in this domain enables students to come to value their personal and propositional knowledge by recognizing its value through the perspectives of others.

Domain "*Connecting experience trough interaction*" according to M. Savin-Baden (2002), is characterized by individual being facilitated through the group process in making sense, through reflection, of his own reality in confronting dilemmas and problems within that reality. Thus, in this domain students use the group to make sense of their world as it appears to be and will use the group to resolve dilemmas and discover the meaning in their lives, which helps them to become autonomous.

Transactional dialogue (after Brookfield, 1985) is used to capture the idea that the group serves as an interactive function for the individual. Thus, individual students by making themselves and their learning the focus of reflection and analysis within the group are able to value alternative ways of knowing. Individuals within transactional dialogue domain will use dialogue and argument as an organizing principle in life so that through dialogue they will challenge assumptions make decisions and rethink goals (M. Savin-Baden, 2002, p. 66).

Problem based learning also provides decision making strategies due to its structural elements and characteristics, which, as it was mentioned, create the conditions for the education of career decision making competence, which are revealed through the Dimensions of Problem Based Learner experience and stances

The very problematic situation involves the development of critical thinking competence, which is vital for any decision making. Critical thinking enables the analyses of choice contexts, which might be presented in the frame of problematic situation as well. The analysis of choice contexts leads to the forming of problem solving strategy, which is presented in the adequate interaction revealing choice possibilities. At the end the quality of problem solution influences the final decision making. These steps of PBL exercising the Learner's Experience dimensions may serve as an educational context for the formation of Informed Intuition, the later being the final result of decision-making competence.

Still, K.S. Meredith (2001, p. 26) asserts that in democratic classes students and teachers may see positive and negative consequences of changing positions, perspectives and attitudes with minimum harm for society. Therefore this type of PBL class is a real learning society allowing mistakes and corrections of negative consequences. PBL may be treated as a liberal education (M. Savin-Baden, 2002), which is characterized by the goal of autonomy and attention to critical thinking development, despite the related difficulties (T. McLaughlin, 1997, p. 169).

As the common concern of the educators is to develop competent school graduates ready for the challenges they are going to encounter as adults, G. Kaminskienė (2002) suggests each school to implement "*Preparation to choose a competence model*" in such a way that the students could evaluate their desires and studying abilities from grade 1 to grade 10, and in the senior grades 11-12 would be able to perceive their further path in the education system or professional activity. In this case PBL might be one of the recommended ties. According to the researchers these collaborative actions are necessary between the secondary and higher schools, branches of industry and labour exchange. Therefore contemplating about autonomy and decision-making competence dilemma it is very important that social units interact without dominating over another party and postulating mutual influence and understanding while seeking a decision, acceptable for both parties.

This principle encourages defining educational dimensions enabling the person for competent choice of the life path. As K.S. Meredith (2001, p. 24) highlights, school graduates have to be ready not just to survive, but to be happy, live a fruitful and meaningful life.

Conclusions

The solution of personal autonomy dilemma and striving to reach the quality of decision-making is a very complex and wide problem. Ensuring personal autonomy during considerations, of choosing a life path and making decision means ensuring the acquisition of decision-making competence. Thus, while summing up the results of the investigation it is reasonable to point out the following conclusions:

- Autonomy is not important in itself and it does not exist for itself, but is needed for enabling a person to act independently in a social context presenting the quality of decision-making competence.
- As learners become aware of the social context they gradually become independent, dispel myths, and can be thought of as "authors of their own worlds". Therefore the school and parents task is not to be a decision maker, even with the best knowledge, what the child or the society needs, but to be the best enabler of student's autonomy, providing the competence for making the decision.
- Based on the concept of liberal education, autonomy of the person is one of the social morality values which instilment is a part of the moral task of education task, raised by the secondary school.
- Problem based learning orientated towards the learner experience dimensions directly deals with the decision making competence, providing the models of career decision-making which include both rational and alternative–to–rational perspectives.
- When developing the personal autonomy, the basis of competence as well as the competence to make the right decision, all individuals will reach their growth and society with school will nurture the culture of consensus, seek bigger openness of society and political maturity.

References

ARENDT, H.: Tarp praeities ir ateities. Aidai, ALK.

BROWN, D.: The Role of Work and Cultural Values in Occupational Choice, Satisfaction, and Success: A Theoretical Statement. Journal of Counselling and Development. 80(2002)1, pp. 48-56.

EVANS, R.N./MCCLOSKEY, G.: Rationale for Career Education. Journal of Industrial Teacher Education 39(2001)1, pp. 30-42.

GRIGAS, R.: Sociologinė savivoka. Specifika, metodai, lituanizacija. Vilnius 2001 (Rosma).

HARTUNG, P.J./BLUSTEIN, D.L.: Reason, Intuition, and Social Justice: Elaborating on Parsons's Career Decision-Making Model. Journal of Counselling and Development 80(2002)1, pp. 41-47.

JOHNSON, P.N/BUBOLTZ, C.N./WALTER, C./RIEDESEL, B.: Assessing a Holistic Trait and Factor Approach to Career Development of College Students. Journal of College Counselling 5(2002)1, pp. 4-14.

KAMINSKIENĖ, G.: Diferencijuoto mokymosi aplinka ir jos prielaidos Lietuvos bendrojo lavinimo vidurinės mokyklos aukštesnėse klasėse. Daktaro disertacija. Kaunas 2002 (KTU).

KERKA, S.: Parenting and Career Development. ERIC Digest 214(2000). Available from: <http://ericacve.org/fulltext.asp>

MCLAUGHLIN, T.: Šiuolaikinė ugdymo filosofija: demokratiškumas, vertybės, įvairovė. Kaunas 1997 (Technologija).

MacMillian English Dictionary for Advanced Learners. International Student Edition. 2002 (Bloomsbury Publishing PLc.).

MEREDITH, K.S.: Laisvė, atsakomybė ir demokratinė mokykla. Permainos 3(2001). pp. 24-27.

RAZ, J.: The Morality of Freedom. Oxford 1986 (Clarendon Press).

SAVIN-BADEN, M.: Problem-based Learning in Higher Education: Untold Stories. Great Britain 2002 (St. Edmundsbury Press, Bury St Edmunds, Suffolk).

TATSUNO, R.: Career Counselling in Japan: Today and in the Future. The Career Development Quarterly 50(2002)3, pp. 211-17.

THAGARD, P.: How to make decisions: Coherence, emotion, and practical inference. In: MILLGRAM, E. (Ed.): Varieties of practical inference. Cambridge, MA 2001 (MIT Press), pp. 355-371.

THANASOULAS, D.: Autonomy and Learning: An Epistemological Approach. Applied Semiotics. Sémiotique appliqué 10(2000). Available from: <http://www.chass.utoronto.ca/french/assa/ASSA-No10/No10-A4.html>

TOFFLER, A.: Valdžių kaita. Vilnius 2001 (Tyto Alba).

TRACTENBERG, L./STREUMER, J./VAN ZOLINGEN, S.: Career Counselling In The Emerging Post-Industrial Society. Arbeitsamt online. Inhalt Kongress A. 2000. Available from: <http://www.arbeitsamt.de/laa_bb/international/InhaltKongressA2/A33Berlin2000Leonel_pdf>

ŽYDŽIŪNAITĖ, V.: Slaugytojų kompetencijų charakteristikos: edukologijos ir slaugos požiūriai. Sveikatos mokslai 4, Vilnius 2002 (UAB „Sveikata“), pp. 53-58.

Vilma Žydžiūnaitė

Fostering the Transcultural Sensitivity of Nurse Students: Curriculum Development through the Incorporation of Religiousness as an Element of Transcultural Health

Introduction

The factor that influences the health of an individual, families and even communities and their functioning is spirituality and religion. Indeed, spirituality and religion can influence every aspect of an individual/patient's life. Transitions in the individual life cycle (ex., marriage, birth, death) are marked by religious ceremonies, observances and traditions. Many patients report new meaning in life and increased depth of spirituality after experiencing a significant illness. *Religion, religiousness, spirituality* and *spiritual well – being* are terms frequently used in discussions about spiritual issues.

> *Religion* has been defined as "any source of transcendent meaning from which people derive a sense of purpose and values that become the standards by which people derive a sense of purpose and values that become the standards by which they direct actions and define, interpret, and judge the activities of human life" (Brigman, 1992, p. 39).

Religion thus provides the beliefs and values that help person determine how he/she will act in various situations and how he/she will evaluate the actions of others. *Religiousness*, originally associated with church attendance and involvement in formal religious activities, has now been defined as adherence to the beliefs and practices of organized practices of an organized church or religious institution (Shafranske & Maloney, 1990). Religiousness reflects how closely a person believes in and practices the beliefs of a particular religious group such as Roman Catholicism, Judaism, Islam, or various Protestant Christian denominations. Thus, both religion and religiousness provide a framework of values, beliefs, and actions that can influence the functioning of the person and the way the individual responds to illness.

Spirituality is a broader term that has been defined and those more personal practices of a religious nature that may or may not emanate from a particular religious institution (Shafranske & Maloney, 1990). Spirituality encompasses a sense of relation or connectedness within oneself, with others, and with a higher, unseen power of God (Reed, 1992). *Spiritual well – being* is defined as a sense of inner peace, compassion for others, reverence for life, gratitude, and appreciation of both unity and diversity (Vaughn, 1986). It includes the conviction that there is a purpose and meaning in life, a relationship with God, and diversity of realistic views and loss (Moberg, 1974).

Spirituality and spiritual well – being may not be linked to a specific religious institution or group. The aspects of religion and spirituality that contribute to more positive personal functioning include religious beliefs related to love, faith, hope, forgiveness, reconciliation and the worth of human beings; a lifestyle that emphasizes commitment, responsibility, giving and caring for others; shared beliefs and values; shared religious and church activities; reverence for marriage and family; and affiliation with a support group or network that provides a sense of belonging (Brigman, 1992). According to Hanson & Boyd (1996), for persons whose situation is not congruent with the beliefs and values of their religion, this can be a source of stress rather that support and comfort. Also, a religious emphasis on sin and judgment may produce feelings of guilt and low self – esteem (Brigman, 1992).

The fit between the patient's religious beliefs and values and the individual situation is critical. Although the inclusion of religious values and beliefs in the health care assessment may seem to be an invasion of the patient's privacy, it is important for a nurse to be sensitive to how these beliefs and values influence a patient's response in illness situations. Even if the individual has no formal religious affiliation, it is important to know her/his worldview and to work with it when delivering nursing care. *A holistic approach suggests that religiousness and related topics should be included in curriculum of nursing education. It is evident that religiousness is a dimension of health, so religiousness should be a dimension of health care (Banks, 1980). The aim of this article is to illuminate the religiousness as an element of curriculum in nursing education.*

Religio – specific nursing care

Rogers (1990) describes human beings as patterns of energy inseparable from environment. Newman (1994) identifies health as the expansion of consciousness. She sees health as harmony of the mind, body, emotions and spirit including environmental context and relationships with others and with nature. People establish their realities through integration of self within the person – environment - culture context, without actual lines of demarcation. Their worlds are continuously being reshaped by new meanings and relationship. Changes in health behavior accompany expansion of consciousness and the specific changes are dictated by inner – directed (related to cultural aspects as well) healing processes.

Thus the *mutual connectedness* is valued in the interaction between the patient and the nurse in the cross-cultural context. What does it include? According Schubert & Lionberger (1996), the mutual connectedness includes the four elements:

1. ***A caring environment,*** which is created when the nurse's behavior conveys a stable caring attitude toward the patient. I could interpret this element as a health care aspect, which is a state of being when the nurse is motivated to serve the best interests of the patient, to promote health and support inner healing process. In this context patient and nurse are getting to know each other by sharing personal and private worlds in this environment. This aspect prepares the way for a healing connection between patient and nurse. Also a caring environment supports growth and self – actualization of those in such a relationship.
2. ***Mutual connectedness*** is joining the nurse and the patient in a relationship committed to the health and healing of the patient. I think, that the nurse remains constant in caring, listening to, and focusing on the patient. The patient's trust may fluctuate until a feeling of joining or bonding occurs in the relationship.
3. ***Nursing therapy*** constitutes work done in the patient – nurse relationship. My comment would be, that the nursing knowledge and skills nourish the holistic therapies (that include cultural sensitive care as well) specifically chosen to enhance healing as it is defined from the patient's cultural standpoint on the consensus base with the nurse.
4. ***Healing*** is the process of the patient and the nurse self – transformation during which latent potential is realized.

Religiousness is closely related to interaction process between the nurse and the patient and to the mutual connectedness between them as well. Religiousness is described as an important factor that influences health care by allowing the patient to accept, treat and cope with suffering and illness (Mickley & Soeken, 1993). These authors describe religiousness as an adaptive coping mechanism that is used by the patient in such a way as to maintain health or effectively deal with suffering and illness. According to Spector (1991), it may be difficult to distinguish those aspects of a patient's beliefs and values that are derived from religion and those derived from culture and ethnicity. Some aspects of religiousness are shared behaviors, beliefs, and values that are passed from one generation to the next and the religiousness behaves sociodynamically like culture.

According to Rooda (1992), the cultural background of an individual is composed of the learned norms, values, customs and beliefs: "Cultural identification gives a background from which nurses can anticipate differences in values, religiousness, dietary practices, lines of authority and beliefs and practices related to health and illness" (p. 340). It becomes clear from Rooda's study that culture is the global concept of which religiousness is a component.

The health care establishment has long been aware of the positive correlation between religiousness and recovery from or coping with disease and illness (Mickley & Soeken, 1993). Now that the concept of cultural competence has been recognized as a necessary part of nursing education and acute, long-term care (Rooda, 1992), the role of religiousness needs investigating because of its interconnectedness with culture and ethnicity. It means, that religiousness gives the nurse a possibility to deliver care that is even more patient oriented than it would be using the dictates of cultural specificity alone.

As health care threatens to redefine the health delivery system, religio-specific nursing and care may have important implications for the future in terms of health promotion. Identification of religiousness provides a background from which the nurse can anticipate differences in values, religious orientation, dietary practices, family life patterns, beliefs and practices related to health and illness (Tripp – Reimer, Brink & Saunders, 1984). The tenets of one's faith can have a positive effect on the major determinants of health, such as a clean and safe environment, adequate nutrition, reduction of the risk factor (e.g., smoking, drug abuse and etc.). In this respect, religio-specific nursing practice is proactive and translates into a healthier population, with the potential for tremendous savings to the health care system. Religio-specific nursing and care can also become another avenue that enables the nurse to understand the patient's orientation to health and illness.

Nursing education and nursing practice and care are the processes by which ordinary people with their own culture – bound views of health, illness and healing are "programmed into the health provider culture" (Spector, 1991, p. 59). Nurses may systematically shed old notions of health, illness and healing for the more scientifically based nursing education. In doing so the nurse may reject alternative forms of healing, not realizing that the patient may not hold the science of health care in the same high esteem. Many times it is religiousness that is responsible for differences in the patient's orientation to health, illness and healing. Therefore, the goal of the nurse is to find a common ground between provider and patient, with maximum benefit for the latter. Religio-specific nursing practice can initiate the search for common ground because of its ability to touch every aspect of the patient – his/her mind, body and spirit. If the nurse is knowledgeable in the area of religiousness and is willing to relate to the patient on the patient's terms, the possibility of positive outcomes is substantially increased.

Here it is important to define the terms related to religio-specific nursing. Religion has reference to the formal religious organizations of the world. Religiousness refers to formal organizations that are important to an individual and the extent to which the individual's life is influenced by what they teach.

Spirituality is often used synonymously with religion, but the terms define different phenomena. In relation to nursing and caring practice and assessment of the patient's needs, spirituality implies a much broader perspective of how an individual goes about searching for the meaning in life.

Religiousness is related to an individual's beliefs and reactions to God, Supreme Being, or the supernatural and may be described in a variety of ways. The trait approach (Mickley & Soeken, 1993) looks at certain characteristics that a person has. One trait that is often studied is religious commitment, which is the cognitive, motivational, and behavioral expression of one's faith. Allport (1967) has described religiousness from two orientations – *intrinsic* and *extrinsic*. It means, that people who are intrinsically oriented attempt to live out the principles of their faith. Extrinsic religiousness is viewed as something to be used and not lived.

Religiousness, as it relates to health, may also be defined as spiritual well – being. As a good indicator of spiritual well – being connotes satisfaction with life and a relationship with God. Thus it has an existential as well as a religious component. Spector (1991) states that culture is composed of beliefs and behaviors that are learned and shared by members of a group. These shared beliefs and behaviors are passed from one generation to the next and are related to ethnicity as well; an ethnicity is the condition of belonging to a particular ethnic group (Spector, 1991). It is important that cross – cultural content and experiences about cultural diversity should be included in nursing education. As religiousness is a component of culture, the content of nursing curriculum should also extend to religiousness.

Religiousness provides a way for the nurse to fine – tune the nurse – patient interaction and relationship. For example, the nurse may encounter two patients who are from the same cultural and ethnic background, and yet embrace different orientation to religiousness. On the other hand, two patients could be of different ethnic and cultural backgrounds and yet share the same orientation to religiousness. The common denominator in each case is religiousness, which is highly individualized in terms of the patient's perspective. When the nurse incorporates aspects of the patient's religiousness into health care delivery, the result may be greater compliance with regimes, care plans and strategies to improve the patient's health. This may be true because the patient firstly has a chance to participate in his/her nursing and care. But certain behavioral objectives identified by the nurse may also coincide with behaviors that are strongly tied to the patient's religiousness. For example, knowledge of the Jewish dietary law is essential to provide religion – specific nursing care to the Jewish patient. It is not unusual for a Jewish patient to refuse to eat hospital foot because of not be kosher (Giger & Davidhziar, 1990). Therefore, consideration of the Jewish patient's attitude toward the importance of dietary laws is an aspect of religio-specific nursing and caring practice.

According to Highfield (1992), people respond to the threat of illness as total beings. Therefore, the nurse should anticipate the influence of religiousness, along with its assets and limitations. The research of Mull, Cox and Sullivan (1987) revealed that effective interventions are based on the whole individual, taking into account more than just physical needs. Religiousness, a resource that can facilitate coping and adaptation to health and illness, should be explored as a component of health care interventions. Raleigh (1992) notes, that a large number of patients report that religiousness is helpful in coping with illness and nurses need to give greater consideration to aspects of religiousness. This idea could be proved by the results of Sodestrom and Martinson (1987) research results, which have shown, that 92% of the patients surveyed felt that "people fall back on their religion in times of crisis" (p. 45).

Thus the religio-specific nursing and caring practice includes the nurse's and patients own cultural backgrounds and religiousness to the health care setting. By the process of assessment the nurse gains knowledge of differences in religious orientation. With these differences in mind, the nurse can fine – tune the interaction with the patient. Respect for the patient's religiousness allows the nurse to be flexible in using aspects of the patient's religiousness to maximize positive patient outcomes.

Transcultural nursing

As our world conflates through the use of technology and travel, we can no longer maintain a single cultural perspective (Leininger, 1994; Meleis, 1996). Transcultural nursing emphasizes liberty over equality, the individual over the structural and the cultural over the political. Leininger (1985, 1995), the key proponent of transcultural nursing, set the agenda in defining culture as the values, beliefs, norms and practices of a particular group. The task for nurses, according to Leininger, is to study designated cultural groups and elicit descriptive information specific to those groups. The underlying assumption that the identity of a person is culturally determined above all else is explicit in her work.

Thus the transcultural nursing, according Leininger (1995, p. 58), is *"focused on comparative care, values, beliefs and practices of individuals or groups of similar or different cultures with the goal of providing culture specific and universal care practices in promoting health or well – being or to help people to face unfavorable human conditions, illness, or death in culturally meaningful ways."*

Leininger has warned her readers that unless transcultural nursing is used to challenge the universalistic assumptions (an aspect of liberalism), which underpin nursing in general, nursing is risking cultural imposition and ethnocentric practices in multicultural societies. In her writings Leininger emphasizes the dangers of ethnocentrism, cultural imposition and cultural conflict (Leininger, 1985, 1991). In order to avoid decontextualization and cultural imposition of people deemed "other", Leininger emphasizes the contextualization of individual, family, group and institutional system (s) in time, place, culture, economy, technology, politics and religion. I would mention, that Leininger constructed a theory based on care which she regards as the essence and the central, unifying and dominant domain to characterize nursing (Leininger, 1984).

In giving primacy to care the author does not assume that care constructs and expressions are universal or that they represent a universal value. On the contrary, Leininger insists that care constructs and expressions are diverse, complex and embedded in the complex and interactive systems of all societies. Through this lens of primacy of care in nursing, Leininger analyzes complex interactive inter/intra – cultural, educational, religious etc. human creations and practices within which care and nursing are constructed and expressed. The researcher argues that care concepts and practices are not universal, but are diverse within the cultural groups and across cultural groups. According to Leininger (1985, 1991, 1995), care and caring practices are the driving force for transcultural nursing, and are diverse and embedded in sociocultural systems of particular nations, families, groups and individual experiences, beliefs and values. Leininger points out that she expects nurses to examine the interactive values of their own culture and social systems and to strive to influence change, when, as Bourdieu (1990) explicates, that culture in practice is found to be non – caring and non – therapeutic. Thus transcultural nursing concepts and insights in their broadest sense can be used to examine and influence change within the nursing[1].

This perspective (mentioned above) need to be reflected and integrated into the basic nursing training and education programs and nurtured in graduates who prepare for specialist roles. According to DeSantis (1991, 1994), to achieve competence in transcultural care there must be recognition of one's own values, openness to cultural differences, possession of a person-oriented learning style, and use of the cultural resources available. Papadopoulos *et al* (1994) agree with this idea and ad, that seeking to achieve such outcomes the curriculum should address course content in themes, which lend themselves to transcultural issues.

Modelling the transcultural curriculum elements

Achieving transcultural nursing, which includes a culturally sensitive care is a process rather the end product (Camphina-Boyte, 1994) and requires emotional and cognitive preparation (Spitzer et al, 1996). False awareness of stereotypes needs to be resolved before knowledge and skills can be effectively learned (Pederson, 1988). Doing this requires a supportive environment in which students and teachers/lecturers can openly and frankly discuss the issues (McGee, 1992) and explore ethnocentric ways of knowing about science and nursing (Abdullah, 1995). Although the classroom discussions on the issues are important to prime the stu-

[1] And the same time I do think, that Leininger's work gives the functionalistic and apolitical definition of culture and inadequate attempts to consider issues of gender, class and power.

dent, they have to be consolidated through experiential learning (DeSantis, 1991; Abdullah, 1995; Meleis, 1996).

Tilki *et al* (1994), Gerrish *et al* (1996) argue, that learning about other cultures can be achieved by exposure to ethnic groups within our own society. Garvey (1997) suggests, that this can be most effectively achieved by traveling abroad and immersion into another culture for no less than three months. Total immersion in the culture allows the individual time to come to terms with that culture (Williamson, 1994) and gives students the opportunity to work with staff as well as patients from alternative cultural backgrounds. It establishes important connections: 1) the nurses share a common profession; 2) the students thrust into an alternative culture, experience minority status and thus disadvantage:

a. ***Struggling with a second language.*** The phenomenon of translation is no more considered an activity of mere transference between lexicon and grammar, as simple substitution of literal material of a language into another one. Nowadays, it is seen as a transcultural event in which the translator (e.g., nurse student from foreign country, who learns the new language, experiences and percepts the foreign language through her/his translation to the herself/himself), earning visibility, is considered an active element, speech and text producer. Thus the subjectivity of translation is very present in the choice of the terms that the nurse uses, as well as in his/her personal style of writing the translated phrases from the foreign language (Garcia & Lima da Nóbrega, 2003).

b. ***Working is an environment, which is relatively strange.*** Because of the striving to give the culturally relevant health care, the nurse needs to know not only foreigners' country of origin, but also their perception of their ethnic identity (May, 1996). Through this experience nurse students make connections with individuals, which transcend stereotypical clustering, i.e. from the standpoint of gender, power, sexuality. In the context of working as a nursing student, various nursing concepts such as *suffering, care, hope, personal dignity, healing, knowing the patient as a person* and etc. transcendent any single cultural orientation and enable the nurse student to make connections with patients through the act of caring (Leininger, 1991). Making connections with people opens the possibility to learn, to listen, to value, to reflect, to empower that open up the new forms of understanding. Thus the students have the possibility to make connections with the situation of patients, which influences their construct of the nurse – patient relationship. Comparison and critical reflection are seen as key ingredients to achieve these outcomes in an educational process of transcultural nursing (DeSantis, 1991; Abdullah, 1995; Meleis, 1996).

To become transculturally oriented the nurse students need to temporarily suspend their own cultural traditions in order to perceive the situation of others (DeSantis, 1994). To do this nurse students have to be conscious of their own values and beliefs. Then they can explicitly compare those with the views of the patient (DeSantis, 1994; Tripp-Reimer *et al*, 1984). For the nurse student this is made up of three elements:

1) the nurse's personal beliefs, values and expectations;
2) the nurse's professional beliefs, values and expectations;
3) the context in which the interaction between the nurse and patient takes place.

Sometimes these values are so close they do not lend themselves to critical scrutiny. Transcultural nursing may expose the nurse student to an alternative culture and language and such experience creates a disjuncture for the nurse student who has to reconsider what has been taken for granted in their own system, alongside analysis of the system that is novel. Therefore, not only do the students become sensitive to the perspective of others, but they also become increasingly self-aware of the multiple factors influencing their everyday interaction with the patients. This is important as the nurse student can take account of suffering that may stem from cultural distress, and thereby relieve cultural pain (Leininger, 1995).

Thus the curriculum, which is oriented to transcultural nursing, needs to incorporate the elements that foster the nurse students' transcultural sensitivity development:

- ***Personal/individual*:** the nurse student is able from the transcultural standpoint – a) to recognize personal limitations and strengths; b) to reflect into her/his capacity to cope with difficult and stressful situations that are related to increased confidence and self-reliance; c) develop personal linguistic competence in using foreign languages; d) to develop continuously transcultural and professional knowledge, skills and abilities.
- ***Interaction with patients*:** the nurse student is able from the transcultural standpoint – a) to develop the won empathy with the vulnerability of patients; b) to recognize, acknowledge and value the differences; c) to enhance ability to find the right questions to ascertain the patient's perspective; d) to involve relatives in the care of patients; e) to verify that mutual understanding has been achieved; f) to help patients achieve what they want from the health care system; g) to evaluate the patient's acculturation, if the patient is an immigrant (ex., what languages the patient reads and speaks; what languages the patient used in his/her childhood; what languages the patient usually speaks and thinks; what languages the patient usually speaks with his/her friends etc.).
- ***Interaction with colleagues*:** the nurse student is able from the transcultural standpoint – a) to increase confidence to challenge and explore the meaning behind practice; b) to accept the constructive criticism; c) to develop a greater insight into the real meaning behind actions and words; d) to increase personal confidence in examining issues with colleagues; e) to recognize, that there is not any right or wrong way of doing things, only suitable responses to the situation.
- ***Professional contextuality:*** the nurse student is able from the transcultural standpoint – a) to change her/his personal view on how the nurse student wishes to nurse and work as a nurse; b) to develop broader understanding about the multidisciplinary teamwork and factors that influence power relationships within the team; c) to develop greater confidence in using the transcultural nursing research and rational to support actions; d) to develop the greater insight into their own value system of nursing, which enhanced the value nurse students placed in the nursing profession; e) to enhance understanding about non-verbal communication and the meaning ascribed to the spoken word; f) to develop reflexivity about the practice.

- ***Educational process:*** such process needs to include the elements that will foster nurse students to acquire the transcultural sensitivity – a) to establish realistic expectations and learning objectives; b) to use in the learning-teaching process the analysis of simple and complex situations, which develop students' objective, reflective and sensitive analysis; c) to discuss more analytical of things that happen in practice and prepared to consider how environmental factors influence events; d) to increase student nurses' open-mindedness, flexibility, self-reliance and confidence to self-directed learning; e) to use different learning strategies a learning-teaching process such as learning from comparison, critical reflections (in written form), writing essays on a theme related to transcultural nursing situations from nursing practice, theory and research standpoints.

Thus from the practical standpoint there are two ways in experiential learning of the transcultural nursing: *a) when the nurse student is going abroad and works as a nurse-"foreigner"; b) when the nurse student has a possibility to work with the people with the different cultural backgrounds (e.g., language, religion etc.).*

Conclusions

- The diverse mix of cultures, languages and points of view on nursing that is evident within the context of different cultures means that an ambition to generate a universally acceptable conceptual nursing framework from the transcultural standpoint is a chimera. However, extreme caution needs to be exercised to ensure that dominant cultural orientations or notions of science do not suppress locally effective ways of nursing that would ultimately result in culturally incompetent caring. A way to ensure this does not happen through learning, which makes connections with individuals in a manner that acknowledges and celebrates cultural difference, but finds a mutually acceptable way of working.
- To fulfill the dictates of religio-specific nursing and caring practice, nurses will have to be aware of their own religiousness and the religiousness of their patients. At some point during their educational process, nurses will need to become familiar with the various religious denominations, but only as a foundation to anticipate those aspects of religiousness that might lead to a positive response to health care.
- The success of religio-specific nursing practice depends largely on how well it is received by the patient and on the nurse's willingness to use this approach to health care delivery. To determine the efficiency of religio-specific nursing practice, evaluation tools must be developed to reveal weaknesses, strengths, inconsistencies and generalizability in the delivery of such nursing and caring.

References

ABDULLAH, S.: Towards as individualized client's care: implications for education. The transcultural approach. Journal of Advanced Nursing 22(1995), pp. 715-720.

BANKS, R.: Health and spiritual dimensions: relationships and implications for professional preparation program. Journal of School Health 4(1980), pp. 199.

BOURDIEU, P.: The Logic of Practice. Cambridge 1990 (Polity Press).

BRIGMAN, K.: Religion and family strengths: implications for mental health professionals. Topics in Family Psychology and Counseling 1(1992), pp. 39-52.

CAMPHINA-BOYTE, J.: The process of cultural competence in health care: a culturally competent model of care. (2nd ed.). USA, Wyoming, OH 1994 (Perfect Printing Press).

DESANTIS, L.: Developing faculty expertise in culturally focused care and research. Journal of Professional Nursing 7(1991), pp. 300-309.

DESANTIS, L.: Making anthropology clinically relevant to nursing care. Journal of Advanced Nursing, 20(1994), pp. 707-715.

GARCIA, T./LIMA DA NÓBREGA, M.: Transcultural adaptation of the Brazilian Portuguese translated version of terms included in ICNP® and identified in the CIPESC. In: OUD, N. (ed.): Acendio 2003: Making nursing visible. 4th European conference of Acendio. Bern 2003 (Verlag Hans Huber), pp. 124-129.

GARVEY, J.: The benefits of transcultural nurse education. Proceedings of the 1st International nursing Conference on Connecting Conversations: Nursing Scholarship and Practice. Reykjavik, Iceland, 20-22 June, 1995. Nursing Research Paper Series, 1(1997), pp. 159-64.

GERRISH, K./HUSBAND, C./MACKENSIE, J.: Nursing for a multi-ethnic society. England, Buckingham 1996 (Open University Press).

GIGER, J./DAVIDHZIAR, R.: Contextual care: Religious considerations for culturally appropriate nursing care. Advancing Clinical Care 5(1990)4, pp. 48-51.

HANSON, S./BOYD, S.: Family health care nursing. Philadelphia 1996 (F.A. Davis Company), pp. 110-111.

HIGHFIELD, M.: Spiritual health of oncology patients" nurse and patient perspectives. Cancer Nursing 15(1992)1, pp. 1-8.

LEININGER, M.: Care: the Essence of Nursing and Health. New York 1984 (Thorofare, Slack).

LEININGER, M.: Qualitative Research Methods in Nursing. Orlando 1985 (Harcourt Brace Jovanovich Group).

LEININGER, M.: Transcultural care diversity and universality: a theory of nursing. Nursing and Health Care 6(1985), pp. 199-202.

Cultural Care Diversity and Universality: A Theory of Nursing. LEININGER, M. (ed.). New York 1991 (National League for Nursing Press).

LEININGER, M.: Transcultural nursing education. A worldwide imperative. Nursing and Health Care 15(1994), pp. 254-257.

LEININGER, M.: Transcultural nursing concepts, theories, research and practices. (2nd ed.). New York 1995 (McGraw – Hill Inc.).

MAY, K.: Middle - Eastern immigrant parents' social networks and help - seeking for child health care. In: SMITH, J.P. (Ed.): Nursing care of children. London 1996 Blackwell Science, pp. 41-55.

MCGEE, P.: Teaching transcultural care: a guide for teachers of nursing and health care. London 1992 (Chapman & Hall).

MELEIS, A.: Culturally competent scholarship: substance and rigor. Advances in Nurse Science 19(1996), pp. 1-16.

MICKLEY, J./SOEKENN, K.: Religiousness and hope in Hispanic and Anglo-American women with breast cancer. Oncology Nursing Forum 20(1993)8, pp. 1171-1177.

MOBERG, D.: Spiritual well – being in late life. In: GUBRIUM, J. (ed.): Late life: Communities and environmental policy. Springfield, IL 1974 (Charles C. Thomas).pp. 256-279.

MULL, C./COX, C./SULLIVAN, J.: Religion's role in the health and well – being of elders. Public Health Nursing 4(1987)3, pp. 151-159.

NEWMAN, M.: Health as expanding consciousness. New York: 1994 (National League for Nursing).

PAPADOPOULOS, I./ALLEYNE, J./TILKI, M.: Promoting transcultural care in a college of health studies. British Journal of Nursing 3(1994), pp. 1116-1118.

PEDERSON, P.: A handbook for developing multi cultural awareness. Alexandria, VA 1988 (American Association for Counselling and Development).

RALEIGH, E.: Sources of hope in chronic illness. Oncology Nursing Forum 19(1992)3, pp. 443-448.

REED, P.: An emerging paradigm for the investigation of spirituality in nursing. Research in Nursing and Health 15(1992), pp. 349-357.

ROGERS, M.: Nursing: Science of unitary, irreducible human beings: Update 1990. In: BARRETT, E. (ed.): Visions of Rogers' science – based nursing. New York 1990 (National League for Nursing), pp. 5-11.

ROODA, L.: The development of a conceptual model for multicultural nursing. Journal of Holistic Nursing 10(1992)4, pp. 337-347.

SHAFRANSKE, E./MALONEY, H.: Clinical psychologists' religious and spiritual orientations and their practice of psychotherapy. Psychotherapy, 27(1990), pp. 72-78.

SCHUBERT, PH./LIONBERGER, H.: A study of client – nurse interaction using the grounded theory method. Journal of Holistic Nursing 13(1996)2, pp. 102-116.

SODESTROM, K./MARTINSON, I.: Patient's spiritual coping strategies: a study of nurse and patient perspectives. Oncology Nursing Forum 14(1987)2, pp. 41-46.

SPECTOR, R.: Cultural diversity in health and illness. Norwalk, CT 1991 (Appleton & Lange).

SPITZER, A./KESSELRING, A./RAVID, C./TAMIR, B./NOAM, R.: Learning about another culture: project and curricular reflections. Journal of Nursing Education 35(1996), pp. 323-328.

TILKI, M./PAPADOPOULOS, I./ALLEYNE, J.: Learning from colleagues of different cultures. British Journal of Nursing, 3(1994), pp. 1118-1124.

TRIPP-REIMER, T./BRINK, P./SAUNDERS, J.: Cultural assessment: content and process. Nursing Outlook 32(1984), pp. 78-82.

VAUGHN, F.: The inward arc: Healing and wholeness in psychotherapy and spirituality. Boston 1986 (New Science Library).

WILLIAMSON, J.: ERASMUS cultural exchanges in Europe. British Journal of Nursing, 3(1994), pp. 1124-1128.

Ernesta Račienė

Sprachen und Kulturen im Globalisierungsprozess. Ein Vergleich zwischen Deutschland und Litauen

Einleitung

Wir leben im Zeitalter der Globalisierung, offener Grenzen, intensiver Integrationsprozesse, was neue Kommunikationsmöglichkeiten eröffnet, Kultur- und Sprachkontakte aktualisiert. Politische, wirtschaftliche und sozial-kulturelle Veränderungen, die durch diese neue Situation in unseren Ländern hervorgerufen werden, beeinflussen die Kultur unserer Völker und vor allem die Sprache als ihr wichtigstes Ausdrucksmittel. Jede Sprache repräsentiert eine besondere und eigene kulturelle, regionale, nationale Befindlichkeit und ermöglicht unsere gesellschaftliche Identifikation. Die Übernahme von Elementen fremder Kultur, die sich in der Sprache als fremdsprachliche Infiltration manifestiert, kann eine gewisse Gefahr für die einzelnen kulturellen Identitäten und für die Sprachen selbst darstellen. Sprachkontakte als eine Art der Kulturkontakte rufen Veränderungen der betroffenen Kontaktsprachen in Phonetik, Grammatik und Wortschatz, die nicht immer positiv bewertet werden.

Zu den Sprachproblemen, die gegenwärtig die Linguisten sowie die Öffentlichkeit verschiedener Länder besonders bewegen, gehört die Frage nach dem Verhältnis der Nationalsprachen zum Englischen. Englisch nimmt die Vorrangsstellung weltweit und auch in Europa. H. Haarman meint, dass die Ära der Globalisierung bedeutet, dass wir in der Zeit sprachökologischer Umwälzungen leben, „in der traditionelle Rollenmuster von Sprachen neu definiert werden und in der ein neuartiges Szenarium für Sprachkontakte Profil gewinnt. Ein Faktor, der den Gesamteindruck der sprachlichen Globalisierung entscheidend prägt, ist die Rolle des Englischen in der Interkommunikation. Das Englische, so scheint es, fördert in besonderem Maße Prozesse kultureller Assimilation und Nivellierung“ (Haarmann 2002:9). Andererseits kann das Wesen der Globalisierung als Dynamik, mit der lokale Kulturen und Sprachen auf internationalistische Trends reagieren, aufgefasst werden (vgl. ebda.:10).

In der letzten Zeit ist auch in Litauen der Einfluss der englischen Sprache besonders groß, ihre Autorität und soziales Prestige sind gewachsen. Ein großer Teil des neuen fremden Wortguts im Litauischen sind Anglizismen, ein anderer Teil stammt aus anderen Sprachen, aber auch sie gelangen in die litauische Standardsprache über das Englische, gewöhnlich über Übersetzungen. Also wird der Gebrauch der fremdsprachlichen Lexik hauptsächlich von der englischen Sprache beeinflusst. Auf Grund der Bedeutungen der entsprechenden englischen Wörter werden den seit langem im Litauischen heimischen Entlehnungen neue Bedeutungen verliehen, angestammte litauische Wörter werden oft nach dem Vorbild des Englischen verdreht. Außerdem ist die litauische Gesellschaft dem Einfluss anderer Sprachen infolge des langen Russifizierungsprozesses nicht mehr widerstandsfähig. Oft hat man kein Sprachgefühl mehr, deshalb wird nicht mehr zwischen dem sinnvollen Eigenen und dem Fremden unterschieden. Diese Tatsache trägt dazu bei, dass man nicht mehr fähig ist, ein Mittel zu finden, das uns bei der Begegnung mit fremder Kultur und Sprache stärken könnte.

Einen anderen negativen Faktor stellt die Unterschätzung der eigenen Kultur und der eigenen Sprache dar. Das geschieht meistens als Folge der schlechten Kenntnis der Muttersprache und ihrer Ausdrucksfähigkeiten. Der allzu häufige Gebrauch der fremdsprachlichen Elemente gefährdet die kulturellen Interessen des Volkes, die durch die Muttersprache erzogen werden. Überhaupt ist man nicht mehr imstande, die Offenheit für andere Kulturen mit dem nationalen Bewusstsein, mit der Bewahrung und Pflege der Muttersprache, zu verbinden. Es ist außerdem zu beachten, dass der Einfluss einer Fremdsprache nur bis zu einer gewissen Grenze als positiv zu betrachten ist, und zwar nur solange er keine Gefahr für das System der aufnehmenden Sprache darstellt und es nicht zerstört. Die Destabilisierung des Litauischen kommt auch durch die unmotivierte Ersetzung der eigenen Elemente durch fremde zum Ausdruck. Damit entsteht eine Gefahr für das Gleichgewicht der Sprachstruktur, das soziale Ansehen der Sprache sinkt, eigene Sprache wird nicht mehr geschätzt.

Für das Deutsche spielt das Englische auch eine immer wichtigere Rolle. Der englische Einfluss auf das Deutsche ist schon im 19. Jht. spürbar, er ist für das gesamte 20. Jht., besonders aber für die letzten 60 Jahre charakteristisch. Was für die litauischen Sprachforscher- und Pfleger ein relativ junges aber sehr aktuelles Problem ist, gehört seit Jahrzehnten zum festen Bestandteil der Forschungen von deutschen Sprachwissenschaftlern.

Die deutschen Linguisten stellen fest, dass der englische Einfluss in der zweiten Hälfte des 20. Jht. in besonderer Weise in Wechselwirkung mit den gesellschaftlichen und kommunikationstechnischen Veränderungen steht. Das ist mit der Entwicklungstendenz der letzten 10 Jahre in Litauen vergleichbar. Aber der Prozess ist in Deutschland viel weiter fortgeschritten: der englische Einfluss „wird nicht mehr nur von einer kleineren oder größeren Minderheit wahrgenommen, sondern betrifft in Beruf und Alltag mittlerweile den weitaus größten Teil der Sprachgemeinschaft" (Bär 2000:18). Das Englische wird nicht nur als Lingua franca, sondern auch als Gegenstand und Mittel der kulturellen Identifikation verstanden. Die Orientierung am Vorbild der USA führte zu einer Modifikation des Deutschen vor allem auf der Ebene des Wortschatzes und der Redewendungen. Die deutschen Wissenschaftler neigen aber dazu, die Tendenz der Anglizismenverwendung als einen Ausdruck lebendiger Sprachentwicklung, ihrer Verjüngung zu betrachten (vgl. ebda.:20, Elitz 2000:143-145). Sie sind der Meinung, dass die Verwendung der Fremdwörter und Anglizismen insbesondere in bestimmten Diskursen sogar sinnvoll ist, wenn sie als Ausdrucksmittel sprachlicher Handlungsmuster die ins Auge gefasste Zielgruppe zu erreichen helfen (z.B. in Werbetexten, im fachsprachlichen Diskurs). Und obwohl die Zahl der Anglizismen in den Jahrzehnten nach 1945 erheblich zugenommen hat, sind deutsche Sprachforscher zur Schlussfolgerung gekommen, dass die Anglizismen keine Bedrohung für die deutsche Sprache darstellen, weil im Verhältnis zum gesamten Wortschatz der deutschen Gegenwartssprache die Zahl der Wörter aus dem Englischen immer noch sehr gering ist. Eine größere Gefahr für das Deutsche bildet der immer zunehmende latente Einfluss des Englischen. Genauso wie die litauischen Linguisten befürchten die deutschen Sprachwissenschaftler, dass durch dieses Einwirken des Englischen die deutsche Sprache viel gravierender umstrukturiert und der englischen angepasst werden kann als durch die Übernahme einzelner Wörter. Es geht hier vor allem um Veränderungen der grammatischen Strukturen und Bedeutungsverschiebungen bei deutschen Wörtern und auch Fremdwörtern, die schon länger im Deutschen heimisch sind (vgl. Hoberg 2000:306-308).

Im vorliegenden Beitrag wollen wir die wichtigsten Tendenzen der Verwendung von Entlehnungen in der litauischen Standardsprache analysieren und mit den Tendenzen der deutschen Standardsprache vergleichen, den Einfluss des Englischen auf den Gebrauch der Entlehnungen in beiden Sprachen untersuchen. Da dieser Einfluss besonders auffallend im Jugenddiskurs ist, werden auch Beispiele der Jugendsprache in Deutschland und in Litauen herangezogen. Die litauischen Belege dazu wurden meistens der Pressesprache entnommen, weil die Pressesprache am schnellsten auf die Prozesse, die auch in anderen Sprachdiskursen vor sich gehen, reagiert und sie offenbart. Einige Beispiele stammen aus anderen Quellen - aus verschiedenen Anzeigen, Werbetexten, Einladungen, Prospekten. Sie sind mit rš vermerkt. Beispielsätze werden manchmal verkürzt angeführt. Die deutschen Belege für den Vergleich wurden aus Wörterbüchern und der Pressesprache entnommen.

Folgende Wörterbücher wurden verwendet: Duden. Deutsches Universalwörterbuch. Mannheim 1996; Duden. Das Fremdwörterbuch. Mannheim 1966; 1974; 1982; 1990; Duden. Das große Fremdwörterbuch. Mannheim 2000. Für Beispielsätze aus der deutschen Presse wurden teilweise die Korpora der geschriebenen Sprache des Instituts für Deutsche Sprache, Mannheim benutzt, die unter www.ids-mannheim.de zu finden sind.

1. Vergleich der Tendenzen der Verwendung von Entlehnungen: Litauen - Deutschland.

Man kann die Tendenz feststellen, seit langem verwendete, in der litauischen Sprache eingebürgerte Entlehnungen durch neue Fremdwörter mit derselben oder ähnlicher Bedeutung zu ersetzen. Manchmal werden dazu auch die Entlehnungen aus derselben Sprache verwendet. Es geschieht oft unter dem Einfluss des Englischen. Einige Beispiele dazu: das alte Lehnwort genocidas wird durch gr.-lat. holokaustas ersetzt, das in das Litauische aus dem Englischen übernommen wurde. Die Bedeutung ist gleich, doch holokaustas wird eher verwendet, wenn man über die Massenvernichtung von Juden spricht, z.B.: Nacių etninis fanatizmas pasaulį sukrėtė holokausto siaubais LA 1997:115. Sovietų komunizmas yra surengęs tautoms tokį pat nusikalstamą holokaustą kaip ir nacizmas mūsų broliams žydams LA 1998:8. Im Medizinbereich wird anstatt des Wortes der griechischen Herkunft profilaktika das lateinische Wort prevencija gebraucht, z.b.: Suvažiavime buvo nutarta, kad didelę reikšmę širdies kraujagyslių prevencijai turi ir sveika mityba LR (GB) 2000:206. Širdies ir kraujagyslių ligų prevencija – ne toks įspūdingas dalykas, kaip širdies persodinimas, tačiau sąmoningai keičiant gyvenseną [...] nebereikės ir širdies operacijų LA 1999:66.

Das Wort lateinischen Ursprungs *eksperimentinis* wird durch das französische Wort *pilotinis, -ė*, das über das Englische gekommen ist, ersetzt, z.B.: Bus parodyta naujo animacinio filmo *pilotinė* serija rš. Ne tik Briuselis, bet ir Maskva iškėlė vadinamojo *pilotinio* Rusijos regiono plėtojant santykius su Vakarų valstybėmis, idėją LR (RV) 2001 181.

Anstelle der deutschen Entlehnung grupė tritt immer öfter der Anglizismus bendas, z.B.: Muzikos *bendas* „Lietus" sukūrė jau pirmą dainą, pelniusią jiems šlovę LR (GB) 1999:200.

Ähnliches lässt sich auch im Deutschen beobachten. Die ältere gr.-lat. Entlehnung Genozid kennzeichnet allgemein „Mord an nationalen, rassischen od. religiösen Gruppen" DU. Das Wort Holocaust wird im DU als die aus dem Englischen übernommene Entlehnung bezeichnet, die vor allem in der Bedeutung „vom Nationalsozialismus betriebene Vernichtung des jüdischen Volkes" verwendet wird, z.B.: Nach dem Holocaust in Litauen hängt auch die Sowjetmacht für Jahrzehnte den Mantel des Schweigens über die Geschichte FAZ 27.07. 2001.

Die lat.-frz. Entlehnung *Prävention* hat die Bedeutung „vorbeugende Maßnahmen". Das davon abgeleitete Adjektiv präventiv bildet mehrere Komposita, die zum Wortfeld „Medizin" gehören, z.B.: *Präventivbehandlung*, *Präventivmedizin, Präventivmittel* DU.

Die lat. Entlehnungen *Experiment, experimentieren* und Adjektive *experimental*, *experimentell* werden im Deutschen durch das französische Wort *Pilot* verdrängt, das besonders produktiv bei der Bildung der Komposita ist und in diesen Bildungen etwas als vorausgeschickten Versuch, als Test zur Feststellung oder Klärung wichtiger Aspekte, Schwierigkeiten o.Ä. kennzeichnet, z.B.: Pilotanlage, Pilotprojekt, Pilotveranstaltung, Pilotfilm, Pilotsendung, Pilotstudie DU. Diesen Zusammensetzungen entsprechen die litauischen attributiven Wortgruppen mit dem Adjektiv *pilotinis*.

Das Wort Band wurde ins Deutsche nach 1945 übernommen und kennzeichnet „Gruppe von Musikern, die vorzugsweise moderne Musik wie Jazz, Beat, Rock usw. spielt" DU. Also verbindet man im Deutschen (genau wie auch jetzt im Litauischen) mit dem Wort *Band* moderne Tanzmusik, z.B.: Harter Rock und wildes Revolutionsgeschrei: mit dieser Mixtur brüllt sich die US-*Band* Rage Against the Machine in die Hitparaden (Sp 22/1996:213).

Mit der immer aktiveren Verwendung von Entlehnungen in der litauischen Standardsprache ist auch die Tendenz der Bedeutungserweiterung dieser Wörter verbunden. Den Lehnwörtern werden auch übertragene Bedeutungen verliehen, d.h. sie werden metaphorisch in einer ihnen ursprünglich nicht üblichen Bedeutung benutzt. Diese neuen Bedeutungen werden in vielen Fällen nach dem Vorbild der entsprechenden Wörter in den westlichen Herkunftsprachen angenommen. Dieselben Tendenzen wurden auch in der deutschen Standardsprache festgestellt (vgl. Yang 1990:112; Hoberg 2000:303-309; Fink 1980:205). Es ist zu bemerken, dass verschiedene Völker bei der Übernahme derselben Entlehnungen ihnen unterschiedliche Bedeutungen verleihen. Oft werden mit diesen Entlehnungen auch sogar andere Begriffe bezeichnet. Also werden neue Wörter aus derselben Quelle, z.B. aus dem Englischen, übernommen und assimiliert, doch werden sie auf eigene Art interpretiert. Folglich kann ein Wort auf seiner Reise durch die Welt einige Bedeutungen annehmen. Ein krasses Beispiel dazu kann die deutsche Sprache liefern: das Wort Handy wurde ins Deutsche aus dem Englischen übernommen und bedeutet bekanntlich „Mobiltelefon". Im Englischen heißt aber *Mobiltelefon mobile phone* (vgl. dazu *mobilusis telefonas* im Litauischen) und das deutsche *Handy* geht auf das englische Adjektiv *handy*, das „griffbereit", „greifbar" bedeutet, zurück (vgl. Bär 2000:19).

Es werden die Bedeutungen der sog. Internationalismen erweitert. Wir wollen einige Beispiele dazu anführen. Das Wort der lateinischen Herkunft *legionierius* ist seit langem in der litauischen Sprache in der Bedeutung „Soldat einer Legion" bekannt (diese und alle weiteren Definitionen der im Wörterbuch fixierten Entlehnungen stammen aus dem litauischen Fremdwörterbuch „Tarptautinių žodžių žodynas" 1985). Jetzt hat das Wort die neue Bedeutung „ausländischer Sportler" bekommen (vgl. Rudaitienė/Vitkauskas 1998. Aus diesem Buch sind Bedeutungserklärungen aller neueren Entlehnungen der litauischen Sprache entnommen). Z.B.: Klaipėdos „Žuvėdros“ moterų krepšinio komandos *legionierė* iš JAV J.T. pasibaigus LMKL rungtynėms po uostamiesčio „Žalgirio“ sporto rūmus vaikšto šiltai apsirengusi LR (SA) 2000:18.

Per lietuvių ir *legionierių* rungtynių pertrauką du krepšinio aistruoliai galės laimėti naujutėlį automobilį „Land Rover Freelander“ LR (TVA) 2000:7, 57.

Man neigt dazu, dieses Wort auch in der abwertenden Bedeutung „Eindringling“, „Ankömmling“ (= atėjūnas, įsibrovėlis) zu verwenden, z.B.: Būtina atlikti kruopščią analizę, kas ir kodėl apvogė kaimą, kas leido kaimo skurdinimo *legionieriams* (= atėjūnams, įsibrovėliams) jaustis nebaudžiamiems bei visagaliams vargų iškankintame postkomunistiniame kaime VL 1996:95.

Das Wort *Legionär* wird im DF 1990 als „Mitglied einer Legion (z.B. der franz. Fremdenlegion)“ definiert. Im DU wird neben der ersten Bedeutung „Soldat einer Legion, der Fremdenlegion“ auch die weitere Bedeutung „[Fußball] spieler, der für einen Verein in einem anderen Land spielt“ angeführt, z.B.: Ich vermisse ihn auf keinen Fall, sagte der 40-Jährige über den *England-Legionär* vom FC Liverpool MM 23.03.2001. Neben dem *Finnland-Legionär* Jan Benda hat der Bundestrainer auch vier Neulinge eingeladen MM 17.04.2001.

Das Wort lateinischer Herkunft *moderatorius* ist in der litauischen Standardsprache als Physik- und Musikterminus bekannt. Jetzt wird dieses Wort nach dem Vorbild des englischen Wortes *moderator* auch in der Bedeutung „Vorsitzender, Leiter einer Versammlung Sitzung, Konferenz, Fernseh - oder Radiosendung“ (= pirmininkas, vedėjas) benutzt, z.B.: Diskusijos moderatorius (= pirmininkas, vedėjas) akcentavo, jog įdomiausia būtų pažvelgti į nematomą kultūros žurnalų pusę LA 1996:201.

Į televiziją lituanistės specialybę turinti 32 metų N.S. atėjo iš žinių radijo, [...] ji čia dirbo redaktore, o pastaruoju metu – moderatore (= vedėja) LR (TVA) 2000:47, 6.

Ähnliche Bedeutungsentwicklung hatte das Wort *Moderator* unter dem Einfluss des Englischen im Deutschen. Im DF 1966 werden zwei Bedeutungen verzeichnet: Moderator [lat.]: 1. Vorsteher eines Moderamens, 2. Verzögerer, Bremsvorrichtung in Kernreaktoren. Das DF 1974 führt als die dritte Bedeutung „Redakteur einer Rundfunk- oder Fernsehanstalt, der durch eine Sendung führt und dabei die einzelnen Programmpunkte ankündigt, erläutert und kommentiert“ an. Jetzt wird die Bedeutung des Wortes *Moderator* im DF 2000 wie folgt erklärt: Moderator < aus lat. Moderator „Mäßiger, Leiter“ > 1. leitender Redakteur einer Rundfunk- od. Fernsehanstalt, der durch eine Sendung führt u. dabei die einzelnen Programmpunkte ankündigt, erläutert und kommentiert. 2. Stoff, der Neutronen hoher Energie abbremst (Kernphysik). 3. Vorsteher eines Moderamens. Z.B.: Temperamentvoll nahm er seine Aufgabe als *Moderator* und Conferencier wahr und gewann das

Publikum rasch zu seiner Unterstützung MM 08.01.2001. Marco Rima ist bekannt durch zahlreiche Fernsehauftritte, auch als Gaststar oder *Moderator* MM 05.01.2001.

Aber auch in der Bedeutung „Vorsitzender, Leiter einer Versammlung, Sitzung, Konferenz" ist dieses Wort im Deutschen sehr häufig anzutreffen, z.B.: Hofmann ist der evangelische *Moderator* beim christlich-jüdischen Gesprächskreis MM 26.01.2001. Ebenso freute sich Präsident und *Moderator* Hans-Jörg Ebert [...] über das Kommen vieler treuer und verdienter Ehrenkappen-Träger MM 01.02.2001. Schriesheimer Realschul-Rektor ist *Moderator* einer Podiumsdiskussion zur Bildungspolitik MM 01.03.2001.

Unter dem Einfluss des englischen Wortes *stylist* werden mit dem Internationalismus lateinischer Herkunft *stilistas, - ė* im Litauischen nicht nur Beherrscher des Stils, des sprachlichen Ausdrucks, jemand, der den Stil, die Sprache korrigiert, sondern auch jemand, der Haare frisiert, Kleider entwirft u.ä. bezeichnet, z.B.: LRT pasiryžo ir kitoms išlaidoms – visą ketvertą nusiuntė pas *stilistę*, kuri pakoregavo ir jų šukuosenas LR (TVA), 1999:52. Paskutinį dešimtmetį *stilistas* stengėsi suvienyti itališkos mados kūrėjus, kad galėtų konkuruoti su žymiausiais pasaulio dizaineriais LA 1999:73.

Das Gleiche gilt für das Lehnwort *Stilist* auch im Deutschen. Es bedeutet „jemand, der die sprachlichen Ausdrucksmittel beherrscht, jemand, der den Stil [einer Sportart] beherrscht" DU 1996. Nach dem Vorbild der englischen Sprache (im Deutschen gibt es auch das Verb stylen zu engl. *to style*) wurde die Bedeutung dieses Substantivs im Deutschen erweitert, *Stylist* wird besonders in der sog. Szenensprache verwendet, wo es Folgendes kennzeichnet: „Der Stilist gehört zu einer Spezies, die sich im Biotop der Medien- und Modeszene bewegt und bei Foto- und Filmproduktionen an Star und Model Hand anlegt. In erster Linie besorgt er die Kleidung, bei größeren Produktionen organisiert und gestaltet er auch die Dekoration".

Das aus dem Griechischen stammende Wort *problema* ist in der litauischen Standardsprache schon lange bekannt. Es bedeutet „theoretische oder praktische Aufgabe, die zu lösen ist", 10. Jetzt erfährt es nach dem Vorbild der Bedeutungen des englischen Wortes *problem* eine Bedeutungserweiterung und ersetzt die üblichen litauischen Wörter *reikalas, dalykas* (= Sache), vargas, rūpestis (= Sorge) u.a., z.B.: JAV medikų duomenimis, 71 proc. nereguliariai miegančių žmonių turi su miegu susijusių *problemų* LR (GB) 1999:230. Vaikų *problemos* kur kas paprastesnės, o ir nuotykiai šviesesni, nerūpestingi LŽ 1999:228.

Auch die Verwendung dieses Wortes wird erweitert. Besonders beliebt sind die Begriffe *problematiška šeima* (= Problemfamilie) R 1999:68, *problemiškas vaikas* (= Problemkind) LŽ 2000:26. In diesem Fall werden aus dem Litauischen die alten heimischen Varianten *sunki šeima* (= schwierige Familie), *sunkus vaikas* (= schwieriges Kind) oder *asociali šeima, asocialus vaikas* verdrängt.

Vergleichen wir die Bedeutungen und den Gebrauch des Wortes *Problem* im Deutschen, so sehen wir, dass es dieselbe Bedeutungserweiterung erfahren hat: Problem [lat. Problema < griech. Problema = das Vorgelegte; gestellte (wissenschaftliche) Aufgabe, Streitfrage]: 1. Schwierige [ungelöste Aufgabe, schwer zu beantwortende Frage, komplizierte Fragestellung; 2. Schwierigkeit. DU. Z.B.: Ein Jahr später als erwartet hat bei der norwegischen Eisenbahngesellschaft das *Jahr-2000-Problem* zugeschlagen MM 03.01.2001. Jetzt hat Sonja aber ein *Problem*: ihr Kleiderschrank platzt aus allen Nähten MM 05.01.2001.

Das Wort *Problem* drückt in sehr zahlreichen Bildungen mit Substantiven aus, dass jemand oder etwas Probleme aufwirft, Schwierigkeiten bereitet DU. Also ist der Gebrauch dieses Lehnworts besonders erweitert, und man spricht im Deutschen von *Problemfamilie, Problemgruppe, Problemkind, Problemfall, Problemmüll* u.a., z.B.: Vor allem Ecstasy und Haschisch seien die *Problem-Drogen*, sagte der Richter MM 10.01.2001. Vier hessische *Problem-Stadtviertel* kommen in diesem Jahr neu ins Förderprogramm „Soziale Stadt“ MM 04.01. 2001.

Unter dem Einfluss des Englischen wird der Gebrauch auch anderer Internationalismen immer häufiger. Ein solches Wort im Litauischen stellt das aus dem Griechischen stammende *nostalgija* dar. In der litauischen Standardsprache wurde es seit langem in der Bedeutung „starkes Heimweh“ verwendet. Jetzt ist die Gebrauchshäufigkeit dieses Wortes besonders gestiegen und zwar in der Bedeutung „sehnsuchtsvolle idealisierte Erinnerung“ oder überhaupt „Sehnsucht“, z.B.: *krepšinio nostalgija* (= Baskelttball-Nostalgie) LR 1997 268, *meilės nostalgija* (= Liebes-Nostalgie) LR (TVA) 1999:263,*nostalgija kinui* (= Kino-Nostalgie) LŽ 1999 197, *sovietinių laikų nostalgija* (= Sowjetzeit-Nostalgie) LA 1996: 577, *šlovės metų nostalgija* (= Ruhmjahre-Nostalgie) LR 2000:259. Im Deutschen war dieses Lehnwort auch vor allem in der Bedeutung "Heimweh" bekannt (vgl. FW 1966). In den 70er Jahren registriert man im FW die hinzugekommene Bedeutung „[schwärmerisch romantisierende, mit Sehnsucht, Wehmut verbundene] Rückwendung zu früheren, in der Erinnerung sich verklärenden Zeiten, Erlebnissen, Erscheinungen in Kunst, Musik, Mode u.a.“ (vgl. FW 1974). Heutzutage wird das Wort *Nostalgie* nach dem Vorbild des gleichbedeutenden eng.-am. *Nostalgia* hauptsächlich in der Bedeutung „von unbestimmter Sehnsucht erfüllte Gestimmtheit, die sich in der Rückwendung zu früheren, in der Erinnerung sich verklärenden Zeiten, Erlebnissen, Erscheinungen in Kunst, Musik, Mode u.a. äußert“ verwendet und die Bedeutung „Heimweh“ wird als veraltet betrachtet (vgl. FW 2000), z.B.: In einigen der Restaurants wird *Nostalgie* gepflegt (MM

18.01.2001). In Ludwigshafen kehrt nun die Legende zurück und bringt neben *Nostalgie* viel Musikalität mit (MM 15.03.2001). Die *Nostalgie* fürs Bonner Understatement kann nicht Ausgangpunkt [...] sein (MM 28.04. 2001). Keiner will die umstrittene *Nostalgie*-Truppe zu den Jubelfeiern verpflichten (MM 09.01. 2001). Weitere Beispiele: Nostalgie-Nähkurs St. GT (05. 01.2001), Nostalgie- und Traditionshäuser (MM 23.01.2001), Nostalgie-Fans MM (23.01.2001), Nostalgie-Gefühle (MM 27.01.2001), Nostalgie-Treff (MM 16.02.2001), Kaffeekränzchen-Nostalgie (MM 03.03.2001), Nostalgie-Karussell (MM 05.04.2001).

Nach dem Vorbild der Bedeutungen und des Gebrauchs des Anglizismus Satellite werden die Bedeutung und Gebrauch des Latinismus *satelitas* erweitert. In der litauischen Standardsprache wurde dieses Wort in folgenden Bedeutungen gebraucht: 1. Bewaffneter Diener, der seinen Herrn begleitet; 2. Himmelskörper, der einen Planeten umkreist; 3. künstlicher Flugkörper, der einen Planeten umkreist. In den letzten zwei Bedeutungen wurde häufiger das litauische Wort *palydovas* verwendet. Jetzt wird der Internationalismus *satelitas* auch in der Bedeutung „Staat, der formal unabhängig, aber faktisch von einem stärkeren Staat abhängig ist", z.B. Šiuo metu Rusija nori išlaikyti Lietuvą savo ekonominėje ir politinėje orbitoje, kad vėliau paverstų savo paklusniu *satelitu* (LA 1994:61).

Das Wort tritt häufig in der Bedeutung „Stadt, die formal unabhängig, aber faktisch einer größeren Stadt untergeordnet ist" auf z.B.: Numatyta, sako architektas, kaip sostinės *satelitus* plėtoti Lentvarį ir Naująją Vilnią (LR 1995:15).

Häufiger ist der Gebrauch dieses Wortes in der übertragenen Bedeutung „Komplize, Helfer" geworden, z.B.: Kai nuolat klausaisi viso to netaupaus kalbėjimo apie galimus pavojus, nenorom klausi savęs: ar kartais mes patys nenuleidžiame savęs į elitinių Rusijos *satelitų* rangą? (LR 2000:14).

Besonders oft verwendet man das Wort *satelitas*, wenn man über politische Parteien, Organisationen spricht, z.B.: Jei Valstiečių partija yra LDDP *satelitas*, tai ir Pramonininkų partija bus tokia pat (LA 1995 61).

In derselben Bedeutungen wird auch das Adjektiv *satelitinis, -ė* benutzt, z.B.: *satelitinė* Lietuva (= Satelliten – Litauen) (LA 1998:82); *satelitinės* partijos (= Satelliten-parteien) (LA 1995 215).

Das Adjektiv *satelitinis, -ė* bedeutet auch „technische Einrichtung", z.B.: Pasirašyti susitarimai dėl *satelitinio* ryšio specialistų mokymo (LR 2000:40). Rūsyje įrengtas bunkeris su *satelitinėmis* komunikacijomis (R 2000:12).

Die Bedeutungsstruktur und die Verwendung des Lehnwortes *Satellit* im Deutschen sind fast die gleichen wie im Litauischen. *Satellit* kennzeichnet: 1. Himmleskörper, der einen Planeten auf einer unveränderlichen Bahn umkreist; 2. Flugkörper, der – auf eine Umlaufbahn gebracht [...] die Erde (oder den Mond) umkreist und dabei bestimmte wissenschaftliche od. technische Aufgaben erfüllt, Daten sammelt o.ä; 3. (abwertend) kurz für Satellitenstaat DU. Das Wort ist produktiv bei der Bildung der substantivischen Zusammensetzungen, in denen es die ähnlichen Bedeutungen wie die litauischen *satelitas, satelitinis, -ė* aufweist, z.B.: Satellitenbild (= satelitinė fotografija), Satellitenschüssel (= satelitinė antena), Satellitenfernsehen (= satelitinė televizija), Satellitenstadt (= miestas satelitas).

Sehr deutlich ist die Tendenz, die Bedeutungen der Anglizismen, seltener Romanismen und Germanismen, zu erweitern, weil sie jetzt besonders intensiv in die litauische Sprache eindringen. Ein Teil dieser Wörter waren in Wörterbüchern fixiert, aber in der letzten Zeit sind sie aus dem passiven in den aktiven Gebrauch gekommen. Als Beispiel kann das englische Wort *spikeris* (Speaker) angeführt werden. Im litauischen Fremdwörterbuch wird das Wort als Diplomatieterminus mit der Bedeutung „Vorsitzender des Parlaments" erklärt. Jetzt wird die Bedeutung dieses Wortes ohne Grund erweitet. Es wird anstelle „Sprecher, Redner, Vorsitzender einer Sitzung, Versammlung, Kundgebung u.a." verwendet, z.B.: Susirinkimo *spikeris* E.Š. pareiškė, kad atvyko ne jauni kareiviai, o 144 divizijos dalinių apmokytas papildymas (LR 1993:52). Šį kartą darbo žmonių organizuotas mitingas atrodė skurdžiai, matyti, sunku pasamdyti šiais laikais gerą *spikerį* (R 1993:74).

Ähnliche Bedeutungserweiterung ist auch im Deutschen nachzuweisen, z.B.:

Ebenfalls im Vorderland erst mal eingesetzt wurden versierte *Speaker*, welche das Geschehen am Viehschauplatz kommentierten (St. GT 15.01.2001).

Paul Krähemann schließlich war als *Speaker* einmal mehr dafür besorgt, die Zuschauer zu informieren (ebda. 16.01.2001).

Ein anderer Anglizismus, dessen Bedeutung sehr erweitert wird, ist *servisas* (Service). Das Wort war heimisch in der litauischen Sprache in der Verbindung *automobilių servisas* , Kurzform *autoservisas* (= Autoservice). Jetzt werden durch *servisas* immer mehr litauische Wörter aptarnavimas, paslaugos, priežiūra (= Bedienung, Dienstleistung, Fürsorge) verdrängt, z.B.: Čia [Fidžio salose] be jokių netikėtumų ir nesklandumų galite mėgautis atostogomis: aukščiausios klasės *servisas*, prabanga ir turtinga pramoginė programa (LŽ 2000:187). Kailiniams, kaip ir automobiliams, reikia *serviso* (LŽ 1999:276).

Eine enorme Bedeutungserweiterung und Verwendungshäufigkeit weist der Anglizismus *Service* im Deutschen auf und scheint besonders aus der Pressesprache die deutschen Wörter *Bedienung, Kundendienst, Kundenbetreuung* völlig verdrängt zu haben. Einige Beispiele dazu: Mehr *Service* in der Bibliothek MM 03.01. 2001. Zuverlässigkeit, *Service* und ein charakteristisches Outfit waren die Attribute, mit deren Hilfe der Interregio im ersten Jahr Zuwachsraten von fast 30 Prozent erreichte (MM 04.01.2001). Ferner spricht man von Party-Service, Rathaus-Service, Prospekt-Service, Bus-Service, Ticket-Service, Tele-

Marketing-Service, Reha-Notdienst-Service, Computer-Service-Unternehmen, Service-Aktion, Service-Bus, Service-Angebot, E-Mail-Service, Internet-Service-Station u.a.m. (MM 01.06.2001).

Der Anglizismus *autsaideris* (outsider) wird als Sportterminus mit der Bedeutung „Sportler oder Mannschaft, die nicht gewinnen“ angeführt. Dieser Anglizismus wird jetzt in der litauischen Sprache oft in der übertragenen Bedeutung „Person, die abseits steht, sich nicht einmischt“ verwendet, obwohl es dazu andere passende litrauische Normvarianten *pašalinis asmuo, pašalaitis, -ė, prašalaitis, -ė, pašalietis, -ė, nesuinteresuotas* asmuo u.a. gibt., z.b.: Beveik visi mano herojai – piešėjas, architektas, zoologas - dėl skirtingų priežasčių lieka visuomenės ir kultūros *autsaideriais* (= pašalaičiais) (LA 1993:82). Visi – vargšai, turčiai, teisybės ieškotojai, lyderiai ir *autsaideriai* (= atstumtieji) Vincui Kudirkai nepatiktų (LA 1999:98).

Im Deutschen gibt es eine Lehnübersetzung zu dem Anglizismus *Outsider – Außenseiter*. Das Wort weist dieselben Bedeutungen auf, wie der Anglizismus *autsaideris* im Litauischen, vgl.: Außenseiter [für eng. Outsider]: 1. (Sport) Wettkampfteilnehmer, dessen Gewinnchancen gering zu veranschlagen sind; 2. Abseits der Gesellschaft, einer Gruppe Stehender; jemand, der seine eigenen Wege geht DU. Eben in der zweiten Bedeutung und in der englischen Schreibweise tritt das Wort *Outsider* in der deutschen Pressesprache sehr häufig auf, z.B.: So etwas soll und darfst nicht geduldet werden, so ein *Outsider* kann nicht die Jugend vertreten und in ihrem Namen zu Fragen des Alltags Stellung nehmen (MM 24.03.1998). Der 1939 geborene Komponist gilt als „radikalster *Outsider*“ des Musiktheaters (MM 24.01.1996). Weitere Beispiele: Euro-Outsider (FR 15. 12.1997, Outsider-Staaten (FR 17.07.1998), Outsider des Arbeitsmarktes (FR 23. 07.1998), Insider-Outsider-Ansätze (FR 04.06.1999), Outsider-Werk (St. GT 24. 02.1999), Outsider-Chemie (ZT 02.09.1996), Hollywood-Outsider (ZT 30.01. 1998), TV-Outsider (ZT 21.10.1998), u.a.m. Manchmal stehen die englische und deutsche Varianten im Text nebeneinander, z.B.: In seiner Gruppe sei er ein *Outsider*, ein Außenseiter. Außer ihm bekleidet keiner ein politisches Amt (St. GT 24.03.2000).

Das Wort französischer Herkunft *gurmanas* (Gourmand) wird im schon erwähnten litauischen Fremdwörterbuch mit der Bedeutung „Kenner und Liebhaber von schmackhaften ausgesuchten Speisen“ fixiert. In der litauischen Standardsprache werden in derselben Bedeutung litauische Wörter *smaguris, -ė, smaguriautojas, -a smaližius, -ė, valgių žinovas, -ė* verwendet. Jetzt beobachtet man im Litauischen völlig unmotivierte Bedeutungserweiterung des Romanismus *gurmanas*. Es wird auch in der übertragenen Bedeutung (besonders im Bereich der Kunst) gebraucht und verdrängt aus der Sprache übliche Wörter *mėgėjas, -a, gerbėjas, -a, žinovas, -ė* (= Verehrer, Liebhaber, Kenner), z.B.: *džiazo gurmanas* (= Jazz-Gourmand) (LZ 1999:228), informacijos gurmanas (= Informations-Gourmand) (LR [TVA] 2000:35, 63), kino gurmanas (= Kino-Gourmand) (LŽ 2000:123), meilės gurmanas (= Liebes-Gourmand) (LŽ 1999:84), meno gurmanas (= Kunst-Gourmand) (LR [GB] 2000:40), muzikos gurmanas (= Musik -Gourmand) (LŽ 1998:14), operos gurmanas (= Oper-Gourmand) (LA 1997:186, žinių gurmanas (= Kenntnisse-Gourmand) (LA 1998:210).

Obwohl *Gourmand* in den von uns analysierten deutschen Belegtexten eher in den Bedeutungen „10, der gern gut und zugleich viel isst; jemand, der ein Kenner in Bezug auf Speisen und Getränke ist und gern ausgesuchte Delikatessen ist; Feinschmecker“ FW verwendet wird, lässt sich auch ähnlicher übertragener Gebrauch, wie im Litauischen feststellen, z.B. Das liegt daran, dass der Gourmand unter Amerikas Folkloregitarristen, Ry Cooder, ihn vor drei Jahren endlich zu einer ersten eigenen Plattenproduktion ins Studio holen konnte (FR, 07.04.1998). Sie scheinen ein Gourmand zu sein, was Ihre Expansionsgelüste anbetrifft? (ZA 31.01.1997).

Eine weitere neue Tendenz bei der Verwendung besonders der neueren Entlehnungen besteht in der graphischen Hervorhebung der Fremdwörter in der Schrift. Die litauischen Sprachbenutzer scheinen an der Notwendigkeit mancher Fremdwörter zu zweifeln und keinen richtigen Ersatz dafür zu finden, deshalb heben sie sie graphisch hervor. Die am meisten verbreitete Art ist, das Fremdwort, für das es keine Entsprechung gibt, in Anführungszeichen zu setzen. Oft werden mit Anführungszeichen die in Originalsprache geschrieben Fremdwörter versehen, z.B.: Būdamas šešiolikos metų, jis mėgo grupės „Depeche Mode“ dainas, po dviejų metų susidomėjo pankroku ir *„independent“* grupių kūryba (LR 1997:305).

Esi studentė, moki užsienio kalbą, vairuoti automobilį, tapk *"au-pair"* mergina (LR [S] 2000:212). L.S. darydama *„body flex“*, per penkis mėnesius atsikratė 25 kilogramų (LR 2000:24).

Solche graphische Hervorhebung ist auch im Deutschen anzutreffen, z.B.: Am ganztägigen Seminar wird im Hinblick auf die Liberalisierung des Telekom-Marktes die Problematik der „Cast mile“ diskutiert (MM 09.01.2001).

Als *„special – coaches“* engagierten sich die Fußballprofis (MM 15.02.2001). Bei den „Compus Chances“ steht *„Service pur* im Mittelpunkt“ (MM 09.01. 2001).

Mit Anführungszeichen werden auch mehr verbreitete, phonetisch und morphologisch adaptierte Fremdwörter-Termini versehen; z.B.: „Acoragil“ valdymą išmanantis *„hakeris“* gali persiųsti iš svetimo kompiuterio bet kokius duomenis (LR [V] 1998:11).

Vagys įėjo pro duris ir susišlavė viską iš lentynų – ne tik alkoholinius gėrimus, bet ir *„pampersus“*. (R 2001:15).

Ähnlich geht man auch in der deutschen Pressesprache vor, vgl.: Die superscharfen Sonnenbrillen aus den USA sind Renner bei den „*Beachern*" (MM 26. 08.1998).

Die Japaner begründen diese „*No-sales*" - Politik mit sozialen Überlegungen (CZ 22.09.1999). In dem wohl bekanntesten Modell dieser Art, dem „*Shirking-Modell*" von Shapiro und Stiglitz... (FR 04.06.1999).

Ziemlich oft verwendet man in der litauischen Presse bei der unadaptierten Schreibung der Fremdwörter eine andere Schrift, z.B.: 54 procentai apklaustųjų tvirtina jog elektroniniu paštu ir IRC *on-line* pokalbių metu jie bendrauja laisviau negu kalbant telefonu ar asmeniškai (LR 1998:11). J. filmas taip pat artimesnis šiam žanrui, o ne *science fiction* (LA 1998:25).

In Werbeanzeigen, Prospekten, besonders in der Jugendpresse werden aber neuere Entlehnungen so wie in der Herkunftssprache, weder adaptiert noch durch irgendwelche Zeichen versehen, geschrieben.

In beiden verglichenen Sprachen gibt es auch die Tendenz, die graphisch hervorgehobenen Entlehnungen in Klammern (oder auch ohne) zu erklären oder ihre Entsprechungen anzuführen, z.B.: Kai „*haute couture*" (aukštoji mada) ir „*pret-a-porter*" (gatavi drabužiai) asociacijos prezidentė S.G. čiupinėjo, vartė ir uostinėjo mano suknelių siūles, dėl įtampos mano veidu tekėjo ašaros (LR [GB] 2001:19). Skaitmeniniai palydovinės TV *tiuneriai* (imtuvai) per keletą pastarųjų savaičių vėl pastebimai atpigo (LR [TVA] 2001:436).

Das Problem des Schneemangels möchte Emig mit Hilfe der Kunstschneemaschinen lösen, die sogenannten „hot snow" (heißen Schnee) produzieren MM 08.01.2001. Zum umfassenden Service zählen ein so genannter „*Doorman*", der *Portier und Sicherheitskraft* zugleich ist (MM 09.01.2001).

Eine andere Variante im Litauischen stellt die Vorgehensweise dar, wenn die litauische Entsprechung durch das in Klammern angeführte Fremdwort erklärt wird, z.B.: Aprangos pasaulio aristokrate vadinama *aukštoji mada* („haute couture") vėl keri Paryžių (LR [GB] 2001:22). Asmenims, nutarusiems susikurti savo *interneto svetaines* (home page), taikomi dar kiti įkainiai (LŽ 2001:10).

Im allgemeinen kann man festhalten, dass Stilisten und Redakteure der seriösen litauischen Zeitungen und Zeitschriften Entlehnungen, besonders die neueren, durch passende litauische Entsprechungen ersetzen, aber in der gesprochenen Sprache werden sie regelmäßig verwendet.

2. Zum englischen Einfluss auf den Jugenddiskurs: Deutschland – Litauen

Ein besonders prägnantes Strukturmerkmal der Jugendsprache im Bereich des Wortschatzes sind Entlehnungen (vgl. Neuland 2000). Im Deutschen gehört vor allem der Gebrauch von Anglizismen zur jugendspezifischer Sprechweise. Die Anglizismen dringen in die deutsche Jugendsprache aus verschiedenen Umfeldern. Das Duden-Wörterbuch der Szenensprachen 2000 führt folgende Haupteinfallstore der Anglizismen an: Sport, Musik und Popkultur, Mode, Models, Fashionzones, Computerslang und Cyberspace, Liebe, Sex und Partnerschaft, Freizeit. Z.B.: beachen, boarden, Coping, Gap-Jumping, Local, Charts, Chickband, deep, Gangsta-Rap, jammen, Label; wriggeln, Body-Bag, dreads, Crossdressing, Hipsters, Outfit, scratchen, Sneakers, Stylist, Temptoo, Zipper; andocken, Attachment, ballern, beamen, chatten, Cybercafe, downloaden, Files, Luser, Link, Messi, Mailbox, online, Pic, posten, Provider, scannen, surfen, User; Assel, checken, Date, durchgedatet, Grrrl, kinky, smart, taff, touchy; beamen, briefen, burnen, Connection, gaga, Gosh!, Performance, snaken u.a.m.

Englische Wörter werden problemlos in das deutsche Flexionssystem integriert, Verben können das deutsche Infinitivsuffix bekommen, die von englischen Nomina gebildeten Verben können durch Affixe erweitert werden, z.B: eine coole Sache, ein cooler Typ, ein smartes Image, spoten, forwarden, abpowern, reinmoven, loscutten

Von den 300 analysierten litauischen Belegen sind 63 Wörter und Wortfügungen Entlehnungen, was 21% ausmacht. Darunter sind 11 Anglizismen, 50 Russizismen und 2 Germanismen. Die Zahlen sprechen für sich, obwohl man sie selbstverständlich nicht verallgemeinern kann. Umfangreichere Untersuchungen speziell zu den Entlehnungen in der litauischen Jugendsprache würden sicherlich ein etwas anderes Bild zeigen. Trotzdem kann ein wichtiger Unterschied zwischen beiden verglichenen Sprachen festgestellt werden, der uns noch einmal verdeutlicht, wie groß der Einfluss des Russischen auf das Litauische gewesen und noch bis jetzt geblieben ist, obwohl sich die Umstände ganz geändert haben. Einige Beispiele der von litauischen Jugendlichen verwendeten Russizismen: barakas, šaiba, ant bajerio, laikyti vierchą, pravalas, koroče, trūko plionkė, numesti majaką/namioką, kabakas, nachaliavu, čiudna, navaras, čiotka, blūdyti, bajeris, krūtas čiuvas, babkės, kislotnas, pacanas, biesina u.a. Jam klausimas atrodo labai *tūpas* (P 1998:8). Man labiausiai patinka uždaros *tūsovkės*, kur daug triukšmo ir mažai žmonių (P 2000:11). Padalvojau, kad tai *dochlas* reikalas (P 1997:9). Oi, darykit kaip nors paprasčiau, mažiau tų *navorotų* (P 1999:4). Tereikia nebūti *zanūdai* ir šypsotis (P 1997:8). Dabar *zubrinu* anglų kalbą (P 1997:12). Po šventės net springdamas aiškina, kokios tavo draugės *dūros*, koks tėvas atšokęs (M 1998:99). Norėčiau vyro prie *babkių*, juodo BMW (P 2001:9). Viena ką norėčiau parašyti, tai ne komentarą, bet atsiliepimą į *pacano* komentarus Int.

Eine Art der Entlehnungen aus dem Russischen bilden die Lehnübersetzungen, die sehr häufig sind, z.B.: Linkiu merginoms vieną kartą *užrišti* su tais kalorijų skaičiavimais (P 2000:7). Pameluoji merginai vardą, pasiguodi, jog neturi namų – ji *pastato* man alaus (P 1997:8). Čia tai jau tikrai nėra ko pykti, vis tiek ūgio nepridėsi – *traukia* vos 160 cm M 2000 3. Na, *nesilaužyk*, tėvai nemato, gali atsipalaiduoti (P 1999:1). Atvažiavo gruzinų teatras ir mane jie nugirdė konjaku. Namo grįžau *šiltutėlė* (ebda.:10).

Die übernommenen russischen Wörter passen sich den Regeln des Litauischen an, sie können mit litauischen Flexionen und Wortbildungsaffixen versehen werden. Z.B.: *Pamajačink*! Int. Kartą su draugais nusprendėme *pasibiesinti* (P 1997:8). Prasimanėm įvairiausių žaidimų, *blūdinėjom* (ebda.:3). Aš čia ilgiau nenoriu gyventi gal todėl, kad visi pažįsta ir labai *pliotkina* (ebda.:4). Dešimtetė *įsikliopino* šešeriais metais vyresnį Ričardą (P 2000:4). Daugumos merginų nuomone Marijonas *navorotnesnis* ir *prikolnesnis* (P 1998:8).

Nach der Wiederherstellung der Unabhängigkeit ist starke Invasion der Wörter englischer Herkunft in der litauischen Jugendsprache festzustellen. Die Analyse der neuesten Entlehnungen aus dem Englischen, die im Buch von Rudaitienė/Vitkauskas (1998) verzeichnet sind, zeigt, dass die litauischen Jugendlichen Aglizismen aus denselben thematischen Bereichen wie die deutschen Jugendlichen übernehmen: z.B.: profas, fanas, boksas, fitnisas, klipas, hitparadas, rokas, rapas, raperis, reperis, reivas, popsas, bandas, hororas, klubas, show, party, make-up, body, drinkas, cool, klipmeikeris, klipmeikingas, klipas, kastingas, juzeris, jam'as, bleidai, adapteris, breikeris, dartas, dance party, diskonas, draftas, draivas, džemas, e-mailas, failas, fanas, feelingas, fitnesas, gegas, hakeris, serfingas, skaneris, skeneris u.a. Muzika: gerai veža, geras *draivas*, gerai varo (P 1999:7). Jei neišlaikei egzamino, visi laiko *lūzere* (P 2001:7). Taigi gyvena lenkaitė, mano *friendas* ir šeimininkas Int. Negaliu, užkibo šitas pasakojimas. Dar perskaičiau – *cool*!!! Int. Mergina jam paliko savo *mobiliako* numerį (P 2000:7). Meilė yra mums visiškas *autas* (M 2001:6). O galvoje sukasi dar bent šimtas *crazy* idėjų (P 1998:2). Patinka protingi, mėgstantys *kreizėti*, būtinai su humoro jausmu (P 1996:9).

Im Litauischen bekommen fast alle Anglizismen eine Endung mit Ausnahmen von einigen direkten Transferen aus dem Englischen ins Litauische, z.B. sex-appeal, e-mail, pop-corn, make-up, show, cool, know-how, non-stop, baby, party, free style, body, time-out, soul. Aber auch diese Ausnahmen erscheinen in manchen Fällen phonomorphologisch adaptiert. Im Deutschen dagegen trifft man viel häufiger englische Entlehnungen, die unverändert bleiben.

3. Fazit

In der Sprache widerspiegelt sich der Kulturwandel eines Volkes, denn die Sprache folgt der Kultur, und wenn sich kulturelle Änderungen einstellen, so ändert sich auch Sprache. Im Deutschen und im Litauischen ist die Verbreitung der Fremdwörter festzustellen, die durch viele psychologische und soziale

Gründe bedingt ist. In Litauen verändert sich die traditionelle Lebensweise, neue Technologien entstehen, neue Realien werden entdeckt, und mit ihnen übernimmt man aus der Herkunftssprache ihre Benennungen. Bei der Verbreitung der Fremdwörter spielen Kontakte mit den Ausländern und Integrationsprozesse eine wichtige Rolle. Die Veränderungen im Gebrauch von Entlehnungen hängen genauso wie die gesamte Entwicklung der Sprache auch von interlinguistischen Faktoren ab.

Deutsch und Litauisch werden vom Englischen sehr stark beeinflusst. Beide Sprachen reagieren auf die neue Rolle des Englischen im Globalisierungsprozess ähnlich, man kann auch sagen, das die litauische Sprache teilweise die Entwicklung, die das Deutsche schon hinter sich hat, wiederholt.

Die Analyse der Belege zeigt Verwendungstendenzen von Entlehnungen in der litauischen Standardsprache, die mit den Tendenzen im Standarddeutsch vergleichbar sind. Allgemein übliche Entlehnungen werden durch andere Entlehnungen von derselben oder ähnlicher Bedeutung ersetzt, Bedeutungen oder Verwendung von Entlehnungen werden in beiden Sprachen erweitert, dabei lassen sich viele Übereinstimmungen in den Bedeutungsstrukturen und im Gebrauch derselben Fremdwörter feststellen.

Die neueren Bedeutungen der Entlehnungen sind im Deutschen meistens schon in Wörterbüchern fixiert, im Litauischen aber noch nicht. Neuere Entlehnungen werden graphisch hervorgehoben; es wird versucht die graphisch hervorgehobenen Entlehnungen durch Übersetzung zu erklären oder passende Entsprechung dazu anzuführen; manchmal werden nach den litauischen Entsprechungen in Klammern Fremdwörter in adaptierter Form oder in der Ursprungssprache geschrieben.

Entlehnungen stellen ein wichtiges charakteristisches Merkmal der jugendlichen Sprachvarietäten des Deutschen und des Litauischen dar. Auf die deutsche Jugendsprache übt das Englische einen besonderen Einfluss aus. In der litauischen Jugendsprache sind viele Russizismen zu beobachten, was teilweise durch starke nachhaltige Wirkung des früheren russischen Einflusses zu erklären ist. Im letzten Jahrzehnt nimmt der Umfang der Entlehnung aus dem Englischen auch in der litauischen Jugendsprache sehr zu.

Der Vergleich zeigt, dass die Anglizismen in beide Sprachen aus denselben Themenbereichen eindringen.

Verallgemeinernd sollte man sagen, dass solch ein häufiger, oft gar unmotivierter Gebrauch von Entlehnungen, die sogar auch eigene Wörter verdrängen, doch eine Gefahr für die ethnische Kultur des Volkes und deren wichtigstes Ausdrucksmittel, die Sprache, darstellt.

Abkürzungen

G – Panelės priedas "Top Girl"
Int – internetas
LA – Lietuvos aidas.
LR – Lietuvos rytas.
LR (GB) – Lietuvos ryto priedas "Gyvenimo būdas".
LR (RV) – Lietuvos ryto priedas "Rytai-vakarai".
LR (S) – Lietuvos ryto priedas "Savaitė".
LR (SA) – Lietuvos ryto priedas "Sporto arena".
LR (TVA) – Lietuvos ryto priedas "TV antena".
LR (V) – Lietuvos ryto priedas "Vartai".
LŽ – Lietuvos žinios.
M – Panelės priedas "Maxi"
P – Panelė
R – Respublika.
rš – raštai.
VL – Valstiečių laikraštis.
CZ – Computer Zeitung.
DF – Duden. Das Fremdwörterbuch.
DU – Duden. Deutsches Universalwörterbuch.
FAZ – Frankfurter Allgemeine Zeitung.
FR – Frankfurter Rundschau.
MM – Mannheimer Morgen.
Sp – Spiegel
St. GT – St. Galler Tagblatt.
ZT – Züricher Tagesanzeiger.

Literatur

BÄR, J.A.: DEUTSCH IM JAHR 2000. EINE SPRACHHISTORISCHE STANDORTBESTIMMUNG. IN: VON EICHHOFF-CYRUS, K.M./HOBERG, R. (HG.): DIE DEUTSCHE SPRACHE ZUR JAHRTAUSENDWENDE. MANNHEIM, LEIPZIG, WIEN, ZÜRICH 2000.

DABARTINĖS LIETUVIŲ KALBOS ŽODYNAS. VILNIUS 1993.

DUDEN-WÖRTERBUCH DER SZENENSPRACHEN. MANNHEIM,LEIPZIG,WIEN,ZÜRICH 2000.

ELITZ, E.: Sprache in den Medien – die Wortverdreher GmbH. In: VON EICHHOFF-CYRUSM K.M./HOBERG R. (hg.): Die deutsche Sprache zur Jahrtausendwende. Mannheim, Leipzig, Wien, Zürich 2000.

FINK, H.: Superhit oder Spitzenschlager: Ein Versuch zur Häufigkeit und Funktion von Anglizismen und „Werbeanglizismen" in deutschen Jugendzeitschriften. In: Studien zum Einfluss der englischen Sprache auf das Deutsche. Tübingen 1980.

HAARMANN, H.: Sprachenvielvalt im Globalisierungsprozess. In: VON WERMKE, M./HOBERG, R./EICHHOFF-CYRUS, K.M. (hg.): Deutsch-Englisch-Europäisch. Impulse für eine neue Sprachpolitik. Mannheim 2002.

HOBERG, R.: Sprechen wir bald alle Denglisch oder Germeng? In: VON EICHHOFF-CYRUS, K.M./HOBERG, R. (hg.): Die deutsche Sprache zur Jahrtausendwende, Mannheim, Leipzig, Wien, Zürich 2000.

NEULAND, E.: Jugendsprache in der Diskussion: Meinungen, Ergebnisse, Folgerungen. In: VON EICHHOFF-CYRUS K.M./HOBERG, R. (hg.): Die deutsche Sprache zur Jahrtausendwende. Mannheim, Leipzig, Wien, Zürich 2000.

RUDAITIENĖ, V./VITKAUSKAS, V.: Vakarų kalbų naujieji skoliniai. Vilnius 1998.

YANG, W.: Anglizismen im Deutschen. Tübingen 1990.

Aistė Urbonienė, Irena Leliūgienė

Development of Professional Motivation in the Study Process as Assumption for the Growth of the Knowledge Society

In the knowledge society the importance of permanent education that guarantees successful individual integration and adaptation in the society and labour market is stressed. In the main indications for the future higher education it is emphasized that there is "a need of "more operational" training, nearer the labour market, but at the same time more responsible and more citizen oriented". That raises new challenges for HE in a lifelong learning perspective (STRATA-ETAN expert group, 2002), because individuals need to master the required competencies and knowledge that also stresses the development of professional motivation as assumption for self-realization of individuals in the knowledge society.

According to Rajeckas (1999), the organization of learning process is conditioned by the level of the society development, namely material living conditions, type of social relations, needs of the society, the situation of culture and science, requirements raised to education and other factors. The organization of educational processes as a composite part of the dominating culture influences greatly the activities and expression of the motivation as well as the development of professional motivation.

The organization of education is influenced by interests of various groups of the society while economic and political structures of the society are changing. The learning process based on principles of liberal and democratic education promotes independence, responsibility and activeness of students as well as forms dutifulness, public spirit and motivation of individuals for studies (Rajeckas, 1999). This principle could be applied evidently to the development of professional motivation during studies as the one that helps to self-actualize and evaluate own possibilities for future professional activities.

Undoubtedly the process of studies influences the development and expression of professional motivation of students. It could be considered as various educational conditions, motivating and non-motivating factors. After unsuccessful efforts to recognize exhaustive analysis of elements of the study process allowing the determination of factors promoting the increase of professional motivation, *it remains unclear, which factors influence the professional motivation and the expression of which factors decreases efforts to obtain the chosen profession and stimulates the termination of studies.*

The aim of this article is to base the factors that stimulate the development of professional motivation during studies on a theoretical level.

Differences between motivation for studies and professional motivation will be discussed in this article as well as the theoretical aspects composing the basis for the future possibility analysis of the development of professional motivation will be introduced. Afterwards possibilities of motivation stimulation during studies will be considered.

It is important to define similarities and differences between motivation for studies and professional motivation, which include different spheres of the phenomenon of motivation, while trying to analyze the possibilities of the development of professional motivation during studies.

A close relationship constantly dominates between motivation for studies and professional motivation as these two kinds of motivation are united by striving to master a certain sphere of knowledge, necessary for future performance of a professional role. The main difference is the fact that professional motivation is oriented towards the future and is related to the professional identity of an individual. That is why a student may be motivated only for studies of certain modules. Meanwhile, motivation for studies is directed towards cognitive aims and results of knowledge mastering. Otherwise, final results differ in their type, though the ways of achieving these results are similar.

Further analysis of the development of professional motivation during studies will be performed based on the following study dimensions, emphasized by Carlsson (2001), Hoel (1995), which are significant for the wish of students to learn and develop their own personality: 1) dimension of study contents, 2) dimension of activity, 3) social dimension, 4) dimension of self-realization.

Dimension of study contents

A great part of students who choose a certain dimension of studies in higher education are in many cases poorly familiarized with their future profession. Students familiarize better with a specific sphere of knowledge and their future professional role during studies. The aspect of the benefit of the acquired knowledge and skills acquires special importance in this context (Carlsson, 2001), therefore becomes very significant to the development of professional motivation. Other scientists emphasize the factor of the benefit of the acquired knowledge for the promotion of the process of educational effectiveness as one of the most important factor to stimulate motivation (Hedin & Svensson, 2001). It should be assured that every student feels the meaning of his/her studies and knows that acquired knowledge and skills may be used in the future professional activities and other interests important for students (Marton, Säljö, 1997), allows predicating that not only student's understanding about the relevant professional role but the interest in this role could be increased in such a way.

However, the dilemma could be envisaged as well, namely: whether academic modules should be harmonized with the modules oriented towards career education during studies; or whether studies should be composed only of modules relevant for the future profession? It is not a casual question, as all modules are unexceptionally related with the future professional activities in some Western European higher schools preparing socio-educational employees. Meanwhile, if to consider study programs of social work and social education in Lithuanian universities, the coordination of academic subjects with professional ones could be noticed.

It could be stated while glancing from the perspective of the benefit factor that students who choose the study program of social work will be more motivated for the module of "social work methods", than for mathematics or general philosophy. Notwithstanding, if study programs of higher education were composed only of subjects oriented towards professional activities, the aspect of academic studies, on witch the idea of university education is based, would disappear.

The content of studies should constructively harmonize theory with practice, as practical part of studies creates conditions for the expression of personal experience of students and discussions. The change between theoretical and practical subjects during studies not only varies studying but also increases the interest in the chosen sphere of knowledge. The content of studies and learning tasks, oriented towards aspects of avail and meaningfulness of studies, motivates students to strive for higher study results and increases the motivation of professional choice. Meanwhile the interest in studies and the suppression of professional motivation may be stipulated by studies oriented towards theoretical knowledge contravening requirements of the labor market.

Dimension of activity

The behavior of an individual is quite often determined by the context and provided possibilities for the expression of his/her activity. A certain context predominates in institutions of higher education, in which a student is governed by formal and informal rules and norms. In other words, formed structural and ideological contexts express academic culture of a specific institution. The effectiveness of learning process as well as the expression of professional motivation is directly dependant on the democracy of learning space (Jarvis, 1992).

The abundance of possibilities of choice is typical of the system of higher education as well as for the whole post-modern society. The possibility to choose, evaluate and change not only influences the motivation, but also develops strategic and communicative skills of students. It greatly depends on the fact whether lecturers acknowledge the changed role of students from passive receivers of knowledge to active participants in the study process.

Three conditions are necessary for the increase of professional motivation and activeness of students, namely: the authority of lecturers, effective structure of studies and learning process oriented towards development (Hedin & Svensson, 2001). It is very important to assure that relations between lecturers and students should be based on mutual respect; however, the authority of lecturers among students would be based on the expression of their competence. According to Jarvis (1992), the authority of a lecturer exists as the given, which is created by the educational institution and also formed by the status of a lecturer as an expert of a certain sphere of science.

The study structure that assures the sense of safety and stability, which helps students to acquire better new knowledge and skills, is important to the development of professional motivation and favorable for students. A great study load preconditions the expression of fatigue and stress that stimulates frustration and other obstacles of effective learning. An important condition of the motivation stimulation is a balanced load of studies assuring the possibility for reflections (Hedin & Svensson, 2001). The compatibility of the study schedule, the organization of practical works, teaching and learning as well as the usage of information technologies and other conditions become important while considering material-organizational conditions of studies.

The third condition, namely the learning process oriented towards the development, is used to assure not only the quality of learning but also the development of every student's personality. It should be valued from the general perspective, when not only new knowledge and practical skills are gained but also professional motivation increases while becoming and important part of the development of the personality itself.

While discussing the quality of higher education, the issue of student empowerment gains importance, i.e. students receive higher responsibility for study control and possibilities of choice (Stewart, 1994). Student empowerment should be performed while creating such an academic environment in which democracy, flexibility, openness etc. prevail and which assures the development of critical thinking towards the intellectual emancipation (Jucevičienė, 1999).

The changed function of a lecturer as "a developer of human resources" (Jarvis, 1992) stipulates lecturers to act as student advisors in the study process, promoting the search for new knowledge. It provides lecturers with higher responsibility to help students mastering such features as flexibility and creativity, which contribute to the adaptation to new living situations, as well as increasing their professional motivation as a precondition of their personality development.

The attitude of lecturers towards their subject, methods and relations with students are factors of no less importance to stimulate the professional motivation. The nature of study environment and emotional status of students, which are of high importance to the expression of the development of professional motivation, depend on the fact whether conditions are created for *equal dialogue*, or on the contrary – *anti-dialogue* (Hedin & Svensson, 2001).

Social dimension

University studies are not only an academic or profession acquisition process, but also a social space, in which interaction between lecturers and students is performed constantly. According to Jarvis (1992), learning is an activity of a human being. That is why the most favorable conditions for the personality development are created only during the interaction of relations. Therefore, psychosocial aspect of the environment is very important, as the possibilities of the expression of open and positive interaction depend on it.

The pairs of players, namely "lecturer-student" and "student-student", are the mains in the educational process. Both interactions are important for the expression of professional motivation; however, the relations between a lecturer and a student may significantly influence the attitude of a student towards the future profession. In the event the basis of the above mentioned interaction consists of the provision that every individual comes to study with his/her life experience, it creates conditions for qualitative interrelations, such as "adult-adult" (Hedin & Svensson, 2001) or "me-you" (Buber, 1994). Basing on Buber, a person becomes *Me* only while being *You*, i.e. an individual may experience his/her being only in interaction with another person. It means that the attitude towards a student as a personality allows the expression of a student's *Me*, which creates conditions for his/her development. This qualitative aspect of the interaction "lecturer-student" creates the basis for mutual encouragement and the increase of professional motivation.

Carlsson (2001), while trying to ground this thought, states that it is very important for the individual to feel that he/she is suitable, that things done by him/her are meaningful and he/she is respected, as it stimulates further development of the personality. Such holistic development of a personality includes not only the development of a student to a competitive specialist and educated person, but also the parallel development of self-evaluation.

Basing on Argyris, the interaction among students during studies should convert into *reflecting society* (Argyris et al, 1985) in the academic space in the most ideal situation. Discussions on various theoretical and practical issues, productive only when every member of a group feels safe and acknowledged in it, may start during social interaction. Namely such social generality reinforces self-confidence of a student and increases self-evaluation.

Dimension of self-realization

Every individual wishes to realize himself/herself, while trying to become an active agent in his/her social environment (Bourdieu, 1986) and reach certain internal and external aims in such a way. The development of a personality is related to gaining new knowledge, skills and experience, thus broadening one's own understanding, developing world-view and strengthening value orientation. It means constant search for learning of something new in spheres of professional improvement and personal development. It is approved by Šiaučiukėnienė (1997), that an individual develops the need of self-realization while studying, which becomes an effective motive of activities when perceived.

Constant renewal of knowledge and skills is necessary in contemporary information society, therefore education is not a certain mastering of technical skills, but more equated to the strategy development of personal problem solution. This new attitude raises new aims for higher education, while requiring non-traditional learning methods that stimulate creativity, curiosity, flexibility, self-cognition etc. of students (Gerdman, 1989). Furthermore, the mastering or change of new knowledge and skills determines the changes of thinking of a student (Svensson, 1984). It may also cause changes in world-view, value system or provision, which is unavoidable in the process of personality development.

Studies of higher education are not only gaining knowledge, skills and qualification degree, but also the being of a personality, which may be ambivalent. With the acquisition of *activity competence*, a student acquires certain professional competences, necessary for labor market. Meanwhile, the acquisition of *academic competence* provides a student with mental outfit of subjects, which are little related to the requirements raised in professional activities (Barnett, 1996). Mastering of activity competence is more important for a future professional, as it creates the basis for a successful professional socialization and qualitative performance of professional tasks. However, academic competence not only expands horizons of an individual but also creates basis for the further process of personality development.

The main aim of socialization of an individual performed in educational system is to develop such a personality, which is capable of adjusting to changing social environment, takes over standards of behavior and general values and also changes it actively in accordance to his/her own needs (Šermukšnytė, 1996). Education tries to integrate new generation into the society, enabling it to participate in professional and social activities. Withal it performs the function of selection, while exercising social-structural allocation of the society, resulting in the occupation of a place in professional life of an individual, which corresponds to his/her capabilities and skills and provides with social status.

Education is "activity of social culture, in which cultural treasury of the past, personal and human life of the present and future perspectives meet" (Laužikas, 1993). Thus, university studies should assure professional and personal development of a student, while providing with many-sided basis of knowledge about a human being, society and culture as well as their interaction.

The development of a studying person in a certain scientific sphere is performed because of interest and experience. It follows that the more students feel the consideration in their opinion and needs, the greater their motivation is (Hedin & Svensson, 2001). Interest of students in their future professional activities is directly related to such existential moments, as self-realization and satisfaction of own needs. That is why it is so important to integrate practical elements into theoretical education, which should stimulate the reflection of students and increase the understanding of a professional role.

Dimensions	Factors
Study contents	Usefulness of studies Meaningfulness of studies Harmonization of theory and practice
Activity	Possibilities of choice Democracy Teacher's role Structure of studies
Social dimension	Psychosocial climate Teacher-student interaction Student-student interaction
Self-realization	Personality development Value internalisation

Table 1: Study dimensions and educational factors

To generalize thoughts presented in this article, it could be stated that the development of professional motivation of students creates an important part of general development of a personality as well as stimulates the increase of motivation for studies. These four dimensions have significant importance for the increase of professional motivation during studies.

Efforts to develop high professional motivation during studies should first of all be expressed as a wish to reach aims such as knowledge necessary for the profession and activity competence. Future hopes of a student are quite often related to the integration into the labor market as a condition of financial safety.

Hereby analyzed influence of various educational factors to professional motivation allows considering that its development depends greatly on the level of the relation of studies with motives of self-realization and those acute for future professional activities. However, even after the effect of these factors is assured, professional motivation may decrease, if the attitude stating that every student is a personality with his/her own needs, interests and experience, which creates conditions for social interaction based on respect and effective studies, is not followed.

References:

ARGYRIS, CH./ PUTNAM, R./SMITH, D.M.: Action Science. San Francisco 1985 (Jossey-Bass Publishers).

BARNETT, R.: The Limits of Competence: Knowledge, Higher Education and Society. Buckingham 1994 (The Society for Research into Higher Education, Open Society Press).

BOURDIEU, P.: Kultursociologiska texter. Stockholm 1986 (Salamander).

BUBER, M.: Jag och du. Falun 1994 (Dualis).

CARLSSON, C.: Ungdomars möte med yrkesutbildning. Stockholm 2001 (HLS Förlag).

GERDMAN, A.: Klient, praktikant, handledare: Om att utveckla en egen yrkesteori. Lund 1989 (Wahlström, Widstrand).

Nyklar till kunskap. Om motivation, handling och förståelse i vuxenutbildning. HEDIN, A./ SVENSSON, L. (eds.). Lund 2001 (Studentlitteratur).

HOEL, M.: Yrkestilpasning og yrkesutvikling: En studie av arbeidstilknytning, yrkeskarrierer og yrkesmotiver blant et kull hogskoleutdannede. Oslo 1995 (Oslo universitet).

JARVIS, P.: Paradoxes of Learning. San Francisco 1992 (Jossey Bass Publishers).

JUCEVIČIENĖ, P.: Lietuvos aukštojo mokslo reformos galimas posūkis: edukologinis vertinimas. Konferencija "Lietuvos mokslas ir pramonė: XXI amžiaus technologijas pasitinkant". Kaunas 1999.

LAUŽIKAS, J.: Švietimo integracijos pagrindai. Pedagoginiai raštai. Kaunas 1993 (Šviesa).

MARTON, F./SÄLJÖ, R.: Approaches to Learning. In: MARTON, F./HOUNSELL, D./ENTWISTLE, N. (eds.): The Experience of Learning. Implications for Teaching and Studying in Higher Education. Edinburgh 1997 (Scottish Academic Press).

RAJECKAS, V.: Mokymo organizavimas. Kaunas 1999 (Šviesa).

STEWART, A.M.: Empowering People. London 1994 (Pitman Publishing).

STRATA-ETAN EXPERT GROUP: Higher education and research for the ERA: Current trends and challenges for the near future. Final Report. European Commission. Directorate – General for Research. October 2002.

SVENSSON, L.: Människobilden i INOM-gruppens forskning: Den lärande människan. Göteborg 1984 (Göteborgs universitet).

ŠERMUKŠNYTĖ, L.: Švietimo sistemos vertinimo kriterijai. Švietimo vadybos įvadas. Kaunas 1996 (Technologija), pp. 127-143.

ŠIAUČIUKĖNIENĖ, L.: Mokymo individualizavimas ir diferencijavimas. Kaunas 1997 (Technologija).

Regina Novikienė, Liuda Šaučiukėnienė

Deep Learning Development Possibilities in the Studies of Mathematics

Introduction

The main issue expressed in the documents of the European education (UNESCO) and the works of educational researchers is a problem of high quality education connected with student's conscious activity, understanding, responsibility and the development of constantly changing individual knowledge. These qualities are very important when a person reaches the necessary capabilities: the values of democracy, ability to present an argument, to listen to another opinion, to generalize, to assess, to make responsible solutions. He must be able to ground his position on the problem under discussion, to learn to study throughout the whole life. The process of mathematics studies may be very suitable when trying to reach the above mentioned qualities, because the purpose of mathematics as a subject is connected with the development of the learning capabilities: learning consciously, acting with responsibility.

If the mission of education is being connected with the development of students' capabilities (Barnett, 1990) while seeking for high quality education one should raise a question: *how do the presented information (subject content), organization of learning methods, teaching methods, learning strategies, etc influence personal development of an individual?*

When investigating the differences of students' learning (Ramsden, 2000; Marton, 1998; Lodewijks, Vermunt (according to Letschert, J. (ed.) (2001), two approaches to learning came to light – the deep and the superficial. They influence the organization of further learning activity. It has been proved that the way of learning and learning strategy has a direct influence on the qualities mentioned. Therefore the investigation of the used strategies when organizing the process of learning may reveal the limitations of deep learning and development possibilities. The research problem emerges: what are deep learning development possibilities when studying mathematics at Kaunas University of Technology?

The object of this investigation is deep learning strategies applied by first-year students of Kaunas University of Technology in the studies of mathematics. The purpose of the investigation is to reveal the strategies used by students in the studies of mathematics at KUT, to discover all shortcomings and to plan deep learning development possibilities. The problem has been formulated after the analysis of educational science literature, after having revealed the indications by which the approach to deep learning is expressed. Applying statistical methods has processed the research results.

The article consists of three parts. In the first part the problem is raised and its relevance is grounded. The second part reveals the results of different researchers' work in the field of learning on the basis of which the research instrument was created. The third part presents the results of the research, their analysis in order to find the possibilities of the approach to deep learning development, to promote the insufficiently revealed learning strategies.

Different ways of learning

It has been proved that the way of learning has a direct influence on study results (Bowden, Marton, 1998; Ramsen, 2000; Vermunt, Lodewijks (according to Letschert, J. (Ed) 2001). It is closely connected with the meaning and structure of knowledge research. F. Marton (1998) has carried out research with students. They were requested to read the same scientific paper. Some students were given a task to read it so that they could remember as many details as possible; the others tried to grasp only the essence of the text. The research showed that those who wanted to understand the main idea of the paper, perceived it deeper than those who tried to remember the details. They remembered them, but were not able to find links, were at a loss in the variety of the details and did not have the whole picture of the article (J. Bowden, F. Marton, 1998, p. 50). This is the way students prepare for their examinations. But unfortunately, their knowledge is usually superficial and momentary.

Some research analyzed by Ramsden (2000), prove that there are two approaches that determine deep learning: some people are orientated to the meaning and the process of obtaining conclusions, trying to understand how they were made; others link learning with remembering facts and their reproduction. The aim of the first approach is to understand, whereas the other supports remembering. The deep and the superficial approaches are expressed not only by raising aims, but also by performing actions. Some students begin to solve a problem from perceiving it, try to determine the links between details, and only then perform in the planned way. Others try to apply a well-known method that they have been used to, and solve the problem according to analogy, nether seeing the whole nor looking for alternative ways.

The knowledge of the first group of students is deep. They may work effectively. The knowledge of the other group is superficial: although they succeed in applying the method, but it is the only one, without alternative ways and improvement possibilities.

The structure of the knowledge depends on the way of learning. If we read a text and do not know its structure, we do not see the hierarchic links between details. Then we can only say what it is about in the beginning, what the author writes later and again. Such structure is direct. If we read a text knowing the structure, the problem and the conclusion and are interested in how this conclusion was made and how it matches our prior knowing, then the structure is furcated. As the mentioned readers' structures are different, they perceive the idea of the text differently. The first group of readers overload their memory as details are the most important for them, whereas the other group seeks to grasp the meaning and details are only a kind of illustration for the. They come together with the idea of the text without even trying to remember them.

Bowden and Marton (1998, p. 71) have singled out six learning conceptions: learning as

a) increasing of learning;
b) remembering and reproduction;
c) application according to analogy;
d) understanding;
e) observation of those who act differently;
f) assessment and changing the structure of knowledge.

The stages of the first group (A, B, C) are characteristic of learning directed towards reproduction. Such learning is superficial. It is connected with the direct structure that overloads memory with details. Unfortunately, memory is short. New facts replace the old ones and the final result is poor.

The stages of the other group (D, E, F) are characteristic of learning directed towards search. Such learning alters the structure of knowledge, dissociates from the stereotype of activity, and opens new ways for analysis, critical thinking and effectiveness. Such learning is linked with the furcated structure of knowledge – perceiving the essence, revealing and assessment of new variations of the phenomenon what leads to changing the whole structure, i.e., perceiving a different phenomenon.

Vermunt (1992) (from Letschert J. (Ed), 2001, p. 27) distinguishes learning styles according to how students process information:

1. Analysis and thinking based learning;
2. Application based learning;
3. Reproduction based learning;
4. Non-structuralized learning.

Personal interests motivate students supporting the first type of style. Their activity leads to better understanding of the problem and development of their personal knowledge structure. The information learned is not for overloading memory, but for changing personal structure of knowledge. It is accepted as a stimulus for thinking. The supporters of this style are usually looking for relations between separate parts, their connection. They are critical and do not take for granted what is written. They get deep into the logic of other people's thinking, compare their own activity with others and try to find out whether it is worth. This is the way of finding the most effective methods of activity.

Application based learning is similar. A person who deeply understands the phenomenon itself can only apply it. But the motives of these learners are more associated with practical needs. The motives of the *third learning style* supporters are to acquire knowledge, to pass examinations, to receive a diploma. They also structuralize information, but their structures depend on how to reproduce information easier and to put it in a compact way. Their learning is connected with summarizing and reproducing separate parts of information. But their summarized information keeps the same structure as it was presented.

Non-structuralized learning is connected with absolute incapability to learn. These learners have difficulties both in understanding and reproduction.

Thus, a gifted person is the one who knows how to learn and uses analysis and thinking based learning. H. Rost's seminar presentation (2000) and Narkevičienė's (2000) research prove this statement, as the difference between "gifted" and "normal" students was in favour of the gifted only in activeness, freedom of expressing thoughts. All this is determined by possession of holistic knowledge structure that gives way to expressing one's opinion, making conclusions, being free and self-confident.

Ramsden (2000) states that the deep and superficial approaches towards learning are not concrete qualities of a person. Every human being may apply either of the mentioned approaches depending on the *organization of learning* (p.66). Students usually adjust to the requirements of the study system. They would do what is useful for them in that system. If during the examination theoretical material is required to be repeated, the students' strategy will be directed towards how many details to remember. If examination tasks require demonstration of understanding, students will organize their activity accordingly. Trying to change the approach to learning we are trying not to change our students, but the organization of learning. It will cause changes in students' learning strategies and their experience.

Research organization and analysis of the results

The differences between strategies of learning activity organization revealed by the researchers working on the learning process are the indicators for the development of the instrument of this research.

The questionnaire consists of 14 statements that express the strategies of deep and superficial learning. The first –year students of technical faculties of Kaunas University of Technology (n 385) assessed them by choosing from 5 positions: 1 – I absolutely disagree with the statement, 2 – I disagree, 3 – I do not know, 4 – I agree, 5 – I absolutely agree with the statement. A high number of collected points will characterize a superficial approach in one block of questions and a deep approach – in the other.

The whole situation is reflected in Fig.1. It shows that deep learning is insufficient. The students on the right side of the distribution choose the deep learning strategy, whereas those on the left do not choose it. The distribution of points is not exponential. If it were exponential, the majority of students would use deep learning strategies. Thus another phase of research is associated with finding out the strategies that are least used in the studies of mathematics.

For data interpretation in separate blocks, the indicators of mode, mean and standard deviation expressed the answer tendencies. Mean and mode approaching 1 mean respondents' disagreement with the statements, whereas their approach to 5 means agreement. The answer tendencies with the highest mean and the lowest standard deviation is considered to be the most unanimous assessment tendencies. The consistency of the blocks was based on Kronbach-alfa values. This highlighted the statements that did not "stick" to the block of statements defining deep learning, what shows the limitations of the chosen learning style.

The applied learning strategies were investigated during both theoretical and practical activities of the studies of mathematics.

The research on theoretical studies according to the approach of deep learning showed that mathematical propositions had been accepted as dogmas with no doubt about their propriety. Students do not dare to ask questions or try to find out. The reason might be a large number of people in the classroom. Then a person is shy to ask, being afraid of the reaction of the others, especially if he/she has already got negative experience (Kline, Saunders, 1993)

Students do not know that it is necessary to match their thinking with the others, or they are not given the possibility. This might be because of the peculiarities of teaching theory, when a lecturer chooses the only form of lecture due to a large number of students. If at secondary school new material used to be delivered only by teacher the student may not know other ways of learning.

If one does not question what he/she reads or hears, that shows a lack of critical thinking. This must be associated with the experience at secondary school where material is presented like a dogma that has to be accepted without arguments, explanations. It has been noticed that nowadays secondary school teachers of mathematics reject proof (Novikienė, Petraitienė, 2002). Lack of possibility to make sure of the propriety of propositions creates barriers for the deep attitude to learning. Students do not link propositions, what makes the knowledge system hard to set and leads to reproductive learning.

A wrong presentation of the material in the course books might cause lack of doubt to what is read. The statement formulated in the beginning of the research makes one believe what is written. You have only to remember and reproduce, and that causes reproductive learning.

The analysis of the use of deep learning strategies during practical activities showed that students do not look for alternative ways of problem solution. They are limited by what they know and are not interested in how their colleagues solve the same problem. This reduces effectiveness.

While analyzing the statements describing the superficial approach towards learning, we noticed that students try to concentrate with the aim to remember as much as possible. They read course books thoroughly, without understanding the essence of the phenomena.

The reason is their lack of learning skills. A student often goes into details without understanding the essence of the phenomenon. He gets lost in the abundance of details and his learning becomes harder. Such learning leads to the formation of the linear structure when one remembers what was written. If marking is changed, the student gets lost and cannot explain anything. Lack of new text reading strategy is connected with disability to use textbooks. *This shows that students' prior experience of using textbook material on their own is insufficient.*

The use of superficial learning strategies might be associated with the assessment of the theory of mathematics. If it is enough to repeat the textbook material, the learning strategies will be directed towards remembering and reproduction. Thus, *the hypothesis emerges that student assessment procedures might be oriented towards tolerance of superficial learning.*

The drawbacks of deep learning inspire planning its development possibilities. As students implicitly believe what is written, in order to change this strategy we should assess only the sufficiently grounded statements without assessing the narration of the facts.

Another deep learning possibility might be noticed in order to avoid presentation of the declarative type contents. The organization of the teaching material should join research elements. That might help to construct conceptions and get phenomena definitions and statements during the research.

By encouraging interest in how a phenomenon is understood by other students, we should more often apply the active teaching methods – discussion, seminar, and group work. That would stimulate self-study of textbook material;

develop a possibility to brush up a language, to use mathematical terminology, to express one's thoughts precisely.

By stimulating self-study, we should deliver information not only by the form of lecture, but also by distant learning, using virtual learning environments. This way it is possible not only to present information, but to organize network discussions, have self-control tests, etc. That would encourage students to be more active, self-independent.

Conclusions

1. Only two students from 385 examined ones have collected maximum sum of deep studies points. Thus, other first-year students of KTU in one or another aspect use deep studies strategies.
2. The most often happening deep studies limitations are connected with position receiving as dogmas, not even doubting that they might be right, and not interesting how the expression understand others.
3. Studies are perceived as an individual process. Therefore the development of deep studies possibilities is connected with studies process alternation, accentuating investigation elements, argumentation, increasing student's activity, stimulating intercommunication, applying discussion seminar and remote studies forms, requiring self- dependent study skills.

References

BARNETT, R.: The Idea of Higher Education. London 1990 (University Pres).

BOWDEN, J./MARTON, F.: The University of Learning.-Great Britain 1998 (Biddles Ltd, Guildford, King's Lynn).

Turning the perspective. New outlooks for education.-Consortium of Institutions for Development and Research in Education in Europe/Netherlands Institute for Curriculum Development, Enschede. LETSCHERT, J.: (ed.) 2001

RAMSDEN, P.: Kaip mokyti aukštojoje mokykloje. 2000 (Aidai).

KLINE, P./SAUNDERS, B.: Ten Steps to a Learning Organizarion. Virginia 1993 (Arlington).

NARKEVIČIENĖ B.: A Model of Capable Children Education Conditions, and Its Expression in Lithuania. Doctoral thesis. Kaunas 2000 (KTU).

NOVIKIENĖ, R./PETRAITIENĖ, S.: Mathematics Studies Limitations Passing From Teaching to Studies Model. Lietuvos matematikos rinkinys 42(2002). Vilnius (Institute of Mathematics and Informatics), pp. 1-6.

Ilona Adomavičiūtė

Rolle der Kultur als Kommunikationsmedium im Kontext der Europäischen Union

Einleitung

Die seit langer Zeit angestrebte EU-Erweiterung Europas ist wahr geworden. Damit ist eine der wichtigsten Seiten der Geschichtslektüre aufgeschlagen, die täglich durch neue Inhalte für weitere Spannung sorgt. In letzter Zeit betrachten wir oft die eigentliche Wiedervereinigung eines alten europäischen Kulturraumes in Zahlen, Fakten, Quoten, Tabellen, Graphiken und beleuchten die aktuellen Debatten über das einzigartige historische Gebilde aus verschiedenen Perspektiven mit Hilfe der fassbaren Kriterien. Einer der existenzielleren Frage kommen wir nur zögernd auf den Grund, nämlich, für wen und mit welchem Ziel „Projekt Europa" durchzusetzen wäre. Welche kulturelle Mittel werden für das Erreichen der Ziele relevant sein und welche von denen die Kulturvermittelnde in Prozessen des interkulturellen Austauschs im Endeffekt berücksichtigen werden. Schließlich wäre dabei sinnvoll der Frage nachzugehen – mit welcher Bereitschaft die einzelnen europäischen Gesellschaften Kultur für das einigende Band und Vision der gemeinsamen Europas betrachten und Gewicht bei der Lösung vieler europäischen Probleme auf die kulturelle Kommunikation legen.

Es erscheint mir wichtig, die Funktionen der Kultur für den politischen Aufklärungsprozess der Europäischen Union zu betonen. Einige der aktuellen Themen und Fragestellungen zu den neuen Perspektiven für das erweiterte Europa möchte ich besonders ins Blickfeld rücken. Vor allem gilt festzustellen, mit welchen Instrumenten die Impulse zu der Verständigung der Kulturen durch die effizientere interkulturelle Kommunikation und durch die Aneignung der neuen Kompetenzen auf dem Gebiet der internationalen Kulturarbeit zwischen den Ländern, aufgegriffen werden können. Des Weiteren sollten alle derzeitige EU- Mitgliedssaaten sich Fragen stellen, welche Bedeutung die Kommunikationspotenziale für die notwendige soziokulturelle Vernetzung im Prozess der EU-Erweiterung haben; was ist mit der europäischen Identität gemeint oder wie steht es mit der Europäischen Öffentlichkeit.

An der Kreation des Projektes „Gemeinsames Haus Europa" sollten wir alle an dem transnationalen Dialog aktiv beteiligen, um die kulturelle Annäherung der Kulturnationen auf dem Gebiet der Kommunikation zu beschleunigen.

Große politisch-historische Entwicklungen wie die Erweiterung 2004 rufen als erstes die emotionale Seite in uns hervor. Wir werden täglich mit den Gefühlswelten der Menschen, mit ihren Ängsten und Hoffnungen konfrontiert. Diese sollen kultur-soziologisch dechiffriert werden, um zu verstehen, wo der Ausgangspunkt ist, d.h. wo wir uns befinden und wohin wir wollen. Dies gelingt nur, wenn wir uns mit dem Begriff „Kultur" als gemeinsamen Nenner, auseinandersetzen. Die lateinische Bedeutung des Wortes „cultura" ist zunächst „die Summe der Bestrebungen einer Gemeinschaft, die Grundbedürfnisse der menschlichen Natur (...) zu befriedigen und untereinander auszugleichen".[1] Darunter fällt auch das Bedürfnis nach einer Veredelung, Verfeinerung und Formung der menschlichen Persönlichkeit. Ein Individuum aber macht noch keine Kultur aus. Es ist die Gesamtheit der geistigen und künstlerischen Lebensäußerungen aller Völker und Staaten, eine Gemeinschaft der Gleichgesinnten, die bestimmt, wonach sie strebt, welche kulturelle Entwicklung sie bewusst eingeht und welche Prinzipien dabei gelten.

Die fünf Erweiterungen der EU haben nicht nur die Geographie, sondern auch die politisch-kulturelle Haltung der Länder zueinander verändert. Heute gilt es, der europäischen Einigung Wege in die Zukunft zu weisen. Der Einigungsprozess erfordert dabei ein offenes und transparentes Gegenübertreten in Sachen Kultur und den Willen an einem neuen EU- Gebilde zu bauen.

Zwar ist die Kultur kein Allerweltsheilmittel gegen noch ungelöste gesellschaftliche Unstimmigkeiten in Europa; aber sie lässt Aufmerksamkeit und Sensibilisierung gegenüber den neuen Mitgliedern der erweiterten Europa wachsen. Da wir momentan 25 gleichberechtigte Staaten in der Union haben, bedarf es einer gemeinsamen Sprache. Auch für Jürgen Habermas, einen der bedeutendsten gegenwärtigen Philosophen und Sozialtheoretiker, erweist sich ein gemeinsames sprachliches Kommunikationsmedium dabei als unabdingbar. Dieses Medium sieht er im Englischen, das den Europäern „als zweite Erstsprache"[2] dienen soll. Erscheinen Sprachen, als einer der wichtigsten Erzeuger der Öffentlichkeit in Plural, wie das der Fall der EU ist, ist somit ein multilingualer Kommunikationskontext begründet, das als massive Hürden im Prozess demokratischer Integration zu verstehen ist.

Meiner Meinung nach muss man nicht unbedingt an den physischen Aspekt einer Sprache appellieren, da Kunst bzw. Kultur auch eine vermittelnde Funktion haben und als „Sprache" für die Verständigung in der europäischen Union – für die angestrebte europäische Einigung – genutzt werden können. Die Sprache der internationalen Kulturarbeit und interkulturellen Kommunikation, könnte als

---

[1] Brockhaus Lexikon. 1955. Band VI, S. 692.

[2] J. Habermas. Zeit der Übergänge, Frankfurt a.M., 2001, S. 122 zitiert aus: P.A. Kraus. „In Vielfalt geeint? Europäische Identität in der Zwickmühle". Blätter für deutsche und internationale Politik. Heft 6, 2004, S. 724.

Bindeglied zwischen den einzelnen Nationen verstanden werden und einige Aufgaben in EU Kulturpolitik übernehmen. Die Vielfalt der Sprachen mit einem gemeinsamen Nenner, nämlich –Kultur, kann keinesfalls als Hindernis bezeichnet werden, denn die Zukunftsaufgaben Europas lassen sich, auch ohne Englisch in Vordergrund zu stellen, verständigen. Nicht die Suche nach einer gemeinsamen Sprache verhindert heute viele wichtige Entscheidungen. Vielmehr ist das europäische Zusammengehörigkeitsgefühl, woran es zur Zeit in alten und neuen EU-Staaten mangelt. Daran konnten auch die sattsam bekannten gemeinsamen Werte und das Kennen von Problemen in der EU bislang nichts ändern. Weder haben sie nach innen für den nötigen zusätzlichen Identitätsschub gesorgt, noch bestimmen sie in großem Stil die europäische Außenpolitik. Was könnte die vielen Stimmen der einzelnen Nationalstaaten in der EU verbinden? Dafür ist, meines Erachtens, die Kultur, als ein wichtiges Medium zur transparenteren Gestaltung der EU prädestiniert und fähig, die schnellere Integration der Ost- und Mitteleuropäischen Staaten zu erzielen.

Kultur per se sollte und kann für die Annäherung der Europabürger zueinander viel stärker beansprucht werden. Denn „Kultur ist das eigentliche Leben. Sie liegt der Politik und Wirtschaft, dem lokalen und dem Feuilleton zugrunde und verbindet sie. Sie ist folglich auch die Substanz, um die es in der Politik geht. Der erweiterte Kulturbegriff umfasst vielmehr Werte und Orientierungen alles dessen, was wir mit dem Desiderat vom „anständigen Leben“ verbinden-: mit realistischen Zukunftschancen auch für unsere Enkel“.[3.] Im Jahre 2004 sind es über 450 Millionen Menschen, die zueinander fanden, dabei veralterte Ansichten und kulturelle Vernachlässigung in allen wichtigen Bereichen der menschlichen Begegnungen überwunden haben. Langsam, aber sicher ist es zu verzeichnen, dass die EU-Bürger/innen zunehmend als entscheidendes Merkmal für das Zusammenwachsen Europas die kulturelle und geschichtliche Vielfalt erachten.

Es ist offensichtlich, dass Kultur – und damit impliziert – alle geistig-kulturellen Prozesse wichtig für das europäische „Miteinander“ zu erachten sein sollten, denn auch jeder einzelne Staat in der erweiterten Europäischem Union definiert sich in erster Linie über die Kultur.

Europäische Identitätsfrage

2004 ist ein wichtiges Jahr für Europa: am 1. Mai haben sich zehn neue Staaten sich der bisherigen EU angeschlossen. Kann man Hier und Jetzt über ein gemeinsames kulturelles Fundament und über eine europäische Wertegemeinschaft in dieser nun sehr großen Gesellschaft sprechen? Bietet unsere europäische Kulturgemeinschaft zum jetzigen Zeitpunkt den Anlass zu denken, dass eine Europäische Identität bereits besteht, oder kann sie erst mit der Zeit entstehen?

Dieser gesellschaftliche Diskurs befindet sich dabei in einem Zwiespalt, einer gewissen Dialektik. Ist es nicht gerade unsere Stärke, dass die Kulturen in Europa so unterschiedlich waren bzw. sind und sich gegenseitig über die Jahrhunderte befruchten konnten? Gäbe es Wolfgang Amadeus Mozart ohne den Italiener, wäre der Sturm und Drang ohne William Shakespeare möglich? Interkultureller Wandel durch Übernahme und Entlehnung von Kulturgütern aus fremden Kulturen in die eigene, spielte schon immer eine wichtige Rolle. Wir kennen keine Kultur, die nicht solche ursprünglich „fremden" Elemente in großer Zahl aufwiese und starke Wechselwirkungen mit anderen Kulturen hatte. Was im 21. Jahrhundert gesucht wird, sind die Bemühungen zum globalen Verständnis zwischen den Kulturen, Akzeptanz des Andersseins in unmittelbarer Nähe und Erneuerung der Sichtweisen aufeinander. Neben der Kulturvielfalt suchen die neuen EU-Länder nach ihren eigenen kulturellen Wurzeln und fürchten, dass durch die große Lawine der neuen kulturellen Einflüsse, im Zusammenhang mit der Erweiterung der Union, die Nationalkulturen ihre Konturen leicht verlieren können. Dies könnte z.B. kleinere EU-Staaten wie Litauen oder Lettland betreffen, die an der Grenze der EU liegen. Ich glaube aber, dass diese Gefahr, nämlich das eigene nationales Gesicht zu verlieren, ist nicht mit der neuen Konstellation in der EU gegeben, denn gerade kleinere Kulturnationen über die Jahrhunderte dauernden Repressalien durch die ehemalige Diktatur der Sowjets, gelernt haben, sein kulturelles Erbe gut zu bewahren – und können heute ihre kulturelle Wurzel stolz präsentieren, ohne die Fremdbeeinflussung zu fürchten. Das heißt, momentan verlaufen die Prozesse der Identitätsbildung sowohl für Europa als ein Ganzes und parallel bilden sich die einzelnen Identitäten der neuen EU Länder. Eine turbulente Zeit für alle.

Es ist ein gutes Zeichen, dass in Debatten und Foren über das neue Europa viel diskutiert wird.

> „Das Geheimnis der Sprache ist groß; die Verantwortlichkeit für sie ist symbolischer und geistiger Art, sie hat keineswegs nur künstlerischen, sondern allgemein moralischen Sinn. Sie ist die Verantwortlichkeit selbst, die es niemanden erlaubt, heute am wenigsten, das Geistig-Künstlerische vom Politisch-Sozialen zu trennen und sich gegen dieses im vornehm Kulturellen zu isolieren."[4]

Das heißt, durch das Wort und durch den Dialog finden wir zueinander.

---

[4] Thomas Mann, Brief an den Dekan der Philosophischen Fakultät der Universität Bonn 1937, in: Jürgen Moeller (Hrsg.): Historische Augenblicke. Das 20. Jahrhundert in Briefen. München 1999, S. 128.

„Europa entsteht im Reden wieder (...), es erlebt einen neuen Gründungsmoment, ohne schon dafür einen Begriff dafür zu haben“[5], bemerkt G. Hofmann.

Demnach ist die Europäische Identität noch im Begriff sich zu entwickeln.

„Eine detaillierte Identitätsbestimmung sei heute noch schwer möglich und Europäer hier sich auf einen langen Prozess einlassen müssen, wenn es um die Suche nach dem „Europäischen Wir“ geht.“[6]

Wir haben ein Territorium, eine gemeinsame Geschichte, Erinnerungen und Wurzeln, bereits eine gemeinsame Wirtschaft, grenzüberschreitende Kooperationen und gemeinsame politisch-kulturelle Absichten auf dem Wege zum faszinierenden Projekt “Europa“. Äußerlich mag es schon eine europäische Identität geben. Dieser Prozess darf aber nicht nur auf das Äußere reduziert werden. Vielmehr soll er sich zu einer inneren Identifikation mit Europa entwickeln. Ich bin der Auffassung, dass ein deutlicheres Bewusstsein wachsender Betroffenheit von europäischen Entwicklungen und Entscheidungen ausgebildet werden muss, um über die Selbstfindung Europas sprechen zu können. Die Bürgerinnen und Bürger Europas müssen sich in diesen Prozess noch aktiver einbringen. Zu den gemeinsamen Ansichten in allen Bereichen der Gesellschaft müsste man mit allen 25 gleichberechtigten Partnern, d.h. mit allen Menschen, welche die Europäische Union ausmachen, erst kommen. Eine europäische Zivilgesellschaft, eine europäische Öffentlichkeit, eine europäische Demokratie müssen sich erst noch herausbilden. Man wird dies in die eigene Lebensgestaltung miteinbeziehen müssen, um sich mit den Gedanken des gemeinsamen Europas identifizieren zu können. Vor allem kommt der identitätsstiftenden Funktion der Kultur in einer sich wandelnden Welt, eine immens wichtige Rolle zu. Neben der Voraussetzung- das Interesse aneinander, bedarf es in erster Linie Kenntnisse über die vorhandenen kulturellen, politischen, sozialen und rechtlichen Unterschiede, um gemeinsam antworten zu können, womit und wodurch wir, Europäer, uns identifizieren möchten.

Kommunikation als Voraussetzung für die Europäische Öffentlichkeit

„Das Zeitalter verlangt nach einer Veränderung der Gesellschaft (...). Damit ist nicht nur die Veränderung von Inhalt und Organisation des Wissens gemeint,(...) sondern vor allen Dingen unser Verhältnis zum Wissen.“[7]

---

[5] Gunter Hofmann, Wildes, vages Reden. Europas Intellektuelle probieren ihre Ideen für die neue EU erst aus, in: „DIE ZEIT“ Nr. 19, 2004.

[6] Rebecca Harms: Rede auf der Eröffnungspodiumsdiskussion der Konferenz „Zukunft Europa“ am 19/20.03.04 in Leipzig.

[7] C. Castoriadis, Durchs Labyrinth-Seele, Vernunft, Gesellschaft, Frankfurt a.M. 1981 in: Klaus-Michael Wimmer: Der Andere und die Sprache, Vernunftkritik und Verantwortung, Berlin 1988, S. 1.

Momentan gibt es in der Europäischen Union noch vieles Unübersetzbares. Immer größere Aufgabenfelder in Politik, Wirtschaft, Sozialem und Kultur sind im nationalen Alleingang nicht mehr zu gestalten und erfordert momentan von den Beteiligten eine starke Handlungskompetenz in Sachen der Dialogführung. Darunter versteht man die Fähigkeit, erfolgreich in gemischt-kulturellen Kontexten zu kommunizieren. Aber erst mal muss die Optimierung der Bedingungen im Bereich der Informationsübertragung, -aufarbeitung und allgemeinen Mediatisierung stattfinden, um die notwendige Artikulation der gemeinsamen europäischen Zielen, Visionen zu ermöglichen. Eine Ansprache, wie man weiß, kündigt bereits das Verhältnis zum Anderen an. Hierin liegt Dialektik insofern, dass die Möglichkeiten miteinander zu kommunizieren erst erschlossen werden müssen, um in einen Dialog zu treten.

> „Der Sprache kommt zwar die sinnstiftende Funktion zu, über deren Geltung aber allein die Subjekte mittels ihrer Erfahrung urteilen können."[8]

Diese Erfahrungen auf der „miteinander" Ebene, müssen erst gemacht werden. Der sprachliche Austausch könnte das Entstehen von gemeinsamen Berührungspunkten implizieren. So einfach es klingt, so schwer sind die Resultate festzustellen.

Aktuell, im Kulturvermittlungsbereich zeigt sich des Öfteren, dass es trotz der vielen Bemühungen, die Kontakte zu den neuen EU-Ländern herzustellen sich als äußerst schwierig erweist. Ich finde es kontraproduktiv, dass in den wenigen Fällen, wo die seltene Ansprache der neuen Beitrittsländer im Rahmen der nationalen Öffentlichkeitsarbeit funktioniert, bezieht sie sich meistens nur auf die gesellschaftlichen Kreise der Alten EU-Länder, die ohnehin in die aktuellen politisch-kulturellen Belange eingeweiht sind. Das heißt, ein breiteres Publikum und somit potenzielle Mitgestalter einer europäischen Öffentlichkeit, sind somit bei der Entstehung eines gemeinsamen Diskurses bereits ausgeschlossen. Ein gut aufgebautes, zielgerichtetes und konkurrenzfähiges Kommunikationsnetz und neue Kommunikationstechniken müssen in nahe Zukunft folgen.

Der Entstehung der Europäischen Öffentlichkeit, liegen aber noch weitere Probleme zugrunde. Es gibt wenig aktuelle und vielseitige Informationen über unsere neuen EU-Nachbarn. Wie soll man diese besser kennenlernen, wenn man den informativen und kulturellen Zugang nicht findet, oder der nicht ausreichend geboten ist? Wie soll die Europäische Meinung überhaupt sich langsam bilden können? Viele europäische Foren, zahlreiche europäische Diskussionen und Konferenzen finden unter den geschlossenen Türen, und viele EU-Bürger fühlen sich somit schlicht nicht angesprochen, was für die Öffentlichkeit und für die Identi-

---

[8] K-M. Wimmer, in: Der Andere und die Sprache, Vernunftkritik und Verantwortung, Berlin 1988, S. 3.

tät Europas nicht positive Auswirkungen haben wird. Schon 1935 rief der Deutsche Unternehmer und Stifter Robert Bosch auf, das Kulturerbe Europas zu bewahren und an der Lösung schwieriger Fragen selbstschöpferisch mitzuwirken, indem man sich persönlich einbringt, um somit zur Völkerverständigung in Europa beiträgt. Das heißt- hier geht es auch um engagierte Bürgergesellschaft, die zum europäischen Themenspektrum auch die Stellung beziehen könnte. Ein leichterer Zugriff auf die Wissensbasis, bezüglich der EU, soll verhelfen, möglichst breite Schichten der Gesellschaft in die EU-Themen einzubeziehen. Transparentere Informationen würden eventuell daraus folgende Bereitschaft zur tiefergehenden kulturellen Wahrnehmung einer anderen Kultur, begünstigen. Das wiederum kann die zwischenstaatliche und die Kommunikation unter den Bürgern Europas zumindest auf der kulturellen Ebene erheblich erleichtern.

Eine weitere Voraussetzung für die Entstehung der europäischen Öffentlichkeit, ist das Intensiveren der internationalen transeuropäischen Kommunikation. Leider auch hier ist vieles mit Problemen behaftet; die Abschaffung der Kulturbeilagen in Deutschland, Verkleinerung der Feuilletons führen dazu, dass noch weniger Informationen über die EU und einzelne EU-Mitglieder zumindest in der Presse zu erwarten sind. Auch die Zahl der Auslandskorrespondenten verringert sich stets. Demzufolge können die wenigen Artikel und Essays zu wichtigen Europagesellschaftlichen Themen gar nicht die Aufmerksamkeit der Leser erlangen. Und die vielen Diskurse erreichen nur einen kleinen Teil der ohnehin an Europa oder an europäischen Kulturdebatten interessierte Menschen. Solche in der Regel machen nur 10-15 Prozent der Bevölkerung als Kulturpublikum Europas aus. Problematisch erscheint mir auch, dass die Diskussionsforen nicht ausreichend beworben werden, also sind diese der breiten Öffentlichkeit wenig bekannt und schwer zugänglich.

Europa als Gemeinwesen im Bewusstsein seiner Bürger kann nicht allein durch die Verfassung oder die gemeinsame Wehrung Euro festsetzen.

> „Die wenigen existierenden europäischen Medien wie Financial Times oder International Herald Tribune sprechen (zudem) eher eine Wirtschafts- und Politikelite als den europäischen Normalbürger an“, heißt es im Aufsatz von Katharina Pesche.[9]

Das würde heißen, dass die internationale Kommunikation in den Medien künftig von ihrer abstrakten, zur Zeit noch auf der Expertenebene sich befindendem Diskurs, auf ein einfacheres und zum Disput animierendes Niveau sich begeben soll. Um verschiedene Stellungnahmen zu vielen Themen der EU in den Medien muss der interessierte Europäer über ein Abonnement gleich mehrerer der großen europäischen Zeitungen verfügen oder sich Extraausgaben über Europa be-

---

[9] K. Pesche: Europa Ansichten. Online-Artikel auf der Website der European-Journalists e.V./ Netzwerk europäische Kommunikation-www.neucom.org.

schaffen. Der als Entstehungsbedingung für ein europäisches Bewusstsein und einer europäischen Öffentlichkeit Habermas so eindringlich geforderten Kommunikation, sind also Grenzen gesetzt. Vielleicht sollten die Berichte nicht nur Informationen über die politischen und wirtschaftlichen Belange in EU beinhalten, sondern mehr Kommunikation auf der kulturellen Ebene anspornen und in der Kultur- das einigende Band für EU sehen; die kulturellen Themen öfters in Vordergrund stellen, die mannigfaltige intraeuropäische Kulturdebatte gewichtiger und seriöser in Betracht ziehen.

„Eine attraktive, ja ansteckende „Vision“ für ein künftiges Europa fällt nicht vom Himmel. Wenn das Thema bisher nicht einmal auf die Agenda gelangt ist, haben wir Intellektuelle versagt“[10], stellt erneut mit Bedauern Jürgen Habermas fest. Ich würde begrüßen, dass die Gesellschaften der Alten und Neuen Europas stärker auf die Herausforderungen und Perspektiven der EU auf den medialen Wegen, da der breiteste Zugriff darauf gegeben ist, vorbereitet werden, die EU-Bürger an europäischen Prozessen teilhaben können und die Intensivierung einer interkulturellen Kommunikationskompetenz als Lebenskompetenz in der Europäischen Union gefördert wäre. Denn ohne Informationen und somit den Gelegenheiten gemeinsam nach Lösungen für die entstandene Probleme zu suchen, kann keine europäische Öffentlichkeit entstehen, zumal das Interesse an EU zur Zeit in der Gesellschaften als noch nicht sehr groß einzuschätzen ist.

Eine weitere Herausforderung ist das neue bzw. erweiterte, dialoggestützte Denken. Insbesondere betrifft es die innovativen Wege in der Kommunikation, in der ein neuartiges Geschichtsverständnis, neue Konfliktbewältigungsstrategien sowie Vermittlungs- und Dialogalternativen zwischen den Europäischen Kulturnationen herausgearbeitet werden und zustande kommen könnte. Die Entstehung der Europäischen Öffentlichkeit und somit auch der europäischen Identität wäre durch das Schaffen einer gemeinsamen Handlungs- und Verständigungsbasis aller Staaten nicht ganz so illusorisch. Im Bereich der interkulturellen Kommunikation sehe ich allerdings die Notwendigkeit – die Entwicklung der interkulturellen Kompetenzen deren voranzutreiben, die als Kulturschaffende, Kulturvermittelnde und -rezipienten in Europa das gemeinsame „Projekt Europa“ gestalten.

Diese sollten nach Möglichkeiten mit dem Wissen um Gemeinsamkeiten und Unterschiede zwischen Kulturen, ausgestattet sein. Dies trägt nämlich dazu bei, die Komplexität einer Kultur zu verstehen. Des Weiteren sollte allen Kommunikatoren (Medien-Journalisten, Politikern, Kulturvermittlern, Pädagogen etc.) die interkulturelle Sensibilität abverlangt werden, Eigenschaften wie Bereitschaft zur

[10] J. Habermas in: K. Pesche. Europa Ansichten. Online-Artikel auf der Website der European-Journalists e.V./Netzwerk europäische Kommunikation-www.neucom.org.

Empathie, Offenheit, Toleranz sind hier von zentraler Bedeutung. Zuletzt sind es Fähigkeiten, Fertigkeiten auf der kommunikativen und der Verhaltensebene, um effektiv mit Menschen einer fremden Kultur interagieren und Probleme lösen zu können, die aus kulturellen Unterschieden resultieren.

Europäische Zukunft in der kulturellen Vielfalt zu gestalten, ist bestimmt kein leichtes Unterfangen, das vielen Europäern sichtlich Probleme bereitet. Denn es ist unumstritten, dass sogar 15 Jahre nach dem Fall der Mauer nicht alle in der Gesellschaft genügend kulturell aufgeklärt, für die Selbstverständlichkeit der Kulturvielfalt im gesamten EU-Gebilde vorbereitet sind. Eine ausgeklügelte europäische Kommunikation muss sich damit auseinandersetzen und primär durch die Nutzung kulturspezifischen Stärken der wechselseitigen Kompensation von Defiziten in den EU- Gesellschaften verhelfen, damit diese ihren Weg zueinander finden können.

"Synergie" – griechische Bezeichnung für „Zusammenarbeiten" – ist und bleibt im Kontext der EU-Debatte sehr aktuell. Die bloße Suche nach kulturellen Gemeinsamkeiten im Sinne eines kleinsten gemeinsamen Nenners und ein Verdrängen der Unterschiede erweist sich dabei als ebenso falsch wie eine ausschließliche Fixierung auf die Unterschiede. Erst durch und dank der intensiver interkulturellen Kommunikation würden die hervorgebrachte kulturelle Kooperationen, gemeinsame Projekte, Programme von höherer Qualität sein, als es die Summe der Einzelelementen schaffen kann. Interkultureller kommunikativer Austausch könnte als neue Ressource definiert werden, der eine Zunahme an Handlungsalternativen bietet, die Möglichkeiten Vorgehens- und Sichtweisen, neue Arbeits- und Organisationsformern eröffnet und fördert die miteinander zu gestalten, sowie lädt ein neues Know-how zu erproben ein. Die Synergiepotenziale sind durch die kulturelle Vielfalt der EU eindeutig gegeben. Die Union, mit ihren derzeitigen 25 zahlreichen eigenständigen Kulturen, sollte man als multikulturelles Team in einem Unternehmen auffassen, das, auf die Erfahrungen der internationalen Firmen und Experten F. Herbrand stützend, „meist deutlich effizienter als monokultureller"[11] ist. Denn die Kommunikationsstärke der Franzosen, hohe Fähigkeiten im Organisationsbereich der Deutschen, Flexibilität und der Improvisationstalent der Italiener sind Synergiepotenziale, von denen alle Kulturstaaten in der EU profitieren können.

Wirtschaftlich hat die Ost-Erweiterung schon fast zu zwei Dritteln stattgefunden. Auch Interkulturalität, als Bereicherung für neue Rahmen der EU, sollte mit einem Selbstverständnis angenommen werden, damit wir uns nicht nur ökonomisch näher kommen. Das ist natürlich eine Herausforderung an alle Beteiligten, die Fähigkeit zu erlangen, die verschiedenen Kulturen als gleichberechtigt

[11] F. Herbrand. Fit für fremde Kulturen. Haupt Verlag, Bern, Stuttgart, Wien, 2002, S. 44.

zu betrachten, alle gegenseitige kulturelle Befindlichkeiten zu achten, über die Jahre gemeinsam am neuen Europa zu werkeln und die Bereitschaft viele Schwierigkeiten dabei auf uns zukommen zu lassen– aufrecht zu erhalten. An welchen Mustern wirtschaftlicher, politischer und vor allem kultureller Kreationen und Formen der Zusammenarbeit kann angeknüpft werden, um Zugang zueinander zu finden? In den aktuellen Kulturdebatten zur EU sollten wir die Kultursoziologie ins Auge fassen. Sie deckt die Beziehungen zwischen den einzelnen Kulturen und deren gesellschaftlichen Ordnungen auf. Kultur bietet dem Menschen ein System von Verhaltensnormen, an denen er sein Handeln orientieren kann. Infolge der Vielzahl der Kulturen sind diese Normen oft verschieden oder gar gegensätzlich. Das Verständnis für die andere Kultur kann beginnen, indem man von den eigenen Normen, als einzig richtigen absieht und das Fremde zulässt. Denn das Schöpfertum, egal in welchem kulturellen System, entspringt zunächst aus dem Volk, greift oft über weite Räume, staatliche, ethnische und sprachliche Grenzen überschreitend aus und vermag selbst einen grundstürzenden Wandel der Staaten und Reiche zu überdauern. Unserem aktuellen Ziel im neuen Europa liegen nicht nur politische Ideale und wirtschaftliche Bestrebungen zugrunde, sondern vor allem notwendige soziale, kulturelle sowie humane Dimensionen. Während die Politik noch um Regelungen und Normen ringt, sind Kunst und Kultur in Sachen europäische Einigung voraus (denken wir an die laufenden unzähligen kulturellen Projekte in Europa!). Wir bauen allmählich Vorurteile ab und erleichtern somit die Integration der beigetretenen Kulturlandschaften.

Plädoyer für Kulturdialog und intensiven kulturellen Austausch

Wie bereits erwähnt, umfasst die Kultur nicht nur Geschichte, Traditionen, Künste und Bildungsfragen, welche die notwendigen Brücken innerhalb der erweiterten EU errichten. Die Existenz der kulturellen Vielfalt erfordert Respekt, Toleranz der Unterschiede und angemessene Umgangsformen miteinander. Heute, wo neue Demokratien entstanden sind, die danach streben, in den europäischen Kulturraum wieder einbezogen zu werden, ist es vonnöten, alle Austauschvarianten der Kulturschaffenden zu unterstützen, die die Traditionen der einzelnen nationalen EU-Kulturen präsentieren und gleichzeitig zur Erweiterung des gemeinsamen Europäischen Kulturraumes beitragen.

Auf die Frage, was Europa zusammenhält, kann man entschieden antworten: die Kultur, die Künste und die Literatur die Jahrhunderte, ja Jahrtausende früher entstanden sind als die wirtschaftliche und politische Allianz Europas und auch in Zukunft gelten werden. Im Grunde fand die Verzahnung zwischen den Kulturen schon immer statt (Übernahme und ständige Ergänzung der Kunststile in Musik, Malerei etc.). Durch den Fortschritt der Technik, die daraus resultierende Mobilität und die mentale Auflösung der geographischen Grenzen ist es möglich geworden, trotz der politischen Zwänge, den kulturellen Austausch bewusster in

Szene zu setzen. Auf der Pressekonferenz der Kulturminister der neuen EU-Mitgliedstaaten im Bundeskanzleramt am 6. Mai 2004 in Berlin wurde primär auf die Europäische Kulturgemeinschaft eingegangen. Diese ist sehr vielschichtig, aber politisch und historisch nicht unbelastet. Daher bedarf es einer Neustrukturierung und einem erweiterten Verständnis von Kulturvielfalt.

„Das Europa, das ich meine, wird nicht ein Erinnerungsschrein sein, sondern eine Idee, ein Symbol, ein geistiges Kraftzentrum", schrieb 1946 Hermann Hesse in einem Brief an Thomas Mann.

Der gegebene kulturelle Pluralismus in der europäischen Staatengemeinschaft stellt, meiner Meinung nach, dieses Kraftzentrum dar. Die Kulturen liegen aber noch mosaikartig nebeneinander. Es gilt demnach, nicht nur das Sprechen über- und voneinander zu zelebrieren, sondern verstärkt das Gespräch miteinander zu suchen, um die gemeinsamen Interessen zur Erhaltung und Erschließung der bedeutenden kulturellen Nachlasse der Länder im Kontext der europäischen Geschichte zu erforschen und die Grundlagen möglicher Zusammenarbeit zu legen.

Warum sollen wir für kulturellen Austausch plädieren? Die Kultur sollte als Quelle des intensiven Austauschs von allen Nationen und Staaten angesehen werden und darauf sollte zukunftsorientiert gebaut werden.

Denn „in Wirklichkeit findet kein ökonomischer und kein militärischer, sondern ein geistiger Wettbewerb statt (...) Europa bleibt nichts weiter übrig, als seine Vielfalt zu mögen, denn darin besteht seine Stärke"[12], so G. Konrád, einer der wichtigsten zeitgenössischen ungarischen Literaten.

Auch die deutsche Kultur ist ohne den Kontext Europa undenkbar. Im Unterschied zu China, wo die Vereinheitlichung der Kulturen angestrebt wird, kann und sollte jedes Land in diesem neuen Europa sich zu den Merkmalen der eigenen kulturellen Mentalität bekennen. Die Fragen zur Nationalkultur der einzelnen europäischen Regionen sind sehr spezifisch und die Auseinandersetzung damit würden den Rahmen der vorliegenden Publikation sprengen.

Die Rolle des Kulturmanagements

Trotz der großen Massenmedienrolle, ist es wichtig die Kulturvermittlung zusätzlich durch konkret angelegte Kunst-, Medien-, Journalisten-, Sprach- und Kulturprojekte zu unterstützen, um somit weitere EU-Interessenten in verschiedenen Gesellschaften zu erreichen. Wünschenswert wäre es in Form einer direkten, bzw. „unmittelbaren persönlichen Begegnung, die der Massenkommunikation weit überlegener ist".[13]

---

[12] Geörgy Konrád: Wir werden normal sein, in: „DIE ZEIT" Nr. 19, 2004, S. 6.

[13] Gerhard Maletzke: Kulturverfall durch Fernsehen? Berlin 1988, S. 109.

Die EU hat zum Glück eine sehr hohe Dichte an Künstlerinnen und Künstlern, eine unglaubliche Anzahl Festivals – vom Jazz über Literatur, Film bis hin zu alternativem Tanz und Musik, womit sich die Kulturarbeit noch lange nicht ausschöpfen wird. Dennoch ein Guter Wille und Enthusiasmus, persönlicher Kontakt zur Kulturbegeisterten allein, genügen allerdings nicht, um effizient den Kulturaustausch zu entwickeln und die Resultate unter die Leute mit verschiedenen kulturellen Bedürfnissen und Vorlieben zu bringen.

Damit die Verständigungs-, Lern- und Austauschprozesse besser funktionieren, bedarf es vor allem in den neuen EU-Ländern tieferer Kenntnisse in Bereichen des Kultur- und Medienmanagements, Kulturmarketings und allgemein in der Projektleitung jeglicher Art. Die Ausbildung des Führungsnachwuchses in diesem Bereich schreitet in den neuen EU Ländern, nur langsam voran. Man sollte die Erfahrungen der westeuropäischen Kollegen durch Lehraufträge in die ost- und mitteleuropäischen Ländern kanalisieren – und somit allen das Kulturmanagement-Handwerk für die Gestaltung der Kommunikationswege, für die effizientere kulturelle Vermittlung ermöglichen. Das heißt, der Transfer der Kulturarbeit aus Sicht der Professionalität ist noch in vielen EU Staaten nicht optimal. Die wenigen Kultur- und Medienexperten aus den Beitrittsländern, die im Ausland ausgebildet worden sind und wieder bereits in die Heimatländer zurückgekehrt sind, könnten viel stärker in die Vermittlungsprozesse der EU einbezogen werden, was bis dato nicht ausreichend geschieht.

Momentan liegt die Macht, Akzente zu setzen neben den einzelnen Personen, eher auf Seiten der Institutionen: Kulturabteilungen der Botschaften, nationale Kulturzentren, einzelne Vereine, Bildungsakademien und andere Träger der Kultur. Dabei ist der Filtereffekt, bezüglich der dialogisch ausgerichteten Ansprache mit den kulturell-politischen Informationen, als sehr gering zu beobachten. Diese sollten stärker in die Mediatisierungssysteme und -prozesse aufgenommen werden. Ein Vernetzungssystem von Politik, Medien, Kultur und Gesellschaft kann und muss klarer ausgearbeitet werden, um die bestehende Vielfalt in der EU zeitgemäß und für die Gesellschaft verständlich präsentieren und platzieren zu können. Wir müssen uns der politischen und der kulturellen Kommunikation bedienen, und dabei auf die aktuellsten Querschnittsthemen unserer Gegenwartsgesellschaften durch die Medien eingehen.

Resümee

Mit dem Beitritt der zehn neuen Staaten, hat Europa an politischer Vielfalt und ökonomischem Potenzial gewonnen. Bereits die grenzüberschreitende Zusammenarbeit in allen Bereichen innerhalb der Europäischen Gemeinschaft erfordert eine Auseinandersetzung mit den landesspezifischen Eigenarten der Nachbarländer. Damit wir noch stärker zusammenwachsen und zu der Vision einer „europäischen Identität“ gelangen, müssen wir uns auf einen langwierigen und komplizierten Kulturprozess einlassen. Nun liegt es an uns, dieser Identität eine Gestalt zu verleihen. Um im Bereich der kulturellen Arbeit zum Gefühl des „Miteinander“ zu gelangen, bietet der Kulturaustausch viele Möglichkeiten zur Annäherung, mit dem Ziel, gemeinsam das Gebilde des neuen Europas mitzugestalten. Solch wichtige historische Momente sind zeitintensiv und mit vielen Veränderungen verbunden. Die Formen der Zusammenarbeit auf der Basis der kulturellen Pluralität und kulturellen Differenzen zu initiieren, ist kein leichtes Unterfangen. Die EU sollte sich zu einer Lern- und Kommunikationsgemeinschaft entwickeln. Denn der Stellenwert der politischen Entwicklungen, der Sprachen und Traditionen ist in den einzelnen Beitrittsländern sehr hoch, und man möchte dies nach Außen kommunizieren und der ganzen kulturellen Vielfalt, entgegenhalten. Damit die nationalen Egoismen nicht die Oberhand gewinnen, sollte das Verständnis, Vertrauen in einem offenen Dialog entstehen, der wiederum, bekräftigt durch den regen kulturellen Austausch, die Fähigkeiten einer erfolgreichen Zusammenarbeit der Beteiligten steigern könnte. Deshalb wird versucht mit Hilfe der medialen Mittel das Interesse an Erkenntnissen übereinander zu beschleunigen. Viele der Osteuropäischen Länder hatten die durch die Wirrungen der Geschichte unfreiwillig verlernt und müssen zu sich und zu Anderen in Europa wiederfinden. Bei der Bewältigung der kulturellen Diversibilität ist es wichtig, die junge Generation für die Schlüsselaufgaben, wie die gemeinsame europäische Geschichte mit Hilfe von Kulturprojekten, Austauschprogrammen und medialer Präsenz zu sensibilisieren, gemeinsame Aktivitätsfelder in Kulturbeziehungen zu suchen und diese für die Fortentwicklung der internationalen Beziehungen zu nutzen.

Inga Šukevičiūtė

Wirtschaftslinguistik als Teilbereich der Fachsprachen und Fachkommunikationsforschung

1. Einleitung

Der vorliegende Artikel befasst sich mit der Fachsprache in der Wirtschaftskommunikation und verfolgt das Ziel, die Besonderheit der Fachsprache Wirtschaft unter den anderen Fachsprachen zu zeigen. Dabei wird die Vielfalt von Schichtungsmodellen der Fachsprache Wirtschaft dargestellt und anschließend wird die praktisch-fachliche Schicht von Ohnacker durch die Beispiele aus den deutschen, litauischen und russischen Geschäftsbriefen typologisch illustriert.

Wirtschaftskommunikation, derer Ausdrucksmittel Fachsprache Wirtschaft ist, ist heutzutage eines der meist erforschten Bereiche auf dem Gebiet der Pragmatik und eines der wichtigsten Schwerpunkte in der Kommunikationstypologie.

Die Erforschung der Fachsprachen hat eine reiche Tradition in allen Ländern. Die Fachsprache jedes Landes hat ihre Entwicklungsrealien, widerspiegelt die Besonderheiten der wirtschaftlichen Entfaltung und Mentalität des Volkes. Dabei werden verschiedene Phasen und Konzeptionen der Fachsprachenforschung unterschieden.

Durch den Einfluss der Wirtschaftslinguistik und der funktionalen Sprachbetrachtung der Prager Schule in den 20er und 30er Jahren veränderte sich das wissenschaftliche Interesse der Fachsprachenforschung und der Bereich der Fachsprachen wurde durch Fachtextlinguistik und Fachtextpragmatik erweitert. Heute umfasst Fachsprachenforschung und Fachsprachenpraxis alle Formen beruflicher und institutioneller Kommunikation in ihrer Funktionalität und Interaktionalität und die Wissenschafts- und Praxisbereiche Terminologiewissenschaft, Textlinguistik und Diskursanalyse; Fachlexikographie, Übersetzten und Dolmetschen, Sprachplanung und Sprachpolitik, fachbezogenen Sprachausbildung und jede Form der Wissensvermittlung.

2. Zum Problem der Fachsprache Wirtschaft

Wirtschaftssprache ist im Gegensatz zu den „klassischen" Fachsprachen der Naturwissenschaft oder der Technik eine sehr komplexe Fachsprache und lässt sich als Fachsprache durch eine Vielfalt der sprachlich-inhaltlichen Bereiche, sprachlichen Realisierungsformen, Sprachstile schwer bezeichnen (vgl. v. Böhme 1999: 235, Heuberger 1997:56). Die Feststellung von Buhlmann/Fearns, dass Wirtschaftsdeutsch „ein bisher nicht definierter Arbeitsbegriff" sei, besitzt auch heute eine gewisse Gültigkeit (1987:306).

Die Wirtschaftssprache ist die Komponente der Wirtschaftskommunikation, und die Wirtschaftskommunikation schließt sowohl Lehr- und Wissenschaftskommunikation als auch Unternehmens-, Bank-, Handels- und Dienstleistungskommunikation ein. Wirtschaftskommunikation besitzt unterschiedliche Facetten auch deshalb, weil sie zugleich institutionelle, fachliche und berufliche Kommunikation ist (vgl. v. Hahn 1993:92, Heuberger 1997:57, Chreptavičienė 1997: 55ff., v. Böhme 1999:235, Brünner 2000:17f.).

Es wird immer diskutiert, ob es die Fachsprache Wirtschaft als integratives Ganzes überhaupt gibt, oder ob es nicht sinnvoller wäre, sich mit den Fachsprachen der Einzeldisziplinen, etwa Börse, Finanzwissenschaften, Währungspolitik u.a. zu befassen. Der zweite Weg impliziert den Verzicht auf die komplexe Bestimmung ,Wirtschaftssprache'.

Zu diesem Standpunkt hat man verschiedene Meinungen. Buhlmann vertritt die Ansicht, dass ,Wirtschaftsdeutsch' "als Sammelbegriff für diverse Fachsprachen" bezeichnen lässt. Dabei werden auch unterschiedliche Spezialisierungsgrade der Fachsprachen der einzelnen Wirtschaftsbereiche akzentuiert, sowie werden verschiedene sprachliche Fertigkeiten in ungleichen Sprachverwendungssituationen hervorgehoben, was über die „sogenannte Wirtschaftssprache" zu sprechen erlaubt (vgl. Buhlmann 1989:85ff.).

Ohnacker reformuliert den Fachsprachenbegriff von Hoffman[1], der sich mehr auf die Fachsprache Wirtschaft bezieht:

> „Die Fachsprache Wirtschaft umfasst die Inhalte des Faches Wirtschaft und die sprachlichen Mittel, mittels derer sie realisiert werden. Sie stellt die Summe der Fachsprachen dar, die in der wirtschaftsbedingten Kommunikation benutzt werden. Diese Fachsprachen sind Subsprachen der Standardsprache und umfassen die Gesamtheit aller sprachlichen Mittel, die in dem fachlich begrenzten Kommunikationsbereich Wirtschaft verwendet werden, um die Verständigung über die fachlichen Inhalte dieses Bereichs zu gewährleisten". (1992:36)

Diese Bestimmung zeigt nicht nur eine bestimmte und abgrenzbare Fachsprache Wirtschaft, sondern auch die Möglichkeit, die Subsprachen und ihre Differenzierungen auszusondern.

Bolten vertritt die These „dass das Sprachgeschehen im Wirtschaftsbereich nicht additiv sondern integrativ beschreibbar ist als ein komplex, empirisch annähernd bestimmbares Beziehungsgeflecht verschiedener fach- und– im weitesten Sinne- berufssprachlicher Ebenen" (1991:74).

[1] Unter Fachsprache versteht Hoffmann „die Gesamtheit aller sprachlichen Mittel, die in einem fachlich abgrenzbaren Kommunikationsbereich verwendet werden, um die Verständigung zwischen den in diesem Bereich tätigen Menschen zu gewährleisten. Diese Mittel bilden einen Teil des Gesamtinventars der Sprache im Sinne einer Subsprache" (1987:53).

Dabei werden besonders die einzelnen Subsprachen betont, die den gesamten Begriff der Fachsprache Wirtschaft bilden.

3. Schichtungsmodelle der Fachsprache Wirtschaft

Im Folgenden werden die Modelle der Fachsprache Wirtschaft näher erläutert, weil die Kriterien jeweilige Modelle wichtig für fremdsprachendidaktische Gegebenheiten sind.

Die in der Literatur vorzufindenden Gliederungen von Wirtschaftskommunikation orientieren sich an der vertikalen Schichtung[2] und als vertikale Schichtung der Kommunikation wird Wissenschafts-/Theoriesprache, Werkstatt-/Berufssprache und Verteilersprache/fachbezogene Umgangssprache hervorgehoben. Diese Differenzierungen werden in neueren Untersuchungen immer neu variiert (vgl. Brünner 2000:19f.).

Die Wirtschaftssprache wird in vier Hauptstile eingeteilt:

- wissenschaftliche Wirtschaftskommunikation (z.B. Lehr- und Lernkommunikation, Vorträge, Veröffentlichungen in wissenschaftlichen Wirtschaftszeitungen),
- publizistische Wirtschaftskommunikation (z.B. Artikel in Tageszeitungen und Zeitschriften),
- Kommunikation des schriftlichen (dienstlich-offiziellen) Geschäftsverkehrs (z.B. Geschäftsbriefe, Telefaxen, Verträge),
- Kommunikation des mündlichen Geschäftsverkehrs (z.B. Telefongespräche, Verhandlungen) (vgl. v. Böhme 1999:235).

Buhlmann/Fearns ordnen die Wirtschaftskommunikation in folgende vier Kommunikationsbereiche:

- Kommunikation von Betrieb zu Betrieb (hierzu u.a. Brief, Telefax, Telefongespräch; Zolldokumente, Verhandlungen, Verträge),
- Kommunikation im Betrieb (hierzu u.a. Bericht, Gespräche, Konferenzen, Sitzungen),
- Allgemeine Information (u.a. Gesetzestexte, Kommentare, Verordnungen),
- Aktuelle Information (u.a. Wirtschaftsnachrichten und Kommentare, Jahresberichte) (vgl. Buhlmann/Fearns 1987:307).

Ohnacker (1992:37) nennt drei Kriterien, die wichtig für das Schichtungsmodell der Fachsprache Wirtschaft besonders unter fremdsprachendidaktischen Gesichtspunkten sind:

---

[2] Hoffmann (1987) behandelt die Gliederung der Fachsprachen hinsichtlich ihrer fachsprachlichen Verwendungsbereiche auf horizontaler Ebene und die Schichtung auf vertikaler Ebene. Zur Bestimmung der Schichten nennt Hoffmann folgende Kriterien wie „die Abstarktionsstufe, die äußere Sprachform, das Milieu und die Teilnehmer an der Kommunikation" (Hoffmann 1987:70).

1. Die einzelnen Schichten repräsentieren unterschiedliche fachliche Ausrichtungen, die verwendeten Textsorten und auch Kommunikationspartner.
2. Das Modell soll textoffen sein, um auch außerhalb des Bereichs stehende Texte einschließen zu können, die mit materiellen Produkten verbunden sind.
3. Das Modell soll adressatenoffen sein, um die Texte involvieren zu können, die der Kommunikation mit Laien über fachliche Themen dienen.

Ohnacker bezieht sich in seinem Modell auf die Einteilung der Fachsprache Wirtschaft von Borgulya und erweitert das Modell mit der Dimension der Kommunikation zwischen Betrieb und Konsument und mit didaktischen Überlegungen zu diesem Thema (vgl. Ohnacker 1992:40f., Heuberger 1997:60f.).

Ohnacker (1992) stellt umfassende linguistische Aussagen über die sprachlichen Merkmale der Fachsprache Wirtschaft dar. Die typischen Merkmale der Fachsprache Wirtschaft werden von ihm in den drei Schichten – theoretisch-wissenschaftlicher, praktisch-fachlicher, populär-wissenschaftlicher Schicht – dargestellt. Im Vergleich zur Standardsprache und zu den wissenschaftlichen Fachsprachen insgesamt können für die Fachsprache Wirtschaft Unterschiede wie z.B. sprachökonomische und inhaltlich verdichtete Ausdrucksweise, die klare Dominanz des Vorgangspassivs festgestellt werden (vgl. 49ff., 86).

Die Schichtung der Fachsprache Wirtschaft (nach Ohnacker 1992):

Bolten entwickelt in Anlehnung an die wirtschaftswissenschaftliche Einteilung des gesamtwirtschaftlichen Handlungsfeldes ein zweiteiliges Schema mit den Konstituenten Makroökonomie (Weltwirtschaft, Volkswirtschaft) und Mikroökonomie (öffentlicher/privater Betrieb, öffentlicher/privater Haushalt).

Jedes einzelne Unternehmen wird als „Substrat sämtlicher Bereiche wirtschaftlichen Denkens und Handelns“ definiert.

Die Wirtschaftskommunikation wird von Bolten folgendermaßen resümiert:

> „Die Kommunikationsstruktur des Unternehmens schließt die genannten Bereiche notwendigerweise ein. Sie repräsentiert demzufolge in vertikaler Hinsicht – und ebenfalls in verkleinertem Maßstab – sämtliche Sprachebenen des gesamtwirtschaftlichen Handlungsfeldes“ (1991:76).

Die Einteilung der Wirtschaftssprache erfolgt in drei Ebenen, die jeweils nach Textsorten und Kommunikationspartnern differenziert werden:

1. Theoriesprache (z.B. Monographien, Forschungsberichte, verwendet u.a. von Wissenschaftlern, Unternehmensführern, Wirtschaftspolitikern);
2. Berufssprache (z.B. Geschäftsberichte, Bilanzen, Verträge, verwendet u.a. von Fachjournalisten, Fachübersetzern, und -dolmetschern);
3. fachbezogene Umgangssprache (z.B. Geschäftsbriefe, Werbespots, Prospekte, verwendet u.a. von Mitarbeitern in Unternehmen, Konsumenten).

Bolten schließt auch die Möglichkeit einer Gliederung nach Fächern in horizontaler Hinsicht nicht aus, besonders für ein Großunternehmen. Die Einteilung spiegelt sich auch im Sprachgebrauch wider, z.B. potentiell grundständige Bereiche eines Unternehmens (u.a. Unternehmensführung, Verwaltung, Finanzen, Logistik, Forschung) haben auch in sprachlicher Hinsicht branchenübergreifenden Charakter.

Merkmale	**Theoretisch-wissenschaftliche Schicht**	**Praktisch-wissenschaftliche Schicht**	**Populär-wissenschaftliche Schicht**
Kommunikationsträger	Wirtschaftswissenschaftler, hochqualifizierte Fachleute	Fachleute, Mitarbeiter, Kunden	Interessierte Fachleute, breite Öffentlichkeit
Explizität/ Implizität	hoher Grad an sprachlicher und inhaltlicher Explizität	hoher Grad an sprachlicher und inhaltlicher Implizität	hoher Grad an inhaltlicher Explizität
Unterteilung/ Textsorten	Wirtschaftswissenschaftliche Monographien, Forschungsberichte, wissenschaftliche Veröffentlichungen, theoretische Abhandlungen in Büchern und Fachzeitschriften	Wirtschaftsdokumentation: Wirtschaftspläne, Berichte, Beschreibung von Arbeitskreisen, **Korrespondenz**, betriebsinterne Mitteilungen, Protokolle, Notizen, Formulare usw.	Wirtschaftsberichte der Tagespresse, Wirtschaftsfachzeitschriften
Themenbereiche	Wirtschaftssysteme, Wirtschaftssektoren, Wirtschaftsindikatoren, Wirtschaftsorganisationen, Wirtschaftsgeographie, Währungssysteme	Distribution der Waren, Geld- und Kreditwesen, Versicherungswesen, Personalwesen, Auftragsabwicklung	Wirtschaftspolitik, Beruf, Geld und Kredit, Märkte, Unternehmen, Technik und Innovation, Management.

Schröder hat das Modell der kumulativen Textanalyse entwickelt, das weiter nach Buhlmann entsprechend der Kommunikation in der Berufssituation modifiziert wurde. Das sog. Schrödersche Modell hat vier Betrachtungsebenen:

1. die Ebene der extralinguistischen und pragmatischen Faktoren (hierzu u.a. Disziplin, Kommunikationssituation, -ziel);
2. die Ebene der Textgroßstruktur (u.a. Textsorte, Spezialisierungsgrad, Thema, Darstellungsart, Textverflechtung, Gliederung);
3. die Ebene der Makrostruktur (bezogen auf den Abschnitt) (u.a. Einbettung, Thema, Darstellungsart, Textbauplan, Textverflechtung, Gliederung);
4. die Ebene der Mikrostruktur (bezogen auf den Absatz) (u.a. Kommunikationsverfahren, Kohärenz/Konnexität, Gliederung, Syntax, Lexik, Grammatische Kategorien) (vgl. Schröder 1983:254ff., Buhlmann 1989:100f.).

Baumann unterstreicht auch die enge Beziehung zwischen der Fachsprache Wirtschaft und einer Unternehmung, wobei durch die Fachsprache nicht nur *Corporate Identity* geformt wird, sondern auch Fachlichkeit eine zentrale Kategorie der Unternehmenskommunikation darstellt, die durch qualitative und quantitative Merkmale gekennzeichnet ist.

Unter dem Phänomen der Fachlichkeit von Unternehmenskommunikation versteht man „ein komplexes System von außersprachlichen und sprachlichen Determinationszusammenhängen, das sich in der sprachlichen Äußerung durch inhaltliche (auf den Kommunikationsgegenstand bezogen), formale (auf die Art der (fach-/sprachlichen Realisierung bezogen) und funktionale Elemente und Relationen herausbildet (Baumann 2000:112).

Ausschlaggebend erscheinen hier die Dimensionen der Fachsprache Wirtschaft, die zur Fachlichkeit der Unternehmungskommunikation dienen und die Dimensionen werden anhand fachlicher Textbeispiele illustriert:

- die interkulturelle Ebene der Fachlichkeit von Unternehmenskommunikation (kulturelle Konventionen bestimmen alle Ebenen des Kommunikationsbereiches);
- die soziale Ebene (entscheidende Rolle spielen hier Elemente der fachlichen Tätigkeitssituation, der sozialen Situation der Partner in der Unternehmenskommunikation sowie Elemente der Umgebungssituation der Partner);
- kognitive Dimension (hierzu gehören Denkfertigkeiten und Handlungsschemata, die Metakommunikation kann im Texte verschieden zum Ausdruck kommen, z.B. für die Sicherung des Textverständnisses wird Fettdruck oder Explikation von Begriffen gebraucht, für die Sicherung der inhaltlichen Akzeptabilität des Textes – eingefügte Illustration – und Belegmaterial);
- inhaltlich – gegenständige Dimension (darunter versteht man den auf Inhalte der Unternehmenskommunikation bezogenen Sachverstand, das Wissens- und Könnensniveau der kommunizierenden Individuen);
- funktionale Dimension (diesem Phänomen liegen Modelle menschlichen Verhaltens zu Grunde. Die funktionelle Dimension der Fachlichkeit bezieht sich auf sprachliche und nicht sprachliche Mittel, die das fachliche Denken bzw. die Handlungs- und Erkenntnisfähigkeit des Einzelnen im konkreten Unternehmensbereich auf einem individuellen Entwicklungsstand repräsentieren);
- textuelle Dimension (Kenntnisse über Textstrukturen können die Aufnahme und Speicherung der im Text vermittelten Information nachhaltig beeinflussen);
- stilistische Ebene (diese Ebene kommt mehr in den Texten für Nichtfachleute vor);
- lexikalisch-semantische (die Unternehmenskommunikation ist eng mit der qualitativen Strukturierung der wissenschaftlichen Terminologie verbunden) (vgl. hierzu Baumann 1994: 65ff., 2000:113ff.).

Die Fachsprache Wirtschaft steht mit ihrer Vielfalt der sprachlich-inhaltlichen Bereiche im Mittelpunkt der Diskussionen und beweist die Aktualität des Problems und stellt mit ihrer horizontalen Auffächerung (in Disziplinen und Fachbereiche) und mit ihrer vertikalen Schichtung (in verschiedene Verwendungssituationen, Fachstile und Register) sowie durch die Verflechtung mit anderen Fachsprachen ein sehr komplexes Feld dar.

4. Zur praktisch-fachlichen Schicht

Als Schwerpunkt dieses Artikels wird die praktisch-fachliche Schicht von Ohnacker gewählt, zu der auch u.a. Geschäftsbriefe gehören (vgl. Ohnacker 1992: 87ff.). Dabei werden nur die typischen fachlichen Merkmale dieser Schicht ausgesondert und mit den folgenden Beispielsätzen in den drei Sprachen typologisch illustriert und dadurch die Thesen von Ohnacker auch in der litauischen und russischer Sprach bewiesen oder abgestritten.

Geschäftsbriefe stellen eine Textart dar, die von stark formalisierten Wendungen geprägt ist, und die Redewendungen und Fachvokabular normiert sind. Die Geschäftsbriefe folgen genau festgelegten Argumentationsmustern und Stilnormen. Jedoch nach 1980 erlebt die Korrespondenz den „neuen Stil". Die drei wesentlichen Prinzipien – übersichtlicher Aufbau, Beschränkung auf das Wesentliche und klare, treffende Ausdrucksweise – verändern den Stil eines Geschäftsbriefes.

- TENDENZ ZUM EINFACHEN *HAUPTSATZ* UND *SATZGEFÜGEN*, Z.B.:

Nach Ablauf der Herstellergarantie haben wir ein interessantes Angebot für Sie- unsere Europlus Neuwagenanschlussgarantie von BMW zu einem sehr attraktiven Preis. Das beiliegende Faltblatt soll Ihnen schon einmal einen Überblick über die zu erwartenden Leistungen geben. Wir sind überzeugt, Sie erkennen sofort die Vorteile, die sich für Sie daraus ergeben.

Susipažinę su Jūsų paskelbto atvirojo konkurso pirkti E2 klasės lengvąjį automobilį ir mikroautobusą sąlygomis, pastebėjome keletą netikslumų, kuriuos pašalinus būtų sudarytos labiau konkurencinės sąlygos dalyviams ir padidėtų Jūsų galimybė pigiau įsigyti minėtus automobilius.

Узнав, что в Кишиневе существует фирма АО »AIEA», которая занимается распространением Вашей продукции, нам экономически не выгодно сотрудничать с ними.

Tendenz zu den einfachen Hauptsätzen und Satzgefügen beweist die Tatsache dass die Fachsprache den präzisen und sprachökonomischen Ausdruck darstellt. Genau wie im Deutschen, gebraucht man im Litauischen und Russischen einfache Satzkonstruktionen.

- KEINE SIGNIFIKANTE TENDENZ ZUR *STAFFELUNG VON NEBENSÄTZEN*, Z.B.:

Sie erteilten einen Reparaturauftrag, der mit Rechnung vom Nr. 1143646 vom 04.04.2000 abgerechnet wurde: Sie haben diesen Auftrag wohl in der Absicht erteilt, die Rechnung nie zu bezahlen um auf diesem Wege Ihren vermeintlichen Anspruch zu befriedigen. Das werden wir nicht akzeptieren. Bitte überweisen Sie den offenen Rechnungsbetrag, der bereits ohne Mahnung bis zum 31. August 2000 fällig ist.

UAB „D.A." pakeitė Jūsų automobiliui komplekte visą turbiną į kitą nenaują veikiančią, kuriai nesuteikiama garantija. Atsiradus pakartotinam gedimui turbinoje iki 2001 06 01, „D.A." UAB įsipareigoja šį mazgą pakeisti nemokamai, neskaitant detalės kainos.

Позвольте выразить Вам искреннее уважение и дать сведения о ситуации на рынке холодильников в Казахстане. Ставим Вас в известность о резких изменениях ситуации на рынке холодильников в Республике Казахстан. Изменения произошли в основном, по двум причинам.

Diese Tendenz bricht noch einmal den Mythus, dass die Wirtschaftssprache sehr kompliziert und unzugänglich in ihren vielen Nebensätzen ist. Die Sprache soll in der Korrespondenz klar für den Adressierenden sein, ungeachtet dessen in welcher Sprache man schreibt. Besonders sieht man diese Tendenz in der russischen Korrespondenzsprache.

- *Nebensätze* werden häufig als Relativsätze, Konjunktionalsätze oder Infinitivkonstruktionen angeschlossen; als Konjunktionen stehen häufig: *wenn, falls, sofern, da, weil, zumal*; Konditionalsätze beginnen häufig uneingeleitet, z.B.:

Falls Sie zu dem Thema noch weitere Informationen erhalten möchten, setzen Sie sich bitte mit S.C. aus der PR-Abteilung in Verbindung.

Es würde uns freuen, wenn wir Ihren Mitarbeiter zum Seminar begrüßen dürfen. Wir wären sehr an der Teilnahme von Herrn D. interessiert, da er sicherlich im Rahmen seiner Aufgaben eine weitere fundamentierte Ausbildung benötigt.

Šis gedimas nėra susijęs su remontu, atliktu š.m. balandžio 3 dieną, nes tai yra kitas variklio mazgas. Siūlome nustatyti, kad 16 ir 17 punkte nurodyti išmatavimo kėbulo vidui, nes išoriniai matavimai neturi prasmės.

И что на территории нашего государвства появится выпускаемая вами продукция, которая зарекомендовала себя как высококачественной и конкурентоспособной.

Relativsätze sind typisch für alle drei Sprachen, Konditionalsätze mehr für die deutsche Sprache. Im Litauischen finden wir am häufigsten die Konjunktionen *nes* und *kadangi*. Im Russischen sind folgende Konjunktionen als *который, чей, что, где, когда.*

- *Adverbialsätze* stehen häufig als Kausalsätze; es dominieren Konditionalsätze, die mit den Konjunktionen wenn, falls, sofern angeschlossen oder asyndetisch vorangestellt werden, frequent sind auch Kausalsätze mit *da, weil, zumal*, z.B.:

Wir sind leider nicht in der Lage das Stück käuflich zu erwerben, da unser Museum zu einem gemeinnützigen, eingetragenen Verein gehört. Wenn Sie uns die Sache schenken wollen, lade ich Sie mit einer Begleitperson herzlich zu einem Museumsbesuch ein.

Įsakau V.B.,Prekybos MB atsarginėmis dalimis skyriaus vadovą, išsiųsti į komandiruotę į Štutgartą (Vokietiją), kurioje 2001m. balandžio mėn. 01-06 dienomis imtinai vyks seminaras.

Мы надеемся, что фирма «А», которая, как мы знаем, тоже продает нашу продук-цию в г. А, будет и дальше сотрудничать с АО «С», но мы повтаряем и в этом году хо-телось, что Вы покупали нашу продукцию через нашего генерального дистрибьютора «Т».

Diese Art von Nebensätzen in der Korrespondenz lässt sich leicht erklären: mit Relativ- und Adverbialsätzen wird etwas ausführlicher erklärt und dargestellt. Sehr wichtig erscheint in der Korrespondenz die adverbiale Bestimmung des Ortes und der Zeit. Die häufigen Konditional- und Kausalsätze beweisen die Wichtigkeit der Bedingung und des Grundes für das Geschehen. Jeder Geschäftsbrief, sowohl mit der Werbung im Hintergrund, als auch mit der Mahnung muss gründlich, manchmal auch mit ausgesonderten Punkten begründet werden.

- *Attribute* treten häufig auf, überwiegend anteponierte Partizipien, auch Präpositional- und Genitivattribute kommen gehäuft vor; Appositionen sind nicht frequent; Attribute dienen der Präzisierung und Differenzierung, z.B.:

Ihr Schreiben vom 06.09.1999 habe ich erhalten und teile bezugnehmend hierauf mit.

Leider müssen wir Ihnen mitteilen, dass die per e-mail versandten Dokumente nicht ordnungsgemäß bei uns eingegangen sind.

Wir haben nun eine genaue Überprüfung der eingereichten Unterlagen durchgeführt und müssen feststellen, dass die von Ihnen übermittelte Aufstellung nicht der unfallkausalen Beschädigung entspricht.

Mes pasiruošę suteikti Jums ekspedicijos paslaugą gabenant Jūsų krovinius geležinkeliu, nepriklausomai nuo gabenamų krovinių kiekio, užtikrinant savalaikį, saugų krovinių vežimą palankiomis Jums sąlygomis. Tikimės abipusiai palankaus bendradarbiavimo.

Наше предприятие гарантирует стопроцентную предоплату, т.к. является рентабельным и платежеспособным, с предоставлением высокого сервисного обслуживания на базе нашего предприятия.

- hinsichtlich *Person* und *Numerus* dominiert die 3. Pers. Sing.; im Bereich der Korrespondenz treten auch die 1. Pers. Pl. und ‚Sie' als höflich distanzierte Anredeform auf, z.B.:

Liebe Indre, beigefügt erhalten Sie ein Word-Dokument des BTL 2000 Internetauftritts. Für Fragen stehe ich Ihnen gerne zu Hilfe.

Sehr geehrter Herr B. Vielen Dank für Ihren freundlichen Brief vom 08.06.1993. Über Ihr Buch zum zehnjährigen Jubiläum und die Feikarten habe ich mich sehr gefreut.

Gerbiamas pone Nimz, remdamiesi Jūsų 1999 vasario 19d. neribotu pažadu apmokėti, mes prašome Jūsų dabar dar kartą nurodytą sumą apmokėti iki 1999 09 17 ir prašome tuo pačiu Jūsų supratimo, kad praėjus terminui savo reikalavimus pateiksime per teismą.

Уважаемые господа, направляем в Ваш адрес оригинал заявки участника ярмарки „Heimtextil".

Уважаемый Антанас! Прошу Вас оказать нам помащь в решении данного вопроса.

Im Bereich der Korrespondenz finden wir am häufigsten höflich distanzierte Anredeform *sehr geehrte(r)/gerb/уважаемые(й)*. Die Anredeform mit *liebe(r)* oder nur mit dem Vornamen ist ziemlich selten, und mehr im Deutschen. Überall ist ‚Sie' zu lesen.

- als *Tempus* dominiert das Präsens zum Ausdruck allgemeingültiger Aussagen, z.B.:

Wir befassen uns im Auftrag des C mit dem Schadenfall vom 18.08.1999, welcher sich in K-W/Deutschland ereignet hat. Bitte korrespondieren Sie mit uns zu unserem Aktenzeichen.

Heute habe ich festgestellt, dass wir noch nicht alle abgestempelte Ausfuhranmeldungen von Ihnen zurückbekommen haben. Wir benötigen diese Unterlagen als Nachweis dafür, dass wir die Ware ausgeführt haben.

UAB „V" turi licenciją verstis geležinkelio transporto ūkine veikla ir itin palankiomis sąlygomis ekspedijuoja krovinius geležinkelio transportu.

Norėtume priminti, kad Jūs asmeniškai rekomendavote mums firmą „FM GmbH“, kaip garbingą ir patikimą partnerį ir reikalavote iš mūsų paskutinės partijos prekių atkrovimo be išankstinio apmokėjimo.

В прошлом году к нам обратились покупатели с претензиями на 4 холодильника по поводу резины.

Als Tempus dominiert neben Präsens im Deutschen auch Perfekt und Präteritum, im Litauischen auch eine Form der Vergangenheit, wie būtasis kartinis, und im Russischen прошедшее. In der Korrespondenz bezieht man sich auf das Geschehene, was sich in der Form der Vergangenheit widerspiegelt und man behandelt zu dem gegenwärtigen Punkt aktuelle Probleme.

- DER *INDIKATIV* DOMINIERT, IMPERATIV UND KONJUNKTIV TRETEN SELTEN AUF, Z.B.:

BMW informierte uns über aktuelle Bedrohungssituationen für alle BMW Händler in Deutschland. Mit bisher zwei Anrufen hat eine unbekannte, männliche Person bei der BMW AG angerufen und folgende Nachricht hinterlassen.

Uns wäre interessant, ob H. unsere Teilnahme unterstützen könnte. Vielleicht könnten wir aus H. das Material für die Messe mit Rabat bekommen.

Vilniaus prekybos, pramonės ir amatų rūmai (VPPAR) organizuoja susitikimą su LR ambasadoriumi K. ponu V.B., K-K B.Į „H-D“ prezidentu p.ČDC ir J-R-K korporacijos „I“ pirmuoju viceprezidentu p. A.I. Į susitikimą kviečiame visus, kas norėtų dalyvauti Lietuvos prekybos namų steigime. Apie dalyvavimą susitikime prašytume mus informuoti iki sausio 20d.

Благодарю Вас за предоставленную информацию. Прошу Вас подготовить более пасширенное предложение по условиям поставок продукуии и возможным скидкам.

In der Korrespondenz sind sowohl Indikativ als auch Imperativ und Konjunktiv zu treffen. Imperativ steht gewöhnlich am Ende des Briefes, als Bitte etwas zu erledigen oder als die Fortsetzung der Kommunikation und ist in der Höflichkeitsform gebraucht. Konjunktiv ist als eine außerdem höfliche, ein bisschen unmutige Form einer Bitte und ersetzt in vielen Fällen den Imperativ.

- *MODALVERBEN* DIENEN DEM AUSDRUCK VON MÖGLICHKEIT, NOTWENDIGKEIT UND AUFFORDERUNG; HÄUFIG SIND *KÖNNEN*, *SOLLEN* UND *MÜSSEN* ZUR SACHBEZOGENEN DARSTELLUNG; *DÜRFEN* STEHT HÄUFIG MIT EINER NEGATION, Z.B.:

Zu dem Gespräch mit Debis kann ich leider nicht kommen.

Im wesentlichen möchte ich mit Ihnen die aktuelle Markt- und Absatzentwicklung in Litauen besprechen.

Leider müssen wir Ihnen mitteilen, dass wir keine Möglichkeit sehen, Ihnen die Fahrzeugstudie SLR zur Verfügung zu stellen.

Kokioje situacijoje galima atsidurti, jeigu pagrindinių gamintojų modeliai bus senesni? Mūsų nuomone, vertinant galima įvesti tam tikrus koeficientus dėl šio reikalavimo, bet nelogiška atsiriboti nuo kokybiškų modelių. Automobilis turi atitikti standartų reikalavimus. Jūsų reikalavimai negali prieštarauti standartams.

Зная о Вашем предприятии и о выпускаемой вами высококачественной продукции, наше предприятие хотело бы сотрудничить с вами.

По второму вопросу можем сказать, что соотношение цен с другими популя-рными торговыми марками...

- *Modale Infinitivkonstruktionen* dienen dem Ausdruck der Notwendigkeit; häufig ist die Konstruktion mit ‚sein+zu+Inf.' zur sachbezogenen und agensabgewandten Darstellung, z.B.:

Gern bestätige ich Ihnen, dass wir damit einverstanden sind, die in Ihrem Brief erwähnten Artikel in einer Publikation in litauischer Sprache herauszugeben.

Es ist mir deshalb eine besondere Ehre, Sie zu einem Vortrag von Herrn Professor C.K. von der Universität P einzuladen.

Reikalingi apyvartų skaičiai kiekvienos atskiros prekės.

С радостью отвечаем на Ваше письмо и скажем прямо нам очень было приятно услышать от Вас, что...

Modale Infinitivkonstruktionen sind mehr typisch für die deutsche Sprache. Für die litauischen und russischen Sprachen ist mehr die elliptische Form bezeichnend.

- Das *Passiv* wird häufig zur sachbezogenen, unpersönlichen Darstellung gebraucht; das Vorgangspassiv ist dominant; es kennzeichnet Thema-Rhema-Abfolgen, z.B.:

In den anderen Fragen verweisen Sie bitte darauf, dass in der Sitzung des Aufsichtsrates im Herbst die angesprochene Thematik ausführlich mit dem Aufsichtsrat diskutiert wird.

Der Vortrag wird in deutscher Sprache gehalten.

Kokie produktai ir nuo kada parduodami Lietuvoje? Trumpai praneškite, kokiais vienetais buvo daroma Jūsų paroda.

Продукция атестованна в Германии ... Мы должны известить Вас, что с фирмой „Тп" в 1997 г. был подписан договор...

Das Passiv wird in der Korrespondenz sowohl der deutschen Sprache als auch in der litauischen und russischen gebraucht, allerdings wird im Litauischen und Russischen mehr zum Aktiv tendiert.

- *Funktionsverbgefüge (FVG)* zur sachbezogenen, agensabgewandten und sprachökonomischen Darstellung treten häufig auf mit Funktionsverben, die durch ‚machen' und ‚geben' ersetzt werden können; frequent sind FVG mit Linkserweiterung; in Texten der Dokumentation werden Funktionsverben häufig eliminiert.

Funktionsverben sind mehr für das Deutsche und das Russische typisch. Im Litauischen gilt der Gebrauch von Funktionsverben als Stilfehler. Trotzdem ist diese Art von Verben nicht so oft in meinen gesammelten Geschäftsbriefen zu treffen, z.B. statt *Erlaubnis geben* steht das einfache Vollverb *erlauben*, oder *Mitteilung machen – mitteilen*. Im Russischen: *сделоть предложение – предложить, сделоть заевление – заевить.*

- *Nominalisierungen* treten häufig zur sprachökonomischen, präsizen und sachbezogenen Darstellung auf, z.B.:

Auslöser dafür war die Eingliederung zusätzlicher außereuropäischer Regionen in die Verantwortung dieses Bereiches, welche wir zum Anlass einer grundsätzlichen Aufteilung der Vertriebszuständigkeiten in Mittel- und Osteuropa in zwei Abteilungen R/1 und R/2 genommen haben.

Nominalisierung ist typisch für die Korrespondenz der deutschen Sprache. Im Russischen und Litauischen finden wir eine gleichmäßige Aufteilung von verbalen und nominalen Stilen.

- *PRONOMINA* STEHEN SELTEN; STATT DESSEN STEHEN NOMINALE FÜGUNGEN, Z.B.:

Sehr geehrter Herr W. Vielen Dank für Ihr Schreiben vom 21. August 2000, dessen Beantwortung ich vornehme. Mit Schreiben vom 23. Dezember 1999, dem Sie nie widersprochen haben, teilten wir mit...

Nuo to laiko iki š.d. iš AB Ž buvo išsiusta 10 laiškų ir faksų. Deja atsakymo mes taip ir negavome.

Mes pasiruošę suteikti Jums ekspedijavimo paslaugą...

Господа, мы передали комерческое предложение на продукцию «С» и хотим дальше ...

Die aufgeführten Beispiele streiten die These von Ohnacker ab. In allen drei Sprachen dominieren in der Korrespondenz Pronomina – und zwar entweder Sg. oder Pl. 1. Person. Unter der Form der 1. Person Pl. ist die Firma zu verstehen, und wenn die 1. Person Sg. steht, ist die Korrespondenz entsprechend von einem zuständigen Mitarbeiter geführt.

- *ADJEKTIVE* STEHEN MEIST ATTRIBUTIV; SIE DIENEN DER PRÄZISIERUNG UND DIFFERENZIERUNG, Z.B.:

... herkömmliche Technologie, rein amerikanisches Produkt, ausfallendes Modell.

... werkseitige, bauseitige Versiegelungen, eventuelle Rückfragen, sensationeller Einkaufspreis.

... lengvasis automobilis, konkurencinės sąlygos, automatinė pavarų dėžė.

... garbingas ir patikimas partneris (Ž).

... высококачественная, конкурентоспособная продукция, генеральный дилер, ежегодные выставки, белорусские холодильники.

Adjektive sind in der Sprache der Korrespondenz sachlich-logisch, nicht so häufig, nur zur wichtigsten Information gebraucht.

Die Thesen von Ohnacker behandeln nur die typischen Merkmale auf der Ebene der Syntax und Morphologie. Ziemlich zweifelhaft erscheint die Behauptung von v. Hahn, „dass die Fachsprachen keine exklusive Syntax besitzen, sondern sie eine spezifische Auswahl aus den gemeinsprachlichen syntaktischen Mitteln treffen" (1983:111). Er führt drei typische Merkmale der fachsprachlichen Syntax von Beneš (vgl. ebda. 1983:113ff.): *Anonymisierung,* die mit den Thesen von Ohnacker vergleichen kann, wo er Pronomina und Passiv behandelt. *Explizite Spezifierung*, die auch wie in den Thesen von Ohnacker Attribute und Relativsätze, Adverbialbestimmungen, aber auch die Präpositionen, die häufig durch Modalpartikel ausgedrückt werden, wie *normalerweise* meiner Ansicht nach *mes galvojame, paprastai; скажем прямо, мы надеемся.* Zu den Spezifierungsmethoden gehören daneben auch die Abtönungspartikel, wie *einigermaßen; deja, apgailestaujame; с радостью.* Das dritte Merkmal von Beneš ist die *Kondenzierung*, die den Komplex von Redundanzarmut, Textsortenspezifik und Schriftlichkeit umreißt. Dieser Komplex schließt die Ausdrucksökonomie und Exaktheit, Nominalisierung, Wortbildungstechniken und nichtsprachliche Zeichen ein.

Die anderen Ebenen, die von Ohnacker außer Acht gelassen werden, finden ihren Platz bei v. Hahn 1983, wie die Lexikologie, die sehr oft als eines der wesentlichen Mittel von Fachsprachen angesehen wird. Die Subsprachen in der Wirtschaft haben auch feste Terminologie und die gebrauchten Wortschatzeinheiten bilden terminologisches Feld (vgl. 1983:87).

v. Hahn laboriert auch an dem Textaufbau in der Fachkommunikation, wo er die fachliche Textstruktur hervorhebt. Die Zielorientiertheit von fachlichen Texten führt meist dazu, dass sie explizit kohärent formuliert werden und gewöhnlich folgende Hauptrelationen zwischen Textelementen in der starken referentiellen Bindung der Kohärenz anzuführen sind: Wirkung, Grund, Schlussfolgerungen, Ausführung (Erläuterung), Gegensatz, Parallelismus, Beispiel, Zeitfolge.

Für den Textaufbau sind auch die Texteinbettung und die Metainformation, sowie Text-Bild-Verhältnis von großer Bedeutung (vgl. ebda. 119ff.).

5. Schlussfolgerungen

Die Fachsprachen bilden in der Forschung moderner Pragmatik einen wichtigen Schwerpunkt, denn die mündliche bzw schriftliche Kommunikation basiert oft auf Fachbereiche, z.B. Arzt – Patient – Kommunikation, Sportler – Trainer-Kommunikation, Verkäufer – Kunde. Ein Defizit der bisherigen Untersuchungen ist die Vernachlässigung der schriftlichen Kommunikation, wo die wichtigsten Merkmale die Fachsprachen am besten zu sehen sind.

Umfassende linguistische Aussagen über die sprachlichen Merkmale der Fachsprache Wirtschaft sind auf der Basis der vorliegenden Untersuchungen nicht unproblematisch. Die syntaktischen Merkmale der Fachsprache Wirtschaft zielen vor allem auf eine sprachökonomische und inhaltlich verdichtete Ausdrucksweise. Bevorzugt werden sprachliche Mittel, die eine agensabgewandte, unpersönliche und sachliche Darstellung ermöglichen. Wortschatz hängt weitgehend von bestimmten Wirtschaftszweigen (Industrie, Handel, Dienstleistungen) oder Branchen (Chemie, Textil, Banken, Versicherungen etc.) ab.

Dieser kleine spezifische Fachsprachen-Exkurs soll die Angewandtheit von schon vorliegenden Modellen beweisen und sie in Zukunft erweitern.

Literatur

ARNTZ, REINER/PICHT, HERIBERT (1989): Einführung in die Terminologiearbeit. Hildesheim u.a.: Georg Olms.

BAASNER, RAINER (1999) (Hrsg.): Briefkultur im 19.Jahrhundert. Tübingen: Niemeyer.

BAUMANN, KLAUS-DIETER (1992): Integrative Fachtextlinguistik. Tübingen: Gunter Narr.

BAUMANN, KLAUS-DIETER, HARTWEG KALVERKÄMPER (Hrsg.) (1992): Kontrastive Fachsprachenforschung (FFF) Tübingen: Gunter Narr.

BAUMANN, KlAus-DIETER (1994): Fachlichkeit von Texten. Egelsbach, Frankfurt a.M., Washington: Hänsel-Hohenhausen (Deutsche Hochschulschriften).

BAUMANN, KlAus-DIETER (2000): Unternehmenskommunikation und Unternehmensidentität aus kommunikativ-kognitiver Sicht. In: K. MORGENROTH (HRSG.): Hermeneutik und Manipulation in den Fachsprachen. Tübingen: Narr, S.107-126.

BÖHME, GÜNTER VON (1999):Zur Vermittlung von Deutsch als Wirtschaftssprache – dargestellt am Beispiel von Textsorten der Geschäftssprache. In: BUNGARTEN, THEO (HRSG.): Wirtschaftshandeln. Kommunikation in Management, Marketing und Ausbildung. Tostedt: Attikon, s. 235-248.

BOLTEN, JÜRGEN (1991): Fremdsprache Wirtschaftsdeutsch: Bestandsaufnahme und Perspektiven. In: MÜLLER, BERND (HRSG.): Interkulturelle Wirtschaftskommunikation. München: Iudicium Verlag, S.71-91.

BRÜNNER, GISELA (2000): Wirtschaftskommunikation. Linguistische Analyse ihrer mündlichen Formen. Tübingen: Niemeyer.

BUHLMANN, ROSEMARIE (1989): ‚Fachsprache Wirtschaft'- gibt es die? In: Jahrbuch Deutsch als Fachsprache 15. Heidelberg: Groos; München: Hueber; München: Iudicium, S.82-108.

BUHLMANN, ROSEMARIE/FEARNS, ANNELIESE (1987): Handbuch des Fachsprachenunterrichts. Unter besonderer Berücksichtigung naturwissenschaftlich-technischer Fachsprachen. Berlin u.a.: Langenscheidt.

CHREPTAVIČIENĖ, VIRGINIJA (1997): Dalykinės komunikacijos raštų mokėjimų ugdymas administravimo universitetinėse studijose. Daktaro disertacija. Kaunas: Kauno Technologijos Universitetas.

HAHN, WALTER VON (1983): Fachkommunikation. Entwicklung, linguistische Konzepte, betriebliche Beispiele. Berlin u.a.: Walter de Gruyter.

HEUBERGER, KATHARINA (1997): Wirtschaftsdeutsch und seine Vermittlung. Tostedt: Attikon.

HOFFMANN, LOTHAR (1987): Kommunikationsmittel Fachsprache. Eine Einführung. 2. völl. Neu bearb. Aufl. Tübingen: Gunter Narr.

OHNACKER, KLAUS (1992): Die Syntax der Fachsprache Wirtschaft im Unterricht Deutsch als Fremdsprache. Frankfurt a.M. u.a.: Peter Lang.

SCHRÖDER, HARTMUT (1983): Zur Arbeit mit sozialwissenschaftlichen Fachtexten in Sprachlehrerveranstaltungen Deutsch als Fremdsprache. In: Jahrbuch Deutsch als Fremdsprache 12. Heidelberg: Groos; München: Hueber; München: Iudicium, S.252-263.

Gerd-Bodo von Carlsburg

Zum Nachdenken anregen – Beispiel Schule

Schule ist nicht nur Lern-, sondern insbesondere auch Lebensort. Das haben Reinert/Heyder insbesondere mit ihrem Band „Lebensort: Schule" (1983), einem Dokumentarbericht mit Reportagen, Ratschlägen und Kommentaren aus der Alltagswelt der LehrerInnen und SchülerInnen, aus dem Schulbetrieb (Reinert/Zinnecker 1978), aus der ‚Stressfabrik Schule', sehr anschaulich bekundet. Die Auseinandersetzung mit der Praxis ist nicht einfach, wenn man vom Universitätsstudium voller Enthusiasmus, vollgepfropft mit theoretischem Wissensstoff den Alltag neu entdecken möchte, weil man illusionistisch ja alles besser machen oder das ‚Gute', die Ziele einer von Humanismus geprägten Schule seit Philipp Melanchthon (1497-1560) und Jan Amos Comenius (1592-1670), vertiefen möchte.

Und aller Anfang ist schwer. Das Nachsinnen, im wahrsten Sinne eines Gebrauchs der Sinne, einer ‚sinnorientierten Schule', die Einstellungen zu Schule und Unterricht, die Aktivitäten... weichen mit zunehmenden Alter einer von Enttäuschungen, von Passivität geprägten Lehrerpersönlichkeit, die burnout-geschädigt ihr Arbeitsfeld verlässt und zunehmend dem Charakterbild einer ‚déformation professionnelle' entspricht. Und immer wieder fragen wir uns deprimiert: Wie entsteht nur dieser Circulus vitiosus des Schulalltaglebens, der Lernwelt, der uns am Ende resignieren lässt? Er ist anscheinend schon zur Conditio sine qua non ‚verkommen'.

Was machen wir anders, besser, um diesem Debakel zu entgehen? Wir, als Lehrende, müssen schon bei/in uns beginnen, uns selbst und unsere SchülerInnen bindungspädagogisch stärken, psychohygienisch festigen und empathischen Streit im Miteinander und in der Sache ertragen. Reinert/Wehr (2001, S. 101f.) reflektieren über das Wohlbefinden in Bezug auf den Lebensort Schule:

> „Sich als Lehrer in einem Klima des Wohlbefindens zu wähnen, bedeutet Kompetenzen zu haben und sich gleichzeitig einen Lernprozess zuzugestehen, in dem man Fähigkeiten erwirbt und erweitert. Dies lässt sich in folgenden fünf ***Bereichen*** realisieren:
>
> 1. Selbstkonzept: persönlichkeitsfördernde Einstellungen und Verhaltensweisen;
> 2. Kommunikation: beziehungsfördernde Verhaltensweisen;
> 3. Kooperation: beziehungsfördernde Arbeitsweisen;
> 4. Fachkompetenz: fachliche und didaktische Fähigkeiten;
> 5. Realitätssinn: realistische Sicht- und Handlungsweisen.
>
> Die Konsequenz des Kompetenzzuwachses bedeutet für die Lehrer-Schüler-Beziehung das Folgen von *personenzentrierten* Impulsen (vgl. Rogers 1985). Verwirklicht werden diese unter

Bezugnahme auf Millers Graphik (vgl. 1998, S. 59) in der Person des Lehrers durch eine *ganzheitliche* Faktorenkomplexion, die in Form eines ***Kongruenz-Dreiecks*** verdeutlicht werden soll:

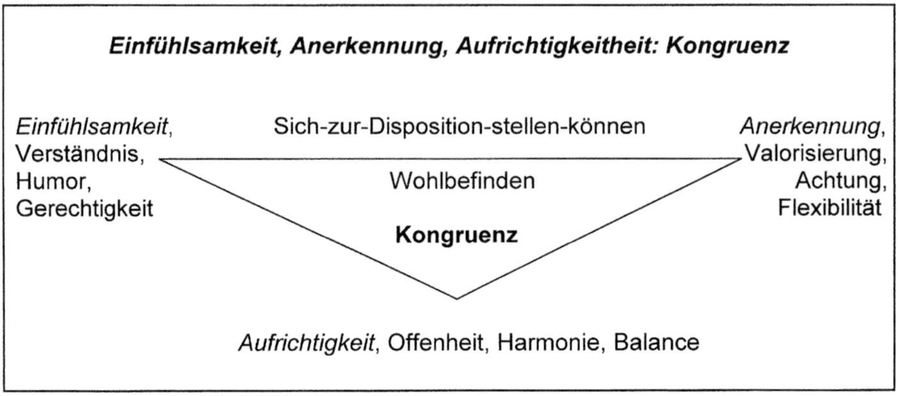

Kongruenz-Dreieck der Lehrerpersönlichkeit“ [i. Orig. kursiv] (Graphik: Reinert von Carlsburg) (S. 101)

Johann Friedrich Herbart (1776-1841)

aus: Herbart 1968, zwischen S. 4/5

„Gleichzeitig wird der Ausbau der transparenten Leistungssteigerung ins Auge gefasst, indem Regeln, Räume, Rituale für Kinder und Jugendliche geschaffen werden, die in einem gemeinsamen Konsens, ein pädagogisches Schulklima gründen. Für Lehrer ist als psychohygienische Basis ein Ausbau der pädagogisch-psychologischen Beratung und Supervision unverzichtbar, denn mit der administrativen Routine, den Lehrplanzwängen und der wenig bildungsfördernden gesellschaftlichen Situation greift letztlich eine Berufsunzufriedenheit (Burnout) um sich und führt somit zu gewalteskalierenden Verhaltensweisen von Lehrkräften, z.B. zu einer ‚gewalttätigen Sprache' (Staudt 1994, S. 17ff.), die als Konsequenz aus Schülerperspektive Gegen-Gewalt und Vandalismus evozieren (vgl. Kaiser 1993; Bade 1994).“ (S. 102)

Im Fokus steht demnach nicht die Schule als Institution, sondern das unterrichtliche und außerunterrichtliche Schulleben, das

Schule gestaltet und lebenswert macht, der „pädagogische Takt“ (Johann Friedrich Herbart [1776-1841]), das gute psychische Milieu, das auch den von SchülerInnen immer wieder bemängelten Humor im Umgang bemängelt, Gerechtigkeit, Autorität, Flexibilität im Umgang und in der Planung, Geduld, Zuneigung, verbunden mit akzeptablem situativen Entscheiden. Gegenseitige Autorität ist geprägt von einer Kongruenz, die gegenseitige *Offenheit* bedingt, Ehrlichkeit sowie das Vermögen, sich dem „Du“, dem Interaktionspartner zur Disposition stellen zu können. Dieses personale Miteinander prägt den Umgang mit und von Schule, das Klima zwischen- und mitmenschlich, die Rahmenbedingungen des Lernens auf der Vorder- und Hinterbühne von Schule (vgl. Reinert/Zinnecker 1978), ist ein Stück Leben, das den Lernalltag zum motivierenden Alltag herbeiführt und SchülerInnen nicht nur das Gefühl der Mitgestaltung an Lernprozessen gibt, sondern realiter Schüler zu selbsttätigem und explorierendem Lernen ermutigt (v. Carlsburg/Heyder 2005, S. 153-170).

Eine lebensfreundlichere Umgestaltung beginnt nicht mit der Modernisierung von Räumen und Inventar, sondern mit der Intensivierung der Bindungen, die Kongruenz, Interaktion und Verhaltensrepertoires beinhalten, wie Reinert/Wehr in ihrer Graphik aufweisen, wobei sie folgende Voraussetzungen postulieren:

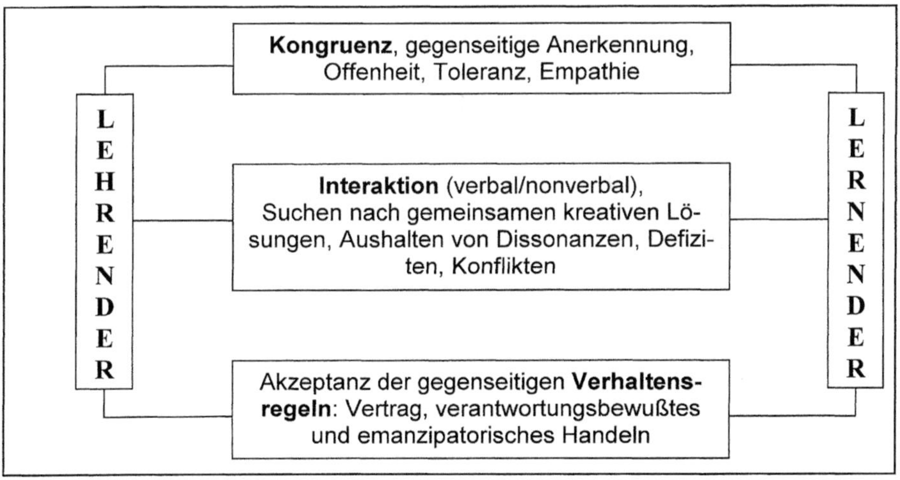

Lösungen zur Konfliktbewältigung: Kongruenz – Interaktion – Verhaltensregeln (Reinert/Wehr 2001, S. 120, Graphik: Reinert von Carlsburg)

- „Fähigkeit des *aktiven Zuhörens* (d.h. Schüler sprechen lassen und nicht unterbrechen, höchstens zustimmende Bemerkungen), um die Schüler zu ermutigen, über ihre momentanen Bedürfnisse zu sprechen. Die Schüler müssen verstehen, dass sie akzeptiert und verstanden werden.
- Lehrer müssen durch *Ich-Botschaften* ihre eigenen Bedürfnisse offen darlegen, damit der Schüler weiß, was den Lehrer belastet. Der Lehrer sollte den Schüler nicht tadeln oder erniedrigen.

- Der Lehrer sollte dem Schüler die drei Konflikt-Löse-Methoden erklären (vgl. Gordon 1977), wobei Anklänge an die Mediation deutlich werden." (S. 121)

These 1:

Schule sieht sich nicht allein in der Funktion als reine Lehranstalt, sondern auch in der Erziehungsrolle einer Unterweisung für das Leben, einer Anleitung für die Internalisierung der Normen und Werte der Gesellschaft.

Für die Erziehung der Kinder ist nicht ausschließlich die Familie zuständig. Schule muss hier Hebammenfunktion/Unterstützung leisten, aber sie ist wiederum nicht der Erzieher unserer Nation.

Lucius Annaeus Seneca (um 4 v.Chr. in Corduba/Spanien geboren – 65 n. Chr. in der Nähe Roms gestorben), Staatsmann (Konsul), Erzieher von Kaiser Nero, Dichter, vor allem Philosoph, postuliert in einem seiner Hauptwerke „Ad Lucilium epistulae morales" (Moralische Briefe an Lucilius), bestehend aus 124 Lehrbriefen, die er zwischen 62-64 n.Chr. verfasst hatte, im 106. Brief (64 n.Chr.):

„NON VITAE; SED SCHOLAE DISCIMUS."
(Für die Schule, nicht für das Leben lernen wir.)

Sokrates
(um 470 v.Chr.-399 v.Chr.)
aus: Lange/Alexander 1982, S. 854

Odenbach (1974, S. 15) beginnt seinen Beitrag „Das Schulleben" mit besagtem Satz Senecas. Auch A. Flitner (1987, S. 7) bezieht sich in seinem Vorwort auf selbigen. Es ist eine Kritik, eine Anklage, der Schule, an deren Lehrern, die nicht auf das Leben vorbereiten, ihr „Ego" nur in ihren Schülern widerspiegeln wollen, nicht wie Sokrates (um 470 v.Chr.-399 v. Chr.) das heuristische Verfahren, das fragend-entwickelnde Gespräch suchten, keine Mahnung an ein mögliches Schulversagen der Schüler seiner Zeit. Allerdings beanstandet Odenbach Senecas Aussage, nimmt sie nicht als Fingerzeig auf Missstände der Polis und ihres Bildungssystems wahr, sieht sie als nicht einsichtige „klare Trennung von Schule und Leben" (S. 15).

„Nun war sich SENECA sicher dessen bewusst, dass auch im Leben gelernt wird – vielleicht auf andere Weise, aber nicht weniger wirkungsvoll; es erhebt sich nur die Frage, ob er auch gesehen hat, dass die Schule Leben ist ...“ (ebda.)

‚Leben' im weitesten Sinne muss beinhalten:

- gegenseitiges Kommunizieren lernen und anwenden;
- Interagieren mit allen Hochstimmungen und Traurigkeiten;
- das Erlernen von selbstbestimmten Lernen im Schulbetrieb;
- das im Schulbetrieb Erlernte, die Schulung des logischen Denkens und verstandesbewussten Handelns, in den außerschulischen Alltag transferieren;
- die Lernwelt der Schule als Lebenswelt der Zukunft begreifen.

Das ist das Leben!

Johann Gottfried von Herder (1744-1803) hielt 3 Jahre vor seinem Ableben seine Schulrede unter der Thematik:

„VITAE; NON SCHOLAE DISCENDUM."
(Für das Leben, nicht für die Schule lernen wir.)

Johann Gottfried von Herder (1744-1803)
aus: Reinert u.a. 1996, S. 25

Will heißen: für die Herzens- und Charakterbildung; für die Abwägung zwischen Seele und Verstand muss Schule prägend wirken, nicht nur der Erwerb von kognitiven Lern-Dimensionen ist Bildungsziel, sondern in erster Linie die Bewusstbarmachung und das Heranwachsen von emotionalen und sozialen Fähigkeiten (vgl. v. Herder 1967, S. 231).

„Was heißt lernen? Man hat davon falsche Begriffe, wenn man glaubt, es heiße: fremde Worte sich einprägen. Worte sind Schälle; ohne Gedanken drücken sie sich zuweilen zumal in der Jugend mit großer Kraft ein; ohne Gedanken hat man sie aber nur als Papagei gelernt: denn bekanntermaßen lernt auch der Rabe, der Papagei Wortschälle und sagt sie zu rechter oder unrechter Zeit wieder.

Worte ohne Gedanken lernen (sic!) ist der menschlichen Seele ein schädliches Opium, das zwar zuerst einen süßen Traum, einen Tanz von Silben und Bildern gewährt, vor dem man sich als vor einer Zauberansicht halbwachend und halbschlummernd fühlt; bald aber spürt man, wie bei dem körperlichen Opium die bösen Folgen dieser Wortträume." (S. 223)

„Endlich (sic!) da das Leben nicht nur Kenntnisse und Gedanken, sondern auch Willen, Triebe, Taten braucht, und in diesen vor allem das Leben besteht, so wendet sich der Spruch ‚nicht der Schule, sondern dem Leben zu lernen,' vorzüglich auf Bildung des Herzens und des Charakters. Was helfe es, tausend Kenntnisse und keinen Willen, keinen Geschmack, keine Lust und trieb zu leben, honett und rechtschaffen zu leben, haben? Im Willen leben wir; das Herz muß uns verdammen oder trösten, stärken oder niederschlagen, lohnen oder strafen; nicht auf Kenntnisse allein, sondern auf Charakter und Triebe, auf die menschliche Brust ist die Wirksamkeit und der Wert, das Glück oder Unglück unseres Lebens gebaut. Leben lernen heißt also seinen Neigungen eine gute Richtung geben, seine Grundsätze reinigen, befestigen, stärken, seine Vorsätze läutern und tapfer begründen, nicht mit dem Kopf allein, sondern auch mit dem Herzen existieren gegen Eltern, Freunde, Lehrer, Mitschüler, Bekannte, Fremde; sich Sitten erwerben, anständige, frohe Sitten, liebenswert machend vor Gott und den Menschen." (S.231)

Georg Kerschensteiner (1854-1932)
aus: Arnhardt/Reinert 2000, S. 195

Auch Georg Kerschensteiner (1854-1932) bemängelt in seiner Festrede am 12. Januar 1908, anlässlich des 162. Geburtstags von Johann Heinrich Pestalozzi (1746-1827), in der Zürcher Peterskirche „Die Schule der Zukunft eine Arbeitsschule" wiederum die Unfähigkeit der Schule zur emotionalen, sozialen Erziehung.

„Die Buchschule des Mittelalters wurde zur Buchschule der Neuzeit. Aber inzwischen ist die Seele des Kindes immer mehr Gegenstand liebevoller Forschung geworden. Inzwischen haben wir immer deutlicher und allgemeiner erkannt, dass die Methoden der passiven Anschauung nicht entfernt dem Seelenleben des Kindes gerecht werden. Inzwischen sind wir uns auch in weiten Kreisen lebendiger als jemals bewusst geworden, dass bauende, schaffende Kräfte von einer bisweilen staunenswerten Größe im Kinde stecken. Das echt Pestalozzische Prinzip der Selbsttätigkeit wurde zum Schlagwort. Der stolze Name der „Erziehungsschule" wurde erfunden, und jene, die sie nach Zillers [Tuiskon Ziller (1817-1882), Herbartianer; G.-B. v.C.] Theorie vom Gesinnungsunterricht zu gestalten vermeinten, nannten sich selbst ‚die wissenschaftlichen Pädagogen'. Und doch, wenn sie die Räume betreten, wo sie arbeiten, die Pädagogen der Selbsttätigkeit, die Pädagogen des Gesinnungsunterrichtes, so finden sie im wesentlichen den gleichen betrieb wie bei den Pädagogen der passiven Anschauung. Sie finden überall die alte Lern- oder, wie man sie vielleicht besser nennen könnte, die alte Buchschule." (Kerschensteiner 1967, S. 397f.)

Was aber unsere heutige Schule dem Kinde, wenn es ins Leben hinübertritt, nicht mitgibt, was sie weit mehr verkümmern lässt als fördert, das sind gewisse aktive Charakterzüge, die

die meisten Kinder im Keime schon hatten, als sie in die Schule eintraten, den Mut der Selbständigeit, den Mut der Selbstbehauptung und der Unternehmungslust, den Mut, Neues und Ungewohntes anzugreifen, die Lust zu beobachten und zu prüfen, vor allem aber nicht nur um seiner selbst willen zu arbeiten, nicht bloß um selbst zu wachsen, um die anderen zu überflügeln, um Sieger zu werden im wilden Kampf des Lebens, sondern auch um die eigenen, reichen Kräfte hilfsbereit allen zur Verfügung stellen zu können, allen, die ihrer bedürfen."

Und so erhebt sich die Frage: Ist es nicht möglich, unsere heutige Schule so umzugestalten, dass sie ihre guten Eigenschaften nicht verliert, gleichwohl aber dem Wesen des Kindes mehr gerecht wird, und dass sie in ihm auch jene aktive Seelenverfassung entwickelt, die sie heute vernachlässigt, ja verkümmern lässt?" (S.403)

„Aus unserer Buchschule muß eine Arbeitsschule werden, die sich an die Spielschule der ersten Kindheit anschließt." (S.404)

„So wollen wir in froher Zuversicht der Zeit entgegenblicken, die uns die neue Schule bringt, die Zeit, deren Morgendämmerung in allen Kulturstaaten bereits unverkennbar ist. Wenn diese Zeit gekommen sein wird, dann werden die Ideen Pestalozzis in die Tat umgesetzt sein. Dann wird wirklich ‚die Anschauung zum Fundament aller Erkenntnis' geworden sein, dann wird wirklich ‚der Grundsatz alles Unterrichtes von der unwandelbaren Urform der menschlichen Geistesentwicklung entnommen' sein (6. Brief an Geßner, Ziff. 34), dann wird der ‚Unterricht zu jener Kunst geworden sein, die lediglich dem Haschen der Natur nach ihrer eigenen Entwicklung Handbietung leistet' (1. Brief an Geßner, Ziff. 34), dann wird ‚Natur und Kunst im Volksunterricht so innig vereinigt sein, als sie jetzt gewaltsam voneinander getrennt sind' (1. Brief an Geßner, Ziff. 39)." (S. 415f.)

Unter Berücksichtigung der historischen Postulate stellt sich für die Schule der Zukunft folgende Aufgaben- und Fragestellung:

These 2:

Schule hat primär die Aufgabe, die Jugend auf veränderte gesellschaftliche Anforderungen vorzubereiten und sich selbst als Institution für Veränderungen zu wappnen, dazu gehört auch die reflektierende Vorbereitung des Lehrers auf den interaktiven Umgang mit dem Schüler im Hinblick auf einen angestrebten Erziehungswandel und eine ‚Atmosphäre gegenseitigen Vertrauens', um erziehend zu unterrichten (Johann Friedrich Herbart [1776-1841]).

Zielstellung:

Bildungsprozesse sind viel stärker als Selbstbildungsprozesse zu entwerfen.

Frage:

Welche Ansprüche ergeben sich an den Lehrer im Hinblick auf eine „edle Form" (Georg Wilhelm Friedrich Hegel [1770-1831] in seiner Schulrede von 1809) des Lehrer-Schüler-Verhältnisses im Hinblick auf den sog. ‚guten' Lehrer?

Kerschensteiner hat diese Zürcher Rede mit eigentlichem Thema „*Die Schule der Zukunft im Geiste Pestalozzis*“ betitelt „und nannte sie eine Arbeitsschule.“ (1930a, III) Genau dieser Anstoß war der Beweggrund, sein Werk „Begriff der Arbeitsschule“ 1911 zu publizieren.

Wie definiert er nun seinen „pädagogischen Begriff der Arbeit“?

> „Das erste, was wir zu beachten haben, ist: Eine manuelle Betätigung mag mit noch so viel Interesse, Eifer, Anstrengung, Übung verbunden sein, Arbeit im pädagogischen Sinn kann sie erst werden, wenn sie Ausfluß einer geistigen Vorarbeit ist, die schon in dieser Vorarbeit zu einem ersten Abschluß kommt, im Fortgang der Ausführung aber immer von neuem aufgegriffen wird und zu neuen Denkprozessen Veranlassung gibt. Rein mechanische, isoliert vom übrigen geistigen Leben ablaufende Arbeit, kann nicht arbeit im pädagogischen Sinn sein.“ (ebda., S. 43f.)

Mit Pestalozzi beginnt sein Vorwort und mit ihm schließt er sein Werk. Dieser Volkserzieher war Kerschensteiners großes pädagogisches Idealbild.

Johann Heinrich Pestalozzi (1746-1827)
aus: Archiv v. Carlsburg

„Vor allem habe ich das Gefühl, dass wir in unseren Bestrebungen ganz im Geiste dessen handeln, der so viel gepriesen und so wenig verstanden wird, der uns in Lienhard und Gertrud, in den Briefen an Heinrich Geßner und besonders im Schwanengesang so oft gelehrt hat, dass nur ‚die Arbeit in der das Kind umgebenden Welt' der elementaren Volksschule ihre Bildungskraft gibt. Ja, ich wage es zu sagen: Ohne dass diese Worte sich genau so in seinen Werken finden, war Pestalozzi der felsenfesten Überzeugung, dass die Berufsbildung die Pforte der Menschenbildung ist.

Drei Generationen sind seit seinem Tode dahingegangen. (...) Der Flugsand der Gedankenlosigkeit hat Berge über Wahrheiten geschüttet, die einst das Herz des unermüdlichen Forschers nach Menschenbildung erfüllten. Aber wirkliche Wahrheiten steigen immer wieder wie Geister aus ihren Grüften auf und wandern umher und beunruhigen die Herzen der Menschen, bis sie endlich Erlösung und Ruhe finden in der Verwirklichung des realen Lebens. Wir alle, die wir mit wissenschaftlichem Ernst und hingebender Energie einer Schulorganisation die Wege bahnen, die dem Vater der Volksschule vor Augen schwebte, bringen diesen Geistern Erlösung. Vielleicht wird dann das Ende des 20. Jahrhunderts das in Vollendung schauen und genießen, was er am Anfang des 19. Jahrhunderts in heißen Kämpfen und in bitterer Not seiner Erkenntnis abgerungen hat.“ (ebda., 167f.)

Die Reformpädagogen begriffen sich als *Gegenbewegung* zu der um die Jahrhundertwende ins 20. Jahrhundert verstandenen Auffassung einer strikten Trennung von Unterricht und Schulleben. Schulleben wird nicht begriffen als additiver Appendix, sondern Schule gilt als Ort der Umsetzung von im Unterricht erworbenem Wissen für Lebensfertigkeiten – kurz formuliert: erst lernen, dann mit dem Erlernten das Gemeinwohl und die Zukunft der Gesellschaft reflektierend, für den humanen Fortschritt *leben* – sowie als nächsten Schritt durch Bildung innerer Kräfte selbsttätig, emanzipiert zu handeln im Sinne einer Erziehung zur Selbsterziehung, eines Trainings von freien Entscheidungen. Nicht der Lehrstoff, der Unterrichtsgegenstand steht allein im Mittelpunkt des Interesses, sondern der Mensch, die SchülerInnen, die explorierend selbstbestimmt lernen sollen, wobei offene Unterrichtsformen bevorzugt werden. Alle SchülerInnen haben ein Recht auf Autonomie.

Hugo Gaudig (1860-1923)
aus: Archiv v. Carlsburg

Alle reformpädagogischen Bewegungen, die sich als Alternative zur ‚Schule der Herbartianer' gründen, verurteilen die strengen Regularien des Unterrichts sowie die Zurückstellung der mitmenschlichen Beziehungen. Im Zentrum des Bildungsgedankens steht das Kind, die Anpassung und Gestaltung des Unterrichts an die kindliche Entwicklung, die freie kindgemäße, natürliche Entfaltung (Bewegungsfreiheit, Spieltrieb, Schaffung von kreativen Freiräumen), die Zusammenarbeit von Schule und Elternhaus. So sieht Gaudig die Schule als „Lebenssphäre" der eigenständigen Persönlichkeitsentwicklung. SchülerInnen dürfen nicht ein „Objekt" pädagogischer Handlungsmodifikationen sein, sondern müssen stets lebendes „Subjekt" bleiben. (Gaudig 1923, S. 67) So setzt Gaudig „arbeiten" gleich „leben" im Sinne von Erleben des Bildungsguts, Zeitgeistes; dies alleine sei Motivation genug, Kraft und Antrieb zu entfalten, was eine „Selbstgesetzgebung" auf der Basis von Eigenständigkeit und eigenem Willen impliziert. (ebda., S. 68f.) Schulleben impliziert als „Gemeinschaftsleben" sowohl eine gleichstufige LehrerInnen-SchülerInnen- als auch SchülerInnen-SchülerInnen-Interaktion. (S. 96) Nur so kann sich eine feste Schulgemeinschaft gründen und fortpflanzen sowie Früchte tragen.

Und in diesem Ansatz einer Entwicklung zur Freiheit herrscht Übereinstimmung mit Kerschensteiner: die Befreiung des Kindes aus erzwungener Rezeptivität, dem Gerechtwerden der geistigen Produktivität. (vgl. S. 73)

Der Lehrer ist Partner und Freund des Kindes. Das setzt eine gegenseitige Akzeptanz und Interesse aneinander voraus, gegenseitiges Tolerieren und Einräumen von Spielraum zur Selbstentfaltung des Einzelnen. Der Aufbau eines „Wir-Bewusstseins", von Wertschätzung ist eine Conditio sine qua non für das Wachsen und Gedeihen eines spannungsfreien Klimas.

So fordert Kerschensteiner „Persönlichkeit" vom Lehrer (1930b, S. 117f.):

> „Von allen, die sich mit dieser Forderung beschäftigt haben, hat sie vielleicht Gaudig in seinem zweibändigen Werk ‚Die Schule im Dienste der Persönlichkeit' ... am eingehendsten behandelt. Die schöne Definition, die ergibt (Bd. I, S. 87) unterschreibe ich gern: ‚Persönlichkeit', sagt er, ‚ist der seiner selbst mächtige, die Kräfte seiner Natur zur Verwirklichung des Ideals seiner Individualität zusammenfassende, auf den Gebieten des Lebens sich frei aus sich heraus bestimmende Mensch.'"

Kompetenzen des ‚guten' Lehrers

Personal:

Flexibilität, Engagement, Bereitschaft zur Selbsterkenntnis und Selbstkritik, Autorität, Empathie (Einfühlungsvermögen)

Tugenden: Humor, Geduld, Gerechtigkeit, Aufrichtigkeit

Fachlich:

Engagement, Aufgeschlossenheit, Fortbildungswillen

Sozial:

Zuneigung, Achtung, Anerkennung, Valorisierung (Wertschätzung) seiner SchülerInnen, Verständnisbereitschaft, Rollen-/Perspektivenübernahme

Wenn, wie zu Beginn der 20er Jahre, Kontakte unter SchülerInnen nicht erwünscht waren, sich alle Aktivitäten unter strengen Regeln vollzogen, Abweichungen vom Kodex Bestrafungen auslösten, dann ist es nicht verwunderlich, dass die SchülerInnen nach Möglichkeiten des Abreagierens suchten, was wiederum strenge Tadel durch das Lehrpersonal zur Folge hatte. Und ein großer Teil der Zöglinge stand der ‚Institution Schule' feindselig gestimmt und gesinnt gegenüber. Stigmatisierungen von der ‚Paukanstalt' bis zur ‚Schule des Vergessens' gehörten bis in die sechziger Jahre zum Vokabular der SchülerInnen.

Bild des ‚guten' Lehrers

Schulfunktionen	Anforderungen an den Lehrer
Sozialisation	Normen verdeutlichen, Aushalten von Dissonanzen und Konflikten
Bildung	Kenntnisse und Einsichten anregen, Weltbegegnung herausfordern
Qualifikation	Wissen vermitteln durch Neugier und Offenheit, Explorieren
Selektion	Nach gerechten und fairen Leistungskriterien zuordnen
Erziehung	Verstehen, Takt, Humor, Wertschätzung, Rollen- und Perspektivenübernahme, Einfühlungsvermögen
Allokation	Fairness
Legitimation	Lehrertugenden, Autorität
Autonomie	Selbstkritik, verantwortungsbewusstes und mündiges Handeln

Das gegenseitige Bedingungsfeld zwischen Schule und Gesellschaft zeigt folgende Abbildung (Reinert v. Carlsburg/Wehr 2004, S. 182, Graphik: Reinert v. Carlsburg):

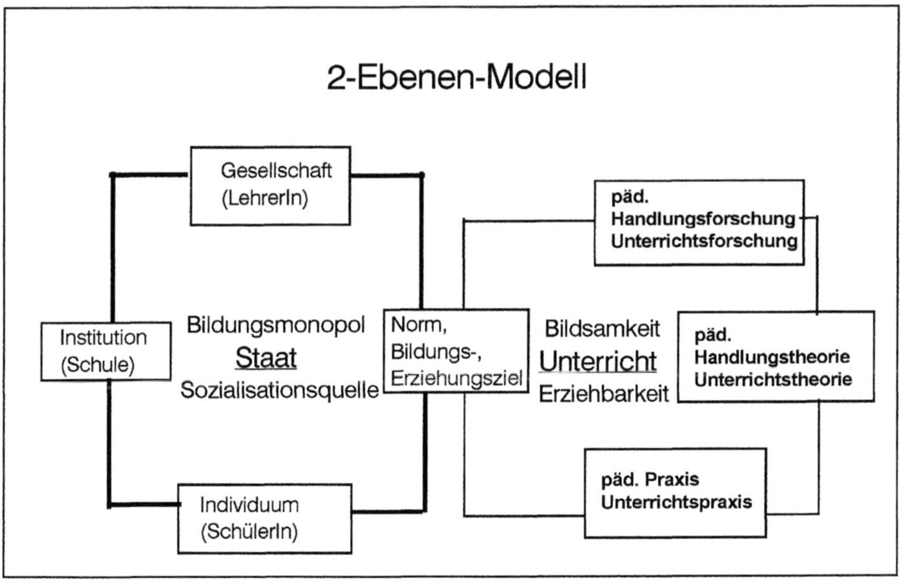

Die ***Funktionen von Schule*** im Hinblick auf ***Gesellschaft*** folgendermaßen darstellen (vgl. Klafki 1994, S. 15-138; Jank/Meyer 2002, insb. Graphik, S. 213):

Funktionen von Schule	→ **Gesellschaft**
Sozialisationsfunktion	Vermittlung von sozialem Verhalten, Sozialkompetenz, sozio-kulturelle Tradierungen (Reproduktion) u. Normen der Gesellschaft
Bildungsfunktion ***formale Bildungstheorie*** ↙ ↘ *funktionale* *methodische* *Bildung*	Entfaltung der geistigen (funktionale Bildung) und seelischen Kräfte/ Persönlichkeit (Begabungen), Bildungsinhalte, Entwicklung kritisch-konstruktiven Denkens, Selbstvertrauen u. personale Handlungskompetenz (methodische Bildung)
Qualifikationsfunktion ***materiale Bildungstheorie*** ↙ ↘ *Objektivismus (Wissen)* *Klassik*	Ausbildung, Aneignung von Wissen u. Kenntnissen, Fertigkeiten, gesellschaftliche, ökonomische Handlungskompetenz
Selektionsfunktion	Auslese nach Leistung, Begabung, Alter..., sozialen Chancen, berufliche Handlungskompetenz, Anforderungen der Berufs- u. Arbeitswelt
Erziehungsfunktion	Erziehung zu Charakterstärke, Toleranz, politische Verantwortung, Moral, Mitmenschlichkeit, Sittlichkeit
Allokationsfunktion	Zuweisung u. Steuerung von sozialen Positionen (s. Reproduktion/Selektion u. Legitimation) durch unterschiedliche Schultypen, Ausleseprüfungen, Leistungsfähigkeit, Internalisierung der Chancen- u. Begabungsungleichheit
Legitimationsfunktion	Stabilisierung der gesellschaftlich-demokratischen Grundordnung (Emile Durkheim – 1858-1917 [arbeitsteilige Gesellschaft, normativer Consensus], Max Weber – 1864-1920 [historische Legitimierung, Tradierungen, Stabilisierung der Gesellschaft]), rationaler Diskurs, Akzeptanz von Chancen- u. Angebotsunterschieden, Akzeptanz von ‚gerechter' sozialer Ordnung, ‚zufriedenes' Selbstbild (Degradierung)
Autonomiefunktion	Sicherung des eigenen ‚freien' Handlungsspielraumes mit eigenen Regeln, Schule als Hinterbühne, „heimlicher Lehrplan" (J. Zinnecker), eigenes Profil, Verwaltung, Finanzierung

Emile Durkheim deutete den Staat – und somit auch den Stellenwert der Bildung und Erziehung – als Produkt einer vom „Kollektivbewusstsein“ gesteuerten Gesellschaft, deren primäre Aufgabe es ist, die gesellschaftlich-demokratische Grundordnung zu stabilisieren. Max Webers Intention war ebenfalls die Stabilisierung der Gesellschaft unter der Prämisse eines von Tradierungen geprägten sozialen Handelns des von insbesondere religiösen Werten bestimmten Staates und des Individuums. Sein positivistischer Ansatz in der Soziologie war fast ausschließlich vom Handeln aus Vernunft geprägt, einer Verhaltensänderung unter den Menschen auf rationaler Basis.

Dies allein würde die Funktion von Schule in Bezug auf das Ganze reduzieren, nicht nur Legitimation, sondern gerade Schulautonomie sollte eine wichtige Schlüsselfunktion im Hinblick auf eine ‚gute' Schule sein, die aber den engagierten, gerechten, humorvollen, flexiblen und leistungs- bzw. einsatzfreudigen ‚guten' Lehrertypus impliziert, der bereit ist, auch außerunterrichtlich seine Schüler zu fordern und fördern, der innerunterrichtlich die Schüler durch schülerorientierten Unterricht, basierend auf einem Arbeitsstil des zwischenmenschlichen Vertrauens unter Schülern und zwischen Lehrern und Schülern zum Explorieren und Transferieren motiviert in einem Klima, das Offenheit und gegenseitiges Sich-zur-Disposition-stellen verträgt, biographisches Verstehen und Akzeptieren untereinander, mit einem Ethos, das ein freundliches Klassenklima mit Freude an der Schule und Spaß am Unterrichten in einem lernbereiten Raum impliziert.

Jean Paul [J.P. Friedrich Richter] (1763-1825)
aus: Jean Paul 1894 (Titelvignette)
nach einem Gemälde von Vogel v. Vogelstein

Es ist faszinierend, wie schon der Schriftsteller und praktizierende Pädagoge Jean Paul [Johann Paul Friedrich Richter] (1763-1825), Ehrendoktor der Universität Heidelberg, vor ca. 200 Jahren dies auf den Punkt gebracht hat:

In seiner Vorrede zur ersten Auflage der „Levana oder Erziehlehre“, geschrieben in „Baireuth, den 2. Mai 1806“, können wir, für uns übersetzt, als pädagogischen Impetus der Neuzeit entnehmen, dass es im „Lande der Pädagogopädien (der Erziehung zu kindlichen Erziehern)“ (Jean Paul 1894, S. 113),eigentlich für jedes Kind einer eigenen Erziehlehre bedürfe.’ Und wenn wir über Erziehung für jedes einzelne Kind *kindgerecht* reflektieren sowie eine *eigene* Erziehlehre womöglich noch schreiben würden, so gerieten wir in ein endloses Unterfangen. Aber hat nicht jeder Pädagoge im Sinne einer Habitualisierung eine eigene Erziehlehre internalisiert, keine Rezeptologie, weil wir Individuen mit unterschiedlichen Biographien und Lernmustern vor uns haben, aber flexible Handlungsmuster, die seiner Denkkraft entsprungen sind?

Wie kann man also individualisieren, ohne das *Generelle*, den ***Bildungs- und Erziehungsauftrag der Schule*** aus den Augen zu verlieren, und umgekehrt?

Erkenntnisvermögen und Anschaulichkeit, wie Jean Paul es noch erlebt hat, sind ein Stück Tradierung, aus dem wir lernen sollten.

Und so heißt es in seiner Vorrede zur ersten Auflage der Levana:

> „Einzelne Regeln ohne den Geist der Erziehung sind ein Wörterbuch ohne Sprachlehre; (...) Allein obgleich der Geist der Erziehung – überall das Ganze meinend – nichts ist als das Bestreben, den Idealmenschen, der in jedem Kinde umhüllt liegt, ... frei zu machen durch einen Freigewordenen, (...) so muss er sich doch, um zu erscheinen, in die bestimmtesten Anwendungen verkörpern.“ (S. 114)

Und aus der Vorrede zur zweiten Auflage, geschrieben in „Baireuth, den 21. November 1811“ formuliert er mit Prägnanz :

> „Leben belebt Leben, und Kinder erziehen besser zu Erziehern als alle Erzieher.“ (S. 118f.)

> „Gewiss bleibt für die Erzieher der Charakter das wahre Elementarfeuer; hab’ ihn nur der Erzieher, so wird dasselbe wenn nicht anzünden, doch wärmen und Kräfte treiben.“ (S. 123)

> „So lasst uns denn mit dem kurzen Kinderarm, d. h. mit dem langen Hebelarm die Zukunft bauen und bewegen und unverdrossen und tapfer das Gute der Zeit erhöhen helfen und das Schlechte unterhöhlen. Ja, sogar der, für dessen Kinder die Fruchtlese zu lange ausbliebe, sage zu sich: ‚Meine Enkel sind ja auch Menschen’, und säe fort.“ (S. 124)

Wie diese ‚Anwendungen’ im Bereich der Lehrerbildung fruchtbar werden sollten und wie sie expressiv werden könnten, das allerdings ist eine Fragestellung, die so leicht nicht zu beantworten sein wird, weil die Ausbildung der künftigen LehrerInnen stets Reformgedanken nachhinkt und nachhinken wird.

Literatur

ARNHARDT, G./HOFMANN, F./REINERT, G.-B.: Der Lehrer. Bilder und Vorbilder. Donauwörth 2000.

BADE, R.: Vandalismus. Was ein Kollegium tun kann. In: PÄDAGOGIK 46(1994), S. 39f.

BÄUERLE, S./MOLL-STROBEL, H./REINERT, G.-B./WEHR, H.: Gewalt in der Schule. 2. erw. Aufl. Donauwörth ^{2}2001.

CARLSBURG, G.-B. VON/HEITGER, M. (HRSG.): Der Lehrer – ein (un)möglicher Beruf. Frankfurt a.M. u.a. 2005, S. 153-170.

CARLSBURG, G.-B. VON/HEYDER, S.: Der „gute“ Lehrer: Impressionen zur Lehrerpersönlichkeit als tragender Determinante der Lehrer-Schüler-Beziehung – Eine Alltags-Studie. In: CARLSBURG, G.-B. VON/HEITGER, M. (HRSG.): Der Lehrer – ein (un)möglicher Beruf. Frankfurt a.M. u.a. 2005.

FLITNER, A.: Für das Leben – Oder für die Schule? Pädagogische und politische Essays. Weinheim/Basel 1987.

GAUDIG, H.: Die Schule im Dienste der werdenden Persönlichkeit. Leipzig 1917.

GAUDIG, H.: Schule und Schulleben. Leipzig 1923.

GAUDIG, H.: Die Schule der Selbsttätigkeit, hrsg. von L. MÜLLER. Bad Heilbrunn 1963.

GAUDIG, H.: Die Schule im Dienste der werdenden Persönlichkeit. In: REBLE, A. (HRSG.): Die Arbeitsschule. Bad Heilbrunn 1979, S. 72-90.

GORDON, Th.: Lehrer-Schüler-Konferenz oder: wie man Konflikte in der Schule löst. Hamburg [3]1977.

HERBART, J.F.: Kleine pädagogische Schriften, besorgt von A. BRÜCKMANN. Paderborn 1968.

HERDER, J.G. VON: Vitae, non scholae discendum. In: FLITNER, W. (HRSG.): Die Erziehung. Pädagogen und Philosophen über die Erziehung und ihre Probleme. Bremen [5]1967, S. 223-232.

HOOF, D. (HRSG.): Die Schulpraxis der Pädagogischen Bewegung des 20. Jahrhunderts. Bad Heilbrunn 1969.

JANK, W./MEYER, H.: Didaktische Modelle. 5. völl. überarb. Aufl. Berlin [5]2002.

JEAN PAUL, BEARB. V. K. FISCHER. Erster Teil. Leben und Lehren Jean Pauls. Levana, 1. Abteilung (Levana oder Erziehlehre). Langensalza [2]1894.

KAISER, A.: Beziehungsfähigkeit statt Gewalt in der Schule. Jungenförderung in der Grundschule. In: päd. extra (1993)1, S. 4-8.

KERSCHENSTEINER, G.: Begriff der Arbeitsschule. Leipzig/Berlin [8]1930a.

KERSCHENSTEINER, G.: Die Seele des Erziehers und das Problem der Lehrerbildung. Leipzig/ Berlin [3]1930b.

KERSCHENSTEINER, G.: Die Schule der Zukunft eine Arbeitsschule. In: FLITNER, W. (Hrsg.): Die Erziehung. Pädagogen und Philosophen über die Erziehung und ihre Probleme. Bremen [5]1967, S. 397-416.

KERSCHENSTEINER, G.: Texte zum pädagogischen begriff der Arbeit und zur Arbeitsschule. Bd. II, besorgt von G. WEHLE. Paderborn 1968.

KERSCHENSTEINER, G.: Die Schule der Zukunft eine Arbeitsschule. In: In: REBLE, A. (Hrsg.): Die Arbeitsschule. Bad Heilbrunn 1979, S. 54-59.

KLAFKI, W.: Neue Studien zur Bildungstheorie und Didaktik. Zeitgemäße Allgemeinbildung und kritisch-konstruktive Didaktik. Weinheim/Basel [4]1994.

KRAFT, P.: Schulleben und soziales Lernen. In: KRAFT, P.: Feste und Geselligkeiten in der Schule. Braunschweig 1979, S. 45-76.

LANGE, E./ALEXANDER, D. (HRSG.): Philosophenlexikon. Berlin 1982, S. 842-848.

LASSAHN, R. (HRSG.): Das Schulleben. Bad Heilbrunn 1969.

MILLER, R.: Gespräche mit gewalttätigen Schülern. In: BÄUERLE, S. (HRSG.): Schülerfehlverhalten. Regensburg 1985.

MILLER, R.: Sich in der Schule wohlfühlen. Weinheim/Basel 1989.

MILLER, R.: Veränderte Schulwirklichkeit – Konsequenzen für die Lehrerfortbildung. In: SCHULMAGAZIN 5 bis 10 (1995)7-8, S. 12-15.

MILLER, R.: Das ist ja wieder typisch. Kommunikation und Dialog in Schule und Schulverwaltung. 25 Trainingsbausteine. Weinheim/Basel 1996.

ODENBACH, K.: Die deutsche Arbeitsschule. Braunschweig ³1963.

ODENBACH, K.: Das Schulleben. In: ODENBACH, K.: Studien zur Didaktik der Gegenwart. Braunschweig ⁵1974, S. 15-36.

REINERT, G.-B.: Schulchroniken als Zeugnisse für Schulleben. In: REINERT, G.-B./GUDJONS, H. (HRSG.): Schulleben. Königstein 1980, S. 167-182 u. 203-203.

REINERT, G.-B./ARNHARDT, G./CORNELIUS, P.: Johann Heinrich Pestalozzi. Anthropologisches Denken und Handeln. Donauwörth 1996.

REINERT, G.-B./GUDJONS, H.: Lehrer ohne Maske? Königstein 1981.

REINERT, G.-B./HEYDER, S.: Lebensort: Schule. Weinheim/Basel 1983.

REINERT, G.-B./WEHR, H.: Gewalt und Gewaltprävention in der Schule. In: BÄUERLE, S. u.a.: Gewalt in der Schule. 2. erw. Aufl. Donauwörth ²2001, S. 68-149.

REINERT, G.-B./ZINNECKER, J. (HRSG.): Schüler im Schulbetrieb. Reinbek 1978.

REINERT VON CARLSBURG, G.-B./WEHR, H.: Selbstbestimmtes Lernen lernen. Donauwörth 2004.

ROGERS, C.R.: Entwicklung der Persönlichkeit. Psychotherapie aus der Sicht eines Therapeuten. Stuttgart 1985.

SANDERS, B.: Der Verlust der Sprachkultur. „Die Pistole ist das Schreibgerät der Analphabeten“. Frankfurt a.M. 1994.

SCARBATH, H.: Träume vom guten Lehrer. Sozialisationsprobleme und dialogisch-förderndes Verstehen. Donauwörth 1992.

SCHÄFER, M.: Aggression unter Schülern. In: REPORT PSYCHOLOGIE 21(1996)9, S. 700-710.

STAUDT, C.: LehrerInnen – Nur Opfer oder auch Täter? Ein Bericht aus der Schülerperspektive. In: PÄDAGOGIK 46(1994)4, S. 17-20.

ZINNECKER, J. (HRSG.): Der heimliche Lehrplan. Untersuchungen zum Schulunterricht. Weinheim/Basel 1975.

ZINNECKER, J.: Die Schule als Hinterbühne oder Nachrichten aus dem Unterleben der Schüler. In: REINERT, G.-B./ZINNECKER, J. (HRSG.): Schüler im Schulbetrieb. Reinbek 1978, S. 29-121.

Authors

Ilona Adomavičiūtė, Dipl.-Kultur- und Medienmanagerin, Ückermünder Str. 11, D-10439 Berlin (Germany) e-mail: ilonaadoms@freenet.de

Giedrė Adomavičienė, Dr., Kaunas College (Lithuania)

Dalia Antinienė, Doctoral Student, Department of Sociology, Faculty of Social Scien-ces, Kaunas University of Technology (Lithuania)

Aldona Augustinienė, Doctoral Student, Department of Educational Systems, Faculty of Social Sciences, Kaunas University of Technology (Lithuania)

Antanas Bagdonas, Dr., Department of Educational Systems, Faculty of Social Sciences, Kaunas University of Technology (Lithuania)

Sandra Balevičienė, Doctoral Student, Department of Chemistry, Faculty of Natural Sciences, Vilnius Pedagogical University (Lithuania)

Nijolė Bankauskienė, Dr., Department of Educational Systems, Faculty of Social Sciences, Kaunas University of Technology (Lithuania)

Viktorija Baršauskienė, Assoc. Prof. Dr., Department of Educational Systems, Faculty of Social Sciences, Kaunas University of Technology (Lithuania)

Gerd-Bodo von Carlsburg, Prof. Dr. phil. Dr. h.c. mult., Institut für Erziehungswissenschaft, Erziehungs- und Sozialwissenschaftliche Fakultät einschließlich Sonderpädagogik, Heidelberg University of Education, Präsident des Weltbund für Erneuerung der Erziehung, Keplerstr. 87, D-69120 Heidelberg (Germany), e-mail: b.v.carlsburg@web.de

Vilija Dabrišienė, Assoc. Prof. Dr., Department of Mathematical research in systems, Faculty of Fundamental Science, Kaunas University of Technology (Lithuania)

Virginija Chreptavičienė, Assoc. Prof. Dr., Department of Educational Systems, Faculty of Social Sciences, Kaunas University of Technology (Lithuania)

Nijolė Čiučiulkienė, Doctoral Student, Department of Educational Systems, Faculty of Social Sciences, Kaunas University of Technology (Lithuania)

Dalija Gudaitytė, Assoc. Prof. Dr., Department of Educational Systems, Faculty of Social Sciences, Kaunas University of Technology (Lithuania)

Brigita Janiūnaitė, Assoc. Prof. Dr., Department of Educational Systems, Faculty of Social Sciences, Kaunas University of Technology (Lithuania)

Rūta Janulevičienė, MSc. in International Business Economics, Aalborg University (Denmark)

Marija Jotautienė, Dr., Kaunas Technical College (Lithuania).

Palmira Jucevičienė, Prof. Dr. habil., Department of Educational Systems, Faculty of Social Sciences, Kaunas University of Technology (Lithuania)

Giedrius Jucevičius, Doctoral Student, Institute of Business Strategy, Kaunas University of Technology (Lithuania)

Robertas Jucevičius, Prof. Dr. habil., Institute of Business Strategy, Faculty of Social Sciences, Kaunas University of Technology (Lithuania)

Eglė Katiliūtė, Doctoral Student, Department of Educational Systems, Faculty of Social Sciences, Kaunas University of Technology (Lithuania)

Violeta Karenauskaitė, Doctoral Student, Department of Educational Systems, Faculty of Social Sciences, Kaunas University of Technology (Lithuania)

Angelė Kaušylienė, Dr., Faculty of Health Care and Social Care, Utena College (Lithuania)

Rūta Klimašauskienė, Doctoral Student, Department of Educational Systems, Faculty of Social Sciences, Kaunas University of Technology (Lithuania)

Irena Leliūgienė, Assoc. Prof. Dr., Department of Educational Systems, Faculty of Social Sciences, Kaunas University of Technology (Lithuania)

Auksė Markevičienė, Doctoral Student, Department of Educational Systems, Faculty of Social Sciences, Kaunas University of Technology (Lithuania)

Gediminas Merkys, Prof. Dr. habil., Department of Educational Systems, Faculty of Social Sciences, Kaunas University of Technology (Lithuania)

Regina Novikienė, Dr., Department of Applied Mathematics, Faculty of Fundamental Science, Kaunas University of Technology (Lithuania)

Ernesta Račienė, Assoc. Prof. Dr., Department of German Philology and Didactics, Faculty of Foreign Languages Vilnius Pedagogical University (Lithuania)

Alicija Ramanauskaitė, Dr., Kaunas Vocational Training Centre for Service Business (Lithuania)

Loreta Rudaitienė, Doctoral Student, Department of Educational Systems, Faculty of Social Sciences, Kaunas University of Technology (Lithuania)

Liuda Šiaučiukėnienė, Prof. Dr. habil., Department of Educational Systems, Faculty of Social Sciences, Kaunas University of Technology (Lithuania)

Edita Štuopytė, Dr., Department of Educational Systems, Faculty of Social Sciences, Kaunas University of Technology (Lithuania)

Inga Šukevičiūtė, Dr. phil., Assistant of the European Parliament, 60 rue Wiertz, B-1047 Brussels (Belgium), e-mail: m.inga@one.lt

Ilona Skersienė, Dr. Department of Educational Systems, Faculty of Social Sciences, Kaunas University of Technology (Lithuania)

Nida Stankevičienė, Assoc. Prof. Dr., Department of Educational Systems, Faculty of Social Sciences, Kaunas University of Technology (Lithuania)

Vida Stasiulionienė, Doctoral Student, Department of Educational Systems, Faculty of Social Sciences, Kaunas University of Technology (Lithuania)

Vilija Stanišauskienė, Dr., Department of Educational Systems, Faculty of Social Sciences, Kaunas University of Technology (Lithuania)

Gintarė Tautkevičienė, Doctoral Student, Department of Educational Systems, Fa-culty of Social Sciences, Kaunas University of Technology (Lithuania)

Rūta Urbanskienė, Assoc. Prof., Institute of Business Strategy, Faculty of Social Sciences, Kaunas University of Technology (Lithuania)

Aistė Urbonienė, Doctoral Student, Department of Educational Systems, Faculty of Social Sciences, Kaunas University of Technology (Lithuania)

Vida Vilkienė, MSc. in Social Sciences (educology), Department of International Relations, Marijampole College (Lithuania)

Ona Visockienė, Assoc. Prof., Institute of Educational Studies, Faculty of Social Sciences, Kaunas University of Technology (Lithuania)

Nijolė Zinkevičienė, Doctoral Student, Department of Educational Systems, Faculty of Social Sciences, Kaunas University of Technology (Lithuania)

Vilma Žydžiūnaitė, Dr., Department of Educational Systems, Faculty of Social Sciences, Kaunas University of Technology (Lithuania)

Addresses of the Editors:

Faculty of Educational and Social Sciences, Heidelberg University of Education (Germany), Keplerstr. 87, D-69120 Heidelberg (Germany), Tel./Fax: +49-6221-477-533

Faculty of Social Sciences, Kaunas University of Technology (Lithuania), Donelaičio 20-400, LT-3000 Kaunas (Lithuania), Tel. +370-7-300-130

Baltische Studien zur Erziehungs- und Sozialwissenschaft

Reihenherausgeber:
Gerd-Bodo von Carlsburg, Irena Musteikienė und Airi Liimets

Band 1 Gerd-Bodo Reinert / Irena Musteikienė (Hrsg.): Litauische Gespräche zur Pädagogik. Humanismus – Demokratie – Erziehung. 1999.

Band 2 Gerd-Bodo Reinert / Irena Musteikienė (Hrsg.): Litauische Gespräche zur Pädagogik II. Staat und Schule. 2000.

Band 3 Jaan Mikk: Textbook: Research and Writing. 2000.

Band 4 Gerd-Bodo Reinert / Irena Musteikienė (Hrsg.): Erziehungswandel und moderne pädagogische Verfahren. 2001.

Band 5 Airi Liimets (Hrsg.): Integration als Problem in der Erziehungswissenschaft. 2001.

Band 6 Gerd-Bodo Reinert / Irena Musteikienė (Hrsg.): Wissenschaft – Studium – Schule auf neuen Wegen. 2002.

Band 7 Palmira Jucevičienė / Gediminas Merkys / Gerd-Bodo Reinert: Towards the Learning Society. Educational Issues. 2002.

Band 8 Gerd-Bodo Reinert / Irena Musteikienė (Hrsg.): Bildung im Zeitalter der Informationsgesellschaft. 2003.

Band 9 Dalia Ohlinger: Argumentation in der Erst- und Fremdsprache. Pragmalinguistische und grammatikalische Aspekte anhand von Argumentationen deutscher und litauischer Studierender. 2003.

Band 10 Inga Šukevičiūtė: Interkulturelle Interferenzen im Bereich Wirtschaftskommunikation. Zur Analyse von Geschäftsbriefen im Vergleich Deutschland, Litauen und Russland. 2004.

Band 11 Gerd-Bodo Reinert von Carlsburg / Irena Musteikienė (Hrsg./eds.): Innovation durch Bildung. Innovation by Education. 2004.

Band 12 Gerd-Bodo von Carlsburg / Palmira Jucevičiene / Gediminas Merkys (eds.): Learning and Development for Innovation, Networking and Cohesion. 2005.

www.peterlang.de

Peter Lang · Europäischer Verlag der Wissenschaften

Gerd Bodo Reinert von Carlsburg / Irena Musteikienė (Hrsg. /eds.)

Innovation durch Bildung
Innovation by Education

Frankfurt am Main, Berlin, Bern, Bruxelles, New York, Oxford, Wien, 2004.
339 S., zahlr. Abb. und Graf.
Baltische Studien zur Erziehungs- und Sozialwissenschaft.
Herausgegeben von Gerd-Bodo Reinert, Irena Musteikienė und Juri Orn. Bd. 11
ISBN 3-631-52820-5 / US-ISBN 0-8204-7336-7 · br. € 56.50*

Der Band dokumentiert eine Reihe von Beiträgen der IX. Internationalen wissenschaftlichen Konferenz zur *Bildungsreform und Lehrerbildung (Švietimo reforma ir mokytoju rengimas).* Das Konferenzthema des Jahres 2003 lautete: Verknüpfung der Hochschul- und Gymnasialbildung: Tradition und Wechsel im Bildungs- und Erziehungsbereich unter Berücksichtigung der Pädagogik der Schule, einschließlich der Allgemeinen Didaktik und Fachdidaktik, abgekürzt wiedergegeben „Innovation durch Bildung". Zu dieser Thematik diskutierten in Vilnius (Litauen) praktizierende Pädagogen, Wissenschaftler und Mitglieder von Bildungsorganisatoren. Behandelt wurden die wichtigsten aktuell anstehenden Fragen, so die schnellere Reformierung des Studiums und die Qualität der Ausbildung im Hinblick auf die Neuorientierung an die Anforderungen der internationalen Standards.

This volume presents contributions of the IX. International scientific meeting on *Educational Reform and Teacher Formation (Švietimo reforma ir mokytoju rengimas).* The meeting topic of the year 2003 was: Combination of the university and high school education: Tradition and change represented in the education and education area, abbreviated under consideration of the education of the school including the general didactics and subject didactics „Innovation by Education". To this topic practicing educationalists, scientists and members of education organizers, discussed in Vilnius (Lithuania). The most important currently upcoming questions, as faster reformers of the studies and the quality of education were treated with a view to the reorientation to the requests of the international standards.

Frankfurt am Main · Berlin · Bern · Bruxelles · New York · Oxford · Wien
Distribution: Verlag Peter Lang AG
Moosstr. 1, CH-2542 Pieterlen
Telefax 00 41 (0) 32 / 376 17 27